Lecture Notes in Computational Science and Engineering

74

For further volumes:
http://www.springer.com/series/3527

Damien Tromeur-Dervout · Gunther Brenner
David R. Emerson · Jocelyne Erhel
Editors

Parallel Computational Fluid Dynamics 2008

Parallel Numerical Methods, Software Development and Applications

Editors
Damien Tromeur-Dervout
Université Lyon 1 - CNRS
Institut Camille Jordan
Bd du 11 Novembre 1918 43
69622 Villeurbanne
France
damien.tromeur-dervout@univ-lyon1.fr

David R. Emerson
Science and Technology Facilities Council
Daresbury Laboratory
Daresbury Science and Innovation Campus
WA4 4AD Warrington Cheshire
United Kingdom
david.emerson@stsc.ac.uk

Gunther Brenner
TU Clausthal
Institut für Technische Mechanik
Adolph-Roemer-Str. 2A
38678 Clausthal-Zellerfeld
Germany
gunther.brenner@tu-clausthal.de

Jocelyne Erhel
INRIA - SAGE
Campus de Beaulieu
35042 Rennes
France
jocelyne.erhel@irisa.fr

ISSN 1439-7358
ISBN 978-3-642-26515-0 ISBN 978-3-642-14438-7 (eBook)
DOI: 10.1007/978-3-642-14438-7
Springer Heidelberg Dordrecht London New York

Mathematics Subject Classification Numbers (2010): 65F10, 65N06, 65N08, 65N22, 65N30, 65N55, 65Y05, 65Y10, 65Y20, 15-06, 35-06, 76-XX 76-06

Cover design: deblik, Berlin

Printed on acid-free paper

Springer is part of Springer Science + Business Media (www.springer.com)

Preface

Parallel CFD 2008, the twentieth in the high-level international series of meetings featuring different aspect of parallel computing in computational fluid dynamics and other modern scientific domains was held May 19 – 22, 2008 in Lyon, France.

The themes of the 2008 meeting included the traditional emphases of this conference, and experiences with contemporary architectures. Around 70 presentations were included into the conference program in the following sessions:
Parallel Algorithms and solvers
Parallel performances with contemporary architectures
Structured and unstructured grid methods, boundary methods
software framework and components architecture
CFD applications (Bio fluid, environmental problem) Lattice Boltzmann method and SPH
Optimisation in Aerodynamics

This book presents an up-to-date overview of the state of the art in Parallel Computational Fluid Dynamics from Asia, Europe, and North America. This reviewed proceedings included about sixty percent of the oral lectures presented at the conference.

The editors.

Parallel CFD 2008 was organized by the Institut Camille Jordan of the University of Lyon 1 in collaboration with the Center for the Development of the Parallel Scientific Computing.

The Scientific Committee and Local Organizers of Parallel CFD 2008 are delighted to acknowledge the generous sponsorship of the following organizations, through financial or in-kind assistance. Assistance of our sponsors allowed to organize scientific as well as social program of the conference.

Scientific Comittee
University Lyon 1

Intel, Germany

SGI, France

Institut Camille Jordan
University Lyon 1

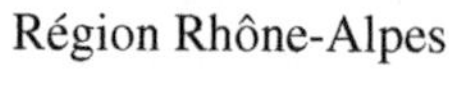

Région Rhône-Alpes

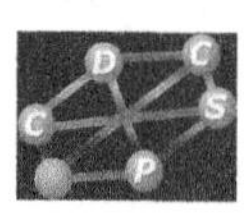

Center for Development of
University Lyon 1

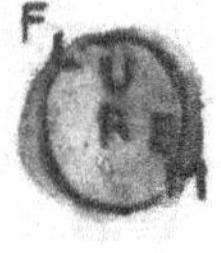

Fluorem

Modelys

Many people worked to organize and execute the conference. We are especially grateful to all members of the international scientific committee. We also want to thank the key members of the local organizing committee David Guibert, Toan Pham Duc, Patrice linel, Simon Pomarede, Thomas Dufaud, Nicolas Kielbasievich, Daniel Fogliani, Fabienne Oudin, Brigitte Hautier, Sandrine Montingy.

We also thank our colleagues Frédéric Desprez from the Laboratoire d'Informatique du Parallélisme (LIP) Ecole Normale Supérieure de Lyon, Michel Lance from the Laboratoire de Mécanique des Fluides et d'Accoustique (LMFA) Ecole Centrale de Lyon, and Patrick Quéré from the Computer Science Laboratory for mechanics and Engineering Sciences (LIMSI) for their help to promote this event.

Damien Tromeur-Dervout
Chairman, Parallel CFD 2008.

Contents

Part V High Order methods

Part VI Parallel Algorithms and Solvers

Part VII Lattice Boltzman and SPH Methods

Part I

Invited speakers

Large Scale Computations in Nuclear Engineering: CFD for Multiphase Flows and DNS for Turbulent Flows with/without Magnetic Field

Tomoaki Kunugi[1], Shin-ichi Satake[2], Yasuo Ose[1], Hiroyuki Yoshida[3], and Kazuyuki Takase[3]

[1] Kyoto University, Yoshida, Sakyo, Kyoto 6060-8501, Japan
[2] Tokyo University of Science, 2641 Yamazaki, Noda 278-8510, Japan
[3] Japan Atomic Energy Agency, 2-4, Shirakata, Tokai-mura 319-1195, Japan
kunugi@nucleng.kyoto-u.ac.jp

Abstract. Large scale computations are being carried out in nuclear engineering fields such as light water reactors, fast breeder reactors, high temperature gas-cooled reactors and nuclear fusion reactors. The computational fluid dynamics (CFD) regarding not only the single-phase flows but also the two-phase flow plays an important role for the developments of advanced nuclear reactor systems. In this review paper, some examples of large scale computations in nuclear engineering fields are illustrated by using a parallel visualization.

Key words: Direct numerical simulation; Multiphase flows; Turbulent flows; Parallel visualization; Magnetohydrodynamics; Nuclear reactors; Fusion reactors.

1 Numerical Simulation of Boiling Phenomena

It is important to remove the heat from industrial devices and nuclear reactors with a high heat flux to insure their safety. In order to enhance the heat transfer, phase change phenomena such as evaporation and condensation have to be utilized, so that it is important to understand the mechanism of boiling to design industrial devices. Although many researchers have experimentally studied the boiling phenomena, it has not been clarified yet because it consists of a lot of complicated phenomena. As for pool boiling experiments, Kenning & Yan measured the spatial and temporal variations of wall temperature in nucleate boiling by using a liquid crystal thermometry. They pointed out the importance of non-uniform wall temperature distribution and suggested that the transient heat conduction model proposed by Mikic & Rohsenow [2] was unrealistic because of the assumption of uniform transient heat conduction. It is relatively difficult to perform numerical analysis of boiling phenomenon because it includes the phase change. On the other hands, a critical heat flux (CHF) is also very important for high heat flux removal. However, the empirical correlation of the CHF is used in most designs. In general, the prediction of CHF is very

D. Tromeur-Dervout (eds.), *Parallel Computational Fluid Dynamics 2008*,
Lecture Notes in Computational Science and Engineering 74,
DOI: 10.1007/978-3-642-14438-7_1,

difficult because of the complexity of relation between the nucleation boiling and the bubble departure due to flow convection. Welch [3] carried out the numerical study on two-dimensional two-phase flow with a phase change model using a finite volume method combined with a moving grid, however it can be applied only to a little deformation of gas-liquid interface. Son & Dhir [4] also carried out the two-dimensional pool boiling simulation using a finite difference method with a moving grid. Recently, Juric & Tryggvason [5] conducted a film boiling with a front tracking method by Unverdi & Tryggvason [6]. They pointed out the importance of the temporal and spatial temperature distribution of the heating plate: heat conduction in the slab. The author developed a new volume tracking method, so-called gMARS: Multi-interface Advection and Reconstruction Solver[7].

This section describes that the MARS is applied to the force convective flow boiling in the channel with an appropriate phase change model: a model for boiling and condensation phenomena based on a homogeneous nucleation and a well-known enthalpy method. This model is good for metal casting problems because of no superheating of liquid. As for the water, it has to be considered the liquid superheat for nucleate boiling phenomena. The direct numerical simulations with this phase change model for pool nucleate boiling and forced convective flow boiling have been performed. The aims of this study are to develop a direct numerical method to simulate boiling phenomena and to simulate the three-dimensional pool nucleate boiling and forced convective flow boiling by the direct numerical simulation (DNS) based on the MARS combined the enthalpy method considered the liquid superheat as the phase change model.

1.1 Numerical Simulation based on MARS

Governing equations.

In this section, the direct numerical multiphase flow solver (MARS) is briefly explained. As for m fluids including the gas and liquid, the spatial distribution of fluids can be defined as

$$\langle F \rangle = \sum F_m = 1.0 \tag{1}$$

The continuity equation of the multiphase flows for m fluids:

$$\frac{\partial F_m}{\partial t} + \nabla \cdot (F_m U) - F_m \nabla U = 0 \tag{2}$$

The momentum equation with the following CSF (Continuum Surface Force) model proposed by Brackbill [9] is expressed as:

$$\frac{\partial U}{\partial t} + \nabla (UU) = G - \frac{1}{\langle \rho \rangle} \nabla P - \nabla \cdot \tau + \frac{1}{\langle \rho \rangle} F_V \tag{3}$$

CSF model :

$$F_V = \sigma \kappa n \langle \rho \rangle / \bar{\rho} \tag{4}$$

here, the mean density at interface, $\bar{\rho} = (\rho_g + \rho_l)/2$, the suffix g denotes vapor and l for water. κ is curvature of the surface, σ the surface tension coefficient and n is the normal vector to the surface.

The momentum equation (3) can be solved by means of the well-known projection method. The Poisson equation for the pressure can be solved by the ILUBCG method. Finally, the new velocity field can be obtained. Once the velocity field can be obtained, it can be transported the volume of fluid by the MARS, i.e., a kind of PLIC (Piecewise Linear Calculation) volume tracking procedure [24] for Eq. (2). The detail description of the solution procedure is described in the reference [7].

The energy equation is expressed as:

$$\frac{\partial}{\partial t}\langle \rho Cv\rangle T + \nabla \cdot (\langle \rho Cv\rangle UT) = \nabla \cdot (\langle \lambda \rangle \nabla T) - P(\nabla \cdot U) + Q, \tag{5}$$

where C_v, T, λ, Q is specific heat, temperature, heat conductivity, heat generation term, respectively. The second term of the right hand side of the equation (5), the Clausius-Clapeyron relation is considered as the work done by the phase change.

Phase Change Model.

As mention in the introduction section, nucleate boiling phenomena need to be modeled. One of ideas for this modeling can be considered:

Nucleate boiling model = Nucleation model + Bubble growth model

Nucleation model: This model gives the homogeneous superheat limit of liquid and the size of nucleus of bubble. A superheat limit T_s is got from the kinetic theory [8] or the usual cavity model. Typically T_s is around 110°C in water pool boiling at an atmospheric pressure. The equilibrium radius r_e of nucleus corresponding to T_s can be calculated by Eq. (6) based on the thermodynamics. The F-value of nucleus corresponding to the ratio of a cell size to a size of nucleus where the shape of nucleus is assumed to be a sphere is given to a computational cell with greater than T_s. In this study, T_s is a parameter. Although T_s varies in spatial on the heated surface from the experiment [1], T_s is assumed to be uniform on the heated surface in the present study. Therefore, particular nucleation sites are not specified.

$$r_e = \frac{2\sigma}{P_{sat}(T_l)\exp\{v_l[P_l - P_{sat}(T_l)]/RT_l\} - P_l}, (T_l \geq T_s) \tag{6}$$

where r_e is a radius of bubble embryo, σ is an interfacial tension, T_l is a temperature of liquid, P_{sat} is a pressure corresponding saturation conditions, P_l is a pressure of liquid, v_l is a volume of liquid per unit mass, and R is the ideal gas constant on a per unit mass basis.

Bubble growth model: Increasing the temperature, liquid water becomes vapor partially if the temperature of liquid is greater than the liquid saturation line (T_l), i.e., liquid-gas mixture (two-phase region) and then eventually becomes the superheated gas phase if the temperature is greater than the gas saturation lime (T_g). This process can be treated by the enthalpy method.

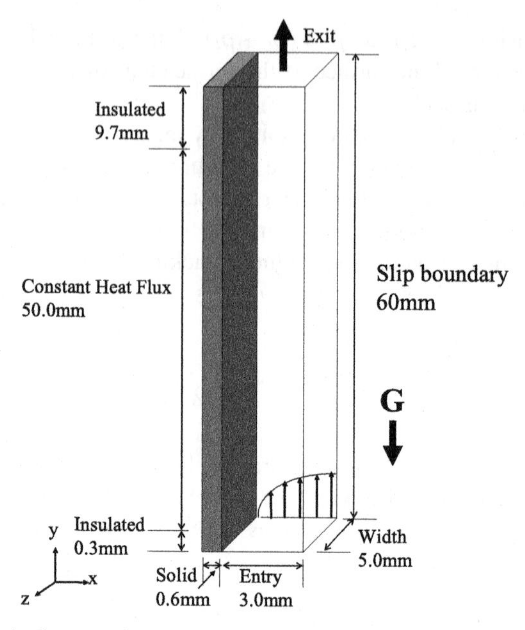

Fig. 1. Computational domain for forced convective flow boiling

1.2 Results and Discussions on Forced Convective Flow Boiling

In order to get some insights into the mechanism of three-dimensional forced convective subcooled flow boiling. Figure 1 shows the computational domain that is a three-dimensional vertical channel and the sidewall is heated with constant heat flux because of considering the spatial and temporal variations of temperature in the solid sidewall. The length of flow channel is 60 mm, 6 mm in height and 5 mm in width. The computational domain is a half channel because of the symmetry. The heating conditions are as follows: 0.3 mm in length from inlet is adiabatic and following 50 mm by heating of a constant heat flux, 2.4 MW/m^2 from the outside of solid sidewall of 0.6 mm in thickness and the remaining 9.7 mm is also adiabatic. The inlet mean velocity is 0.5 m/s with a parabolic profile. The no-slip condition at the sidewall, the slip velocity at the symmetric boundary and constant pressure at the outlet are imposed as the boundary conditions. The periodic velocity and temperature conditions are applied to the spanwise boundaries. The water pressure is atmospheric and the degree of water subcooling is 20 K. The solid wall is assumed to be a stainless steel. The degree of superheat is set to be 50 K. The computational cell is uniform cubic shape and the size of cell is 100 μm and the number of cells is 36(x) 600(y) 50(z)=1,080,000. Time increment is 5 μsec. The fictitious temperature difference used in the enthalpy method is ΔT=0.1 K.

Resulting from the forced convective flow boiling computation as shown in Fig. 2, the series of bubble growth are depicted as black spots, and the gray contours show the temperature distribution at x=0.25 mm every 52.5 msec. From the temperature evolution, the thermal boundary layer has been developing during computation and does not reach to the equilibrium state at the downstream in this computation, so that it is necessary to carry out much longer computation. The maximum size of bubble is around 1 mm at the present stage. It is interesting that the higher temperature stagnation region is formed in the middle of the channel and three higher temperature streaks are observed in the thermal boundary layer. The bubble generated in the upstream region becomes just like an obstacle and makes a high temperature stagnation or recirculation region behind the nucleated bubbles. Eventually, another bubble will generate in that stagnation or recirculation region because the degree of liquid subcooling could be decreased, i.e., it can be saying a "chain generation of boiling bubbles."

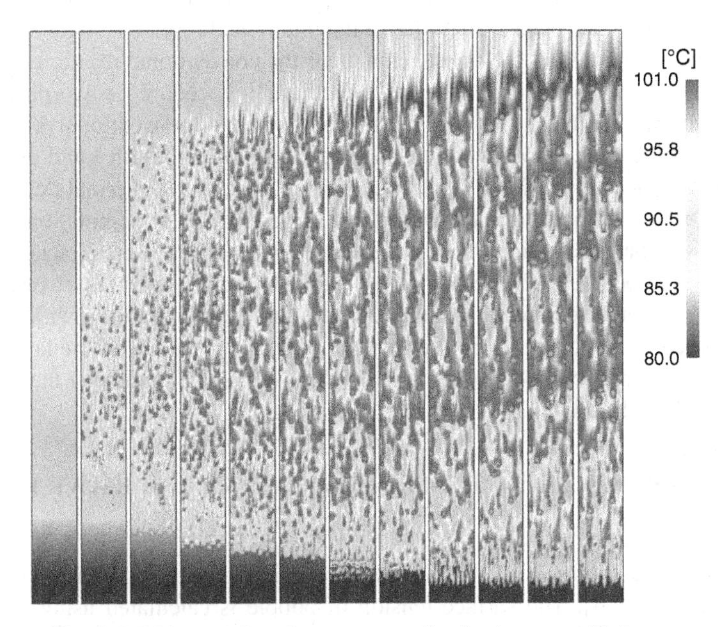

Fig. 2. Bubble growth and temperature distribution every 52.5 msec

2 CFD for Two-Phase Flow Behaviors in Nuclear Reactor

Subchannel analysis codes [11]-[13] and system analysis codes [14, 15] are usually used for the thermal-hydraulic analysis of fuel bundles in nuclear reactors. As for the former, however, many constitutive equations and empirical correlations based on experimental results are needed to predict the water-vapor two-phase flow behavior. If there are no experimental data such as an advanced light-water reactor which has been studied at the Japan Atomic Energy Agency (JAEA) in Japan and named as a reduced moderation water reactor (RMWR), it is very difficult to obtain the precise predictions [16, 17, 18]. The RMWR core has a remarkably narrow gap spacing between fuel rods (i.e., around 1 mm) and a triangular tight lattice fuel rod configuration in order to reduce the moderation of the neutron. In such a tight-lattice core, there is no sufficient information about the effects of the gap spacing and the effect of the spacer configuration on the two-phase flow characteristics. Therefore, in order to analyze the water vapor two-phase flow dynamics in the tight-lattice fuel bundle, a large-scale simulation under the full bundle size condition is necessary. The Earth Simulator [19] enables that lots of computational memories are required to attain the two-phase flow simulation for the RMWR core.

In JAEA, numerical investigation on the physical mechanisms of complicated thermal-hydraulic characteristics and the multiphase flow behavior with phase change in nuclear reactors has been carried out In this numerical research, some of the authors in JAEA pointed out the improvements of the conventional reactor core thermal design procedures and then proposed a predicting procedure for two-phase flow characteristics inside the reactor core more directly than the conventional procedures for the first time in the world by reducing the usage of constitutive and empirical equations as much as possible [20]. Based on this idea, a new thermal design procedure for advanced nuclear reactors with the large-scale direct simulation method (TPFIT: Two-Phase flow simulation code using advanced Interface Tracking) [21] has been developed at JAEA. Especially, thermal hydraulic analyses of two-phase flow positively for a fuel bundle simulated by the full size using the Earth Simulator are performed [22]. This section describes the preliminary results of the large-scale water-vapor two-phase flow simulation in the tight-lattice fuel bundle of the RMWR core by the TPFIT code.

2.1 Numerical Simulation of Two-Phase Flow Behavior in 37 RMWR Fuel Rods

The TPFIT code is based on the CIP method [23] using the modified interface-tracking method [24]. The surface tension of bubble is calculated using the CSF [9]. Figure 3 shows the computational geometry consisting of 37 RMWR fuel rods. The geometry and dimensions simulate the experimental conditions done by JAEA [25]. Here, the fuel rod outer diameter is 13 mm and the gap spacing between each rod is 1.3 mm. The casing has a hexagonal cross section and a length of one hexagonal side is 51.6 mm. An axial length of the fuel bundle is 1260 mm. The water flows upward from the bottom of the fuel bundle. A flow area is a region in which deducted the cross-sectional area of all fuel rods from the hexagonal flow passage. The spacers are installed into the fuel bundle at the axial positions of 220, 540, 750 and 1030 mm from the bottom. The axial length of each spacer is 20 mm. Inlet conditions of water are as follows: temperature 283°C, pressure 7.2 MPa, and flow rate 400 kg/m^{2}s. Moreover, boundary conditions are as follows: fluid velocities for x-, y- and z-directions are zero on every wall (i.e., an inner surface of the hexagonal flow passage, outer surface of each fuel rod and surface of each spacer); velocity profile at the inlet of the fuel bundle is set to be uniform. The present simulations were carried out under the non-heated isothermal flow condition in order to remove the effect of heat transfer due to the fuel rods to the fluid. A setup of a mixture condition of water and vapor at the heating was performed by changing the initial void fraction of water at the inlet of the analytical domain.

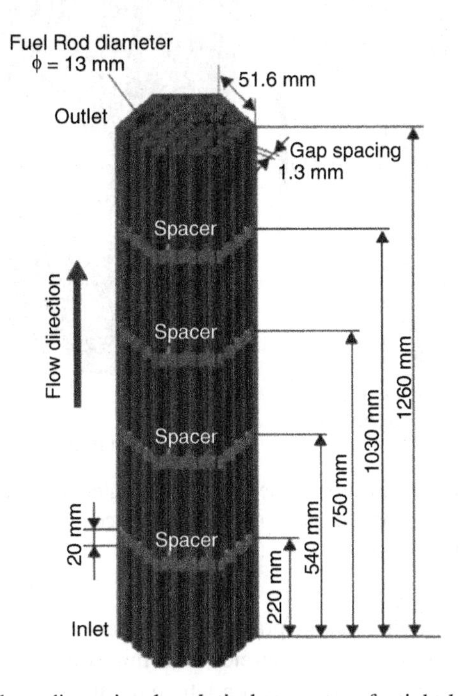

Fig. 3. Outline of three-dimensional analytical geometry of a tight-lattice fuel bundle

2.2 Results and Discussions

Figure 4 shows an example of the predicted vapor structure around the fuel rods. Here, the distribution of void fraction within the region from 0.5 to 1 is shown: 0.5 indicates just an interface between the water and vapor and is shown by green; and 1 indicates the non-liquid vapor and is shown by red. Vapor flows from the upstream to downstream like a streak through the triangular region, and the interaction of the vapor stream to the circumferential direction is not seen. On the other hand, since the vapor is disturbed behind a spacer, the influence of turbulence by existence of the spacer can be predicted.

In order to predict the water-vapor two-phase flow dynamics in the RMWR fuel bundle and to reflect them to the thermal design of the RMWR core, a large-scale simulation was performed under a full bundle size condition using the Earth Simulator. Details of water and vapor distributions around fuel rods and a spacer were

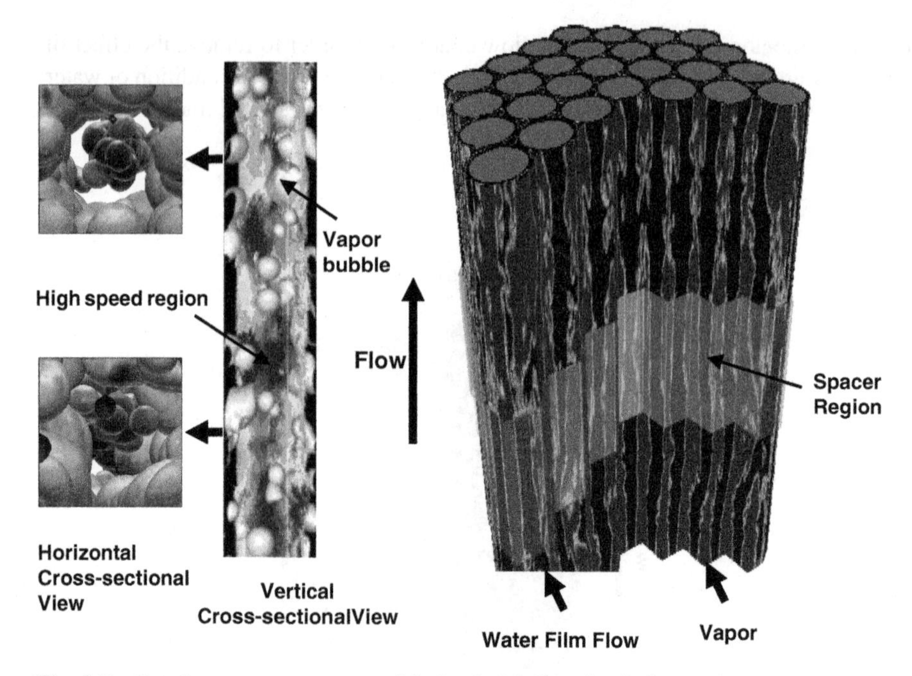

Fig. 4. Predicted vapor structure around fuel rods; black region indicates the water (void fraction is 0), and light grey region indicates an interface between water and vapor which means the void fraction is in between 0 and 1.

clarified numerically. A series of the present preliminary results were summarized as follows: 1) The fuel rod surface is encircled with thin water film; 2) The bridge formation by water film appears in the region where the gap spacing between adjacent fuel rods is narrow; 3) Vapor flows into the triangular region where the gap spacing between fuel rods is large; 4) A flow configuration of vapor shows a streak structure along the triangular region.

3 DNS for Turbulent Flows with/without Magnetic Field

On the other hand, in the gas-cooled reactor and the fast breeder reactors the coolant is a single-phase flow at mostly turbulent situation. Direct numerical simulations (DNSs) for turbulent flows have been carried out to investigate the turbulent structure in the flow passage and around each fuel rod surface at Reynolds number of 78,000 as shown in Fig. 5. A second order finite difference method is applied to the spatial discretization, the 3rd order Runge-Kutta method and the Crank-Nicolson method are applied to the time discretization, and the time advancing scheme is the fractional time step is used for the coupling scheme, so called, Dukowcz-Dvinsky scheme. The number of used central processing unit is 1,152 and it corresponds to 144 nodes. The total memory is 2 terabytes and the total number of computational grid is 7,200 million points [26, 27]. The flow visualization is very important to grasp the entire picture of the flow behavior, so that a parallel visualization technique is applied in this case. As for the thermofluid behavior in fusion reactors, the magnetohydrodynamic (MHD) effect is very important. DNS for turbulent flow in a parallel channel has been performed as shown in Fig. 7 [28] drawn by the parallel visualization. The turbulent structure is suppressed by the magnetic field, i.e., Lorentz force: Hartmann numbers are 0 (non-MHD) and 65 (MHD).

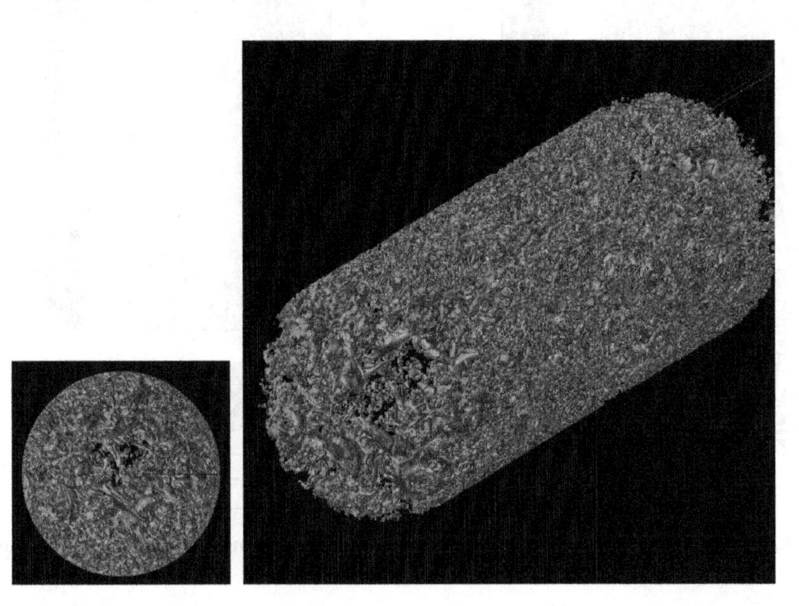

Fig. 5. Second invariant contour surface of velocity tensor of single-phase turbulent flow in a pipe Re=78,000, 7,200 Mega grids, 2 Tera Bytes

4 Conclusion

The computational fluid dynamics (CFD) regarding not only the single-phase flows but also the two-phase flow plays an important role for the developments of advanced nuclear reactor systems. To establish the large-scale simulation procedure with higher prediction accuracy is very important for the detailed reactor-core thermal-design in the nuclear engineering. Moreover, the parallel visualization technique is very useful to understand the detailed flow and heat transfer phenomena.

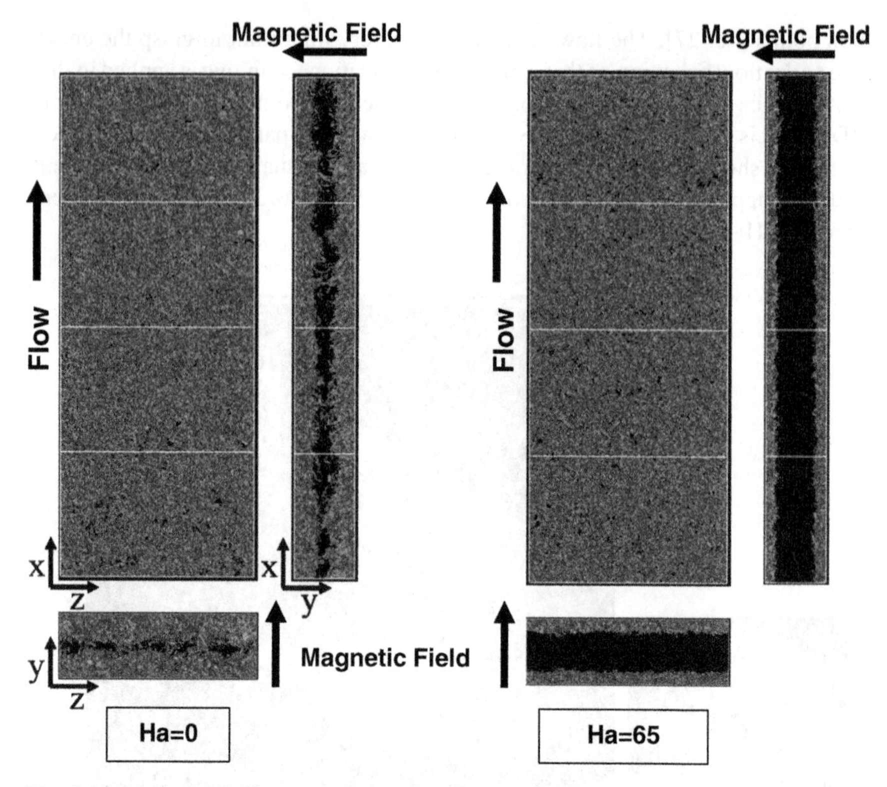

Fig. 6. Magnetic field effect on turbulent flow in a parallel channel: Second invariant contour of velocity tensor

[1] Kenning, D. B. R., Youyou, Y Pool boiling heat transfer on a thin plate: features revealed by liquid crystal thermography *International Journal of Heat and Mass Transfer*, Vol. 39, No. 15, 3117-3137, 1996.

[2] Mikic, B. B., Rohsenow, W. M Bubble growth rates in non-uniform temperature field *ASME J. Heat Transfer* 91, 245-250, 1969.

[3] Welch, S. W. J Local Simulation of Two-Phase Flows Including Interface Tracking with Mass Transfer *J. Comput. Phys.*, vol. 121, 142-154, 1995.

[4] Son, G., Dhir, V. K Numerical Simulation of Saturated Film Boiling on a Horizontal Surface *J. Heat Transfer*, Vol. 119, 525-533, 1997.

[5] Juric, D., Tryggvason, G Computations of boiling flows *Int. J. Multiphase Flow*, vol. 24, 387-410, 1998.

[6] Unverdi, S., Tryggvason, G A front-tracking method for viscous, incompressible, multifluid flows *J. Comput. Phys.*, vol. 100, 25-37, 1992.

[7] Kunugi, T MARS for Multiphase Calculation *Computational Fluid Dynamics Journal*, Vol.9, No. 1, 563-571, 2001
[8] Carey, V. P Liquid-vapor Phase-change Phenomena - An introduction to the Thermophysics of Vaporization and Condensation Processes in Heat Transfer Equipment, Chapters 5 and 6, Taylor & Francis, 1992.
[9] Brackbill, J. U., Kothe, D. B., Zemach, C A continuum method for modeling surface tension *J. Comput. Phys.*, Vol. 100, 335-354, 1992.
[10] Kunugi, T., Saito, N., Fujita, Y., Serizawa, A Direct Numerical Simulation of Pool and Forced Convective Flow Boiling Phenomena *Heat Transfer 2002*, Vol. 3, 497-502, 2002.
[11] Kelly, J. E., Kao, S. P., Kazimi, M. S THERMIT-2: A two-fluid model for light water reactor subchannel transient analysis *MIT-EL-81-014*, 1981.
[12] Thurgood, M. J COBRA/TRAC - A thermal-hydraulic code for transient analysis of nuclear reactor vessels and primary coolant systems, equation and constitutive models *NURREG/CR-3046, PNL-4385*, Vol. 1, R4, 1983.
[13] Sugawara, S., Miyamoto, Y FIDAS: Detailed subchannel analysis code based on the three-fluid and threefield model *Nuclear Engineering and Design*, vol.129, 146-161, 1990.
[14] Taylor, D TRAC-BD1/MOD1: An advanced best estimated computer program for boiling water reactor transient analysis, volume 1 - model description *NUREG/CR-3633*, 1984.
[15] Liles, D TRAC-PF1/MOD1: An advanced best-estimate computer program for pressurized water reactor analysis *NUREG/CR-3858, LA-10157-MS*, 1986.
[16] Iwamura, T., Okubo, T Development of reduced-moderation water reactor (RMWR) for sustainable energy supply *Proc. 13th Pacific Basin Nuclear Conference (PBNC 2002)*, Shenzhen, China, 1631-1637, 2002.
[17] Iwamura, T Core and system design of reduced-moderation water reactor with passive safety features *Proc. 2002 International Congress on Advanced in Nuclear Power Plants (ICAPP 2002)*, No.1030, Hollywood, Florida, USA, 2002.
[18] Okubo, T., Iwamura, T Design of small reduced-moderation water reactor (RMWR) with natural circulation cooling *Proc. International Conference on the New Frontiers of Nuclear Technology; Reactor Physics, Safety and High-Performance Computing (PHYSOR2002)*, Seoul, Korea, 2002.
[19] Earth Simulator Center, Annual report of the earth simulator center (April 2002- March 2003) *Japan Marine Science and Technology Center*, 2003.
[20] Takase, K., Tamai, H., Yoshida, H., Akimoto, H Development of a Best Estimate Analysis Method on Two-Phase Flow Thermal-Hydraulics for Reduced-Moderation Water Reactors *Proc. Best Estimate Twenty-O-Four (BE2004)*, Arlington, Washington D.C., USA, November, 2004.
[21] Yoshida, H., Takase, K., Ose, Y., Tamai, H., Akimoto, H Numerical simulation of liquid film around grid spacer with interface tracking method *Proc. International Conference on Global Environment and Advanced Nuclear Power Plants (GENES4/ANP2003)*, No.1111, Kyoto, Japan, 2003.
[22] Takase, K., Yoshida, H., Ose, Y., Kureta, M., Tamai, H., Akimoto, H Numerical investigation of two-phase flow structure around fuel rods with spacers by

large-scale simulations *Proc. 5th International Conference on Multiphase Flow (ICMF04)*, No.373, Yokohama, Japan, June, 2004.
[23] T. Yabe, T The constrained interpolation profile method for multiphase analysis *J. computational Physics*, vol.169, No.2, 556-593, 2001.
[24] Youngs, D.L Numerical methods for fluid dynamics, Edited by Morton K.W. & Baine, M.J. *Academic Press*, 273-285, 1982.
[25] Kureta, M., Liu, W., Tamai, H., Ohnuki, A., Mitsutake, T., Akimoto, H Development of Predictable Technology for Thermal/Hydraulic Performance of Reduced-Moderation Water Reactors (2) - Large-scale Thermal/Hydraulic Test and Model Experiments - *Proc. 2004 International Congress on Advanced in Nuclear Power Plants (ICAPP 2002)*, No.4056, Pittsburg, Pennsylvania, USA, June, 2004.
[26] Satake, S., Kunugi, T. Himeno, R High Reynolds Number Computation for Turbulent Heat Transfer in a Pipe Flow, M. Valero et al. (Eds.), Lecture Notes in Computer Science 1940 *High Performance Computing*, 514-523, 2000.
[27] Satake, S., Kunugi, T., Takase, K., Ose, Y., Naito, N Large Scale Structures of Turbulent Shear Flow via DNS, A. Veidenbaum et al. (Eds.), Lecture Notes in Computer Science 2858 *High Performance Computing*, 468-475, 2003.
[28] Satake, S., Kunugi, T., Takase, K., Ose, Y Direct numerical simulation of turbulent channel flow under a uniform magnetic field for large-scale structures at high Reynolds number *Physics of Fluid*, 18, 125106, 2006.

Scalable algebraic multilevel preconditioners with application to CFD

Andrea Aprovitola[1], Pasqua D'Ambra[2], Filippo Denaro[1], Daniela di Serafino[3], and Salvatore Filippone[4]

[1] Department of Aerospace and Mechanical Engineering, Second University of Naples, via Roma 29, I-81031 Aversa, Italy,
andrea.aprovitola@tele2.it & denaro@unina.it
[2] Institute for High-Performance Computing and Networking (ICAR), CNR, via Pietro Castellino 111, I-80131 Naples, Italy,
pasqua.dambra@na.icar.cnr.it
[3] Department of Mathematics, Second University of Naples, via Vivaldi 43, I-81100 Caserta, Italy,
daniela.diserafino@unina2.it
[4] Department of Mechanical Engineering, University of Rome "Tor Vergata", viale del Politecnico 1, I-00133, Rome, Italy,
salvatore.filippone@uniroma2.it

Abstract. The solution of large and sparse linear systems is one of the main computational kernels in CFD applications and is often a very time-consuming task, thus requiring the use of effective algorithms on high-performance computers. Preconditioned Krylov solvers are the methods of choice for these systems, but the availability of "good" preconditioners is crucial to achieve efficiency and robustness. In this paper we discuss some issues concerning the design and the implementation of scalable algebraic multilevel preconditioners, that have shown to be able to enhance the performance of Krylov solvers in parallel settings. In this context, we outline the main objectives and the related design choices of MLD2P4, a package of multilevel preconditioners based on Schwarz methods and on the smoothed aggregation technique, that has been developed to provide scalable and easy-to-use preconditioners in the Parallel Sparse BLAS computing framework. Results concerning the application of various MLD2P4 preconditioners within a large eddy simulation of a turbulent channel flow are discussed.

Key words: Preconditioning technique, Schwarz domain decomposition, Krylov methods.

1 Introduction

The solution of linear systems is ubiquitous in CFD simulations. For example, the integration of time-dependent PDEs modelling CFD problems, by using implicit or semi-implicit methods, leads to linear systems

D. Tromeur-Dervout (eds.), *Parallel Computational Fluid Dynamics 2008*,
Lecture Notes in Computational Science and Engineering 74,
DOI: 10.1007/978-3-642-14438-7_2,

$$Ax = b, \tag{1}$$

where A is a real $n \times n$ matrix, usually large and sparse, whose dimension and entries, conditioning, sparsity pattern and coupling among the variables may change during the simulation. Furthermore, because of the high computational requirements of large-scale CFD applications, parallel computers are often used and hence the matrix A is distributed among multiple processors.

Krylov solvers are the methods of choice for such linear systems, but their efficiency and robustness is strongly dependent on the coupling with suitable preconditioners that are able to provide a good approximation of the matrix A at a reasonable computational cost. Unfortunately, among the various available preconditioners, no one can be considered the "absolute winner" and experimentation is generally needed to select the best one for the problem under investigation. Furthermore, developing parallel implementations of preconditioners is not trivial, since the effectiveness and the parallel performance of a preconditioner often do not agree.

Algebraic multilevel preconditioners have received an increasing attention in the last fifteen years, as testified also by the development of software packages based on them [13, 21, 22, 27]. These preconditioners, which approximate the matrix A through a hierarchy of coarse matrices built by using information on A, but not on the geometry of the problem originating A (e.g. on the discretization grid of a PDE), are potentially able to automatically adapt to specific requirements of the problem to be solved [31]. Furthermore, they have shown effectiveness in enhancing the convergence and robustness of Krylov solvers in a variety of applications [25, 24].

In this paper we discuss some issues in the design and develoment of software implementing parallel algebraic multilevel domain decomposition preconditioners based on Schwarz methods. We start from a description of such preconditioners, to identify algorithmic features that are relevant to the development of parallel software (Section 2). Then we present MLD2P4, a package providing parallel algebraic multilevel preconditioners based on Schwarz domain decomposition methods, in the context of the Parallel Sparse BLAS (PSBLAS) computing framework for distributed-memory machines (Section 3). Specifically, we outline the main objectives and the related design choices in the development of this package. Furthermore, we report on the application of different MLD2P4 multilevel preconditioners, coupled with GMRES, to linear systems arising within a Large Eddy Simulation (LES) of incompressible turbulent channel flows, and discuss the results obtained in terms of numerical effectiveness and parallel performance (Sections 4 and 5). We give a few concluding remarks at the end of the paper (Section 6).

2 Algebraic Multilevel Schwarz Preconditioners

Domain decomposition preconditioners are based on the divide and conquer technique; from an algebraic point of view, the matrix to be preconditioned is divided into submatrices, a "local" linear system involving each submatrix is (approximately) solved, and the local solutions are used to build a preconditioner for the whole original matrix. This process often corresponds to dividing a physical domain associated to the original matrix into subdomains (e.g. in a PDE discretization), to (approximately) solving the subproblems corresponding to the subdomains and to building an approximate solution of the original problem from the local solutions. On parallel computers the number of submatrices usually matches the number of available processors.

Additive Schwarz (AS) preconditioners are domain decomposition preconditioners using overlapping submatrices, i.e. with some common rows, to couple the local information related to the submatrices (see, e.g., [29]). We assume that the matrix A in (1) has a symmetric nonzero pattern, which is not too restrictive if the matrix arises from some PDE discretization. By using the adjacency graph of A, we can define the so-called δ-overlap partitions of the set of vertices (i.e. row indices) $W = \{1, 2, \ldots, n\}$ [8]. Each set W_i^δ of a δ-overlap partition of W identifies a submatrix A_i^δ, corresponding to the rows and columns of A with indices in W_i^δ. Let R_i^δ be the (restriction) matrix which maps a vector v of length n onto the vector v_i^δ containing the components of v corresponding to the indices in W_i^δ. The matrix A_i^δ can be expressed as $A_i^\delta = R_i^\delta A (R_i^\delta)^T$ and the classical AS preconditioner is defined by

$$M_{AS}^{-1} = \sum_{i=1}^{m} (R_i^\delta)^T (A_i^\delta)^{-1} R_i^\delta,$$

where m is the number of sets of the δ-overlap partition and A_i^δ is assumed to be nonsingular. Its application to a vector v within a Krylov solver requires the following basic operations: restriction of v to the subspaces identified by the W_i^δ's, i.e. $v_i = R_i^\delta v$; solution of the linear systems $A_i^\delta w_i = v_i$; prolongation and sum of the w_i's, i.e. $w = \sum_{i=1}^m (R_i^\delta)^T w_i$. The linear systems at the second step are usually solved approximately, e.g. using incomplete LU (ILU) factorizations. Variants of the classical AS preconditioners exists; the most commonly used one is the Restricted AS (RAS) preconditioner, since it is generally more effective in terms of convergence rate and of parallel performance [9].

From the previous description we see that the AS preconditioners exhibit an intrinsic parallelism, which makes them suitable for a scalable implementation, i.e. such that the time per iteration of the preconditioned solver is kept constant as the problem size and the number of processors are proportionally scaled. On the other hand, the convergence rate of iterative solvers coupled with AS preconditioners deteriorates as the number of sets W_i^δ, and hence of processors, increases [29]. Therefore such preconditioners do not show algorithmic scalability, i.e. the capability of keeping constant the number of iterations to get a specified accuracy, as the number of processors grows.

Optimal Schwarz preconditioners, i.e. such that the number of iterations is bounded independently of the number of the submatrices (and of the size of the grid, when the matrix comes from a PDE discretization) can be obtained by introducing a global coupling among the overlapping partitions, through a coarse-space approximation A_C of the matrix A. The *two-level Schwarz* preconditioners are obtained by combining a basic Schwarz preconditioner with a coarse-level correction based on A_C. In this context, the basic preconditioner is called *smoother*.

In a pure algebraic setting, A_C is usually built with a Galerkin approach. Given a set W_C of *coarse vertices*, with size n_C, and a suitable $n_C \times n$ restriction matrix R_C, A_C is defined as $A_C = R_C A R_C^T$ and the coarse-level correction operator to be combined with a generic AS preconditioner M_{1L} is obtained as

$$M_C^{-1} = R_C^T A_C^{-1} R_C,$$

where A_C is assumed to be nonsingular. The application of M_C^{-1} to a vector v corresponds to the restriction $w = R_C v$, to the solution of the linear system $A_C y = w$ and to the prolongation $z = R_C^T y$.

The operators M_C and M_{1L} may be combined in either an additive or a multiplicative framework. In the former case, at each iteration of a Krylov solver, M_C^{-1} and M_{1L}^{-1} are independently applied to the relevant vector v and the results are added. This corresponds to the *two-level additive* Schwarz preconditioner

$$M_{2LA}^{-1} = M_C^{-1} + M_{1L}^{-1}.$$

In the multiplicative case, a possible combination consists in applying first M_C^{-1} and then M_{1L}^{-1}, as follows: coarse-level correction of v, i.e. $w = M_C^{-1} v$; computation of the residual $y = v - Aw$; smoothing of y and update of w, i.e. $z = w + M_{1L}^{-1} y$. These steps correspond to the following Schwarz preconditioner, that we refer to as *two-level hybrid post-smoothed* preconditioner:

$$M_{2LH-POST}^{-1} = M_{1L}^{-1} + \left(I - M_{1L}^{-1} A\right) M_C^{-1}.$$

Similarly, the smoother may be applied before the coarse-level correction operator (*two-level hybrid pre-smoothed* preconditioner), or both before and after the correction (*two-level hybrid symmetrized* preconditioner).

An algebraic approach to the construction of the set of coarse vertices is provided by the *smoothed aggregation* [4]. The basic idea is to build W_C by suitably grouping the vertices of W into disjoint subsets (aggregates), and to define the coarse-to-fine space transfer operator R_C^T by applying a suitable smoother to a simple piecewise constant prolongation operator. The aggregation algorithms are typically sequential and different parallel versions of them have been developed with the goal of achieving a tradeoff between scalability and effectiveness [32]. The simplest parallel aggregation strategy is the *decoupled* one, in which every processor independently applies the sequential algorithm to the subset of W assigned to it in the initial data distribution. This version is embarrassingly parallel, but may produce non-uniform aggregates near boundary vertices, i.e. near vertices adjacent to vertices in other processors, and is strongly dependent on the number of processors and on the initial partitioning of the matrix A. Nevertheless, the decoupled aggregation has been shown to produce good results in practice [32].

Preconditioners that are optimal in the sense defined above do not necessarily correspond to minimum execution times. For example, when the size of the system to be preconditioned is very large, the use of many processors, i.e. of many small submatrices, may lead to large coarse-level systems, whose exact solution is generally computationally expensive and deteriorates the implementation scalability of the basic Schwarz preconditioner. A possible remedy is to solve the coarse level-system approximately; this is generally less time expensive, but the correction, and hence the preconditioner, may lose effectiveness. Therefore, it seems natural to use a recursive approach, in which the coarse-level correction is re-applied starting from the current coarse-level system. The corresponding preconditioners, called *multilevel* preconditioners, can significantly reduce the computational cost of preconditioning with respect to the two-level case. Additive and hybrid multilevel preconditioners are obtained as direct extensions of the two-level counterparts; a detailed descrition may be found in [29, Chapter 3]. In practice, finding a good combination of the number of levels and of the coarse-level solver is a key point in achieving the effectiveness of a multilevel preconditioner in a parallel computing setting; the choice of these two features is generally dependent on the characteristics of the linear system to be solved and on the characteristics of the parallel computer.

3 The MLD2P4 software package

The *MultiLevel Domain Decomposition Parallel Preconditioners Package based on PSBLAS (MLD2P4)* [13] implements multilevel Schwarz preconditioners, that can be used with Krylov solvers available in the PSBLAS framework [20] for the solution of system (1). Both additive and hybrid multilevel variants are available; the basic AS preconditioners are obtained by considering just one level. An algebraic approach, based on the decoupled smoothed aggregation technique, is used to generate a sequence of coarse-level corrections to any basic AS preconditioner, as explained in Section 2. Since the choice of the coarse-level solver is important to achieve a trade-off between optimality and efficiency, different coarse-level solvers are provided, i.e. sparse distributed and sequential LU solvers, as well as distributed block-Jacobi ones, with ILU or LU factorizations of the blocks. More details on the various preconditioners implemented in the package can be found in [13].

The package has been written in Fortran 95, to enable immediate interfacing with Fortran application codes, while following a modern object-based approach through the exploitation of features such as abstract data type creation, functional overloading and dynamic memory management. Single and double precision implementations of MLD2P4 have been developed for both real and complex matrices, all usable through a single generic interface.

The main “object” in MLD2P4 is the *preconditioner data structure*, containing the matrix operators and the parameters defining a multilevel Schwarz preconditioner. According to the object-oriented paradigm, the user does not access this structure directly, but builds, modifies, applies and destroys it through a set of MLD2P4 routines. The preconditioner data structure has been implemented as a Fortran 95 derived data type; it basically consists of an array of base preconditioners, where a base preconditioner is again a derived data type, storing the part of the preconditioner associated to a certain level and the mapping from it to the next coarser level. This choice enables to reuse, at each level, the same routines for building and applying the preconditioner, and to combine them in various ways, to obtain different preconditioners. Furthermore, starting from a description of the preconditioners in terms of basic sparse linear algebra operators, as outlined in Section 2, the previous routines have been implemented as combinations of building blocks performing basic sparse matrix computations (for more details see [6, 12]). The PSBLAS library, which includes parallel versions of most of the Sparse BLAS computational kernels proposed in [17] and sparse matrix management functionalities, has been used as software layer providing the building blocks. The vast majority of data communication operations required by MLD2P4 have been encapsulated into PSBLAS routines; only very few direct MPI calls occur in the package. The choice of a modular approach, based on the PSBLAS library, has been driven by objectives such as extensibility, portability, sequential and parallel performance.

The modular design has naturally led to a layered software architecture, where three main layers can be identifed. The lower layer consists of the PSBLAS kernels. The middle one implements the construction of the preconditioners and their application within a Krylov solver. It includes the functionalities for building and applying various types of basic Schwarz preconditioners, for generating coarse matrices from fine ones and for solving coarse-level linear systems; furthermore, it provides the routines combining these functionalities into multilevel preconditioners. The middle layer includes also interfaces to the third-party software packages UMFPACK [14], SuperLU [15] and SuperLU_DIST [16], performing sequential or distributed sparse LU factorizations and related triangular system solutions, that can be exploited at different levels of the multilevel preconditioners. The upper layer provides a uniform and easy-to-use interface to all the preconditioners implemented in MLD2P4. It consists of few black-box routines suitable for users with different levels of expertise; non-expert users can easily select the default basic and multilevel preconditioners, while expert ones can choose among various preconditioners, by a proper setting of different parameters.

A more detailed description of MLD2P4 can be found in [7, 13]; a deep analysis of the effectiveness and parallel performance of MLD2P4 preconditioners, as well as a comparison with state-of-the-art multilevel preconditioning software, can be found in [7, 12].

4 Using MLD2P4 in the LES of turbulent channel flows

MLD2P4 has been used within a Fortran 90 code performing a LES of turbulent incompressible channel flows, in order to precondition linear systems which are a main computational kernel in an Approximate Projection Method (APM). We briefly describe the numerical procedure implemented in the code, to show how MLD2P4 has been exploited; for more details the user is referred to [1].

Incompressible and homothermal turbulent flows can be modelled as initial boundary-value problems for the Navier-Stokes (N-S) equations. We consider a non-dimensional weak conservation form of these equations, involving the volume average of the velocity field $\mathbf{v}(\mathbf{x},t)$ [23]. In the LES approach $\mathbf{v}(\mathbf{x},t)$ is decomposed into two contributions, i.e. $\mathbf{v}(\mathbf{x},t) = \bar{\mathbf{v}}(\mathbf{x},t) + \mathbf{v}'(\mathbf{x},t)$, where $\bar{\mathbf{v}}$ is the resolved (or large-scale) filtered field and $\mathbf{v}'$ is the non-resolved (or small-scale) field. In particular, the so-called top-hat filter (with uniform filter width) is equivalent to the volume-average operator, hence the N-S equations in weak conservation form can be considered as filtered governing equations [28].

In our application, an approximate differential deconvolution operator, $A_{\mathbf{x}}$, is applied to $\bar{\mathbf{v}}$, to recover the frequency content of the velocity field near the grid cutoff wavenumber, which has been smoothed by the application of the volume-average operator [2]. The resulting deconvolution-based N-S equations have the following form:

$$\int_{\partial\Omega(\mathbf{x})} \widetilde{\mathbf{v}} \cdot \mathbf{n}\, dS = s, \qquad A_{\mathbf{x}}^{-1}\left(\frac{\partial \widetilde{\mathbf{v}}}{\partial t}\right) = \mathbf{f}_{conv} + \mathbf{f}_{diff} + \mathbf{f}_{press} + \mathbf{f}_{sgs}, \tag{2}$$

where $\Omega(\mathbf{x})$ is a finite volume contained into the region of the flow, $\widetilde{\mathbf{v}} = A_{\mathbf{x}}(\bar{\mathbf{v}})$, $\mathbf{f}_{conv}$, $\mathbf{f}_{diff}$ and $\mathbf{f}_{pres}$ are the convective, diffusive and pressure fluxes of the filtered equations, and $\mathbf{f}_{sgs}$ contains the unresolved subgrid-scale terms, that can be modeled either explicitly or implicitly. We disregard the source term s and adopt an implicit subgrid-scale modelling, hence $\mathbf{f}_{sgs} = 0$ (see [2] for more details).

The computational domain is discretized by using a structured Cartesian grid. Uniform grid spacings are used in the stream-wise (x) and span-wise (z) directions, where the flow is assumed to be homogenous (periodic conditions are imposed on the related boundaries). A non-uniform grid spacing, refined near the walls, is considered in the wall-normal direction (y), where no-slip boundary conditions are prescribed. The equations (2) are discretized in space by using a finite volume method, with flow variables co-located at the centers of the control volumes; a third-order multidimensional upwind scheme is applied to the fluxes.

A time-splitting technique based on an APM is used to decouple the velocity from the pressure in the deconvolved momentum equation (see [3] for the details). According to the Helmholtz-Hodge decomposition theorem, the unknown velocity field $\tilde{\mathbf{v}}$ is evaluated at each time step through a predictor-corrector approach based on the following formula:

$$\widetilde{\mathbf{v}}^{n+1} = \mathbf{v}^* - \Delta t \nabla \phi^{n+1},$$

where $\mathbf{v}^*$ is an intermediate velocity field, Δt is the time step, and ϕ is a scalar field such that $\nabla\phi$ is an $O(\Delta t)$ approximation of the pressure gradient. In the predictor stage, $\mathbf{v}^*$ is computed using a second-order Adams-Bashforth/Crank-Nicolson semi-implicit scheme to the deconvolved momentum equation, where the pressure term is neglected. The correction stage requires the computation of $\nabla\phi^{n+1}$ to obtain a velocity field $\widetilde{\mathbf{v}}$ which is divergence-free in a discrete sense. To this aim, ϕ^{n+1} is obtained by solving a Poisson equation with non-homogeneous Neumann boundary conditions, which has a solution, unique up to an additive constant, provided that a suitable compatibility condition is satisfied. By discretizing this equation (usually called pressure equation) with a second-order central finite volume scheme, we have

$$
\begin{aligned}
&-\phi_{i,j,k}^{n+1}\left[\frac{2}{\Delta x^2}+\frac{2}{\Delta z^2}+\frac{1}{h_y}\left(\frac{1}{\Delta y_{j+1}}+\frac{1}{\Delta y_j}\right)\right]+\frac{\phi_{i+1,j,k}^{n+1}+\phi_{i-1,j,k}^{n+1}}{\Delta x^2}+\\
&+\frac{\phi_{i,j,k+1}^{n+1}+\phi_{i,j,k-1}^{n+1}}{\Delta z^2}+\frac{\phi_{i,j+1,k}^{n+1}}{h_y\Delta y_{j+1}}+\frac{\phi_{i,j-1,k}^{n+1}}{h_y\Delta y_j}=b,
\end{aligned}
\tag{3}
$$

where $\phi_{i,j,k}^{n+1}$ is the (approximated) value of ϕ^{n+1} in the center of the (i,j,k)-th control volume and the right-hand side b depends on the intermediate velocity field, on the grid spacings and on the time step. The equations (3) are suitably modified near the boundaries to accomplish the divergence-free velocity constraint.

The pressure equations form a sparse linear system, whose size is the number of cells of the discretization grid and hence, owing to resolution needs, increases as the Reynolds number grows. Usual orderings of the grid cells lead to a system matrix, A, which has a symmetric sparsity pattern, but is unsymmetric in value, because of the non-uniform grid spacing in the y direction. The linear system is singular but results to be compatible, since a discrete compatibility condition is ensured according to the prescribed boundary conditions.

The solution of the pressure system at each time step of the APM-based procedure usually accounts for a large part of the whole simulation time, hence it requires very efficient solvers. It is easy to verify that $\mathscr{R}(A)\cap\mathscr{N}(A)=\{0\}$, where $\mathscr{R}(A)$ and $\mathscr{N}(A)$ are the range space and the null space of A; this property, coupled with the compatibility of the linear system, ensures that in exact arithmetic the GMRES method computes a solution before a breakdown occurs [5]. In practice, the Restarted GMRES (RGMRES) method is used, because of the high memory requirements of GMRES, and the application of an effective preconditioner that reduces the condition number of the restriction of A to $\mathscr{R}(A)$ is crucial to decrease the number of iterations to achieve a required accuracy in the solution.[5]

Different MLD2P4 preconditioners, coupled with the RGMRES solver implemented in PSBLAS, have been applied to the discrete pressure equation. Since the original LES code is sequential and matrix free, this has required the matrix A to be assembled and distributed among multiple processors. The sparse matrix management facilities provided by PSBLAS have been used to perform these operations. Note that the cost of this pre-processing step is not significant, since A does not change throughout the simulation and a very large number of time steps is required to obtain a fully developed flow. Similarly, the time for the construction of any preconditioner can be neglected, since this task must be performed only once, at the beginning of the simulation.

5 Numerical experiments

The LES code exploiting the MLD2P4 preconditioners has been run to simulate biperiodical channel flows with different Reynolds numbers. For the sake of space, we

[5] The computed solution has the form x_0+z, where $z\in\mathscr{K}(A,r_0)\subseteq\mathscr{R}(A)$ and $\mathscr{K}(A,r_0)$ is a Krylov space of suitable dimension associated to A and $r_0=b-Ax_0$.

report only the results concerning a single test case, focusing on the effectiveness and the performance of various MLD2P4 proconditioners in the solution of the discrete pressure system.

For the selected test problem, the domain size is $2\pi l \times 2l \times \pi l$, where l is the channel half-width, and the Reynolds number referred to the shear velocity is $Re_\tau = 1050$; a Poiseuille flow, with a random Gaussian perturbation, is assumed as initial condition. The computational grid has $64 \times 96 \times 128$ cells, leading to a pressure system matrix with dimension 786432 and 5480702 nonzero entries. The time step Δt is 10^{-4}, to meet stability requirements.

The experiments have been carried out on a HP XC 6000 Linux cluster with 64 bi-processor nodes, operated by the Naples branch of ICAR-CNR. Each node comprises an Intel Itanium 2 Madison processor with clock frequency of 1.4 Ghz and is equipped with 4 GB of RAM; it runs HP Linux for High Performance Computing 3 (kernel 2.4.21). The main interconnection network is Quadrics QsNetII Elan 4, which has a sustained bandwidth of 900 MB/sec. and a latency of about 5 μsec. for large messages. The GNU Compiler Collection, v. 4.3, and the HP MPI implementation, v. 2.01, have been used. MLD2P4 1.0 and PSBLAS 2.3 have been installed on top of ATLAS 3.6.0 and BLACS 1.1.

The pressure matrix has been distributed among the processors according to a 3D block decomposition of the computational grid. The restarting parameter of RGMRES has been set to 30; to stop the iterations it has been required that the ratio between the 2-norm of the current and the initial residual is less than 10^{-7}. At each time step, the solution of the pressure equation computed at the previous time step has been choosen as starting guess, except at the first time step, where the null vector has been considered. Multilevel hybrid post-smoothed preconditioners have been applied, as right preconditioners, using 2, 3 and 4 levels. Different solvers have been considered at the coarsest level: 4 parallel block-Jacobi sweeps, with ILU(0) or LU on the blocks, have been applied to the coarsest system, distributed among the processors (the corresponding preconditioners are denoted by xLDI and xLDU, respectively, where x is the number of levels); alternatively, the coarsest matrix has been replicated on all the processors and a sequential LU factorization has been used (the corresponding preconditioners are denoted by xLRU). All the LU factorizations have been computed by UMFPACK. RAS has been used as smoother, with overlap 0 and 1 and ILU(0) on each submatrix; RAS has been also applied as preconditioner, for comparison purposes. Similar results have been obtained for each preconditioner with both the overlap values; for the sake of space, we show here only the results concerning the overlap 0.

In Table 1 we report the mean number of RGMRES iterations over the first 10 time steps, on 1, 2, 4, 8, 16, 32 and 64 processors. Although this number of time steps is very small compared with the the total number of steps to have a fully developed flow (about 10^5), it is large enough to investigate the behaviour of RGMRES with the various preconditioners. No test data are available for 2LRU because the coarse matrix at level 2 is singular, and therefore the UMFPACK factorization fails; moreover, on one processor 2LDU is the same as 2LRU, thus it is missing too. The smallest iteration count is obtained by 3LRU, followed by 4LRU; in both cases, the

Table 1. Mean number of iterations of preconditioned RGMRES on the pressure equation

Procs	2LDI	2LDU	3LDI	3LDU	3LRU	4LDI	4LDU	4LRU	RAS
1	42	—	22	18	18	18	18	18	132
2	44	17	21	18	17	18	18	17	142
4	44	21	22	20	18	21	21	20	151
8	46	24	24	23	18	23	23	21	160
16	45	25	25	23	18	22	22	20	159
32	50	33	26	26	19	26	26	22	177
64	49	34	26	26	19	25	25	20	175

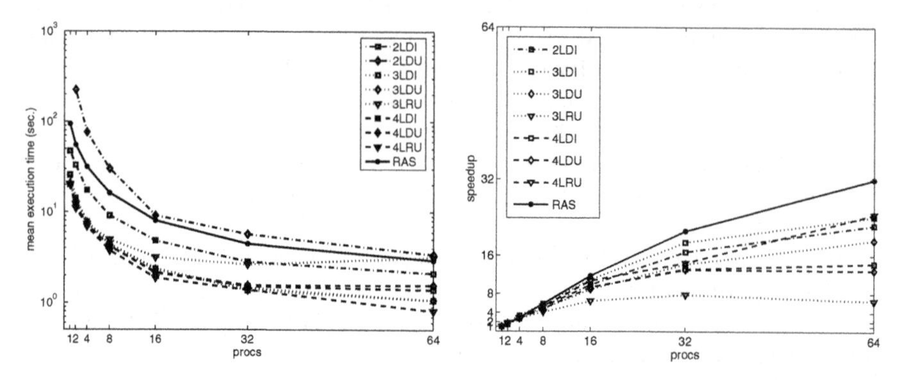

Fig. 1. Mean execution times (left) and speedup (right) of preconditioned RGMRES on the pressure equation.

number of iterations is bounded independently of the number of processors. 3LDI, 3LDU, 4LDI and 4LDU are less effective in reducing the iterations; the main reason is the approximate solution of the corresponding coarsest-level systems through the block-Jacobi method. The number of iterations is about the same with 3LDU, 4LDI and 4LDU, while it is slightly larger with 3LDI, according to the lower accuracy achieved by 3LDI in the solution of its coarsest-level systems (with 4 levels, the local submatrices at the coarsest level are very small and almost dense, and the ILU(0) factorization is practically equivalent to the LU one). For the same accuracy reasons, the two-level preconditioners are the least effective among the multilevel ones. The data concerning RAS confirm the effectiveness of the coarse-level corrections in reducing the iterations almost independently of the number of processors.

In Figure 1 we show the mean execution time, in seconds, of RGMRES with the various preconditioners and the related speedup. The speedup for 2LDU is missing, because 2LDU does not work on 1 processor (using as reference 2LDU on 2 processors gives a misrepresentation of the performance of 2LDU). We see that the smallest execution times are obtained with 4LRU, although 3LRU performs better in terms of iteration count; this is because the small size of the coarsest matrix in 4LRU yields a significant time saving in the solution of the coarsest-level system. Conversely, the execution times concerning 3LRU are greater than the times of the remaining three- and four-level preconditioners, just because of the cost of dealing with the coarsest matrix. The speedup achieved by 4LRU is satisfactory (in particular, it is 14.4 on 32 processors and 24.6 on 64), while the speedup of 3LRU confirms that, for the problem at hand, this preconditioner lacks parallel performance. Using 3LDI, 3LDU, 4LDI and 4LDU leads to close execution times, except on 64 processors, on which the three-level preconditioners are slightly faster. Accordingly, on 64 processors, 3LDI and 3LDU show a better speedup than their four-level counterparts. 2LDI and 2LDU generally require a larger time than the other multilevel preconditioners; this is in agreement with the larger size of the coarsest matrix and with the iteration count of these preconditioners. Furthermore, 2LDU is also more costly than RAS. Finally, RAS achieves the highest speedup values (e.g., 32.0 on 64 processors), despite an increase in the number of iterations of more than 30% when going from 1 to 64 processors; this confirms that a tradeoff between optimality and parallelism must be sought after when using Schwarz preconditioners.

6 Conclusions

In this paper we focused on the design and implementation of scalable algebraic multilevel Schwarz preconditioners, which are recognized as effective tools for obtaining efficiency and robustness of Krylov methods in the solution of linear systems arising in CFD (and other) applications. We described a Fortran 95 package, named MLD2P4, which provides various versions of the above preconditioners through a uniform and simple interface, thus giving the user the possibility of making the most effective choice for his specific problem. Finally, we discussed the application of different MLD2P4 preconditioners in the solution of linear systems arising in the numerical simulation of an incompressible turbulent channel flow. The results obtained demonstrate that this application may benefit from the use of MLD2P4 and show the potential of the package in the development of scalable codes for CFD simulations.

[1] Andrea Aprovitola, Pasqua D'Ambra, Filippo Denaro, Daniela di Serafino, and Salvatore Filippone. Application of parallel algebraic multilevel domain decomposition preconditioners in large eddy simulations of wall-bounded turbulent flows: first experiments. Technical Report RT-ICAR-NA-07-02, ICAR-CNR, Naples, Italy, 2007.

[2] Andrea Aprovitola and Filippo M. Denaro. On the application of congruent upwind discretizations for large eddy simulations. *J. Comput. Phys.*, 194(1):329–343, 2004.

[3] Andrea Aprovitola and Filippo M. Denaro. A non-diffusive, divergence-free, finite volume-based double projection method on non-staggered grids. *Internat. J. Numer. Methods Fluids*, 53(7):1127–1172, 2007.

[4] Marian Brezina and Petr Vaněk. A black-box iterative solver based on a two-level Schwarz method. *Computing*, 63(3):233–263, 1999.

[5] Peter N. Brown and Homer F. Walker. GMRES on (nearly) singular systems. *SIAM J. Matrix Anal. Appl.*, 18(1):37–51, 1997.

[6] Alfredo Buttari, Pasqua D'Ambra, Daniela di Serafino, and Salvatore Filippone. Extending PSBLAS to build parallel schwarz preconditioners. In K. Madsen J. Dongarra and J. Wasniewski, editors, *Applied Parallel Computing*, volume 3732 of *Lecture Notes in Computer Science*, pages 593–602, Berlin/Heidelberg, 2006. Springer.

[7] Alfredo Buttari, Pasqua D'Ambra, Daniela di Serafino, and Salvatore Filippone. 2LEV-D2P4: a package of high-performance preconditioners for scientific and engineering applications. *Appl. Algebra Engrg. Comm. Comput.*, 18(3):223–239, 2007.

[8] Xiao-Chuan Cai and Yousef Saad. Overlapping domain decomposition algorithms for general sparse matrices. *Numer. Linear Algebra Appl.*, 3(3):221–237, 1996.

[9] Xiao-Chuan Cai and Marcus Sarkis. A restricted additive Schwarz preconditioner for general sparse linear systems. *SIAM J. Sci. Comput.*, 21(2):792–797, 1999.

[10] Xiao-Chuan Cai and Olof B. Widlund. Domain decomposition algorithms for indefinite elliptic problems. *SIAM J. Sci. Statist. Comput.*, 13(1):243–258, 1992.

[11] Tony F. Chan and Tarek P. Mathew. Domain decomposition algorithms. In *Acta numerica, 1994*, Acta Numer., pages 61–143. Cambridge Univ. Press, Cambridge, 1994.

[12] Pasqua D'Ambra, Daniela di Serafino, and Salvatore Filippone. On the development of PSBLAS-based parallel two-level Schwarz preconditioners. *Appl. Numer. Math.*, 57(11-12):1181–1196, 2007.

[13] Pasqua D'Ambra, Daniela di Serafino, and Salvatore Filippone. *MLD2P4 User's and Reference Guide*, September 2008. Available from http://www.mld2p4.it.

[14] Timothy A. Davis. Algorithm 832: UMFPACK V4.3—an unsymmetric-pattern multifrontal method. *ACM Trans. Math. Software*, 30(2):196–199, 2004.

[15] James W. Demmel, Stanley C. Eisenstat, John R. Gilbert, Xiaoye S. Li, and Joseph W. H. Liu. A supernodal approach to sparse partial pivoting. *SIAM J. Matrix Anal. Appl.*, 20(3):720–755, 1999.

[16] James W. Demmel, John R. Gilbert, and Xiaoye S. Li. An asynchronous parallel supernodal algorithm for sparse Gaussian elimination. *SIAM J. Matrix Anal. Appl.*, 20(4):915–952, 1999.

[17] Iain S. Duff, Michele Marrone, Giuseppe Radicati, and Carlo Vittoli. Level 3 basic linear algebra subprograms for sparse matrices: a user-level interface. *ACM Trans. Math. Software*, 23(3):379–401, 1997.

[18] Evridiki Efstathiou and Martin J. Gander. Why restricted additive Schwarz converges faster than additive Schwarz. *BIT*, 43(suppl.):945–959, 2003.

[19] Salvatore Filippone and Alfredo Buttari. *PSBLAS: User's and Reference Guide*, 2008. Available from http://www.ce.uniroma2.it/psblas/.

[20] Salvatore Filippone and Michele Colajanni. PSBLAS: A library for parallel linear algebra computation on sparse matrices. *ACM Trans. Math. Software*, 26(4):527–550, 2000. See also http://www.ce.uniroma2.it/psblas/.

[21] Michael W. Gee, Christofer M. Siefert, Jonathan J. Hu, Ray S. Tuminaro, and Marzio G. Sala. ML 5.0 smoothed aggregation user's guide. Technical Report SAND2006-2649, Sandia National Laboratories, Albuquerque, NM, and Livermore, CA, 2006.
[22] Van Emden Henson and Ulrike Meier Yang. BoomerAMG: A parallel algebraic multigrid solver and preconditioner. *Appl. Numer. Math.*, 41:155–177, 2000.
[23] Randall J. LeVeque. *Finite volume methods for hyperbolic problems*. Cambridge Texts in Applied Mathematics. Cambridge University Press, Cambridge, 2002.
[24] Paul T. Lin, Marzio G. Sala, John N. Shadid, and Ray S. Tuminaro. Performance of fully-coupled algebraic multilevel domain decomposition preconditioners for incompressible flow and transport. *Int. J. Numer. Meth. Eng.*, 67:208–225, 2006.
[25] Gérard Meurant. Numerical experiments with algebraic multilevel preconditioners. *Electron. Trans. Numer. Anal.*, 12:1–65 (electronic), 2001.
[26] Yousef Saad. *Iterative methods for sparse linear systems*. Society for Industrial and Applied Mathematics, Philadelphia, PA, second edition, 2003.
[27] Yousef Saad and Masha Sosonkina. pARMS: A package for the parallel iterative solution of general large sparse linear systems user's guide. Technical Report UMSI2004-8, Minnesota Supercomputing Institute, Minneapolis, MN, 2004.
[28] Pierre Sagaut. *Large eddy simulation for incompressible flows. An introduction*. Scientific Computation. Springer-Verlag, Berlin, third edition, 2005.
[29] Barry F. Smith, Petter E. Bjørstad, and William D. Gropp. *Domain decomposition. Parallel multilevel methods for elliptic partial differential equations*. Cambridge University Press, Cambridge, 1996.
[30] Marc Snir, Steve Otto, Steven Huss-Lederman, David W. Walker, and Jack J. Dongarra. *MPI: The Complete Reference. Vol. 1 – The MPI Core*. Scientific and Engineering Computation. The MIT Press, Cambridge, MA, second edition, 1998.
[31] Klaus Stüben. A review of algebraic multigrid. *J. Comput. Appl. Math.*, 128(1-2):281–309, 2001.
[32] Ray S. Tuminaro and Charles Tong. Parallel smoothed aggregation multigrid: Aggregation strategies on massively parallel machines. In *Proceedings of the 2000 ACM/IEEE conference on Supercomputing*, Dallas, TX, 2000. CDROM.

Acceleration of iterative solution of series of systems due to better initial guess

Damien Tromeur-Dervout[1] and Yuri Vassilevski[2]

[1] University of Lyon, University Lyon 1, CNRS, Institut Camille Jordan F-69622 Villeurbanne Cedex, France dtromeur@cdcsp.univ-lyon1.fr
[2] Institute of Numerical Mathematics, Gubkina str. 8, 119333 Moscow, Russia vasilevs@dodo.inm.ras.ru

Abstract. Efficient choice of the initial guess for the iterative solution of series of systems is considered. The series of systems are typical for unsteady nonlinear fluid flow problems. The history of iterative solution at previous time steps is used for computing a better initial guess. This strategy is applied for two iterative linear system solvers (GCR and GMRES). A reduced model technique is developed for implicitly discretized nonlinear evolution problems. The technique computes a better initial guess for the inexact Newton method. The methods are successfully tested in parallel CFD simulations. The latter approach is suitable for GRID computing as well.

Key words: Krylov method, Newton method, Proper Orthogonal Decomposition, client-server architecture, grid computing

1 Introduction

Large systems of equations are solved iteratively. Any iterative technique consists of three basic procedures: the choice of the initial guess, the computation of the next iterate, and the stopping criterion. Each procedure influences the efficiency of the iterative solution in its own way: smart start, fast convergence, and prevention of oversolving. The fast convergence is conventionally achieved by a combination of an appropriate Krylov type method and a preconditioning technique. The stopping criteria have been considered both for nonlinear (sufficient reduction of residual for inner solves) [5] and linear solves (reliable estimate of true errors) [7, 8, 3].

In this paper we address the choice of the initial guess when a series of systems produced by time stepping schemes has to be solved [15]. In particular, we consider series of linear systems with the same nonsymmetric matrix and with different nonsymmetric matrices. For Newton type solvers of nonlinear systems generated by fully implicit discretizations, we present a new method of the choice of an initial guess based on the model reduction [14, 15, 16]. The solution of the reduced model provides more efficient initial guess than the solution from the previous time step. In

D. Tromeur-Dervout (eds.), *Parallel Computational Fluid Dynamics 2008*,
Lecture Notes in Computational Science and Engineering 74,
DOI: 10.1007/978-3-642-14438-7_3,

spite of an extra work spent on the solution of the reduced model, the total complexity of each time step decreases considerably, due to the super-linear convergence of the inexact Newton solver.

The considered methods are parallel since they use basic linear algebra operations only. Numerical results presented below show actual acceleration in parallel CFD simulations and demonstrate the applicability of the reduced model method in GRID computing [16]. The series of systems are generated by time stepping methods for different formulations of unsteady Navier-Stokes equations.

The paper outline is as follows. In section 2 we consider series of linear systems with the same matrix generated by the projection method for the unsteady 3D Navier-Stokes equations. In section 3 we consider series of linear systems with different matrices produced by the projection method for the unsteady Low Mach number equations. In section 4 we present a new algorithm INB-POD for series of nonlinear systems which appear in a fully implicit scheme for the unsteady Navier-Stokes equations in the streamfunction-vorticity formulation.

2 Linear systems with the same matrix

We consider a series of linear systems appearing in the numerical simulation of unsteady incompressible fluid flow. The projection scheme for the unsteady Navier-Stokes equations with Dirichlet boundary condition is based on the pressure correction. Its idea is to project a predicted velocity field onto the divergence-free space by solving an elliptic equation for the pressure correction. In discrete form, the projection reduces to a series of linear systems

$$Ax = b^k \tag{1}$$

with different right hand sides b^k. The standard choice of the initial guess is the zero vector since the unknown solution x represents a pressure increment.

The matrix A is the product of finite difference discretizations of the divergence and gradient operators, and is singular and stiff. Symmetry of the matrix depends on whether the divergence and gradient mesh operators are conjugate with respect to Euclidian scalar product; in our discretization they are not conjugate due to particular finite difference stencils and the matrix A is nonsymmetric.

For the iterative scheme we use the Generalized Conjugate Residual (GCR) method

1. Compute $r_0 = b - Ax_0$. Set $p_0 = r_0$.
2. For $j = 0, 1, 2, \dots,$ until convergence Do:
3. $\alpha_i = \dfrac{(r_j, Ap_j)}{(Ap_j, Ap_j)}$
4. $x_{j+1} = x_j + \alpha_j p_j$
5. $r_{j+1} = r_j - \alpha_j Ap_j$
6. Compute $\beta_{ij} = -\dfrac{(Ar_{j+1}, Ap_i)}{(Ap_i, Ap_i)}$, for $i = 0, 1, \dots, j$
7. $p_{j+1} = r_{j+1} + \sum\limits_{j=0}^{j} \beta_{ij} p_i$

8. EndDo

Although GMRES method requires less vectors to be accumulated, in comparison to the GCR method, and provides similar convergence rate, the additional data available in GCR allow us to construct a very good initial guess. Indeed, besides the Krylov subspace vectors $\mathscr{K} = \{p_j\}_{j=1}^{k}$ which are A^TA-orthogonal, we possess extra data, vectors $\{Ap_j\}_{j=1}^{k}$. Since the projection of b onto the space $A\mathscr{K}$ is equal to $\sum_{j=1}^{k}(b,Ap_j)(Ap_j,Ap_j)^{-1}Ap_j$, the projection of the solution $x = A^{-1}b$ onto $\mathscr{K}$ is [17]

$$\hat{x} = \sum_{j=1}^{k} \frac{(b,Ap_j)}{(Ap_j,Ap_j)} p_j. \tag{2}$$

This observation implies that the projection of an unknown solution onto the accumulated Krylov subspace may be easily computed (k scalar products). The accumulation of the subspaces $\mathscr{K}$ and $A\mathscr{K}$ may be continued for several right hand sides, if the initial guess for each subsequent solve is computed by (2), and x_0 is set to $\hat{x}$. It is evident that the larger is the subspace $\mathscr{K}$, the better is the approximation $\hat{x}$ to x. The accumulation of vectors p_j and Ap_j is limited by a practical capacity of the computer memory: as soon as it is exhausted, the data p_j and Ap_j are erased and the new accumulation process begins. Another restriction for the number of accumulated vectors is imposed by the expense of computation of k scalar products. It should be smaller than the cost of the iteration stage.

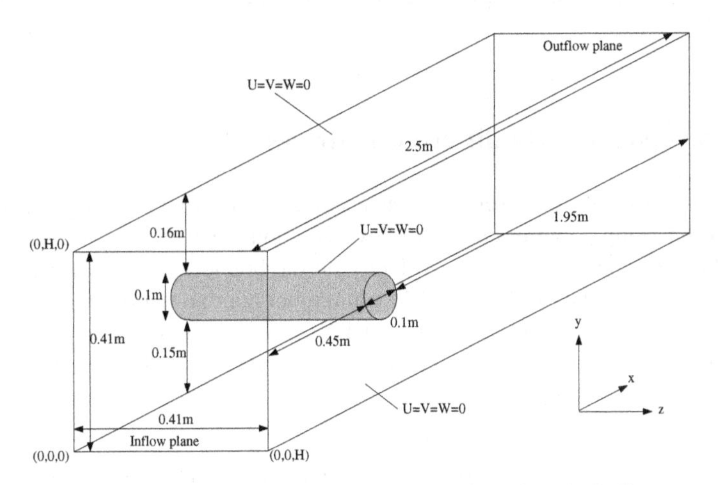

Fig. 1. Computational domain in the case of quasi-periodic flow

We illustrate the method on the test case **3D-2Z** of quasi-periodic flow described in details in [13]. The obstacle (thin cylinder) is lifted 1 mm above the plane of symmetry of the rectangular channel (Fig.1). The unsymmetry produces quasi-periodic flows with vortex separations for $Re = 100$. The inflow and outflow Dirichlet boundary conditions simulate the Poiseuille flow, the other part of the boundary represents the no-slip condition. We consider the mesh $80 \times 72 \times 72$ and 800 time steps with $\Delta t = 0.01$. The size of system (1) is $3.5 \cdot 10^5$, the stopping criterion is $\|r_j\| < 10^{-5}$ which implies 10^{11}-fold reduction of the initial residual for $x_0 = 0$. The system (1) is solved by the preconditioned GCR method on 16 processors of a COMPAQ cluster of alpha ev6 processors (667MHz). In Table 1 we show the total number of GCR iterations, n_{tot}, and the total GCR time t, for different maximal dimensions of the accumulated Krylov subspace $\max\#\mathscr{K}$. We observe that the minimal computation time is achieved for moderate (50-80) dimensions of Krylov subspaces. The reduction of the total number of iterations for higher values of $\max\#\mathscr{K}$ does not compensate the increase of the projection (2) expense. We remark that for the fully developed flow the projection with even $\max\#\mathscr{K} = 50$ provides $10^5 - 10^6$-fold reduction of the initial residual due to the trivial initial guess. This results in 4-5 GCR iterations per time step, in contrast to 11-12 iterations for $x_0 = 0$. However, the actual speed-up is 1.5 due to the overhead of the projection computation.

Table 1. Total number of GCR iterations, the iterative solution time and maximal dimension of the accumulated Krylov subspace

$\max\#\mathscr{K}$	0	50	80	125
n_{tot}	10385	5272	4525	4187
t (s)	1668	1108	1107	1217

3 Linear systems with different matrices

In this section we consider a series of linear systems appearing in the numerical simulation of unsteady compressible fluid flow. The non reactive flow of compressible fluid with prescribed values of the velocity on the boundary obeys the full system of Navier-Stokes equations. The low-Mach number approximation [6] reduces it to a system of two momentum-type equations for velocity and temperature fields and a constrained divergence equation. The projection scheme generates a series of linear systems

$$A^k x = b^k \tag{3}$$

with different matrices A^k and different right hand sides b^k. The matrix A^k is the product of a diagonal matrix with temperature entries and finite difference discretizations of the divergence and gradient operators, and is singular and stiff. Symmetry of the matrix depends on whether the divergence and gradient mesh operators are conjugate; in our discretization they are not conjugate and, therefore, A^k is nonsymmetric. The principal feature of the series $\{A^k\}$ is that the matrices (and their eigenvectors and eigenvalues) vary slowly with the growth of k since the temperature changes slowly at each time step. The standard choice of the initial guess for (3) is the zero vector since the unknown solution x represents a pressure increment.

Since the matrices A^k are different, their common Krylov subspace cannot be generated and the advantages of the GCR method may not be used. The proposed approach is to accumulate the sequence of independent Krylov subspaces and associated images of matrices A^k produced by the GMRES method. The Krylov subspace $\mathscr{K}$ consists of mutually orthonormal vectors $\{v_j\}_{j=1}^m$ such that

$$AV_m = V_{m+1}H_{m+1,m},$$

where matrix V_m is composed of the vectors v_j and $H_{m+1,m}$ is an upper Hessenberg matrix containing the projection of A^k onto V_m. The error $e_0^k = x_0 - x$ satisfies the system $A^k e_0^k = A^k x_0 - b^k$ which may be projected onto $\mathscr{K}$. This projection is used for the correction of the initial guess x_0:

$$\hat{x} = x_0 - V_m \hat{H}_{m,m}^{-1} \hat{G}_m V_m^T (A^k x_0 - b^k) \tag{4}$$

where $\hat{G}_m$ is a sequence of Givens rotations such that $\hat{G}_m^{-1}$ reduces $H_{m+1,m}$ to an upper triangular matrix $\hat{H}_{m,m}$. All the operations of the projection (4) are implemented in the standard realization of the GMRES method.

In the framework of a series (3), the choice of a better initial guess may be stated as follows. Assume that $k-1$ systems have been solved and for the i-th system, $i = k-l, ..., k-1$, the following data are accumulated: m_i, V_{m_i}, G_{m_i}, H_{m_i,m_i}. Then the initial guess x_0^k for the k-th system is computed by the sequence of projections (4)

$$x_0^{i+1} = x_0^i - V_{m_i} \hat{H}_{m_i,m_i}^{-1} \hat{G}_{m_i} V_{m_i}^T (A^i x_0^i - b^i),\ i = k-l, ..., k-1 \tag{5}$$

with $x_0^{k-l} = 0$. In contrast to the projection (2), the number of accumulated data l is restricted not only by the practical capacity of the computer memory, but the discrepancy between the eigenvectors and eigenvalues of A^{k-l} and A^{k-1}.

The method was tested for a model chemical reactor [15]. The projection (4) with $l = 3$ reduces the norm of the initial residual (due to the trivial initial guess) by four orders of magnitude. However, the computed initial guess causes slow down of the convergence rate. As a result, the computational gain is as much as 15% of the solution time.

If the matrices in the series (3) are not "close" to each other, the projection (4) may even increase the initial residual, in comparison with the trivial initial guess. For example, our numerical evidence shows that the projection (4) is useless for the series of systems produced by the Newton method (rf. Table 2).

We conclude this section with the remark that the projection (4) may save certain amount of computations, if the eigendata of matrices of systems (3) are relatively close to each other, and the number of GMRES iterations and the computer memory allow to accumulate the projection data. In this respect, we mention a Ritz's value based strategy [4, 12] for the choice of appropriate vectors from a series of Krylov subspaces.

4 Series of nonlinear systems

Implicit discretizations of unsteady nonlinear problems generate series of nonlinear systems

$$F^i(u^i) = 0. \tag{6}$$

The nonlinear system may be solved by the Inexact Newton Backtracking (INB) method [1, 2, 9] with the finite difference approximation of the Jacobian-vector multiplication. The INB method offers global convergence properties combined with potentially fast local convergence. The method has outer (nonlinear) and inner (linear) iterations. The inner linear solve uses a preconditioner. The effective choice of the initial guess reduces the number of outer iterations due to the super-linear convergence of the outer iterations.

The basic step of the INM method is the approximate solution of linear system

$$(F^i)'(u_k)s_k = -F^i(u_k) \tag{7}$$

with relative reduction of the residual (for trivial initial guess) η_k. The forcing term η_k is chosen dynamically so that to avoid oversolving the systems (7). Backtracking is used to update s_k in order to globalize the convergence. The iterative solution of (7) requires only evaluation of $(F^i)'(u_k)$ on a vector. This allows to replace $(F^i)'(u_k)v$ by its finite difference approximation, e.g.,

$$(F^i)'(u_k)v = \frac{1}{\delta}[F^i(u_k + \delta v) - F^i(u_k)]. \tag{8}$$

Hereinafter, the GMRES method restarted after each 30 iterates (GMRES(30)) method is used for the iterative solution of (7). The arithmetical complexity of the INB method is expressed as the total number of function evaluations n_{evF} and the total number of preconditioner evaluations n_{evP} (if any); the remaining overheads are negligible.

We remark that the method (4) of computation of the initial guess for the sequence of *linear* systems (7) is ineffective: the values n_{evF}, n_{evP} do not decrease in comparison with the trivial initial guess. To illustrate the assertion, we consider the steady counterpart of the unsteady nonlinear problem discussed in the next section. The problem is discretized on a square mesh with $h = 2^{-6}$ and the parameter ν is set to 1. The nonlinear system with 3969 unknowns is solved by the algorithm INB (for other details we refer to the end of the section). The stopping criterion for the INB algorithm is $\|F(u_k)\| < 10^{-7}\|F(0)\|$. In Table 2 we show the total number of linear iterations n_{lit}, n_{evF} and n_{evP} for different strategies of choosing the initial guess for GMRES iterations: the trivial vector corresponding to $l = 0$ in (5), or $l \neq 0$ in (5) corresponding to l consequent projections on the l newest Krylov subspaces.

The data of Table 2 indicate that the complexity of the INB algorithm increases with l growing. The inefficiency of the strategy (5) may be explained by essential differences in spectral properties of subsequent Jacobian matrices. In contrast, an appropriate choice of the initial guess for *nonlinear* iterations reduces considerably the arithmetical complexity of the method. To this end, we introduce two notions.

Table 2. Complexity of the algorithm INB for different numbers of Krylov subspaces used in the projection.

l in (5)	0	1	2
n_{lit}	306	290	363
n_{evF}	344	350	442
n_{evP}	325	329	420

Proper orthogonal decomposition (POD) provides a way to find optimal lower dimensional approximations of the given series of data. It produces an orthonormal basis for representing the data series in a certain least squares optimal sense [10, 11]. Combined with the Galerkin projection, the POD is a tool for generation of reduced models of lower dimension. The reduced models may give a better initial guess for the Newton solution at the next time step.

The POD gives the solution to the problem: find m-dimensional subspace $S \subset R^N$ most close to the given set of n R^N-vectors $\{u^i\}_{i=1}^n$:

$$S = \arg \min_{S \in R^{N\times m}} \sum_{i=1}^{n} \|u^i - P_S u^i\|^2.$$

Here P_S is the orthogonal projection onto S. In order to solve this problem, we define the correlation matrix $R = XX^T$, $X = \{u^1 \dots u^n\}$, and find m eigenvectors of the problem

$$Rw_j = \lambda_j w_j, \quad \lambda_1 \geq \dots \geq \lambda_N \geq 0 \tag{9}$$

corresponding to m largest eigenvalues $\lambda_1 \geq \dots \geq \lambda_m$. Then

$$S = \text{span}\{w_j\}_{j=1}^m, \qquad \sum_{i=1}^{n} \|u^i - P_S u^i\|^2 = \sum_{j=m+1}^{N} \lambda_j. \tag{10}$$

The computational cost of finding m largest eigenvalues of symmetric matrix R is modest. For $m \sim 10$, the application of the Arnoldi process requires a few tens of R-matrix-vector multiplications in order to retrieve the desirable vectors with a good accuracy. The matrix-vector multiplication is cheap and parallel due to the factored representation $R = XX^T$. Alternatively, the solution of (9) may be reduced to the solution of n-dimensional eigenvalue problem

$$\tilde{R}\tilde{w}_j = \tilde{\lambda}_j \tilde{w}_j, \quad \tilde{\lambda}_1 \geq \dots \geq \tilde{\lambda}_n \geq 0$$

with explicitly calculated matrix $\tilde{R} = X^T X$.

A reduced model is generated on the basis of POD for a sequence of n solutions of (6) $\{u^i\}_{i=i_b}^{i_b+n-1}$. The eigenvectors $\{w_j\}_{j=1}^m$ form the basis of the m-dimensional subspace and matrix $V_m = \{w_1 \dots w_m\} \in R^{N\times m}$ is the projector onto this subspace. The reduced model is the Galerkin projection of (6) onto this subspace:

$$V_m^T F^i(V_m \hat{u}^i) = 0, \qquad \text{or} \qquad \hat{F}^i(\hat{u}^i) = 0, \tag{11}$$

where the unknown vector $\hat{u}^i \in R^m$ and $\hat{F}^i : R^m \to R^m$.

The reduced model is the nonlinear equation of very low dimension m. For its solution, we adopt the same INB algorithm with the finite difference approximation of the Jacobian-vector multiplication. Being the formal Galerkin projection, each evaluation of function $\hat{F}^i(\hat{u}^i_k)$ in the kth INB iterate is the sequence of the following operations: $u^i_k = V_m \hat{u}^i_k$, $f^i_k = F^i(u^i_k)$, and $\hat{f}^i_k = V_m^T f^i_k$. Therefore, the overhead is matrix-vector multiplications for V_m and V_m^T, i.e., $4Nm$ flops which is negligible compared to the evaluation of function $F(u)$. Another important consequence of low dimensionality of (11) is that the INB algorithm may be applied without any preconditioner.

Coupling POD and reduced model gives a powerful tool for acceleration of the fully implicit schemes [14, 15]. Let n, the length of data series, be defined, as well as the desirable accuracy ε for F^i: $\|F^i(u^i)\| \leq \varepsilon$. For any time step $i = 1, \ldots,$ perform:

ALGORITHM INB-POD

IF $i \leq n$, SOLVE $F^i(u^i) = 0$ BY PRECONDITIONED INB
WITH THE INITIAL GUESS $u^i_0 = u^{i-1}$ AND ACCURACY ε

ELSE

1. IF$(mod(i,n) = 1)$:
   A) FORM $X = \{u^{i-n} \ldots u^{i-1}\}$;
   B) FIND SO MANY LARGEST EIGENVECTORS w_j OF $R = XX^T$ THAT $\sum_{j=m+1}^{N} \lambda_j \leq \varepsilon$;
   C) FORM $V_m = \{w_1 \ldots w_m\}$
2. SET $\hat{u}^i_0 = V_m^T u^{i-1}$
3. SOLVE $\hat{F}^i(\hat{u}^i) = 0$ BY NON-PRECONDITIONED INB
   WITH THE INITIAL GUESS $\hat{u}^i_0$ AND ACCURACY $\varepsilon/10$
4. SET $u^i_0 = V_m \hat{u}^i$
5. SOLVE $F^i(u^i) = 0$ BY PRECONDITIONED INB
   WITH THE INITIAL GUESS u^i_0 AND ACCURACY ε

The reduced model is slightly oversolved, this provides better initial guess u^i_0. The number of eigenvectors is chosen adaptively in the above algorithm: it allows to form a reduced model that approximates the original model with the desirable accuracy ε.

The appealing feature of the method is its modularity: computation of the reduced model solution is separated from the original solver and is based on its simple algebraic modification. It makes the INB-POD algorithm easy to implement in codes with a complex architecture.

We tested the method for the backward Euler approximation of the unsteady 2D Navier-Stokes equations [15]. The classical lid driven cavity problem in the streamfunction-vorticity formulation was considered. The computational domain was the unit square, and the Reynolds number was 1000. The velocity of the lid was periodic, $v(t) = 1 + 0.2\sin(t/10)$, and the flow demonstrated the quasiperiodic behavior. Indeed, in the case of $v(t) = 1$, the unsteady solution is stabilized within $t_s \sim 150$. Therefore, to get a quasi-periodic solution, we need the periodic forcing term with the period $T < t_s$ but comparable with t_s, $T \sim t_s$. If $T \ll t_s$, the inertia of the dynamic system will smear out the amplitude of the oscillations; if $T > t_s$, the dynamic system will have enough time to adapt to the periodic forcing term and to demonstrate periodic behavior. The function $\sin(t/10)$ has the period $T = 20\pi$ which fits perfectly the above restrictions for the quasi-periodicity. It is well known that the feasible time step Δt for approximation of periodic solutions satisfies $12\Delta t = T$, and we set $\Delta t \sim 5$.

In Table 3 we compare the performance of the standard INB implicit solver with the initial guess equal to the solution at the previous time step $u_0^i = u^{i-1}$ and the proposed INP-POD solver on the square mesh with mesh size 2^{-8}.

Table 3. Performance of the algorithms INB and INB-POD for the lid driven cavity problem.

	INB			INB-POD		
time step, i	10	20	30	32	42	52
$\lVert F^i(u_0^i)\rVert$	0.36	0.79	0.09	22e-6	1e-6	2.6e-6
n_{evF}	166	186	189	44+55	45+11	44+19
n_{evP}	160	180	183	0+51	0+ 9	0+16
CPU time	13.4	15.3	16.1	1.2+4.2	1.1+1.1	1.1+1.2

The parameters of the algorithm INB-POD were as follows. The data (solutions) series are $\{u^{20k-10}\dots u^{20k+9}\}$, $k = 1,2,\dots$, i.e., $n = 20$, and the dimension of the reduced model is fixed to $m = 10$. In Table 3 we present the arithmetical complexity of certain time steps, in terms of n_{evF}, n_{evP} and the CPU time, as well as the quality of the initial guess $\lVert F^i(u_0^i)\rVert$ due to $u_0^i = u^{i-1}$ or due to the reduced model, $u_0^i = V_m\hat{u}^i$. The first entry of each sum in the Table corresponds to the contribution of the reduced model, the second one is due to the original model. The first observation is that the acceleration is significant, ~2-6-fold in comparison with the standard algorithm INB. The reason is in much better initial guess for the original model solver (cf.$\lVert F^i(u_0^i)\rVert$). Due to the super-linear convergence of the INB algorithm, this results in smaller values of n_{evF}, n_{evP}. The price to be paid for this reduction is the cost of the reduced problem solution. As it was mentioned before, the complexity of function $\hat{F}$ evaluation only slightly exceeds that for F, whereas the number of preconditioner evaluations is zero for the reduced model. Since in the considered application (as well as in the absolute majority of applications), the complexity of the preconditioner evaluation dominates over the complexity of the function evaluation, the speed-up is attributable to the ratio of n_{evP} for the standard algorithm and the accelerated one. As it is seen from the Table, this ratio depends on the time moment. At the time step 32, the quasi-periodic flow is not yet very well stabilized, and the prediction of the reduced model is not as good ($\lVert F^i(u_0^i)\rVert \sim 10^{-5}$) yielding only 2-fold acceleration. At the time steps 42 and 52, the initial guess due to the reduced model is very good, ($\lVert F^i(u_0^i)\rVert \sim 10^{-6}$) and the acceleration is 6-fold.

We note that the solution of the eigenvalue problem may be performed asynchronously with the implicit solution: as soon as V_m is formed, the reduced model becomes the active substep. Moreover, the POD and the implicit INB solution may be performed on different computers: the data to be exchanged are X and V only, and the communication delay does not block the simulation. The underlying client-server architecture of the method makes it very appealing in GRID computing applications [16]. The target architecture is represented by a set of low cost computational resources (clusters) connected via a standard slow Ethernet communication network with high latency time.

The POD acceleration can be used wherever POD data are available. The asynchronous non-blocking communications between the POD generator and the solver resource provide computation of the time step without idling. Therefore, a slow network with high latency time is affordable for the proposed technology. Also, the POD generator task is waiting for data from the solver resource and computing the POD basis when sufficient data are gathered. For the sake of more efficient use of the POD generation resource, it may work with other tasks as well. For instance, the POD generator can be used by several solvers and perform the POD on different sets of solutions tagged by the generating solver. In addition, the reduced basis may be used for other postprocessing tasks such as data visualisation or a posteriori error estimation.

Our GRID experiments involve a computer A (SGI Altix350 with Ithanium 2 processors, 1.3Gb/s network bandwidth) and a computer B (6 nodes Linux cluster with AMD BiAthlon 1600+ MP processors, 100Mb/s Ethernet internal network). A latency of 140 μs and a maximum bandwidth of $71 Mb/s$ have been measured for the communication network between the computers A and B. Figure 2 presents the elapsed time of the INB solver for the quasi-periodic flow with and without POD acceleration, on a homogeneous parallel computer and in the GRID architecture context. The Figure shows that the INB-POD algorithm gives quite similar results when it performs on the GRID context A-B or on the homogeneous computer A. Consequently, the computer B can be used with no penalty on the performance.

Acknowledgement

This work has been supported by the Région Rhône-Alpes in the framework of the project: "Développement de méthodologies mathématiques pour le calcul scientifique sur grille".

[1] S. Eisenstat and H. Walker, Globally convergent inexact Newton methods. *SIAM J. Optim.* 4, 1994, 393-422.

[2] S. Eisenstat and H. Walker, Choosing the forcing terms in an inexact Newton method. *SIAM J. Sci. Comput.*, 17, 1996, 16-32.

[3] G. Golub and G. Meurant, Matrices, moments and quadrature II: how to compute the norm of the error in iterative methods. *BIT*, 37-3, 1997, 687-705.

[4] P. Gosselet and Ch. Rey, On a selective reuse of Krylov subspaces in Newton-Krylov approaches for nonlinear elasticity. *Domain decomposition methods in science and engineering*, Natl. Auton. Univ. Mex., Mexico, 2003, 419-426.

[5] C. T. Kelley, *Iterative Methods for Optimization*. Frontiers in Applied Mathematics 18. SIAM, Philadelphia, 1999

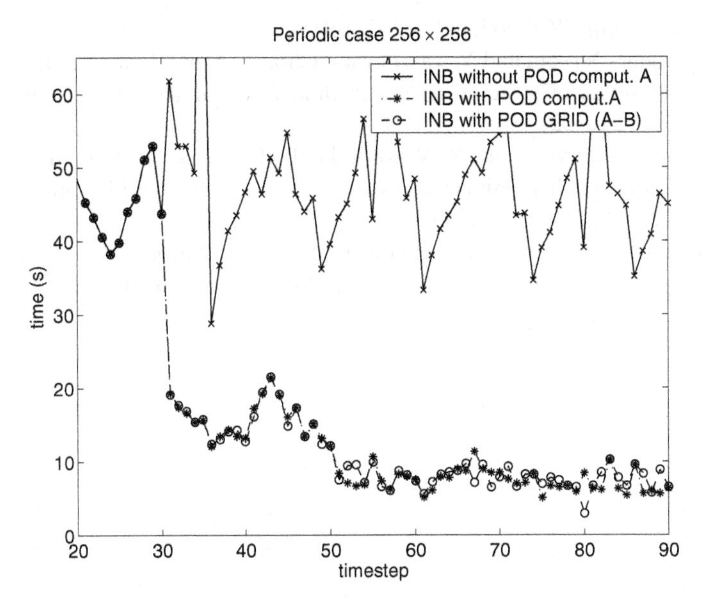

Fig. 2. Elapsed time of GRID computation A-B and computation on homogeneous parallel computer A.

[6] I. Keshtiban, F. Belblidia and M. Webster Compressible flow solvers for Low Mach number flows – a review. *Technical report CSR2*, Institute of Non-Newtonian Fluid Mechanics, University of Wales, Swansea, UK, 2004.

[7] G. Meurant, *Computer solution of large linear systems.* Amsterdam, North-Holland, 1999.

[8] G. Meurant, The computation of bounds for the norm of the error in the conjugate gradient algorithm.*Numerical Algorithms*, 16, 1998, 77-87.

[9] M. Pernice and H. Walker, NITSOL: a Newton iterative solver for nonlinear systems. *SIAM J. Sci. Comput.*, 19, 1998, 302-318.

[10] M.Rathinam and L.Petzold, Dynamic iteration using reduced order models: a method for simulation of large scale modular systems. *SIAM J. Numer.Anal.*, 40, 2002, 1446-1474.

[11] M. Rathinam and L. Petzold, A new look at proper orthogonal decomposition. *SIAM J. Numer.Anal.*, 41, 2003, 1893-1925.

[12] F. Risler and Ch. Rey, Iterative accelerating algorithms with Krylov subspaces for the solution to large-scale nonlinear problems. *Numer. Algorithms* 23, 2000, 1-30.

[13] M. Schefer and S. Turek, Benchmark Computations of Laminar Flow around a Cylinder. In: Flow Simulation with High-Performance Computers II (E.H.Hirschel ed.), *Notes on Numerical Fluid Mechanics*,52, Vieweg, 1996, 547-566.

[14] D. Tromeur-Dervout and Y. Vassilevski, POD acceleration of fully implicit solver for unsteady nonlinear flows and its application on GRID architecture.

Proc. Int. Conf. PCFD05, 2006, 157-160.

[15] D. Tromeur-Dervout and Y. Vassilevski, Choice of initial guess in iterative solution of series of systems arising in fluid flow simulations. *J.Comput.Phys.* 219, 2006, 210-227.

[16] D. Tromeur-Dervout and Y. Vassilevski, POD acceleration of fully implicit solver for unsteady nonlinear flows and its application on grid architecture.*Adv. Eng. Softw.* 38, 2007, 301-311.

[17] D. Tromeur-Dervout, Résolution des Equations de Navier-Stokes en Formulation Vitesse Tourbillon sur Systèmes multiprocesseurs à Mémoire Distribuée. Thesis, Univ. Paris VI, 1993.

Part II

Optimisation in Aerodynamics Design

Aerodynamic Study of Vertical Axis Wind Turbines

Merim Mukinović[1], Gunther Brenner[1], and Ardavan Rahimi[1]

Institute of Applied Mechanics, Clausthal University of Technology, Germany
merim.mukinovic@tu-clausthal.de,
gunther.brenner@tu-clausthal.de,
ardavan.rahimi@tu-clausthal.de

1 Introduction

With respect to wind energy, there is an increasing interest in decentralized, small systems with a nominal power output of 5-25 kW. This has motivated the resurgence of interest in Vertical-Axis Wind Turbines (VAWT), that offer several advantages to the more conventional Horizontal-Axis (HAWT) machines. The VAWT is inherently omnidirectional, and hence obviates the need to provide a yawing mechanism. Due to their simpler configuration the productions costs and service effort is potentially lower than for HAWT. In one sense, the price paid for structural simplicity is aerodynamic complexity: VAWT aerodynamics is inherently unsteady, and highly nonlinear. However, recent development in CFD methods capable of prediction in detail the unsteady aerodynamics has greatly increased the understanding of VAWT aerodynamics.

The present paper is organized as follows: In the section a short overview of the computational model is presented. Subsequently, parameter models used to optimize VAWT and the CFD model is presented. In order to present most important aerodynamic effects, some of the current two dimensional and three dimensional CFD simulation results are presented together with VAWT parameter model.

2 Computational models

Figure 1 presents a simplified cross-section of the rotor. The rotor consists of two straight NACA-4418 airfoils which are at both ends connected with the shaft. During one revolution of the rotor the relative velocity U_{rel} at the airfoil is changing due to large variations of the angle of attack α, often as much as $\pm 20° - 30°$. The blades are then operated in a stalled condition most of the time. If the rate of variation of the angle of attack is large enough, a hysteresis effect known as dynamic stall may occur. Such nonlinear phenomena require that the simulations be carried out in the time domain.The respective parameters as well as free-stream conditions are presented in the Table 1.

D. Tromeur-Dervout (eds.), *Parallel Computational Fluid Dynamics 2008*,
Lecture Notes in Computational Science and Engineering 74,
DOI: 10.1007/978-3-642-14438-7_4, © Springer-Verlag Berlin Heidelberg 2010

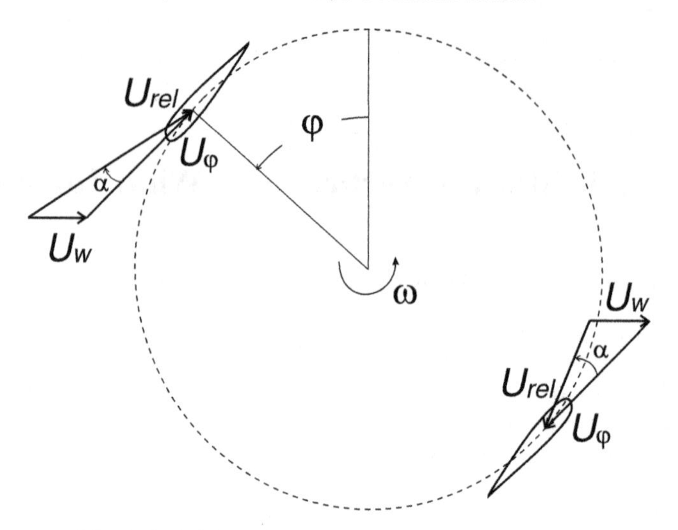

Fig. 1. Definition of Coordinate Systems and velocities.

airfoil	cord c [m]	span H [m]	rotor radius R [m]	wind velocity $U_w\left[\frac{m}{s}\right]$
NACA 4418	0.5	2	1.73	10 m/s

Table 1. Parameters of the rotor and free stream conditions.

2.1 Aerodynamic parameter model

One of the most successful approaches for predicting VAWT is the Double Multiple Streamtube (DMST) model [1]. This approach is capable to predict the periodic loads and average output power of the rotor. The loading in a blade element depends on the local relative velocity, which is the result of the local wind speed and the rotational motion. The local wind speed depends on the retardation of the wind speed past the upwind blade element, which depends in turn on the loading of the downwind blade element. This interaction is modeled using the DMST approach. Since the focus of the present paper is on the CFD simulations, details regarding this approach are omitted here. Crucial input parameters of the model are the aerodynamic coefficients of the profile which are typically available for static variations of angle of attack up to about $\pm 20°$ [2]. In present investigation aerodynamic coefficients for $Re =$ 850000 and $\alpha = \pm 90°$ are used [3]. Larger variations of angle of attack as well as the dynamic effects have to be modeled based on empirical ideas, such as the Gormont model for the hysteresis effect [4]. This parameter model, once validated, allows a fast analysis of the flow, forces and loads acting on the blades and is therefore the base for further optimization of the rotor. However, in order to qualify this approach, a more details analysis of the 2D and 3D flow around the rotor is requires. This may be obtained using CFD calculations.

2.2 CFD Analysis

For the two-dimensional CFD analysis, representing the flow in one plane though the rotor, the commercial package ANSYS CFX is used. The flow is assumed to be incompressible and fully turbulent. The shear stress transport model is used together with scalable wall functions. A single processor is sufficient to run these cases with the overall grid size of about 200 000 cells. In the rectangular stationary domain, a rotational domain is embedded using the sliding mesh interface. Critical issues with respect to the setup of the problem are the size of the domain which has to be large enough in order to avoid interference in particular with the outflow boundaries (see Figure 2). The spatial resolution along the airfoils has to be reasonably fine. For the

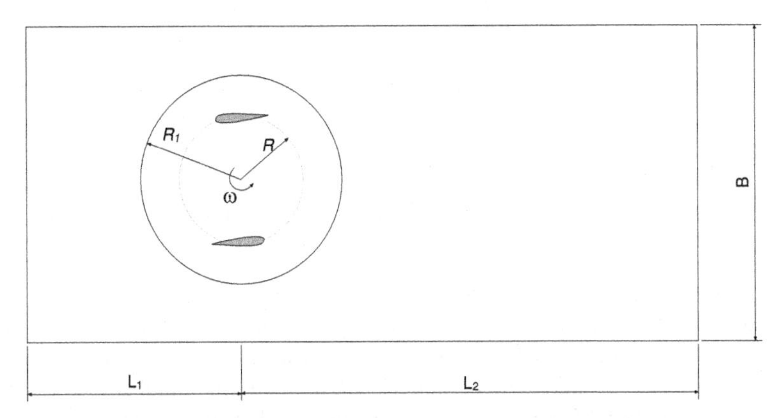

Fig. 2. Domain geometry. $L_1 = 2B = 30R$, $R_1 = 2R$, $L_2 = 50R$.

three dimensional, transient analysis, the DLR TAU code [5] is used. In that code, a Chimera overlapping grid technique is used for the variable interpolation between the rotating and the stationary domain. This code solves the compressible RANS equations with preconditioning to improve the convergence for the present low Mach number flows. The 3D results presented in the following section are obtained on a massive parallel system installed at the German Aerospace Center.

3 Results

One of the most important parameters of the VAWT is the solidity $\sigma = Nc/R$ where N is number of airfoils. Beside a strong influence on the power coefficient of the VAWT [6], for $\sigma < 0.5$ the DMST model have a poor agreement with experiment. For the current setup $\sigma = 0.578$. One of the measure for efficiency of VAWT is the power coefficient C_P which is usually plotted versus tip speed ratio $\lambda = U_\varphi/U_w$. Figure 3 shows quite good agreement of of DMST and CFD results for λ up to the 1.5. However, for large λ, larger deviations are observed. This may be explained with the dynamic stall effects [4] which are not included in the current implementation of the DMST model.

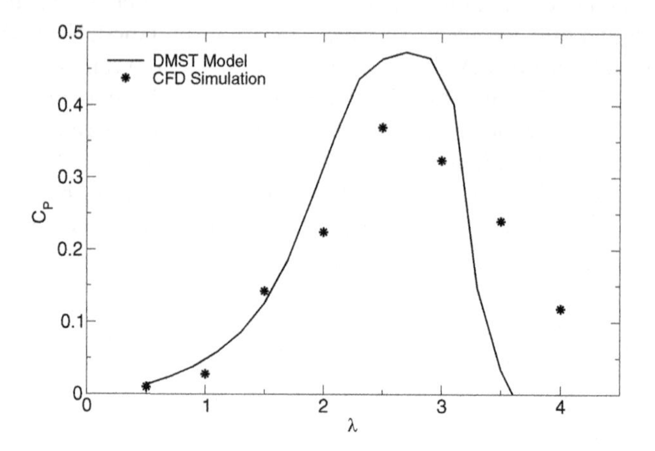

Fig. 3. Power coefficient of the VAWT $C_P = P/(\rho U_w^3 HR)$. 2D-CFD simulations versus DMST model. For the parameters see table 1.

The dynamic stall effects become obvious in Figure 4 where the moment of the single airfoil is compared. Underestimated static aerodynamic coefficients in the DMST model leads to significantly smaller peak moment (maximal angle of attack) as compared with the CFD simulation. Both results predict significantly smaller moment in the downstream positions of the rotor because a large part of the wind energy is exerted in the upstream area.

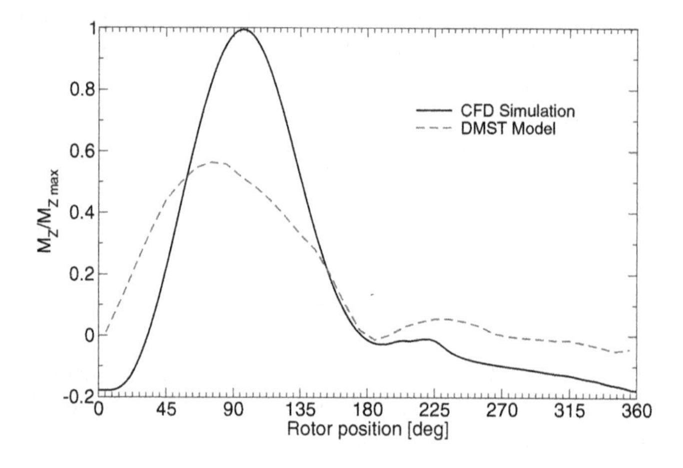

Fig. 4. Normalized moment of the single airfoil with respect to the rotor position. $\lambda = 2.5$

One of the aerodynamic effects which is not included in the present DMST model is the moment of the aerodynamic forces around the aerodynamic center of the airfoil (pitching moment). Figure 5 shows the overall moment of aerodynamic forces Mz and the moment of the resulting aerodynamic forces acting on the single airfoil Mz_{Fxy}. The pitching moment $(Mz - Mz_{Fxy})$ obviously gives negative contribution to the resulting rotor moment.

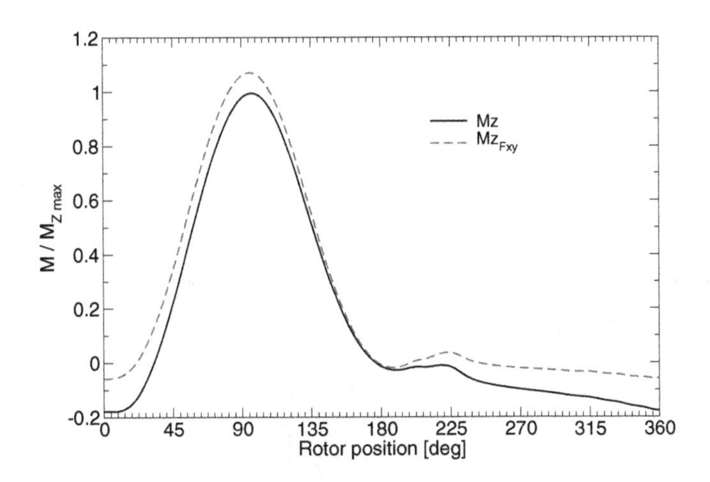

Fig. 5. The effect of the pitching moment,$\lambda = 2.5$.

The above presented CFD results are obtained for a simplified 2D configuration which does not include the rotor shaft. The real configuration contains also and the parts which connects the airfoils and the shaft what requires full 3D simulation. In this case, 3D simulations are only possible using parallel computers. For $\lambda = 2.5$ the full 3D model is simulated on the above mentioned cluster using 96 processors. The model consists of 14.5 million cells and a computational time of about 40 hours is spent per single rotor revolution.

Figure 6 shows the comparison of 2D and 3D model results where the difference in the peak moment is remarkable. This is mainly due to the influence of the finite span and the drag of the rotor connecting parts. Pressure difference contours presented in Figure 7 illustrate the influence of the finite span. Non-continuous distribution in spanwise direction results in decreased moment as compared to the 2D simulation. 3D simulation results shows also that the drag moment of the rotor connecting parts is about 15% of the maximal moment.

4 Conclusion

In the present paper the incompressible, time dependant and turbulent flow past a vertical axis wind turbine (VAWT) was simulated in two and three dimensions. The

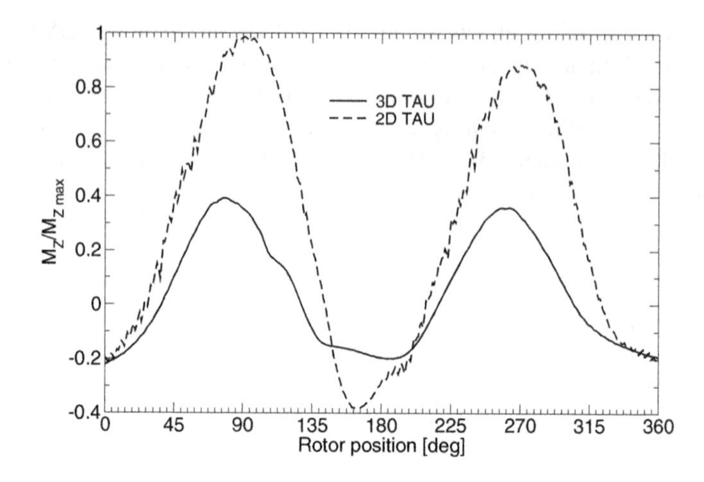

Fig. 6. Normalized resulting moment of the rotor during the second rotor revolution.

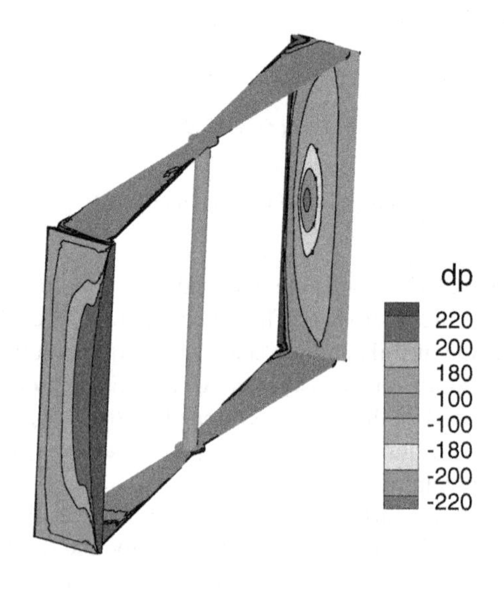

Fig. 7. Pressure difference distribution. $\lambda = 2.5$

results serve to validate parameter models (such as DMST) which are used to optimize such wind turbines. A quite large disagreement between the DMST model and the two dimensional CFD results is observed. This discrepancy motivates the improvement to the parameter models in order to take into account e.g. for dynamic stall effects, which is suspected to be the predominant cause for uncertainties.

Additionally, large differences between the two- and three-dimensional CFD simulations are observed. This leads to the conclusion that 2D CFD simulation may not be sufficient for the aerodynamic analysis of VAWTs due to a large variation of lift along the airfoil span. Thus, 2D approaches may be very useful in order to demonstrate the effect of constructional changes of the rotor in a more qualitative way. For a quantitative analysis of the aerodynamic loads as well as to provide reliable data for parameter models three dimensional simulations of the time dependant flow field past the VAWT are indispensable.

Acknowledgement. The authors would like to thank Arbeitsgemeinschaft industrieller Forschungsvereinigungen "AIF" for the financial support and German Aerospace Center for the code and computational resources.

[1] I. Paraschivoiu. Aerodynamic Loads and Performance of the Darrieus Rotor. *Journal of Energy*, 6(6):406–412, 1981.
[2] I.H. Abbott and Albert E. von Doenhoff. *Theory of Wing Sections*. Dover Publications, Inc., New York, 1959.
[3] U.S. Paulsen. Aerodynamics of a full-scale, non rotating wind turbine blade under natural wind conditions. Technical Report Risø-M-2768, Risø National Laboratory, Denmark, 1989.
[4] R.E. Gormont. An analytical model of unsteady aerodynamics and radial flow for application to helicopter rotors. Technical Report 72-67, U.S. Army Air Mobility Research and Development Laboratory, 1973.
[5] D. Schwamborn, T. Gerhold, and R. Heinrich. The DLR TAU-CODE: Recent Applications in Research and Industry. In P. Wesseling, E. Oñate, and J. Périaaux, editors, *European Conference on Computational Fluid Dynamics, ECCOMAS CFD 2006*. TU Delft, The Netherlands, 2006.
[6] M. Fallen and J. Ziegler. Leistungsberechnung für einen Windenergiekonverter mit vertikaler Achse. *Brennsoff-Wärme-Kraft*, 33(2):54–59, 1981.

Parallel Shape Optimization of a Missile on a Grid Infrastructure

Erdal Oktay[1], Osman Merttopcuoglu[2], Cevat Sener[3], Ahmet Ketenci[3], Hasan U. Akay[4]

[1] EDA – Engineering Design & Analysis Ltd. Co., Ankara, Turkey
eoktay@eda-ltd.com.tr
[2] ROKETSAN – Missile Industries, Inc., Ankara, Turkey
omerttopcuoglu@roketsan.com.tr
[3] Dept. of Computer Engineering, Middle East Technical University, Ankara, Turkey
{sener, ketenci}@ceng.metu.edu.tr
[4] Dept. of Mechanical Engineering, Indiana University-Purdue University Indianapolis, Indiana, USA hakay@iupui.edu

Abstract. A computational tool is developed to be used in the preliminary design of an air vehicle. This tool parametrically optimizes the airframe shape. In order to search the entire solution space thoroughly, a genetic algorithm is used. Code parallelization is utilized to decrease the convergence time of the airframe shape design of a realistic missile geometry on a Grid infrastructure to further improve the search quality. In this work, a generic missile geometry is taken as a test case for a design application. The problem is to maximize the weighted average of lift-to-drag ratio for given mass and propulsion unit.
Keywords: Design Optimization, Shape Optimization, Genetic Algorithms, Parallel Computing, Grid Computing.

1 Introduction

Conventional methods that use gradient information have been used many times for engineering design problems in the past [1, 11, 3]. But, they often fall into difficulty of handling the high dimensional problems. It is hard to search the entire design space by these methods, especially if the problem includes disjoint solution sets. Besides, they require gradient information which is hard to obtain, or nonexistent.

Heuristic methods have become widespread [4, 5, 6, 7]. They can deal with such stiff problems with high dimensionality and with large number of local solutions, and are able to avoid unfeasible regions and handle inequality constraints.

In this paper, genetic algorithms are used, because they are well suited for parallel computing, which is an important property for engineering design problems with large number of parameters. There has been an increasing tendency in the use of

D. Tromeur-Dervout (eds.), *Parallel Computational Fluid Dynamics 2008*,
Lecture Notes in Computational Science and Engineering 74,
DOI: 10.1007/978-3-642-14438-7_5,

parallel genetic algorithms during the last decade [8, 9]. In this study, such a parallel implementation is realized on a Grid infrastructure.

2 Problem Definition

2.1 The Cruise Flight

The aim is to obtain an airframe geometry that provides the longest flight possible. Consider the following cruise equations, for $\gamma = 0$:

$$\dot{x} = V \,, \tag{1}$$

$$\dot{z} = 0 \,, \tag{2}$$

$$\dot{V} = (F_D + F_T)/m \,, \tag{3}$$

$$V.\dot{\gamma} = F_L/m - g \,. \tag{4}$$

where V and γ are speed and flight path angles, respectively. F_T is thrust force, m is mass, and they are functions of time. F_D and F_L are the drag and lift forces, respectively, and are calculated as follows:

$$F_D = \frac{1}{2}.\rho.V^2.S_{ref}.C_D \,, \tag{5}$$

$$F_L = \frac{1}{2}.\rho.V^2.S_{ref}.C_L \,. \tag{6}$$

Here the drag and lift coefficients C_D and C_L are functions of Mach number M and angle of attack α. These coefficients correspond to the trim values; therefore, it is not necessary to deal with the short term variations. Since angle of attack is arbitrary, it is chosen so as to satisfy the condition $\dot{\gamma} = 0$. The atmospheric variables density ρ, and speed of sound c are found from the standard atmosphere model. The problem demands that, for a given set of initial and terminal values of V, the final value of range x_f be maximum. Additionally, different configurations are checked for a static stability requirement that the minimum value of the stability margin should be greater than a reference value.

2.2 Shape Optimization

In this work, it is assumed that propulsion, warhead and instruments sections are previously defined. This way, the length and diameter of the body, and the nose shape are already determined. Thus, the properties related to the body are taken as constants, and are not considered into optimization process. The configuration is built with a body with circular cross section and two sets of fins. The first fin set (wing section), consists of two panels whereas the second set (tail section) consists of cross-oriented four fins. The following parameters are selected as design variables on cross-sectional plane of each fin set:

1. Position of leading edge on the body l_l
2. Leading edge sweep angle Λ_l
3. Root chord length c_r
4. Aspect ratio $A = b^2/2S$, S: plan-form area
5. Taper ratio $\lambda = c_t/c_r$

Each plane is of double wedged cross section. It is assumed that thickness-to-chord ratios are fixed along the half span. The following parameters are selected as design variables for airfoil geometries, for each set:

1. Thickness-to-chord ratios of upper and lower wedges t/c
2. Cone angles for leading edges of upper and lower wedges γ_l
3. Cone angles for trailing edges of upper and lower wedges γ_t

A typical airframe geometry, as well as the design parameters are illustrated in Figure 1.

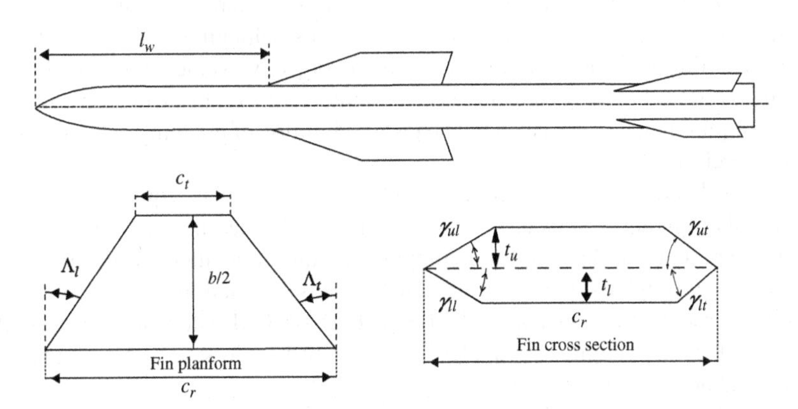

Fig. 1. Airframe geometry.

3 Method

3.1 Genetic Algorithms

The genetic algorithms technique is used frequently in non-linear optimization and constraint satisfaction problems. For a given problem, it maintains a population of candidate solutions; then it evaluates these solutions and assign objective values as the fitness values (in maximization). Based on the fitness values of the candidates, a new generation of candidate solutions is reproduced by means of operators inspired from evolutionary biology, such as mutation, selection, and crossover. This reproduction process favors for the better candidates (principle of the survival of the fittest), thus it reduces the probability of appearance of non-fitted solutions in the new generation.

The process is iterated until solutions converge to an optimum fitness value; hence it is a simulated evolution process which typically requires a large number of fitness evaluations.

3.2 Shape Optimization of a Missile

At each step of iteration, the master (i.e., the genetic algorithm's engine) generates configuration parameters as the solution set, which are combined to form the individual. All these values are bounded with both upper and lower limits, which are physical or geometrical constraints. These design variables are represented by integers called genes, thus the desired resolution of each parameter must also be specified.

The total set of solution proposals (individuals) is called population, which is divided into sub-populations. Each sub-population, in the form of a set of configuration parameters, is sent by the master to a separate worker. The sub-population size is set to the population size divided by the number of workers configured.

Each worker calculates a series of fitness values belonging to a group of individuals. The fitness value of an individual is the objective value being equal to the final range covered at the end of the cruise flight Moreover, the violations of static stability requirement are punished by means of inhibition of breeding, which reduces useless search space.

In the evaluation, an aerodynamic preliminary-design tool is required. This refers to the family of industry-standard codes, which combine the linear aerodynamic theory with experimental data, and use component build-up methods to handle whole configuration through its parts. The simulation needs drag and lift values. The semi-empirical aerodynamic analysis tool MISSILE DATCOM calculates these coefficients; and they are supplied in tabular form with respect to several Mach numbers and angle of attack α values [10].

All the fitness values calculated by workers are transferred back to the master where the population is updated as the best performances are rewarded with higher reproductive rates. That is, the master finds the fittest (best) solutions, out of the ones provided by workers, using the user supplied parameters for mutation, selection, and crossover genetic operators. The flow of this mentioned process is illustrated in Figure 2.

3.3 Implementation of Problem on GridAE

In this study, the genetic algorithms technique is implemented on the Grid environment since the search space is extremely large. This enables to obtain a higher level of diversification, and it becomes possible to find better solutions within this large solution space by the use of a Grid infrastructure. The method is experienced using the Grid-based Artificial Evolution framework (GridAE)[5] deployed on the South Eastern

[5] The development of GridAE is supported by the SEE-GRID project, funded by the European Commission under the contract FP6 RI-002356.

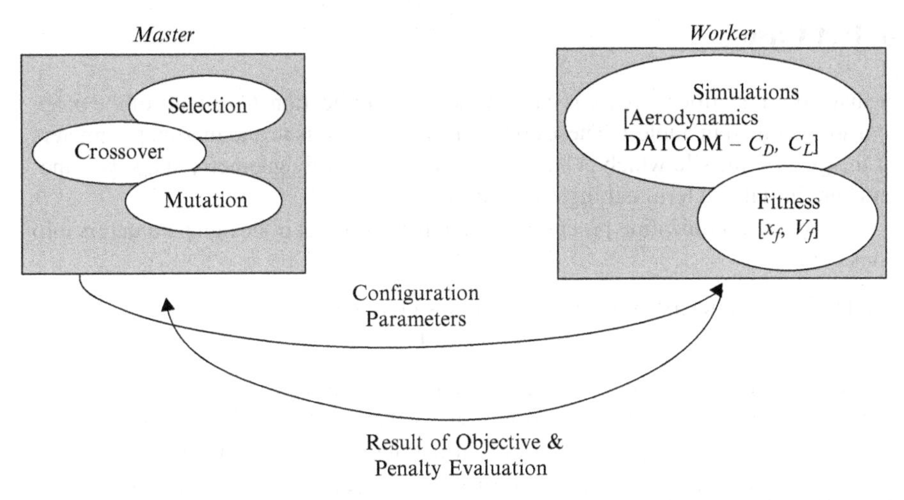

Fig. 2. Illustration of an overall solution flow where the master acts as the genetic engine and each worker calculates the fitness values.

European Grid (SEE-GRID) infrastructure [11, 12], which, like the other EGEE-like Grids, consists of a set of middleware services deployed on a broad collection of computational and storage resources, plus the services and support structures put in place to operate them [13]. It is a stable infrastructure, running the gLite Grid middleware on Scientific Linux Operating System instances [14, 15], for any scientific research, especially for projects where the time and resources needed for running the applications are considered impractical when using traditional IT infrastructures. Currently, it has about 40 sites (clusters) over 13 countries in the region with around 2000 various types of CPUs.

The GridAE framework aims to create a transparent interface for the user which would distribute the execution of artificial evolution applications (through the genetic algorithms technique) onto the Grid. Other than using its built-in genetic library, a user may develop and use his/her own genetic operators. This framework is designed to fit to EGEE-like Grid infrastructures, currently supporting the gLite middleware.

GridAE employs the master-worker paradigm, where it is theoretically possible to submit thousands of workers. As messaging is not supported among the sites by the infrastructure, it reserves and uses some temporary area on the Grid storage elements to simulate a shared memory region for the GridAE tasks to communicate with each other. Hence, submitting all the tasks to a single site and using a storage element located nearby obviously increase the evolution performance; however, one may need to wait for a very long time before that many processors in one site become available. Thus, GridAE makes use of the resource brokers for multi-site Grid execution of its tasks, where load-balancing is provided implicitly.

4 Test Case

A shape optimization test case is carried out as an application. Constants of the problem are specified in Table 1. These constants include, ballistic and inertial parameters of the generic missile which is handled in this work, body geometry shape parameters, and initial and terminal flight conditions.

The shape optimization process takes the following set of design parameters into consideration:

1. Design parameters for the wing section – 11 values
2. Design parameters for the tail section – 11 values

Resultant geometry parameters for each planform section are given in Table 2.

In order to give an insight into how the genetic algorithms work, some steps from the convergence history of a design parameter are given in Figure 3, as an example. The figures illustrate how the diversity in the population reduces and finally converges to an optimum value. Each dot in the figures represents an individual for corresponding AR (aspect ratio of the tail) value. The figure shows four ensembles belonging to different stages of the iteration. This c uses a 125-member population, and the final values of the design parameters are presented in Table 2.

In addition, the effects of number of generations and population size on the maximum fitness achieved are illustrated in Figure 4. It can be observed that the population size is directly proportional to maximum fitness values when the number of generations is fixed. Thus, it can be claimed that increasing the population size will result in less number of generations for the parameters to converge their optimum. Also, the maximum fitness values tend to increase as population evolves as one may expect. The drops on the graph can be avoided by using improved parameters for the genetic algorithm or introducing elitism.

Full Mass	2100 kg	Body Length	5.0 m	$Z0$	2000 m
Empty Mass	1010 kg	Diameter	0.6 m	$V0$	240 m/s
Full Com	3.0 m	Exit Diameter	0.4 m	zr	2000 m
Empty Com	2.3 m	Nose Length	0.8 m	Vr	280 m/s
Thrust	40 kN	Nose Shape	Ogive	γr	30 deg
Burnout Time	70 sec			αlim	8 deg

Table 1. Constant parameters used in the test case

5 Conclusions

Genetic algorithms prove to be reliable engineering design tools, especially in problems with a high degree of parameters space. It can safely be used in aeronautical design applications, in which the number of design variables are high, the sensitivity

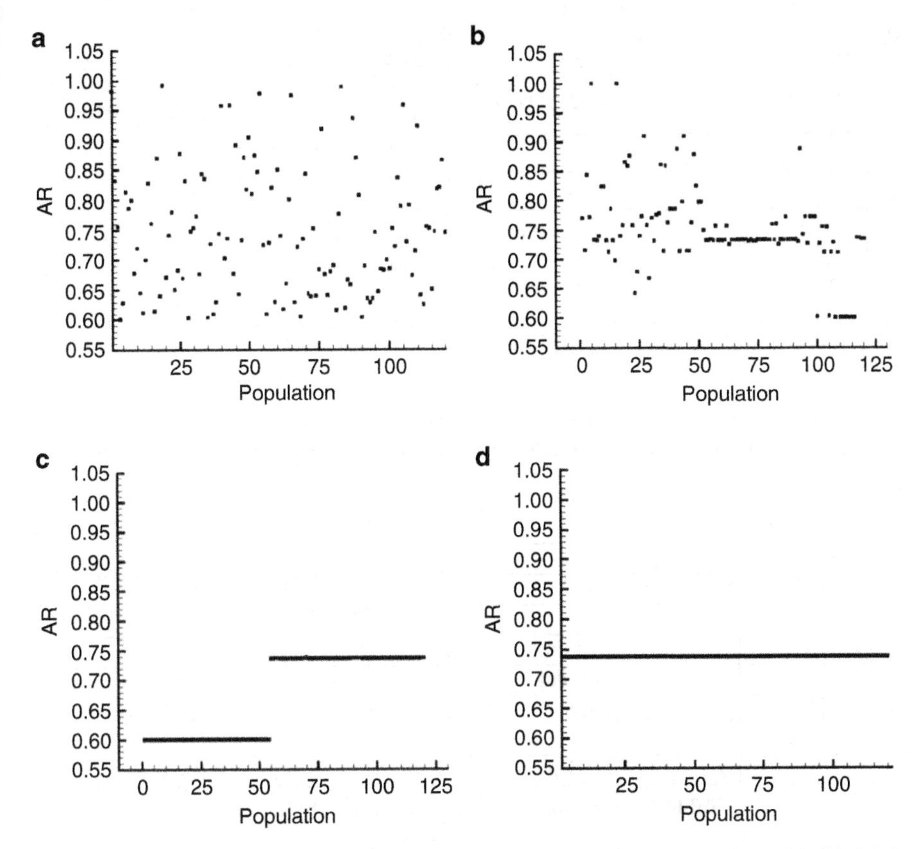

Fig. 3. Selected steps in the convergence history of aspect ratio of the tail **(a)** Initial, **(b)** After 30 generations, **(c)** After 50 generations, **(d)** After 80 generations

functions are inaccurate or unavailable, and the design space includes discontinues parts.

More importantly, the genetic algorithm technique is very suitable for parallelization. In current technology, this aspect is invaluable and even crucial. Parallel genetic algorithm applications fit very well to the loosely-coupled nature of the Grid infrastructures. This makes it possible to submit genetic algorithm applications onto Grid environments with many ten thousands of processors. These advantages encourage the inclusion of the more advanced methods or handling of the more stiff problems in design.

[1] O. Baysal,M.E. Eleshaky, Aerodynamic Design Optimization Using Sensitivity Analysis and Computational Fluid Dynamics. AIAA 91-0471 (1991)
[2] Ide, H., Abdi, F., Shankar, V.J.: CFD Sensitivity Study for Aerodynamic/Control Optimization Problems. AIAA 88-2336 (1988)

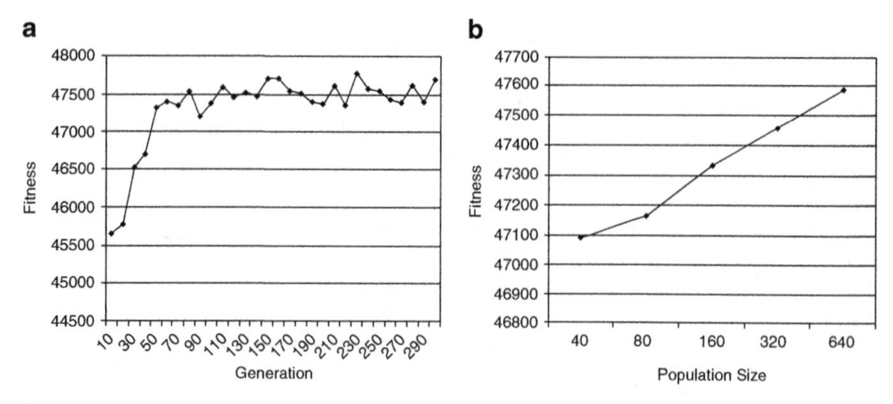

Fig. 4. (a) Maximum fitness values achieved at each 10th generation where population size is fixed at 40. (b) Maximum fitness values achieved for different population sizes after 100 generations.

(a) wing case					(b) tail case			
l_w	2.3 m	t_u	0.037		l_w	4.48 m	t_u	0.041
c_r	1.74 m	t_l	0.077		c_r	0.75 m	t_l	0.071
c_t/c_r	0.42	γ_{ul}	15.5 deg		c_t/c_r	0.67	γ_{ul}	24.8 deg
$b^2/2S$	0.46	γ_{ut}	11.0 deg		$b^2/2S$	0.74	γ_{ut}	25.5 deg
Λ_l	10.9 deg	γ_{ll}	7.8 deg		Λ_l	26.6 deg	γ_{ll}	23.8 deg
		γ_{lt}	7.9 deg				γ_{lt}	24.4 deg

Table 2. Shape optimisation (a) for wing , (b) for tail

[3] Reuther, J., Jameson, A., Farmer, J., Martinelli, L., Suanders, D.: Aerodynamic Shape Optimization of Complex Aircraft Configurations via an Adjoint Formulation. AIAA 96-0094 (1996)

[4] Anderson, M.B., Burkhalter, J.E., Jenkins, R.M.: Missile Aerodynamic Shape Optimization Using Genetic Algorithms. AIAA 99-0261 (1999)

[5] Foster, N.F., Dulikravich, G.S.: Three-Dimensional Aerodynamic Shape Optimization Using Genetic and Gradient Search Algorithms. J. Spacecraft and Rockets 34, 36–41 (1997)

[6] Blaize, M., Knight, D., Rasheed, K.: Automated Optimal Design of Two Dimensional High Speed Missile Inlets. AIAA 98-0950 (1998)

[7] Tekinalp, O., Bingl, M.: Simulated Annealing for Missile Optimization: Developing Method and Formulation Techniques. J. Guidance, Control, and Dynamics 27, 616–626 (2004)

[8] Lee, J., Hajela, P.: Parallel Genetic Algorithm Implementation in Multidisciplinary Rotor Blade Design. J Aircraft 33 (1996)

[9] Jones, B.R., Crossley, W.A., Lyrintzis, A.S.: Aerodynamic and Aeroacoustic Optimization of Rotorcraft Airfoils via a Parallel Genetic Algorithm. J Aircraft 37 (2000)
[10] Buns, K.A., Stoy, S.L., Vukelich, S. R.: Missile DATCOM: User Manual – 1997 Fortran 90 Revision (1998)
[11] GridAE: Grid-based Artificial Evolution Framework, http://GridAE.ceng.metu.edu.tr
[12] South Eastern European Grid-enabled Infrastructure Development Grid Project, http://www.see-grid.eu
[13] Enabling Grids for E-sciencE, http://www.eu-egee.org
[14] Lightweight Middleware for Grid Computing, http://glite.web.cern.ch/glite
[15] Scientific Linux, https://www.scientificlinux.org

Analysis of Aerodynamic Indices for Racing Sailing Yachts: a Computational Study and Benchmark on up to 128 CPUs.

Ignazio Maria Viola[1], Raffaele Ponzini[2], Daniele Rocchi[3], Fabio Fossati[3]

[1] Research Unit, the University of Auckland, New Zealand
im.viola@auckland.ac.nz
[2] High-Performance Computing Group, CILEA Consortium, Milan, Italy
[3] Mechanical Department, Politecnico di Milano, Milan, Italy

Abstract. This work presents a feasibility study for trustable and affordable CFD analysis of aerodynamic indices of racing sailing yachts. A detailed reconstructed model of a recent America's Cup class mainsail and asymmetrical spinnaker under light wind conditions has been studied using massive parallel RANS modeling on 128 CPUs. A detailed comparison between computational and experimental data has been performed and discussed, thanks to wind tunnel tests performed with the same geometry under the same wind conditions. The computational grid used was of about 37 millions of tetrahedra and the parallel job has been performed on up to 128 CPUs of a distributed memory Linux cluster using a commercial CFD code. An in deep analysis of the CPU usage has been performed during the computation by means of Ganglia and a complete benchmark of the studied case has been done for 64, 48, 32, 16, 8 and 4 CPUs analyzing the advantages offered by two kind of available interconnection technologies: Ethernet and Infiniband. Besides to this computational benchmark, a sensitivity analysis of the global aerodynamic force components, the lift and the drag, to different grid resolution size has been performed. In particular, mesh size across three orders of magnitude have been investigated: from 0.06 million up to 37 million cells. The computational results obtained here are in great agreement with the experimental data. In particular, the fully tetrahedral meshes allow appreciating the beneficial effect of the increasing of the grid resolution without changing grid topology: a converging trend to the experimental value is observed. In conclusion, the present results confirm the validity of RANS modeling as a design tool and show the advantages and costs of a large tetrahedral mesh for downwind sail design purposes.
Keywords: CFD, Parallel Computing, Benchmarks, Yacht Sail Plans, Downwind Sails, Wind Tunnel Tests.

1 Introduction

RANS analysis is playing a central role in the recent America's Cup (AC) races for both hydrodynamic and aerodynamic design aspects. In the last 30 years computational analysis capabilities and affordability have grown so much that in the last AC (2007, Valencia, Spain) all the twelve syndicates had invested a comparable amount

D. Tromeur-Dervout (eds.), *Parallel Computational Fluid Dynamics 2008*,
Lecture Notes in Computational Science and Engineering 74,
DOI: 10.1007/978-3-642-14438-7_6,

of money in experimental tests and in computational resources. It is only in last few years that RANS has become a trustable design tools, in particular in the sail design field. In fact, in some sailing condition the flow around the sails are largely separated and a large computational effort is required to accurately compute the resultant aerodynamic forces. The aerodynamics of sails can be divided in three branches: the aerodynamic of upwind sails, reaching sails and running sails. Upwind sails are adopted when sailing at small apparent wind angle (AWA), typically smaller than 35, where AWA is generally defined as the angle between the yacht course and the undisturbed wind direction at the 10m reference height above the sea surface. Single mast yachts, namely sloop, adopt a mainsail and a jib or a genoa, which are light cambered airfoils designed to work close to the optimum efficiency, i.e. to maximize the lift/drag ratio. The flow is mainly attached and consequently un-viscous code has been adopted with success since sixties to predict aerodynamic global coefficients [1], [2], and in the last decades several RANS applications have shown a good agreement with wind tunnel tests [3], [4]. Reaching sails are adopted when sailing at larger AWA, typically from 45 to 160. Sloop modern racing yachts often adopt the mainsail and the asymmetrical spinnaker, which are more cambered airfoils designed to produce the maximum lift [5], [6]; in fact sailing at 90 AWA the lift force component is aligned with the course direction. The flow is attached for more of the half chord of the sail and separation occurs on the trailing edge of the asymmetrical spinnaker. In particular, the flow field is strongly three dimensional because of the increasing of the vertical velocity component, the tip and root vortexes are strongly connected to the trailing edge vortex. Reaching sail aerodynamics requires the capability to correctly compute the separation edge on the leeward spinnaker surface, hence un-viscous code are not applicable and Navier-Stokes code might be adopted. The first RANS analysis has been performed by Hedges in 1993 [7], [8] with limited computational resources. More recently, in 2007 [9] and 2008 [10] two works performed with less than 1 million of tetrahedral cells show good agreement with wind tunnel data: differences between computed and measured force components are between 11Running sails are adopted at larger AWA and sloop yachts generally adopt a mainsail and a symmetrical spinnaker. The flow is mostly separated and sails work as bluff bodies. Separation occurs on the sail perimeters and the drag has to be maximized [5], [6]. In the AC races, the racing curse is around two marks positioned along the wind direction, in such a way that half of the race has to be sailed upwind and half downwind. In the leeward leg, yacht sails at closer AWA to increase the apparent wind component (due to their own speed) in light air, and sails al larger AWA to reduce the sailed course in stronger breeze. In the recent AC races a wind speed limitation lead to sail mainly reaching than running and for this reason particular focus has been placed on asymmetrical spinnakers.

In the present work, an America's Cup Class, version 5 [11], are studied in a downwind reaching configuration sailing at 45 AWA with mainsail and asymmetrical spinnaker and a RANS analysis has been performed to investigate the benefits in the global force computation accuracy with a very large mesh. A 37 millions of cells mesh has been performed with the commercial codes Gambit and Tgrid by Ansys Inc., which adopt a bottom-down approach: meshes are generated from lower to higher topology, hence from edges to surfaces and than to volumes. Only tetrahedral cells have been adopted. The computation has been performed with Fluent 6.3.26 (Ansys Inc.) solving the uncompressible Navier-Stokes-equations. In Figure 1 a visualization of the mathematical model is showed. The herein obtained computational results on the 37 million-cell mesh have been compared with both computational (previously obtained on smaller meshes and under the same fluid dynamics conditions) and experimental data acquired in the Politecnico di Milano Twisted Flow Wind Tunnel. In the following of the paper the experimental set-up is described, then the computational aspects are highlighted together with the hardware and the interconnection technologies used in the parallel run of the numerical simulations, finally numerical results are discussed and compared with experimental measurements in terms of aerodynamics indices such as lift and drag global coefficients.

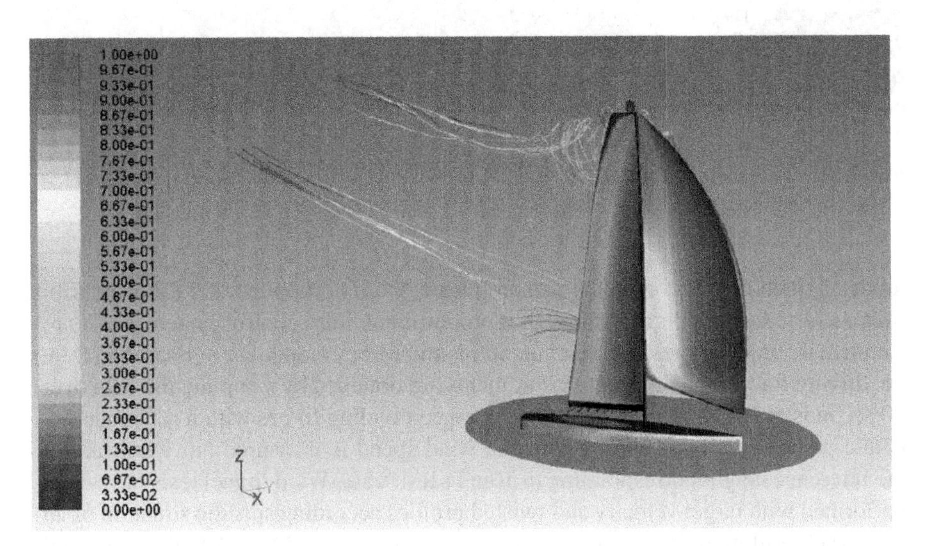

Fig. 1. static pressure coefficient distribution on sails and hull (Cp=(p-p0)/q, where p0 is the outflow undisturbed reference pressure and q is the inflow undisturbed reference dynamic pressure). Path lines colored by time show the boom and the mainsail tip vortexes. The yacht is sailing at 45 of apparent wind angle (i.e. the angle between the hull longitudinal axes and the incident wind at the reference height of 10m full scale) and is 5 leeward heeled. The America's Cup Class (version 5) mainsail and asymmetrical spinnaker for light wind are trimmed to produce the maximum driving force in the boat direction.

2 Experimental measurements

Experimental test has been performed in the Politecnico di Milano Twisted Flow Wind Tunnel. It is a closed circuit wind tunnel with two test sections respectively designed for civil and aerospace applications. On the left of Figure 1 the wind tunnel rendering is presented, airflow is running anti clockwise. On the lower side, aerospace low turbolence test section is showed. On the upper side, the long civil boundary layer test section, 36m length, 14m wide and 4m high, where sail plan tests are performed, which is showed on the right of Figure 2. The 1:12,5 scaled

Fig. 2. left: Politecnico di Milano Twisted Flow Wind Tunnel closed circuit; right: sail plan test in the boundary layer test section.

model is fitted on a 6-component dynamometer and it is supplied of 7 drums to trim sails as in real life, operated through a proportional radio control system. Sails are trimmed to produce the maximum aerodynamic force component in boat direction, i.e. driving force. Then actual measurements are obtained by sampling the data over 30 seconds at 100Hz. Coefficients are obtained dividing forces with a reference dynamic pressure and sail area. Reference wind speed is measured 5m windward at the reference height corresponding to 10m in full scale. Wind tunnel tests have been performed with target velocity and twisted profiles according specific situation of an ACC yacht sailing in Valencia atmospheric boundary layer. More details about wind tunnel tests can be found in [12].

3 Numerical analysis

The commercial code Fluent (Ansys Inc.) with a segregated solver strategy has been used to solve the equations of the flow around the sailing boat without considering time dependence (i.e. steady state), volume forces (i.e. gravity) and density variations and therefore energy equation hasnt been solved. SIMPLE scheme has been solved and first discretization order has been adopted. None turbulence model has been adopted. All the computations were performed on a Linux Cluster equipped with 74 CPUs AMD Opteron 275 dual-core (2.2 GHz, 2 GB/core) interconnected with Infiniband 4x (10GB/s) and Gigabit Ethernet. Due to the lack of information for such kind of models we launched the execution of the computation on 128 CPUs according to the maximum degree of parallel processing permitted by the license. The overall computation together with all the input and output operations and file writing took about one week. During the computation we monitored and analyzed the usage of the CPU using Ganglia (a system able to monitoring and store data concerning the usage of network and CPU in clusters computers); observing that the usage of the CPU was sub-optimal, we decide to perform an accurate benchmark in order to find out the optimal CPU usage. In particular we perform a descending benchmark on 64, 48, 32, 16, 8 and 4 CPUs testing two type of interconnection network, Infiniband and Ethernet Gigabit, and performing 100 iterations starting from the archived data. The benchmarking could not be performed on less then 4 CPUs since it was not possible to allocate in memory the 37 million elements mesh on less then 32GB of memory and wondering to take advantage of the multi-core architecture. For this reason all the results concerning the speed-up evaluation and the efficiency are referred to the 4 CPUs test case. In Figure 3 (left) the total wall time is plotted against the number of the used CPU for the two interconnections considered. In Figure 3 (rigth) the speed-

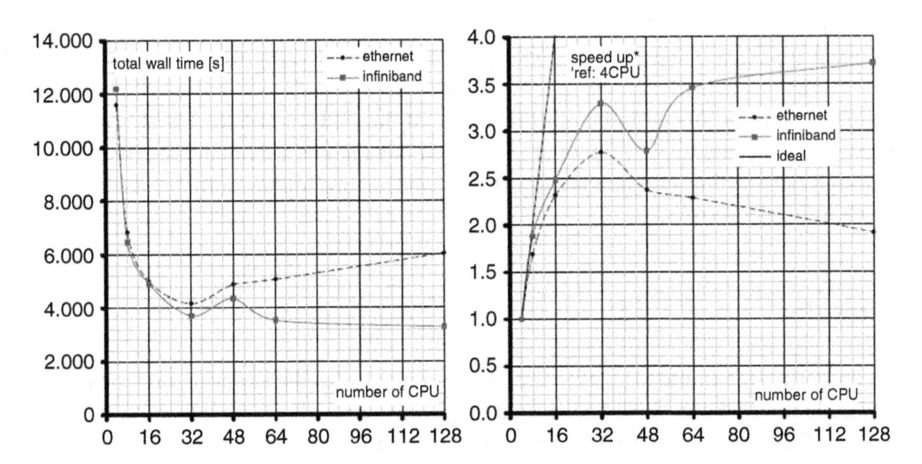

Fig. 3. Results on 100 iterations, left: Total wall time; right: speed-up.

up with respect to the 4 CPUs test is plotted against the number of the used CPU for the two interconnections considered, in Figure 4 the efficiency, again with respect to the 4 CPUs test case, is plotted against the number of the used CPU for the two interconnections considered. In order to better appreciate the gain on using a more performing interconnection and due to the fact that we did not had the possibility to compute the bench on less then 4 CPUs, in Figure 4 we plot the relative speed-up computed at a fixed value of used CPU according to:

$$\text{Relative speed-up} = (\text{Infiniband speed-up})/(\text{Ethernet speed-up}) \quad (1)$$

The relative speed-up shows that Infiniband speed-up rise up to about double the Ethernet speed-up in the case of 128 CPUs. All the benchmark results are consistent

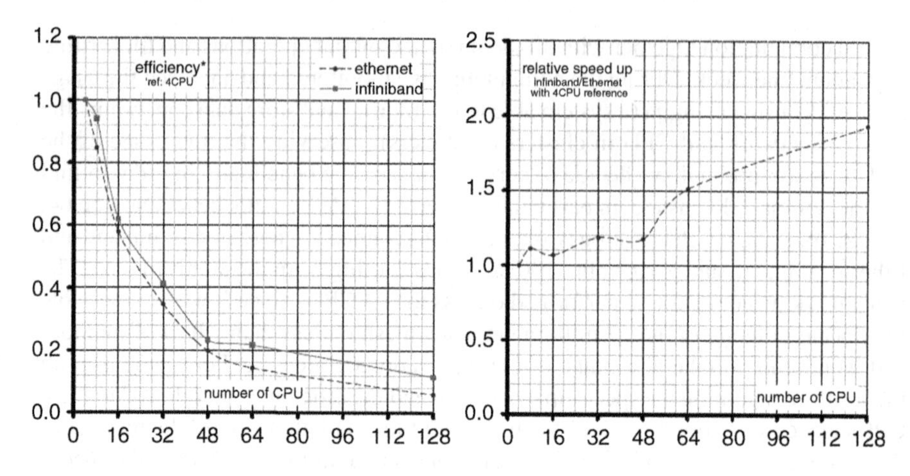

Fig. 4. Results on 100 iterations, left: efficiency; right: gain on speed-up computed as in equation (1).

to the fact that for this case the optimal usage of the CPU is obtained with a degree of parallelism equal to 32, moreover significant advantages are obtainable by means of a high performing interconnection (Infiniband) using higher number of CPU as shown in Figure 4 right.

4 Results

The numerical simulations showed a good agreement with the experimental data, the 37M cells mesh shows differences smaller than 3% in both the global aerodynamic force coefficients lift and drag, defined as following:

$$CD = \frac{drag}{\frac{1}{2}\rho V^2 S}, \; CL = \frac{lift}{\frac{1}{2}\rho V^2 S} \tag{2}$$

Where *drag* and *lift* are forces along the wind and perpendicular to the wind, respectively, in the horizontal plane acting on the yacht model above the water-plane (included hull rigging and sails), ρ air density, V undisturbed incoming reference wind speed measured at 10m height full-scale, S sail area (sum of the two sail surfaces).

The converging criteria is based on the drag and lift coefficients, which are monitored every iteration until the average values become stable. Figure 5 shows the sensitivity analysis to the mesh dimension: the drag (left) and lift (right) coefficients divided by the experimental values are plotted for the 37M cells mesh together with three other meshes of 0.06M, 1M, 6.5M respectively, obtained in previously validated studies and under the same fluid dynamics conditions. Circle and square marks show the average value and the error bars show the standard deviation of the coefficient oscillations. Increasing the mesh size of about three order of magnitude an increasing accuracy is obtained: the maximum differences between computed lift and drag with respect to the experimental values is smaller than 8% for the coarser mesh and becomes smaller than 3% for the finer mesh. By the way, the lift coefficients trend comes across the experimental values: lift is overestimated with coarser mesh and underestimated with finer mesh. Increasing mesh size both drag and lift curve are decreasing monotone. Further researches will be aimed to explore mesh larger than 100 million-cells, which has not been performed up to now because of the computational requirements that would be larger than 100GB of memory usage. Nevertheless the herein discussed work shows that some kind of large scale parallel approaches to RANS code applications in this filed can be a valid candidate to overcome these technical limits. The four meshes are fully tetrahedral and with similar

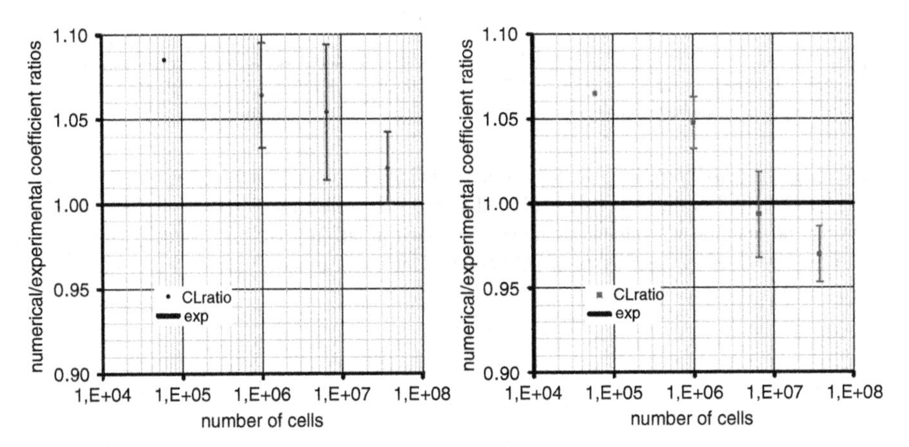

Fig. 5. numerical/experimental coefficient ratios are plotted versus the overall number of cells. Circle and squared marks show drag CD (left) and lift CL (right) average values, respectively, and error bars show the standard deviation of the coefficient signals.

grow rate (the linear dimension ratio between two adjacent cells in the wall-normal direction), hence they are all topologically similar. The wall adjacent tetrahedron dimension, and hence the distance between the tetrahedron centre and the wall (namely the first cell-centre-height y1) have a dramatic impact onto the resultant overall cells number. In Figure 6 on the left, the ratio between the first cell-centre-height of each mesh and the cell-centre-height of the coarser mesh are plotted versus the resultant overall number of cells. In Figure 6 on the right, the y+ values are plotted versus the overall number of cells. An horizontal section at 1/3 height of the yacht model from the water-plane has been considered and the y^+ values are referred to the asymmetric spinnaker leeward edge intersecting the plane.

Values are collected from the cells placed on the asymmetrical spinnaker at 1/3 height of the yacht model from the water-plane at the last iteration stage. In figure the maximum, minimum and average y^+ values are plotted for each mesh.

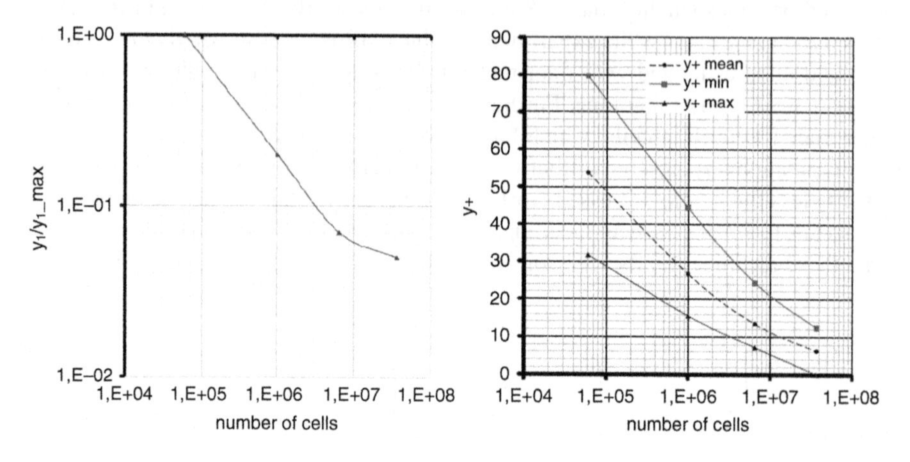

Fig. 6. left: first cell-centre-height of each mesh divided by the first cell-centre-height of the coarser mesh of 0.06 million cells are plotted versus the overall mesh size; right: maximum, minimum and average y^+ values computed at the last iteration stage and collected from the cells of the asymmetrical spinnaker at 1/3 height of the yacht model from the water-plane are plotted versus the overall mesh size.

5 Conclusion

In the present work a detailed study of feasibility of CFD approaches on the study of aerodynamics indices in racing sailing yachts is discussed. The main purpose of the study was to understand the usefulness of parallel computational approaches on the evaluation of several typical aerodynamic indices used to design and test in a synthetic manner the performance of a racing sailing yacht. In order to reach this scope a 37 million cells computational model of a ACC-V5 yacht model have been studied on 128 CPUs at the CILEA computer centre, using the parallel version of the commercial code Fluent (Ansys, Inc.) and all the typical aerodynamic factors, such as lift and drag coefficients, as been computed under steady state condition. Computed coefficients have been compared with experimental measurements performed at the Politecnico di Milano Twisted Flow Wind Tunnel, showing very good agreement: differences in both lift and drag smaller than 3%. In order to evaluate the usefulness of such approach (i.e. using a 37 million mesh) with respect to smaller discretization, we compared the herein obtained results with other pre-computed ones obtained respectively with a 0.06M, 1M, and 6.5M elements and under the same fluid dynamics conditions. An increase in force coefficient computed accuracy has been observed increasing the mesh size. Finally wondering to understand the better balancing between number of processors, mesh dimension and CPU usage, we performed a benchmark of 100 iteration of the same computational model using 64, 48, 32, 16, 8 and 4 CPUs and with two king of interconnection technologies. In this sense the best configuration is obtained using Infiniband interconnection and 32 CPUs. In conclusion this work show the feasibility of very large parallel CFD processing with a concrete gain in accuracy that confirm the usefulness of computational approaches as trustable and affordable tools for design and hypothesis testing today more and more complementary to the necessary experimental analysis.

Acknowledgement

The Authors are grateful to the Fluent Italia (Ansys) personnel and in particular in the person of Marco Rossi for their support and the confidence accorded during this experimentation.

[1] J.H. Milgram: The Aeodynamic of Sails; proceedings of 7th Symposium of Naval Hydrodynamic, pp. 1397-1434, 1968.
[2] Arvel Gentry: The Application of Computational Fluid Dynamics to Sails; Proceedings of the Symposium on Hydrodynamic Peformance Enhancement for Marine Applications, Newport, Rhode Island, US, 1988.
[3] H. Miyata, Y.W. Lee. Application of CFD Simulation to the Design of Sails. Journal of Marine Science and Technology, 4:163-172, 1999.
[4] A.B.G. Querard and P.A. Wilson. Aerodynamic of Modern Square Head Sails: a Comparative Study Between Wind-Tunnel Experiments and RANS Simulations. In the Modern Yacht, Southampton, UK, 11-12 Oct 2007. London, UK, The Royal Institution of Naval Architects, 8 pp. 107-114, 2007. http://eprints.soton.ac.uk/49314/.
[5] P.J. Richards, A. Johnson, A. Stanton: America's Cup downwind sails - vertical wings or horizontal parachutes?; Journal of Wind Engineering and Industrial Aerodynamics, 89:1565–1577, 2001.
[6] William C. Lasher, James R. Sonnenmeier, David R. Forsman, Jason Tomcho: The aerodynamics of symmetric spinnakers; Journal of Wind Engineering and Industrial Aerodynamics, 93:311-337, 2005.
[7] K.L. Hedges: Computer Modelling of Downwind Sails; MF Thesis, University of Auckland, New Zealand, 1993.
[8] K.L. Hedges, P.J. Richards, G.D. Mallison: Computer Modelling of Downwind Sails; Journal of Wind Engineering and Industrial Aerodynamics 63 95-110, 1996.
[9] William C. Lasher and Peter J. Richards: Validation of Reynolds-Averaged Navier-Stokes Simulations for International America's Cup Class Spinnaker Force Coefficients in an Atmospheric Boundary Layer; Journal of Ship Research, Vol. 51, No. 1, pp. 22–38, March 2007.

[10] William Lascher & James R. Sonnenmeier: An Analysis of Practical RANS Simulations for Spinnaker Aerodynamics; Journal of Wind Engineering and Industrial Aerodynamics, 96 143–165, 2008.
[11] Challenger of Record and Defender for America's Cup XXXII: America's Cup Class Rule, Version 5.0; 15th December 2003;
[12] F. Fossati, S. Muggiasca, I.M. Viola, A. Zasso: Wind Tunnel Techniques for Investigation and Optimization of Sailing Yachts Aerodynamics; proceedings of 2nd High Performance Yacht Design Conference; Auckland, NZ, 2006.

Parallel implementation of fictitious surfaces method for aerodynamic shape optimization

S. Peigin[1], B. Epstein[2], and S. Gali[3]

[1] Israel Aerospace Industries, Israel
[2] The Academic College of Tel-Aviv Yaffo, Israel
[3] R&D IMoD, Israel

Abstract. An efficient approach to optimization of real-life 3D aerodynamic shapes for minimum drag is presented. The method allows for substitution of fictitious surfaces for those parts of aircraft which are not subject to modification in the process of design. The approach essentially reduces the geometrical complexity of computational models thus making the optimization of complete aerodynamic configurations practically feasible and, eventually, increasing the accuracy and reliability of optimization. The optimization framework is that of OPTIMAS, the in-house IAI optimization tool based on Genetic Algorithms and full Navier-Stokes computations. The method is illustrated by example of wing shape optimization for the full configuration of generic business jet with close coupled wing-body-nacelle, and vertical/horisontal tails. The results indicate the applicability of the method to practical aerodynamic design.

1 INTRODUCTION

The objective of reducing the design cycle of an aircraft prompted a major research and industrial effort into CFD driven optimization methods. Based on this effort, a certain progress has been achieved in the incorporation of automatic CFD-based tools into the overall technology of aerodynamic design [1]- [2].

As the technology for high-speed flight had matured, the development of competitive business jets requires the application of advanced design methods able to substantially reduce the design cycle and to improve the aerodynamic performance [3]. This leads to the necessity of a greater exploitation of automatic CFD driven optimization tools in the process of aerodynamic design of business jets.

In particular, a new technology of aerodynamic design, based on CFD driven constrained optimization for minimum drag, has been recently developed by the authors in Israel Aerospace Industries (IAI). This technology makes use of the in-house optimization tool OPTIMAS driven by Genetic Algorithms and accurate drag prediction (see Ref. [4]-[6]). With this technology, wing shape was optimized separately and then incorporated into the whole aircraft ([5]) or optimized as an integral part of wing-body configuration as in Ref. [6].

The applicability of automatic optimizers to more complex aerodynamic configurations such as wing-body-nacelle-pylon combination is generally plagued by

D. Tromeur-Dervout (eds.), *Parallel Computational Fluid Dynamics 2008*,
Lecture Notes in Computational Science and Engineering 74,
DOI: 10.1007/978-3-642-14438-7_7, © Springer-Verlag Berlin Heidelberg 2010

several reasons. Even if the design is restricted to only lifting surface(s), aerodynamic accuracy considerations demand to sufficiently resolve at least those parts of the configuration which directly influence the optimized parts of the aircraft or/and may change the design conditions such as the total lift.

This poses the following problems. Firstly, sufficiently resolved complex aerodynamic shapes require the huge computational volume for their optimization making the full-scale optimization impractical. This problem is not likely to be fully solved in the near future even with the advent of more powerful computer resources. Secondly, the optimization of complex shapes requires the use of fine computational grids not only for the initial configuration but for the whole family of tested geometries. These grids must be consistent in the following sense. The computational error prompted by the change in a numerical grid (which is due to the change in a tested shape) must be negligible compared to the accuracy of CFD computations. The latter represents a small value of about 0.0001-0.0004 in terms of drag coefficient. The problem looks intractable where unstructured grids are employed since in this case a change in shape results in grid restructuring, which frequently leads to significant changes in the value of drag coefficient. Even with structured grids (used in OPTIMAS), this still may present a problem especially where important parts of the designed aircraft are closely coupled and/or include "dire straits" between them. In this case, a mesh movement is performed in a very bounded space region which increases the mesh generated computational error. Body installed nacelle-pylon closely coupled with a wing gives a realistic example of the above problem.

The existing practice is to reduce the design geometry to a simpler configuration (e.g. by omitting nacelle) or, alternatively, to drastically reduce the search space.

In this work we show that a simple "naive" reduction of geometrical complexity may yield low optimization accuracy (that is, to significantly reduce the potential gain in aerodynamic performance). It is proposed to overcome the above described obstacles by replacing a part of optimized aircraft (which by itself is not subject to modification) by an infinitely thin surface in the way which allows to reproduce the needed aerodynamic effect.

The method is exemplified by optimizing the wing of a generic business jet with body installed nacelle, which greatly influences the flow around the wing. In this example, the nacelle is replaced by an infinitely thin surface which simulates the needed blockage effect. The geometrical reduction of this kind allowed for sufficiently representative search spaces keeping the overall accuracy to a good level.

The results indicate the applicability of the method to practical aerodynamic design in engineering environment.

2 OPTIMIZATION METHOD

In this section, we briefly describe the optimization method recently developed by the authors. Two-dimensional applications of the method may be found in [4], the optimization of isolated 3D wings was considered in [5], while the optimization of wing-body configuration was reported in [6].

The driver of the optimization search is a variant of Genetic Algorithms (GAs). The main features of the method include a new strategy for efficient handling of nonlinear constraints in the framework of GAs, scanning of the optimization search space by a combination of full Navier-Stokes computations with the Reduced Order Models (ROM) method and multilevel parallelization of the whole computational framework which efficiently makes use of computational power supplied by massively parallel processors (MPP).

2.1 Design Space and Search Algorithm

In this work it is assumed that:

1) The geometry is described by the absolute Cartesian coordinate system (x, y, z), where the axes x, y and z are directed along the streamwise, normal to wing surface and span directions, respectively;

2) Wing planform is fixed;

3) Wing surface is generated by a linear interpolation (in the span direction) between sectional 2D airfoils;

4) The number of sectional airfoils N_{ws} is fixed;

5) The wing-body boundary (the wing root airfoil) is not subject to change in the optimization process;

6) Shape of sectional airfoils is determined by Bezier curves of order N.

In the absolute coordinate system, the location of the above profiles is defined by the corresponding span positions of the trailing edge on the wing planform, twist angles $\{\alpha_i^{tw}\}$ and dihedral values $\{\gamma_i^{dh}\}$ (relatively to the root section).

Thus the design space includes the coordinates of the Bezier control points and twist and dihedral values. The dimension N_D of the search space is equal to:

$$N_D = (N_{ws} - 1) \cdot (2N - 3)$$

and a search string S contains N_D floating point variables a_j $(j = 1, ..., N_D)$. The string components are varied within the search domain D. The domain D is determined by values Min_j and Max_j, which are the lower and upper bounds of the variable a_j.

The optimization tool OPTIMAS uses Genetic Algorithm as its search engine. Genetic Algorithms became highly popular as optimization methods in the last two decades. The basic idea behind Genetic Algorithms is to mathematically imitate the evolution process of nature. They are semi-stochastic optimization methods that are conveniently presented using the metaphor of natural evolution: a randomly initialized population of individuals (set of points of the search space at hand) evolves following a crude parody of the Darwinian principle of the survival of the fittest. The main point is that the probability of survival of new individuals depends on their fitness: the best are kept with a high probability, the worst are rapidly discarded.

As a basic algorithm, a variant of the floating-point GA is employed [7]. We used the tournament selection, which enables us to increase the diversity of the parents, single point crossover operator, the non-uniform mutation and elitism principle.

The constraints handling can be basically outlined as follows (for more detail see [8]):

- Instead of the traditional approach where only feasible points may be included in a path, it is proposed to employ search paths through both feasible and infeasible points
- With this end in view, the search space is extended by evaluating (in terms of fitness) the points, which do not satisfy the constraints imposed by the optimization problem. A needed extension of an objective function may be implemented by means of GAs due to their basic property: contrary to classical optimization methods, GAs are not confined to only smooth extensions.

2.2 Computational Efficiency

Low computational efficiency of GAs is the main obstacle to their practical use where the evaluation of the cost function is computationally expensive as it happens in the framework of the full Navier-Stokes model.

To resolve this problem, we use Reduced-Order Models approach, where the solution functionals are approximated by a local data base. The data base is obtained by solving the full Navier-Stokes equations in a discrete neighbourhood of a basic point (basic geometry) positioned in the search space. Specifically a mixed linear-quadratic approximation is employed. One-dimensionally, the one-sided linear approximation is used in the case of monotonic behaviour of the approximated function, and the quadratic approximation is used otherwise.

In order to ensure the accuracy and robustness of the method a multidomain prediction-verification principle is employed. That is, on the prediction stage the genetic optimum search is concurrently performed on a number of search domains. As the result each domain produces an optimal point, and the whole set of these points is verified (through full Navier-Stokes computations) on the verification stage of the method, and thus the final optimal point is determined.

In order to additionally improve the computational efficiency of the tool, we employ an embedded multilevel parallelization strategy [9] which includes: Level 1 - Parallelization of full Navier-Stokes solver; Level 2 - Parallel CFD scanning of the search space; Level 3 - Parallelization of the GAs optimization process; Level 4 - Parallel optimal search on multiple search domains and Level 5 - Parallel grid generation.

2.3 Incorporation of fictitious surfaces

The proposed method of fictitious surfaces consists in replacing certain parts in the geometrical model of an aircraft by infinitely thin surfaces. It is assumed that the replaced part of the configuration is not subject to shape modification.

In the present framework, the technical implementation is especially simple. First, the initial geometry is reduced by excluding the part which is to be simulated by an infinitely thin surface, and a multi-block structured computational grid for the reduced configuration is constructed. In the next stage, an appropriate face of an "air-block" is chosen and declared to be a fictitious surface. The major requirement to the choice of the fictitious surface is, in terms of location and size, to simulate (in a very approximate way) the major aerodynamic effect caused by the deleted part of the configuration.

Mathematically, fictitious surfaces are regarded as surfaces of zero thickness, and the surface boundary condition is implemented on the both sides of them. Since we solve the viscous Navier-Stokes equations, the no-slip condition is applied in exactly the same way as for "real surfaces".

Specifically in this work we replace the nacelle junction by an appropriately chosen 2D surface. The main idea behind such a replacement is to simulate the blockage effect induced by the flow around and through the nacelle. It appeared that the exact size and location of the fictitious surface make little effect on the results of optimization.

From the other side, the optimization with totally omitted nacelle suffers from low aerodynamic accuracy (apparently due to the unjustified elimination of the blockage effect). Which is even worse, the nacelle elimination produces inconsistent aerodynamic results. This means that the problem of low accuracy could not be solved by simply adjusting the design conditions or improving the resolution of computations. We remind that optimization based on CFD computations for the complete configuration including the real nacelle junction appeared infeasible due to the huge expenditure of computer resources.

3 ANALYSIS OF RESULTS

The optimized configuration presents an example of a generic business jet and includes body, wing with winglets, nacelle-pylon, and vertical and horizontal tails.

As a CFD driver, the IAI in-house full Navier-Stokes code NES was employed. The code allows for high accuracy lift/drag computations in subsonic and transonic regime (see [10]-[11]). For the considered business jet configuration, the computations by the code NES also exhibited good accuracy.

The planview of pressure distribution for the original geometry (full aircraft configuration) and for the wing-body combination without nacelle and tails (at the same cruise flight conditions) is respectively presented in Fig.1 and Fig.2. The analysis of these pictures demonstrates that the body and inboard wing pressure contours significantly differ from one case to another, which is mainly due to the blockage effect produced by the nacelle.

Two optimization cases were considered. In the first case (labeled *Case_GBJNP_1*), the geometry was reduced to the combination of wing-body, and the wing was optimized in the presence of the body only. In the second test case (*Case_GBJNP_2*) the nacelle influence was modeled by including a fictious surface in the optimized wing-body geometry. For both optimization cases the design point was $M = 0.80$, $C_L = 0.46$. The geometrical constraints were imposed on the maximum thickness, leading edge radius and trailing edge angle of the wing sectional airfoils, which were kept to the original level. The number of sectional airfoils subject to design was equal to three.

The first "no-nacelle" optimization *Case_GBJNP_1* showed a significant gain in drag (about 9 aerodynamic counts at the design point) in comparison with the original wing-body configuration. This is illustrated by Fig.3, where lift/drag was computed for the stand alone wing-body combination without nacelle.

However, for practical design, the success of optimization should be estimated in terms of drag reduction for the complete business jet configuration (including nacelle-pylon). Unfortunately the analysis of the complete aircraft, in the presence of the nacelle (Fig.4), showed that the real gain at the design point falls to nearly zero. Note that this "naive" optimization nevertheless yields a better off-design lift/drag behaviour at higher than the design lift and Mach values (see Fig.4 and Fig.5).

We must emphasize that attempts were made to improve the above optimization by a better adjustment of design conditions (which may be regarded as a "cheap" simulation of the nacelle influence) but they did not yield consistent results.

Thus it became clear that the nacelle influence must be simulated in the optimization process in order to increase the gain in the aerodynamic performance. In principal, the ultimate way to do it is to use as an objective function the total drag of the complete aircraft geometry. However, a simple estimation shows, that this requires the huge computational resources which makes the optimization infeasible from practical viewpoint.

In order to overcome this we propose to simulate the blockage effect caused by nacelle through the incorporation of a fictitious surface into the computational model. With this end in view, in the second test case *Case_GBJNP_2*, the computational grid previously used for the wing-body combination, was modified by choosing a joint face of two computational "air-blocks" as the fictitious surface. Technically this means that the transparent block boundary condition (which previously merged the neighboring blocks, see Ref. [10]) was replaced by the Navier-Stokes no slip condition. Note that this boundary condition was applied on the both sides of the face (which belongs to two original computational blocks). Aerodynamically this means that the face was treated as a fictitious infinitely thin surface. Thus, in the process of optimization, the actual complete geometry was replaced by the above described computational model (Fig.6), and the wing was optimized as an integral part of this composite computational model. Finally, the optimal shape was analyzed in the framework of the complete configuration which included the real nacelle junction and the vertical/horizontal tails.

The optimization yielded a significant gain in drag in terms of the actually optimized geometry (wing-body-fictitious-surface). An important point is that, contrary to the first (no-nacelle) optimization, the most of gain was retained in terms of the full aircraft configuration analyzed a posteriori. The total drag of the complete configuration was reduced by 10 aerodynamic counts at the design conditions. The corresponding pressure distrubution for the optimized configuration at the design conditions is given in Fig.7 while for the original shape it was presented in Fig.1. A more detailed analysis demonstrates that the optimization resulted in a smoother pressure distribution on the wing upper surface.

The off-design behaviour of the optimized business jet may be assessed from Fig.8-9, where lift/drag polars at $M = 0.80$, $M = 0.82$, and Mach drag rise at

$C_L = 0.46$ are respectively compared with the original curves. It is clearly seen that the gain due to optimization increases at higher free-stream lift and Mach values.

It is interesting to compare the difference between the considered optimization cases in terms of the resulting wing shapes. The corresponding airfoil shapes at $2y/b = 0.10$, $2y/b = 0.26$ and $2y/b = 1.0$ are respectively depicted in Fig.10-12. As aerodynamically expected, the "nacelle influence" is more significant in the inboard part of the wing gradually diminishing towards the tip region. At the tip, the both optimized wing sections are practically identical.

Finally, it appeared that the exact size and location of the fictitious surface make little effect on the results of optimization. It is important from the practical viewpoint that, as the analysis showed, the fictitious surface may be determined approximately with the only requirement to roughly simulate the blockage effect caused by the nacelle.

4 CONCLUSIONS

Fictitious surface method for reducing the geometrical complexity of full aircraft configurations has been developed. The method was successfully applied to design of a complete transonic business jet with close coupled wing-body-nacelle-pylon, and vertical/horisontal tails. The results indicate good accuracy and robustness of the method which make it suitable for practical engineering design.

[1] Kroo, I., and Manning, V., "Collaborative Optimization: Status and Directions" Optimization in Aerodynamics", *AIAA paper,* AIAA-2000-4721, 2000.

[2] Piperni, P., Abdo, M., and Kafyeke, F., "The Application of Multi-Disciplinary Optimization Technologies to the Design of a Business Jet", *AIAA paper,* AIAA-2004-4370, 2004.

[3] Jameson, A., Martinelli, L., and Vassberg, J., "Using Computational Fluid Dynamics for Aerodynamics - a Critical Assessment", *Proceedings of ICAS 2002*, Paper ICAS 2002-1.10.1, Optimage Ltd., Edinburgh, Scotland, U.K., 2002.

[4] Epstein, B., and Peigin, S., "Robust Hybrid Approach to Multiobjective Constarined Optimization in Aerodynamics", *AIAA Journal,* Vol. 42, No.8, 2004, pp. 1572–1581.

[5] Epstein, B., and Peigin, S., "Constrained Aerodynamic Optimization of 3D Wings Driven by Navier-Stokes Computations" *AIAA Journal*, Vol. 43, No. 9, 2005, pp.1946-1957.

[6] Peigin, S., and Epstein, B., "Multipoint Aerodynamic Design of Wing-Body Configurations for Minimum Drag", *Journal of Aircraft*, Vol. 44, No. 3, 2007, pp.971-980.

[7] Michalewicz, Z., *Genetic Algorithms + Data Structures = Evolution Programs* (New-York: Springer Verlag), 1996

[8] Peigin, S. and Epstein, B., "Robust handling of non-linear constraints for GA optimization of aerodynamic shapes", *Int. J. Numer. Meth. Fluids*, Vol. 45, No.8, 2004, p.1339–1362.

[9] Peigin, S., and Epstein, B., "Embedded parallelization approach for optimization in aerodynamic design", *The Journal of Supercomputing*, Vol.29, No. 3, 2004, p.243–263.

[10] Epstein, B., Rubin, T., and Seror, S., "Accurate Multiblock Navier-Stokes Solver for Complex Aerodynamic Configurations. *AIAA Journal*, Vol. 41, No. 4, 2003, pp.582–594.

[11] Seror, S., Rubin, T., Peigin, S. and Epstein, B., "Implementation and Validation of the Spalart-Allmaras Turbulence Model for a Parallel CFD Code", *Journal of Aircraft,* Vol. 42, No.1, 2005, pp.179–188.

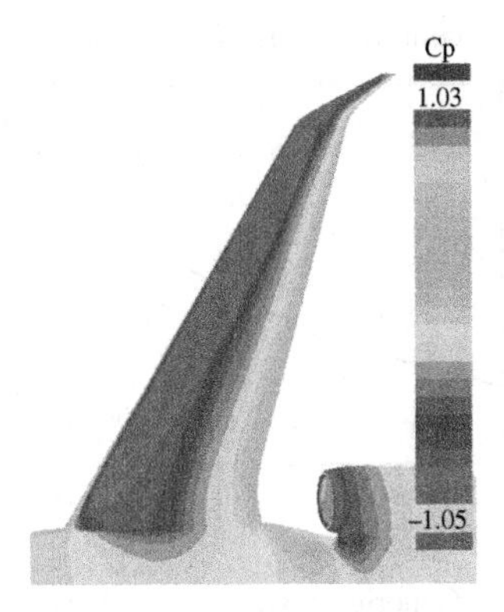

Fig. 1. Generic business jet. Pressure distribution for full aircraft configuration: original geometry.

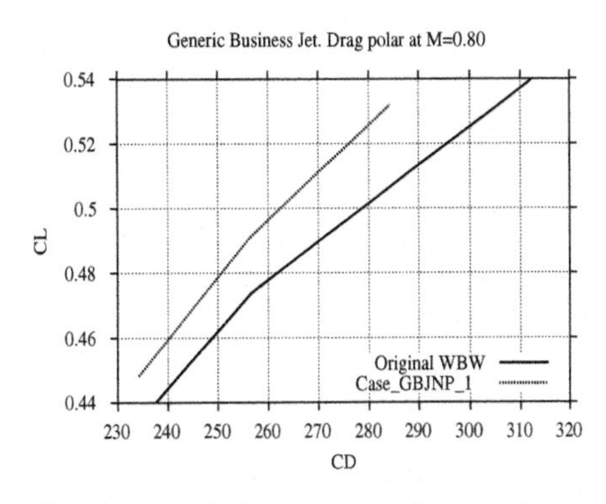

Fig. 3. Generic business jet. Drag polar at $M = 0.80$ for wing-body configuration. Original wing vs. optimal one (*Case_GBJNP_1*).

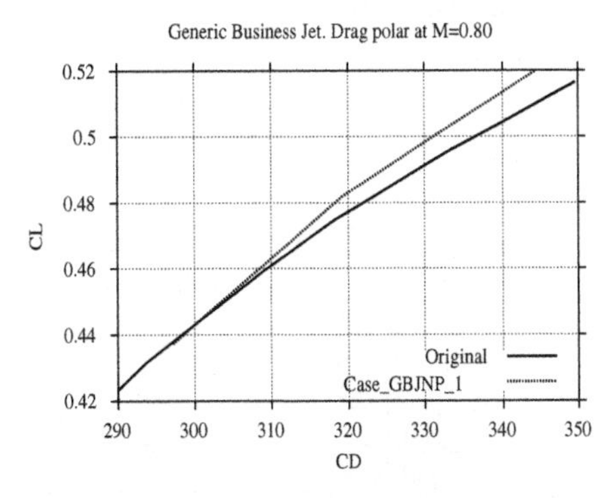

Fig. 4. Generic business jet. Drag polar at $M = 0.80$ for full aircraft configuration. Original wing vs. optimal one (*Case_GBJNP_1*).

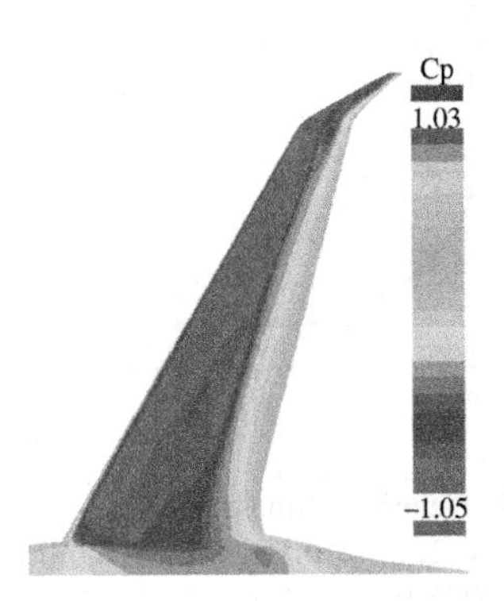

Fig. 2. Generic business jet. Pressure distribution for wing-body configuration: original geometry.

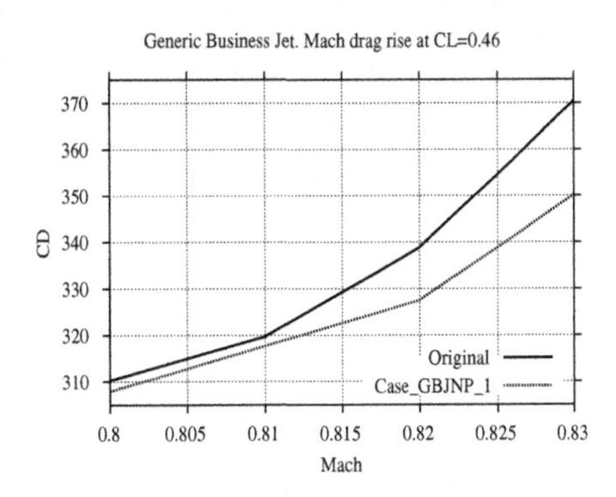

Fig. 5. Generic business jet. Mach drag divergence curve for full aircraft configuration. Original wing vs. optimal one (*Case_GBJNP_1*).

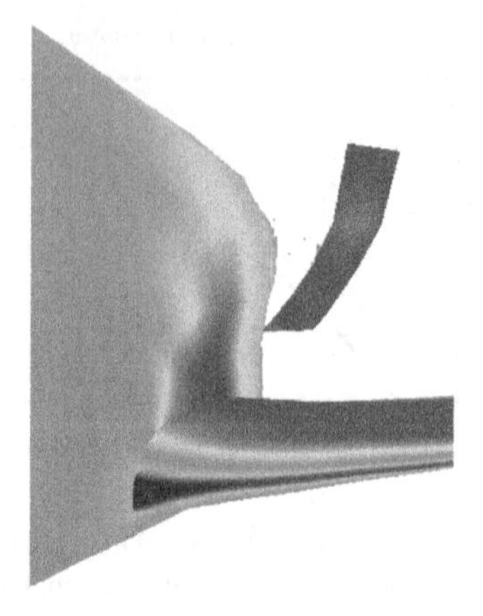

Fig. 6. General view of a generic business jet: computational model.

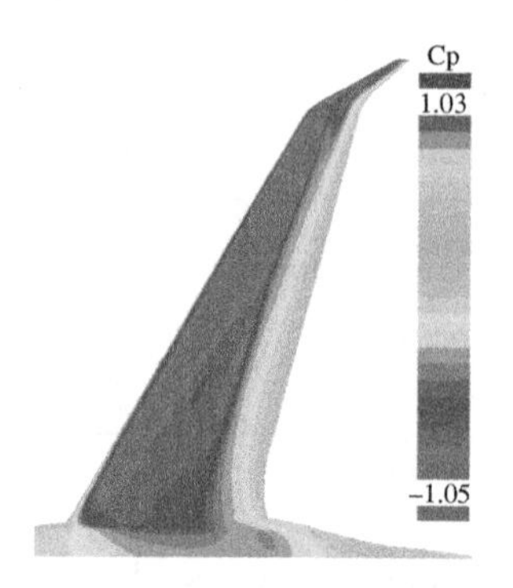

Fig. 7. Generic business jet. Pressure distribution for full aircraft configuration: *Case_GBJNP_2*.

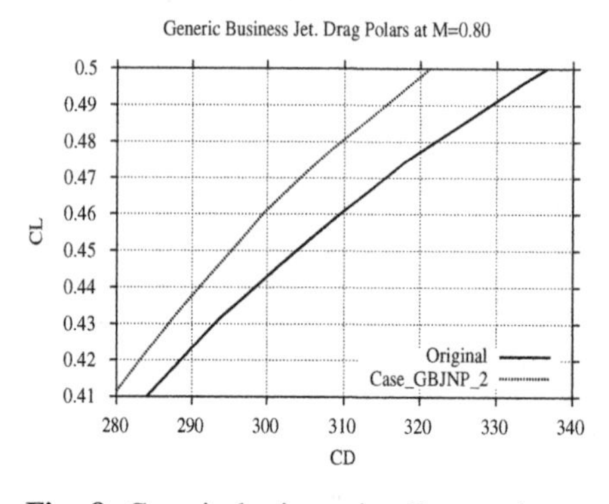

Fig. 8. Generic business jet. Drag polar at $M = 0.80$ for full aircraft configuration. Original wing vs. optimal one (*Case_GBJNP_2*).

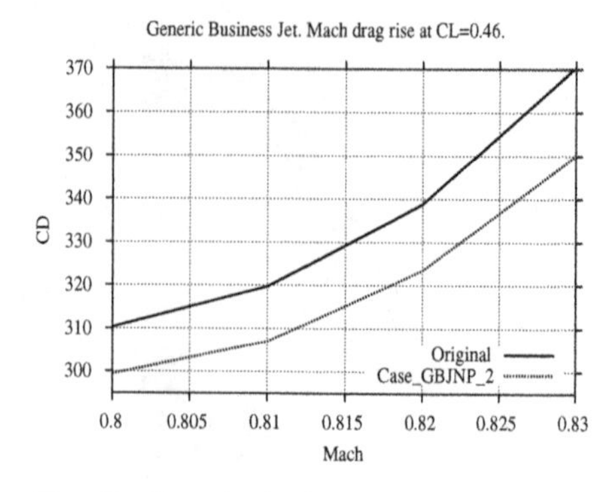

Fig. 9. Generic business jet. Mach drag divergence curve for full aircraft configuration. Original wing vs. optimal one (*Case_GBJNP_2*).

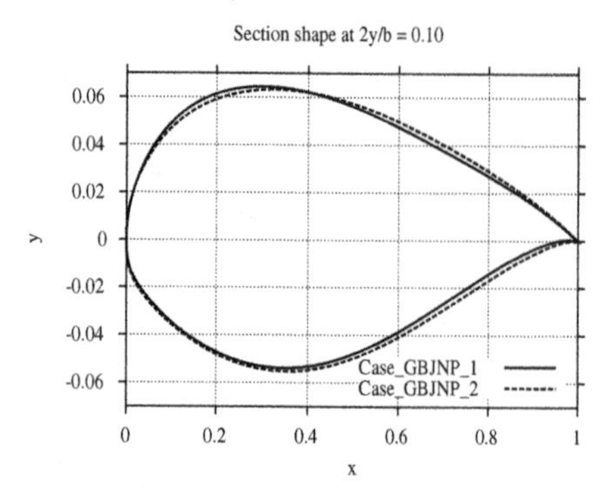

Fig. 10. Generic business jet. Wing sectional shape at $2y/b = 0.10$. *Case_GBJNP_1* vs. *Case_GBJNP_2*.

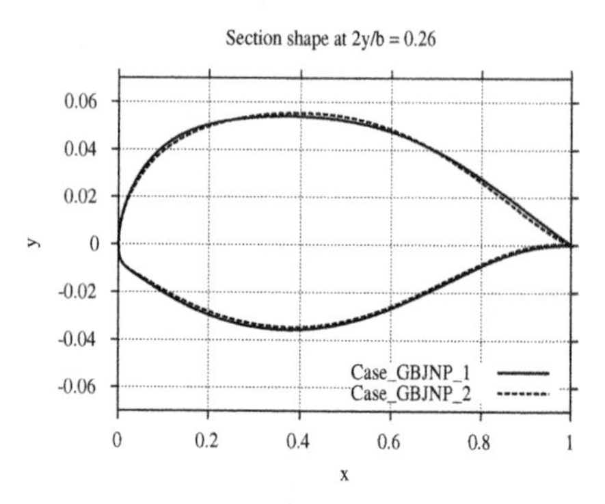

Fig. 11. Generic business jet. Wing sectional shape at $2y/b = 0.26$. *Case_GBJNP_1* vs. *Case_GBJNP_2*.

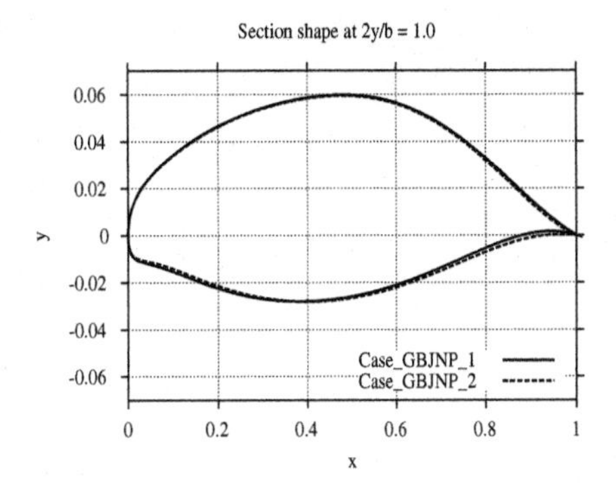

Fig. 12. Generic business jet. Wing sectional shape at $2y/b = 1.0$. *Case_GBJNP_1* vs. *Case_GBJNP_2*.

Path Optimization of Dual Airfoils Flapping in a Biplane Configuration with RSM in a Parallel Computing Environment

Mustafa Kaya[1] and Ismail H. Tuncer[2]

[1] Middle East Technical University, Ankara, Turkey mkaya@ae.metu.edu.tr
[2] Middle East Technical University, Ankara, Turkey tuncer@ae.metu.edu.tr

Abstract. The path of dual airfoils in a biplane configuration undergoing a combined, non–sinusoidal pitching and plunging motion is optimized for maximum thrust and/or propulsive efficiency. The non–sinusoidal, periodic flapping motion is described using Non-Uniform Rational B-Splines (NURBS). The Response Surface Methodology (RSM) is employed for the optimization of NURBS parameters in a parallel computing environment. A gradient based optimization algorithm, steepest ascent method is started from the optimum point of response surfaces. Unsteady, low speed laminar flows are also computed in parallel using a Navier-Stokes solver based on domain decomposition. It is shown that the parallel optimization process with RSM suggests a quick and accurate initial guess for a gradient based optimization algorithm.

1 INTRODUCTION

Flow characteristics of flapping wings are currently investigated experimentally and numerically to shed some light on the lift, drag and propulsive power considerations for a MAV flight[1, 2]. It should be noted that in order to maximize the thrust and/or the propulsive efficiency of a flapping airfoil, its kinematic parameters, such as the flapping path, the frequency and the amplitude of the flapping motion, need to be optimized.

In earlier studies, the present authors employed a gradient based optimization of sinusoidal and non–sinusoidal flapping motion parameters of flapping airfoils[3, 4]. These optimization studies with a limited number of optimization variables show that the thrust generation and efficiency of flapping airfoils may be increased significantly. However, the gradient based global optimization process becomes computationally expensive as the number of optimization variables increases in the non–sinusoidal flapping motion definition with NURBS.

Response surface methodology (RSM) is mainly employed for the construction of global approximations to a function based on its values computed at various points[5]. The method may also be employed for the optimization of a function when the objective function is expensive in terms of computational resources[5, 6].

D. Tromeur-Dervout (eds.), *Parallel Computational Fluid Dynamics 2008*,
Lecture Notes in Computational Science and Engineering 74,
DOI: 10.1007/978-3-642-14438-7_8,

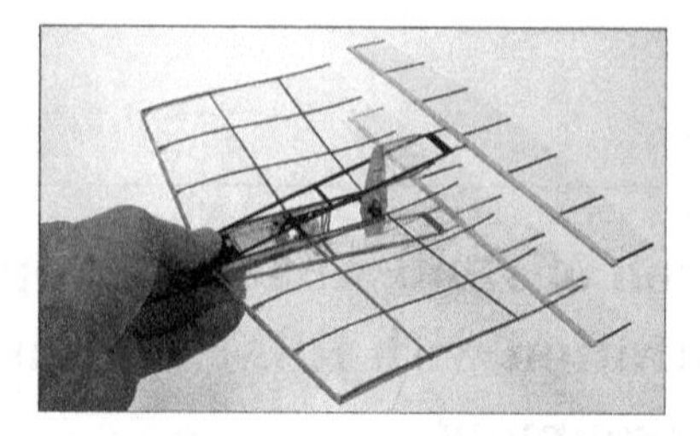

Fig. 1. Flapping-wing MAV model (Jones and Platzer)

In the present study, the thrust generation of dual airfoils flapping in a biplane configuration undergoing a combined non–sinusoidal pitching and plunging motion is optimized using RSM. First, a single airfoil undergoing a non-sinusoidally flapping motion is considered. RSM for 3 optimization variables is assessed and optimization data are compared to the gradient based optimization method in terms of the performance and the accuracy. Next, the non–sinusoidal flapping motion of dual airfoils with seven optimizationvariables is considered.

2 Response Surface Methodology, RSM

RSM is based on building approximate models for unknown functional relationships between input and output data. In this study, the function is the average thrust coefficient, C_t, which is based on the integration of the drag coefficient over a flapping period. It is a function of flapping parameters, V_i, in a given flight condition and can be written as

$$C_t = \eta(V_1, V_2, V_3, \dots) \tag{1}$$

The function $\eta(\mathbf{V})$ is in fact the solution of the Navier-Stokes equations. An approximate response surface, $g(\mathbf{V}) \cong \eta(\mathbf{V})$ may then be constructed over some $\mathbf{V}$ region [5]. In this study, $g(\mathbf{V})$, is chosen to be a quadratic function of V_i's:

$$g(V_i) = a_{11}V_1^2 + 2a_{12}V_1V_2 + 2a_{13}V_1V_3 + \dots + a_{22}V_2^2 + \dots \tag{2}$$

The constants, a_{ij}, are evaluated throught a least-square minimization of the error between the response surface and a certain number of the Navier-Stokes solutions based on a design of experiment (DoE). In this study, the Box-Behnken[7] DoE method is employed.

3 Periodic Path defined by NURBS

A smooth curve S based on a general n^{th} degree rational Bezier segment is defined as follows[9]:

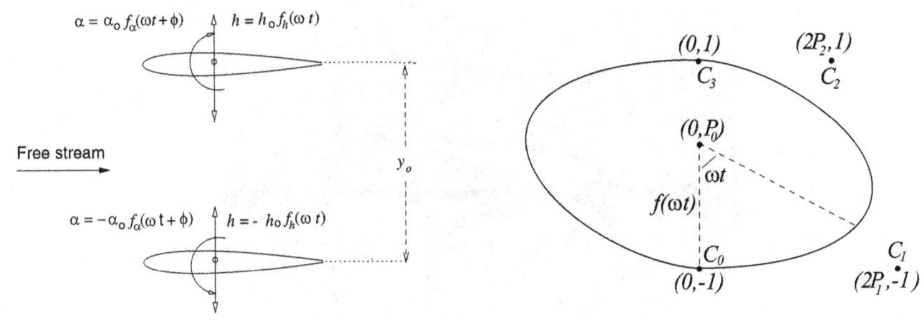

Fig. 2. Out-of-phase flapping motion of two airfoils in a biplane configuration

Fig. 3. Flapping path defined by a 3rd degree NURBS

$$S(u) = (x(u), y(u)) = \frac{\sum_{i=0}^{n} W_i B_{i,n}(u) C_i}{\sum_{i=0}^{n} W_i B_{i,n}(u)} \qquad 0 \leq u \leq 1 \tag{3}$$

where $B_{i,n}(u) \equiv \frac{n!}{i!(n-i)!} u^i (1-u)^{n-i}$ are the classical n^{th} degree Bernstein polynomials, and $C_i = (x_{pi}, y_{pi})$, are called control points with weights, W_i. Note that $S(u=0) = C_0$ and $S(u=1) = C_n$. A closed curve which describes the upstroke and the downstroke of a flapping path is then defined by employing a NURBS composed of two 3rd degree rational Bezier segments. The periodic flapping motion is finally defined by 3 parameters. The first parameter P_0 defines the center of the rotation vector on a closed curve. The remaining two points, P_1 and P_2 are used to define the x coordinates of the control points, which are $C_1 = (2P_1, -1)$ and $C_2 = (2P_2, 1)$ (Figure 3).

The x and y coordinates on the periodic NURBS curve may be obtained as a function of the parameter u:

$$x(u) = \frac{2P_1 u(1-u)^2 + 2P_2 u^2 (1-u)}{2u^2 - 2u + 1} \qquad y(u) = \frac{2u - 1}{2u^2 - 2u + 1} \tag{4}$$

A non-sinusoidal periodic function, f, is then defined by $y(u)$. For a given ωt position, the Equation 5 is solved for u. Once u is determined, $y(u) \equiv f(\omega t)$ is evaluated using Equation 4.

$$\tan(\omega t) = -\frac{x(u)}{y(u) - P_0} \tag{5}$$

4 Numerical Method

Unsteady, viscous flowfields around flapping airfoils in a biplane configuration are computed by solving the Navier-Stokes equations on moving and deforming overset grids. A domain decomposition based parallel computing algorithm is employed. PVM message passing library routines are used in the parallel solution algorithm. The computed unsteady flowfields are analyzed in terms of time histories of aerodynamic loads, and unsteady particle traces.

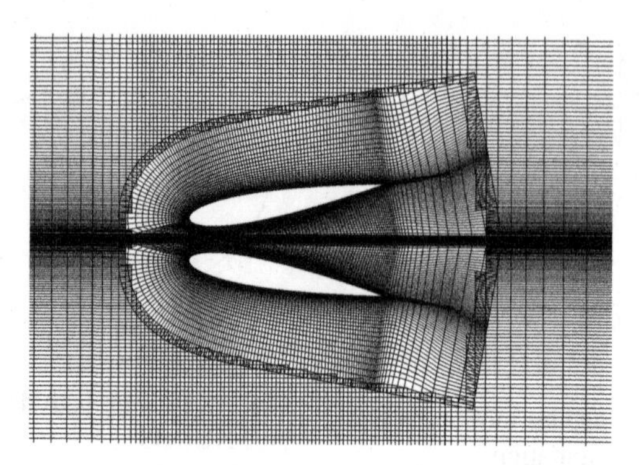

Fig. 4. Moving and deforming overset grid system

4.1 Flow Solver

The strong conservation-law form of the 2-D, thin-layer, Reynolds averaged Navier-Stokes equations is solved on each subgrid. The convective fluxes are evaluated using Osher's third–order accurate upwind–biased flux difference splitting scheme. The discretized equations are solved in parallel by an approximately factored, implicit algorithm. The overset grid system (Figure 4) is partitioned into subgrids. The holes in the background grid are excluded from the computations by an *i-blanking* algorithm. The conservative flow variables are interpolated at the intergrid boundaries formed by the overset grids[10].

4.2 Flapping Motion

The flapping motion of the upper airfoil in plunge, h, and in pitch, α, is defined by:

$$\begin{aligned} h(t) &= h_0\, f_h(\omega t) \\ \alpha(t) &= \alpha_0\, f_\alpha(\omega t + \phi) \end{aligned} \tag{6}$$

where h_o and α_o are the plunge and pitch amplitudes, f is a periodic function based on NURBS, ω is the angular frequency which is given in terms of the reduced frequency, $k = \frac{\omega c}{U_\infty}$. ϕ is the phase shift between plunge and pitching motions. The pitch axis is located at the mid-chord. The flapping motion of the lower airfoil is in counter-phase. The average distance between the dual airfoils is set to $y_0 = 1.4$.

The flapping motion of the airfoils are imposed by moving the airfoil grids over the background grid (Figure 4). The airfoil grids are deformed as they come close to the symmetry line between the airfoils.

4.3 Optimization based on the Steepest Ascent Method

Optimization based on the Steepest Ascent is also performed for validation. The gradient vector of the objective function is given by

$$\nabla O(\mathbf{V}) = \frac{\partial O}{\partial V_1}\mathbf{v_1} + \frac{\partial O}{\partial V_2}\mathbf{v_2} + \cdots$$

where V_i's are the optimization variables. The components of the gradient vector are evaluated numerically by computing an unsteady flow solution for a perturbation of the optimization variables one at a time.

4.4 Parallel Processing

The parallel solution algorithm is based on the domain decomposition. The moving and deforming overset grid system is decomposed into its subgrids first, and the solution on each subgrid is computed in parallel. Intergrid and overlapping boundary conditions are exchanged among subgrid solutions at each time step of the unsteady solution. The unsteady flow solutions needed for the RSM and the gradient vector components, are also carried out in parallel. PVM (version 3.4.5) library routines are used for inter–process communication. Computations are performed in a cluster of Linux based computers with dual Xeon and Pentium-D processors.

5 Results and Discussion

A validation study is first performed. The optimization result with RSM is compared against the gradient based optimization. The path optimization of the flapping airfoils in a biplane configuration with 8 optimization variables is performed next.

5.1 Validation Study

The optimization of a single airfoil flapping on a non-sinusoidal path is studied first. The flapping motion is a combination of non-sinusoidal pitching and sinusoidal plunging. The values for the reduced flapping frequency, $k \equiv \frac{\omega c}{U_\infty}$, and the plunge amplitude, h_0, are fixed at $k = 1.0$ and $h_0 = 0.5$. The optimization variables are the pitch amplitude, α_0, the phase shift between plunging and pitching, ϕ, and the NURBS parameter, $P_{0\alpha}$ (Table 1). The flowfields are computed at Mach number of 0.1 and a Reynolds number of 10^4.

Table 4 gives the design points used for constructing the response surface. Design points are chosen based on Box-Behnken matrix of runs[7]. The P_1 and P_2 values are constrained within the range 0.2 to 5.0, and P_0 in the range -0.9 to 0.9, in order to define a proper flapping motion which does not impose excessively large accelerations.

The parallel computations for 13 unsteady flow solutions take about 2 hours of wall clock time using 40 processors. The cross sections of the constructed response surface are shown in Figure 5. The same optimization case is also studied using the steepest ascent method. The initial conditions required for this method are given in Table 2. The parallel computations, which required 58 unsteady flwo solutions, take about 20 hours of wall clock time using 15 processors.

Table 1. Fixed parameters and optimization variables in validation study

Case	k	h_0	$P_{1\alpha}$	$P_{2\alpha}$	P_{0h}	P_{1h}	P_{2h}	$P_{0\alpha}$	α_0	ϕ
	1.0	0.5	1.0	1.0	0.0	1.0	1.0	V	V	V

Table 2. Initial conditions for steepest ascent method

Case	k	h_0	$P_{1\alpha}$	$P_{2\alpha}$	P_{0h}	P_{1h}	P_{2h}	$P_{0\alpha}$	α_0	ϕ	C_t
	1.0	0.5	1.0	1.0	0.0	1.0	1.0	**0.0**	**20°**	**90°**	0.09

Table 3. Optimization results for the validation study

Case	$\alpha_o(^o)$	$\phi(^o)$	$P_{0\alpha}$	C_t
RSM	9.3	90.6	0.03	**0.17**
Steepest Ascent	9.2	90.7	−0.01	**0.15**

Table 4. RSM design points for the validation study

DoE	$\alpha_o(^o)$	$\phi(^o)$	$P_{0\alpha}$
1	5.0	30.0	0.0
2	5.0	150.0	0.0
3	35.0	30.0	0.0
4	35.0	150.0	0.0
5	5.0	90.0	−0.9
6	5.0	90.0	0.9
7	35.0	90.0	−0.9
8	35.0	90.0	0.9
9	20.0	30.0	−0.9
10	20.0	30.0	0.9
11	20.0	150.0	−0.9
12	20.0	150.0	0.9
13	20.0	90.0	0.0

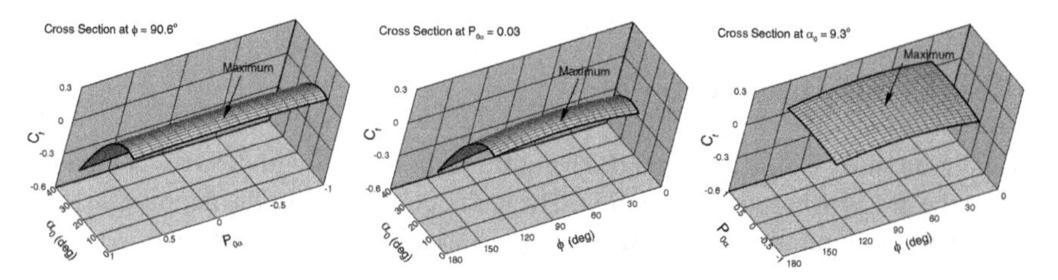

Fig. 5. Response surfaces for the validation case

Table 3 gives the optimization results based on the RSM and the steepest ascent method. It is observed that while the optimum solution is about the same in both cases, the number of unsteady flow solutions is significantly smaller in the RSM than in the steepest ascend method.

5.2 Optimization for Dual Airfoils

Dual airfoils flapping in a biplane configuration is studied next with 8 optimization variables, namely, the NURBS parameters defining the plunging path, P_{0h}, P_{1h}, P_{2h}, the NURBS parameters defining the pitching path, $P_{0\alpha}$, $P_{1\alpha}$, $P_{2\alpha}$, the pitching amplitude, α_0 and the phase shift between plunging and pitching, ϕ. The values for the reduced flapping frequency, $k \equiv \frac{\omega c}{U_\infty}$, and the plunge amplitude, h_0, are fixed at $k = 1.5$ and $h_0 = 0.53$.

These values are from an earlier study which optimized the sinusoidal flapping motion of dual airfoils in biplane[8]. At this study, the maximum thrust is computed to be $C_t = 0.45$.

Design of experiment due to Box-Behnken is summarized in Table 5. A total of 113 unsteady flow solutions are computed in about 10 hours of wall clock time using

Table 5. RSM design points and the computed thrust values

Desing No.	P_{0h}	P_{1h}	P_{2h}	$P_{0\alpha}$	$P_{1\alpha}$	$P_{2\alpha}$	$\alpha_0(^o)$	$\phi(^o)$	C_t
1	0.0	0.5	0.5	0.0	1.0	1.0	9.0	90.0	0.42
2	0.0	0.5	2.0	0.0	1.0	1.0	9.0	90.0	0.49
...	...	...	...	...	...	...	...	...	...
112	0.0	1.0	1.0	0.0	1.0	1.0	12.0	105.0	0.46
113	0.0	1.0	1.0	0.0	1.0	1.0	9.0	90.0	0.42

Table 6. Optimization results

	P_{0h}	P_{1h}	P_{2h}	$P_{0\alpha}$	$P_{1\alpha}$	$P_{2\alpha}$	$\alpha(^o)$	$\phi(^o)$	C_t
RSM	−0.30	2.00	2.00	−0.30	1.13	2.00	10.1	90.1	**0.89**
Navier-Stokes	−0.30	2.00	2.00	−0.30	1.13	2.00	10.1	90.1	**0.85**

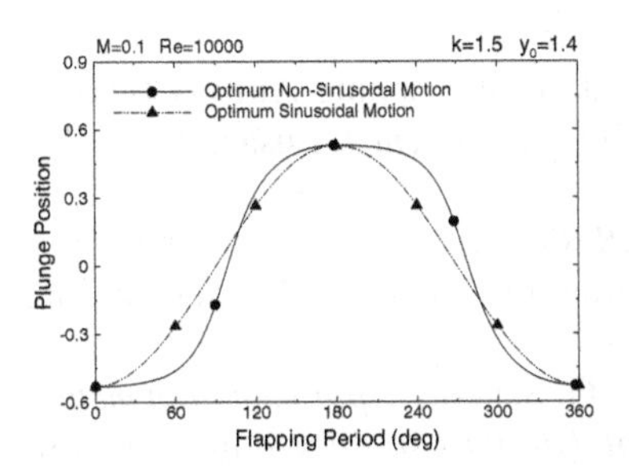

Fig. 6. Variation of optimum plunge position

Fig. 7. Variation of optimum pitch position

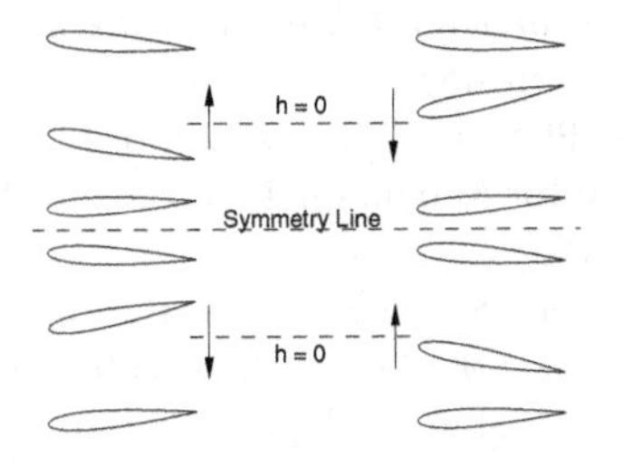

Fig. 8. Optimum non-sinusoidal flapping motion

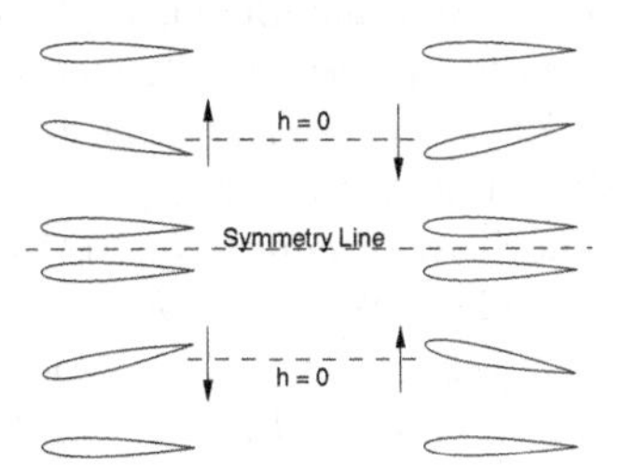

Fig. 9. Optimum sinusoidal flapping motion

64 processors. The response surface has 45 parameters (Eqn. 2). The optimization results are given in Table 6. The flow solution performed at the optimum conditions produces a thrust value of $C_t = 0.85$, which is about 4% off from the RSM prediction. The optimum sinusoidal and non-sinusoidal flapping motions are given in Figures 6, 8 and 9.

6 Conclusion

The path optimization of flapping airfoils with 8 optimziation variables is successfully performed using RSM in a parallel computing environment. In a validation stufy it is shown that RSM is highly efficient in comparison to a gradient based method.

[1] T.J. Mueller (editor), *Fixed and Flapping Wing Aerodynamics for Micro Air Vehicles*, AIAA Progress in Aeronautics and Astronautics, Vol 195, Reston, VA, 2001.

[2] W. Shyy, M. Berg and D. Lyungvist, "Flapping and Flexible Wings for Biological and Micro Air Vehicles", *Pergamon Progress in Aerospace Sciences*, Vol 35, p: 455-505, 1999.

[3] I.H. Tuncer and M. Kaya, "Optimization of Flapping Airfoils For Maximum Thrust and Propulsive Efficiency", *AIAA Journal*, Vol 43, p: 2329-2341, Nov 2005.

[4] M. Kaya and I.H. Tuncer, "Path Optimization of Flapping Airfoils Based on NURBS", *Proceedings of Parallel CFD 2006 Conference*, Busan, Korea, May 15-18, 2006.

[5] Roux, W.J., Stander, N. and Haftka, R.T., *Response Surface Approximations for Structural Optimization*, International Journal for Numerical Methods in Engineering, Vol. 42, 1998, pp. 517-534.

[6] D. H. Van Campen, R. Nagtegaal and A. J. G. Schoofs, *Approximation methods in structural optimization using experimental designs for for multiple responses*, in H. Eschenauer, J. Koski and A. Osyczka (eds.), Multicriteria Design Optimization, Springer, Berlin, 1990, pp. 205-228.

[7] Box, G. E. P. and Behnken, D. W., *Some new three level designs for the study of quantitative variables*, Technometrics, Vol. 2, pp. 455475.

[8] M. Kaya, I.H. Tuncer, K.D. Jones and M.F. Platzer, "Optimization of Flapping Motion of Airfoils in Biplane Configuration for Maximum Thrust and/or Efficiency", *AIAA Paper*, No 2007-0484, Jan 2007.

[9] L. Piegl and W. Tiller, *The NURBS Book*, 2nd ed., Springer-Verlag, Berlin, 1997.

[10] I.H. Tuncer, "A 2-D Unsteady Navier-Stokes Solution Method with Moving Overset Grids", *AIAA Journal*, Vol. 35, No. 3, March 1997, pp. 471-476.

Part III

Grid methods

Convergence Improvement Method for Computational Fluid Dynamics Using Building-Cube Method

Takahiro Fukushige[1] Toshihiro Kamatsuchi[1] Toshiyuki Arima[1] and Seiji Fujino[2]

[1] Fundamental Technology Research Center, Honda R&D Co., Ltd.
1-4-1 Chuo, Wako-shi, Saitama, 351-0193, Japan
takahiro_fukushige@n.w.rd.honda.co.jp

[2] Kyushu University
6-10-1, Hakozaki, Higashi-ku, Fukuoka-shi, 812-858, Japan

1 Introduction

Computational Fluid Dynamics (CFD) has become an important tool for aerodynamics by the improvements of computer performance and CFD algorithm itself. However, the computational time of CFD continues to increase, while progress of computer has been made. One of the reasons is considered that application of CFD has become more complex. For example, CFD is employed to estimate aerodynamics performance for a complex shaped object such as formula one car (Fig. 1) Concerning complex shape, however, the problem of grid generation still remains. It requires so much time and labor. To overcome the problem in meshing for complex-shaped object, we have already proposed an algorithm [3]. The algorithm consists of two approaches. One is Immersed Boundary method [6], and the other is Building-Cube Method (BCM) [4]. The basic idea of Immersed Boundary method is applied to cells in the vicinity of solid boundary, and Cartesian grid method is performed for other cells. These approaches have several advantages except for solution convergence. In this paper, Implicit Residual Smoothing (IRS) [2] is proposed for improvements of solution convergence.

Fig. 1. Formula 1 race car (Honda Racing F1 Team RA106).

D. Tromeur-Dervout (eds.), *Parallel Computational Fluid Dynamics 2008*,
Lecture Notes in Computational Science and Engineering 74,
DOI: 10.1007/978-3-642-14438-7_9,

2 Numerical Methods

2.1 Computational Gird of BCM

The computational grid used in this study is based on the Building-Cube Method. BCM grid generation consists of two steps: the first is to generate 'cube' of various sizes to fill the flow field as shown in Fig. 2. The second step is to generate Cartesian grid in each cube (Fig. 3). Three cells overlap between adjacent cubes to exchange the flow information at the boundaries.

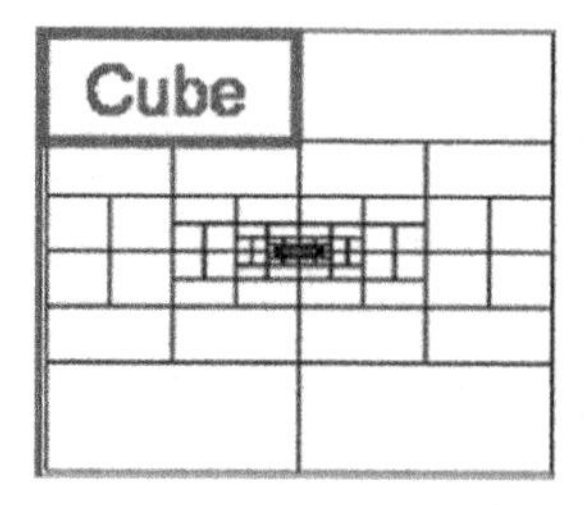

Fig. 2. Cube boundary around airfoil.

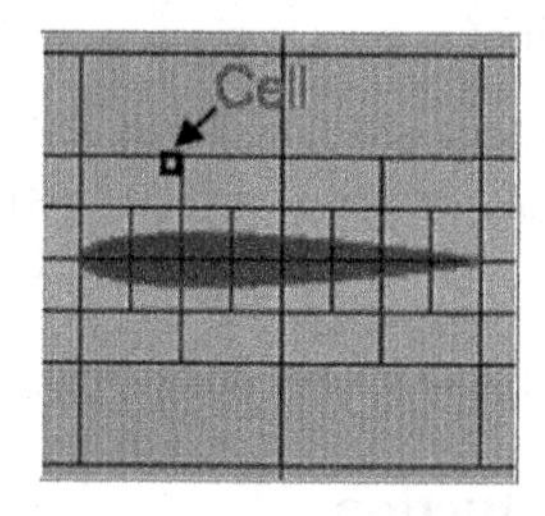

Fig. 3. Cartesian mesh around airfoil.

2.2 Flow Simulation

In this study, the non-dimensionalized Navier-Stokes equations governing compressible viscous flow are used for solving flow field. It can be written in the conservative form as,

$$\frac{\partial Q_j}{\partial t} + \frac{\partial F_j}{\partial x_j} - \frac{1}{Re}\frac{\partial G_j}{\partial x_j} = 0 \quad (j = 1,2,3), \tag{1}$$

where t, x, and Re are time, the coordinate, and the Reynolds number, respectively. The conservation variables vector Q, the inviscid flux vector F, and viscous flux vector G are defined by

$$Q_j = \begin{bmatrix} \rho \\ \rho u_i \\ \rho e \end{bmatrix}, F_j = \begin{bmatrix} \rho u_j \\ \rho u_i u_j + p\delta_{ij} \\ u_j(\rho e + p) \end{bmatrix}, G_j = \begin{bmatrix} 0 \\ \tau_{ij} \\ u_k \tau_{kj} + \lambda(\partial T/\partial x_j) \end{bmatrix}. \tag{2}$$

Here, ρ, p, e, t, and λ are density, pressure, specific total energy, temperature, and thermal conductivity of fluid, respectively. u_i represents the velocity component of the flow in the coordinate direction x_i . It is noted that subscripts i, j, and k represent coordinate indices, and Einstein's summation rule is applied. This set of equations is closed by the equation of state for perfect gas.

$$p = (\gamma - 1)\rho \left(e - \frac{1}{2} u_j u_j \right). \quad (3)$$

where γ is the ratio of the specific heats. The components of viscous stress tensor $\tau_{ij}$are given by

$$\tau_{ij} = \mu \left(\frac{\partial u_i}{\partial x_j} + \frac{\partial u_j}{\partial x_i} \right) - \frac{2}{3} \mu \frac{\partial u_k}{\partial x_k} \delta_{ij}. \quad (4)$$

whrere δ_{ij} is Kronecker's delta symbol and μ is viscosity coefficient, calculate by Surthrerland's relationship. For turbulent flow simulation, the viscosity μ is replaced by eddy viscosity which is evaluated by the Spalart and Allmaras one-equation model.

Specifications of the flow solver in this study are as follows.

- Discretization : FVM
- Convection term : HLLEW (Riemann's solver) [5]
- High-resolution scheme : 4th order compact MUSCL TVD [9]
- Time marching scheme : LU-SGS [10]
- Turbulence model (RANS) : Spalart-Allmaras 1eq. model [7]
- Other numerical technique : Preconditioning method for Mach-uniform formulae [8]

2.3 Parallelization

The flow solver described above is parallelized with Message Passing Interface (MPI) library. In order to achieve optimal parallel performance, an identical number of cubes is assigned to each CPU and only information of boundary cells are exchanged. Figure. 4 shows a overall flow-solution procedure of this flow solver. According to this figure, only once communication is occurred in each iteration. It also contributes to obtain good parallel performance.

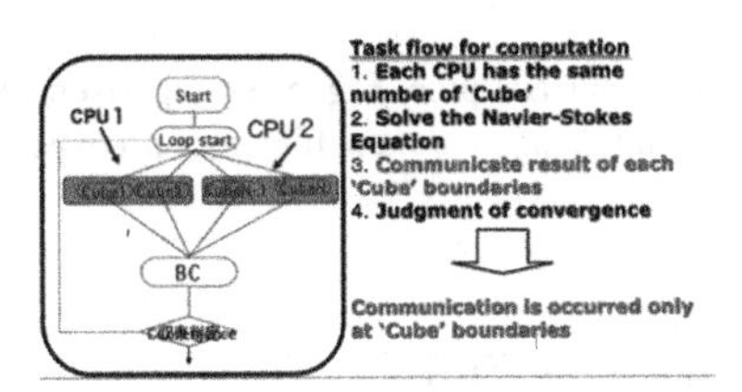

Fig. 4. Overall flow-solution procedure.

2.4 Validation

To validate of this flow solver, flow simulation around a 2D-airfoil (RAE2822) was carried out. The computational grid is shown in Fig. 5. The total number of cubes

is 618 with 16x16 cells in each cube. Therefore total number of computational cells are about 150,000. The inflow Mach number, Reynolds number and angle of attack are 0.73, 6.56 million and 2.70 degrees. Detailed flow conditions can be found in the experiment report by Cook et al. [1]
Figure 6 shows Mach number contour in the flow field. A shockwave is observed at about 55% chord length from the leading edge on the upper surface of the airfoil. The computed surface-pressure and skin-friction distributions are compared with experimental data (Figures 7 and 8). The result of computation shows overall good agreement with the experiment.

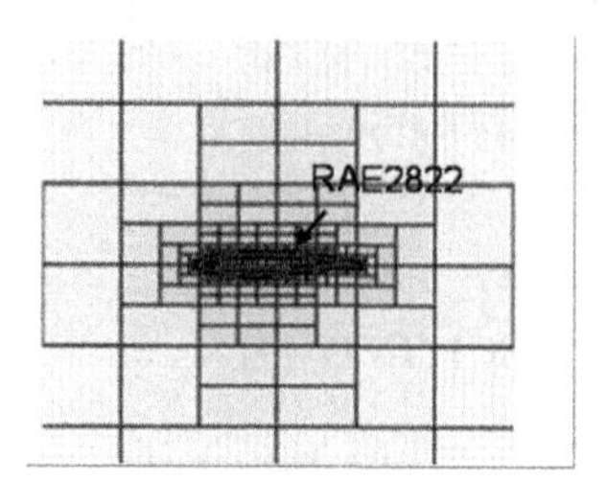

Fig. 5. Computational grid.

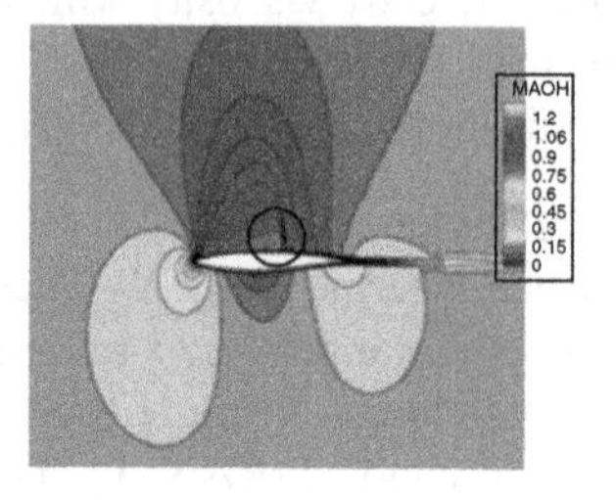

Fig. 6. Mach number contours.

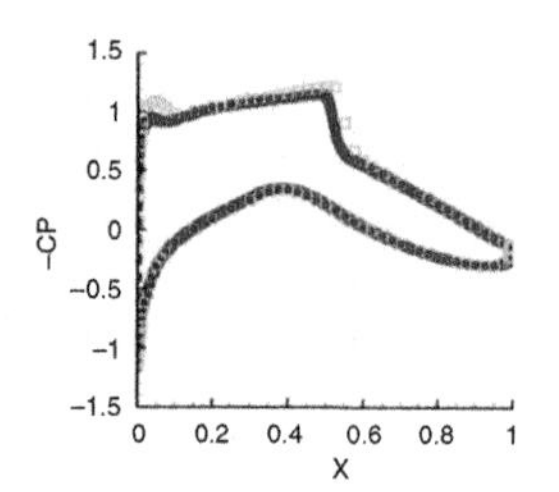

Fig. 7. Comparison of computed surface pressure coefficient with experimental data.

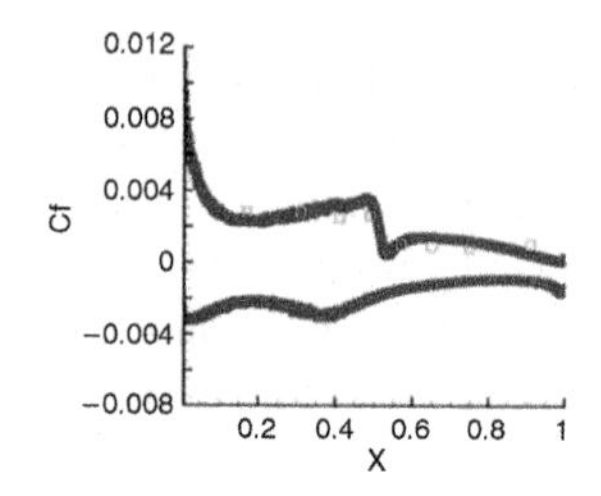

Fig. 8. Comparison of computed skin friction coefficient with experimental data.

2.5 Implicit Residual Smoothing (IRS)

The governing equations discretized by FVM are solved by LU-SGS scheme. To improve solution convergence, the IRS is adopted. R_i is defined as,

$$R_i = \Gamma^{-1}\left[\frac{\partial F(Q_i)}{\partial x} - \frac{1}{Re}\frac{\partial G(Q_i)}{\partial x}\right]. \tag{5}$$

where, Γ^{-1} is preconditioning matrix [8]. In this study the explicit residual $R_i$is modified as,

$$\bar{R}_i = R_i + \varepsilon\Sigma(\bar{R}_j - \bar{R}_i). \tag{6}$$

where $\bar{R}_i$is the smoothed residual, $\bar{R}_j$is residual of cell j adjoining cells i and ε is a positive parameter used to control the smoothing. Equation (6) can be solved by Jacobi iteration described as follows,

$$\bar{R_i^m} = \frac{R_i + \varepsilon\Sigma \bar{R_j^{m-1}}}{1+\varepsilon\Sigma 1}. \tag{7}$$

3 Results and Discussion

To investigate the effect of the IRS on solution convergence, flow simulation around a 3D airfoil was carried out. Computational conditions are shown in Table 1. The computational grids used in the investigation are shown in Table 2. The total number of grids and computational region for all cases is same so that the grid density is uniform. Figure 9 shows these computational grids.

Table 1. Computational conditions.

Condition	Value
Reynolds Number	1.5x10^6
Mach Number	0.19
Number of Surface Element	4.0x10^6
Turbulence Model	On

Table 2. Specification of computational grid arrangements.

Type	Number of cubes	Number of cell per cube	Total number of cells
A	1	2,097,152(=128^3)	2,097,152
B	8(=2^3)	262,144(=64^3)	2,097,152
C	64(=4^3)	32,768(=32^3)	2,097,152
D	512(=8^3)	4,096(=16^3)	2,097,152

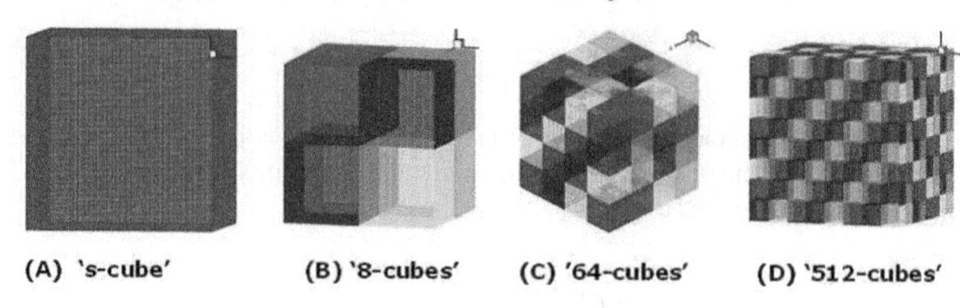

Fig. 9. Computational grids.

- Convergence History of 'single-cube' and 'multi-cube' without IRS

Convergence history is shown in Fig. 10. As the number of cubes increases, the convergence history gets worse with oscillation. One of the reasons is poor information transfer at cube interfaces which might have caused duller solution near the interface. To overcome this problem, the IRS was applied.

- Convergence History of 'single-cube' with IRS

Convergence history is shown in Fig. 11. The convergence history depends on the value of ε . The following 'multi-cube' computations were performed using ε =1.2 determined by a series of numerical tests.

- Convergence History of 'multi-cube' with IRS

Convergence history is shown in Fig. 12. The IRS is used on each cube. Regardless of the number of cubes, significant improvement for convergence is obtained by the IRS. Possible reason is that the IRS promotes propagation and attenuation of residual error across cube interface.

Finally, the IRS is applied to a practical computational grid in which cubes of heterogeneous size were used to represent complex geometries. In this case cell size at interface is different which, in general, makes convergence worse. Figure 13 shows a computational grid. Good solution convergence was obtained as shown in Fig. 14.

- Parallel Efficiency

The computations were performed using NEC SX-8 which is vector parallel computer with 64 GB main memory and 8 processors in each node. The communication device between nodes is Crossbar. To evaluate parallel efficiency, grid D in Table. 2 was used. Good parallel efficiency was obtained as shown in Fig. 15. Moreover, it is unaffected even with the IRS.

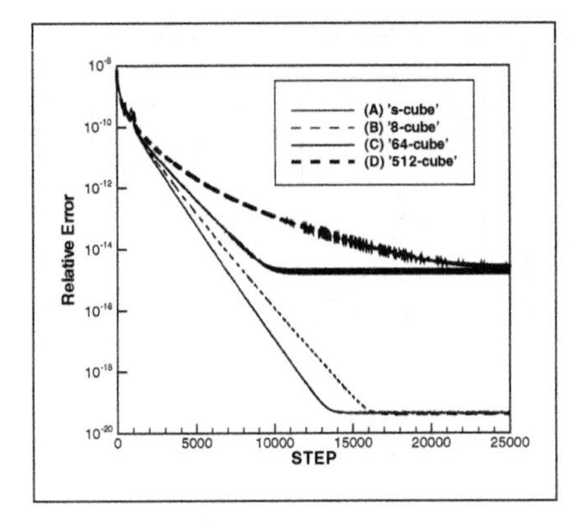

Fig. 10. Convergence history for several cube arrangements w/o IRS.

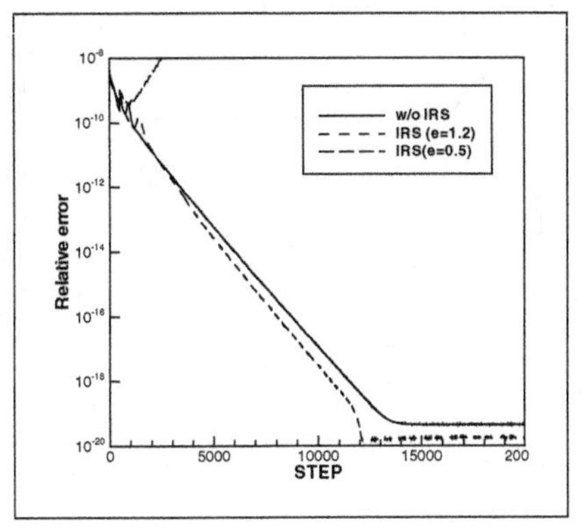

Fig. 11. Convergence history for 's-cube' with and w/o IRS.

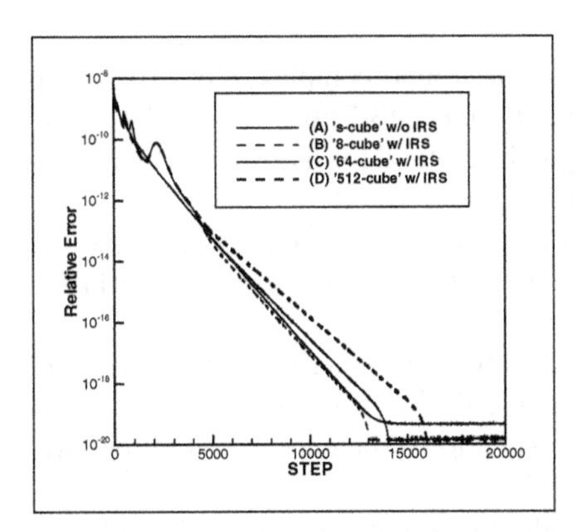

Fig. 12. Convergence history for several cube arrangements w/ IRS.

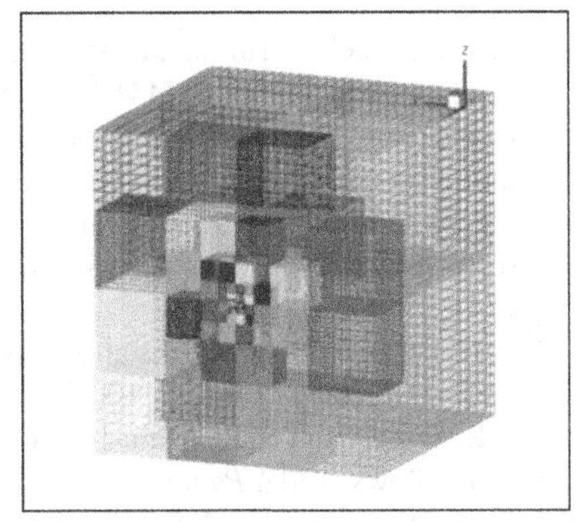

Fig. 13. Computational grid for actual problem.

4 Conclusion

A method was proposed to improve convergence for 'multi-cube' CFD. The basic idea behind the present method was to apply the IRS to 'multi-cube' CFD. Regardless of the number of cubes and their sizes, significant improvement for convergence was obtained without deteriorating parallel efficiency. In the case of heterogeneous cube arrangement good solution convergence was also confirmed.

[1] P. H. Cook, M. A. McDonald, and M. C. P. Firmin. Airfoil rae2822-pressure distributions, and boundary layer and wake measurements. *Experimental Data Base for Computer Program Assessment*, AGARD-AR-138, 1979.

[2] A. Jameson. The evolution of computational methods in aerodynamics. *Journal of Applied Mechanics Review*, 50(2):1052–1070, 1983.

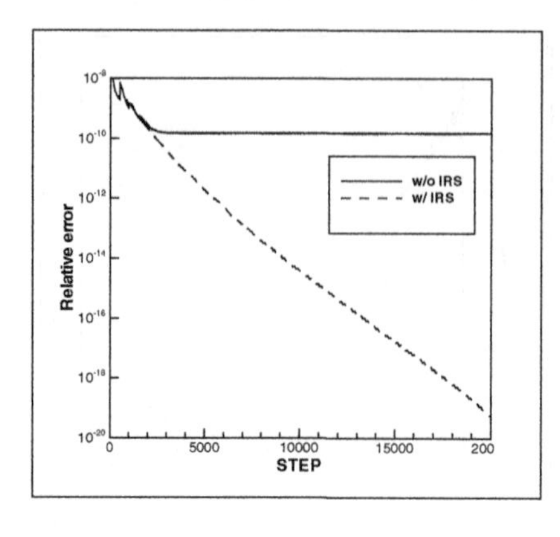

Fig. 14. Convergence history for actual problem w/ and w/o IRS.

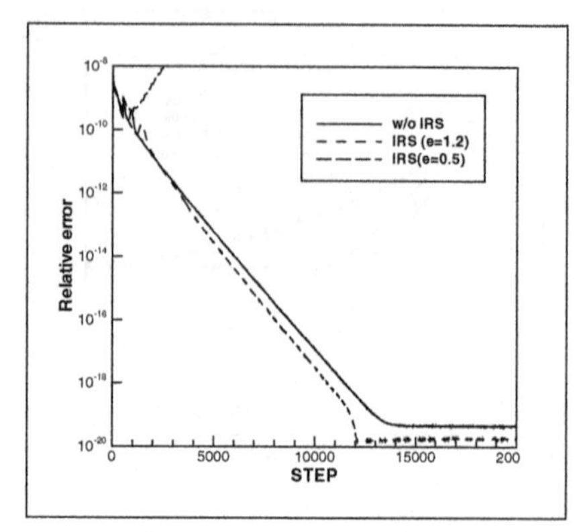

Fig. 15. Speed-up ratio on SX-8.

[3] T. Kamatsuchi. Turbulent flow simulation around complex geometries with cartesian grid method. *SIAM J. Sci. Stat. Comput*, 13(2):631–644, 2007.

[4] K. Nakahashi and L. S. Kim. Building-cube method for large-scale, high resolution flow computations. *AIAA Paper*, 2004-0423, 2004.

[5] S. Obayashi and G. P. Guruswamy. Convergence acceleration of a navier-stokes solver for efficient static aeroelastic computations. *AIAA Journal*, 33(6):1134–1141, 1995.

[6] C. S. Peskin. Flow patterns around heart valves: A numerical method. *Journal of Computational Physics*, 10(2):252–271, 1972.

[7] P. R. Spalart and S. R. Allmaras. A one-equation turbulence model for aerodynamic flows. *AIAA Paper*, 92-0439, 1992.

[8] J. M. Weiss and W. A.Smith. Preconditioning applied to variable and constant density time-accurate flows on unstructured meshes. *AIAA Journal*, 33(11):2050–2057, 1995.

[9] S. Yamamoto and H. Daiguji. Higher-order-accurate upwind schemes for solving the compressible euler and navier-stokes equations. *Computer and Fluids*, 22(2/3):259–270, 1993.

[10] S. Yoon and A. Jameson. Lower-upper symmetric-gauss-seidel method for euler and navier-stokes equations. *AIAA Journal*, 26(9):1025–1026, 1998.

Aerodynamic Analysis of Rotor Blades using Overset Grid with Parallel Computation

Dong-Kyun Im[1], Seong-Yong Wie[1], Eugene Kim[1], Jang-Hyuk Kwon[1], Duck-Joo Lee[1], and Ki-Hoon Chung[2], and Seung-Bum Kim[2]

[1] Dep. of Aerospace Engineering, Korea Advanced Institute of Science and Technology, 373-1 Guseong-dong, Yuseong-gu, Daejeon, 305-701, Republic of Korea
[2] Korea Aerospace Research Institute , 115 Gwahangro, Yuseong, Daejeon, 305-333, Republic of Korea

Abstract. The helicopter aerodynamics is simulated in hovering and forwarding flight using the unsteady Euler equations. As the steady condition, flight test of DLR-F6 and hovering flight test data of Caradonna & Tung's rotor blades were used, and as the unsteady condition, non-lift forwarding flight test data of the rotor blades were used. The parallelized numerical solver was validated with the two of data above. By using this solver, AH-1G rotor blades to forwarding flight numerical test were conducted. In the test of forwarding flight, the numerical trim was applied to decide cyclic pitching angles using the Newton-Raphson method, and the results were good well match to the experimental data, Especially, the BVI effects were well simulated in advancing side in comparison with other numerical results. To consider the blade motion and moving effects, an overset grid technique is applied and for the boundary, Riemann invariants condition is used for inflow and outflow.
Keywords: Overset Grid, Helicopter Rotor, Hovering Flight, Forward Flight

1 Introduction

A helicopter has unique flying characteristics, that are complex aerodynamic phenomena. Continuous changes of aerodynamic environment and loads give rise to the significant difficulty of rotorcraft aerodynamics. Strong tip vortices in the rotor wakes cause the flow-field to became highly unsteady and non-uniform at the rotor disk. These characteristics increase the calculation time and memory size in numerical analysis. Moreover, in forwarding flight, the periodic motions - pitching and flapping - of the rotor blades should be considered with aerodynamic stability. This study is covering the numerical analysis of the rotor blades in hovering and forwarding flight using parallelized overset grid with rotor trim. The parallelized overset grids system uses a static load balancing tied to the flow solver based on the grid size of a decomposed domain [1]. The solver is based on a cell-centered finite volume method [2]. The parallelization is based on a domain decomposition; communication is realized using MPI. The numerical fluxes at cell interface are computed using the Roe's flux difference splitting method and the third order MUSCL (Monotone

D. Tromeur-Dervout (eds.), *Parallel Computational Fluid Dynamics 2008*,
Lecture Notes in Computational Science and Engineering 74,
DOI: 10.1007/978-3-642-14438-7_10,

Upwind Scheme for Conservation Law) scheme. The diagonalized ADI method is used with second order dual time stepping algorithm for the time integration. First, the steady condition was tested for verification of the parallelized solver. The DLR-F6 aircraft configuration was selected for the steady condition. The configuration was provided in the DPW II and III (Drag Prediction Workshop which has been organized by AIAA Applied Aerodynamics Technical Committee) and is a twin engine wide body aircraft which has a wing-body-nacelle-pylon and wing-body-fairing aircraft. The numerical hovering flight test was conducted using Caradonna & Tung's rotor blades that consist of NACA0012 airfoil section and AH-1G rotor blades were used for the forwarding flight with rotor trim. The rotor trim of cyclic pitch angle was made to be zero value of the periodic moments of rotor disk direction in reasonable averaged thrust. And for the iteration method of rotor trim, the Newton-Raphson method was used [7].

2 Validation of Parallelized Solver

First, for the validation of numerical solver of helicopter aerodynamics, the DLR-F6 aircraft configuration was analyzed in steady condition. The structured grid based solver KFLOW3D which has been developed at KAIST ASDI Lab is a parallelized solver. For the high performance parallel computing, 28 Pentium 4 2.13GHz processors are used [3].

2.1 DLR-F6

Computations are implemented on DLR F6 wing-body configuration in Figure 1. The DLR F6 wing body configuration model represents a transonic wing-body transport configuration. The structured grids are provided by the website of Boeing Commercial Airplanes group. In computations, three difference density grids (coarse, medium, medium fine) are used as shown Table 1. The computational conditions are $M_\infty = 0.75$, $Re_\infty = 3 \times 10^6$ at $\alpha = -3.0°, -2.0°, -1.5°, 0.0°, 1.0°, 1.5°$. Fully turbulent boundary layer is assumed in the computations.

	Grid points	Y+	No. of BL cell	Avg. Time (hours)
Coarse	2.1M	1.25	18	6.5
Medium	3.7M	1.0	36	15
fine	13.0M	0.8	36	32

Table 1. Summary of the Grid System

Figure 2 (left) shows the drag polar results using the three grid sets in Table 1. These results are in good agreement with the experimental data, and exhibit the convergence, with respect to the grid density, of numerical results toward the experimental data.. The lift coefficients with the angle of attack are shown in Figure 2 (right), and the results also show good agreement with experimental data.

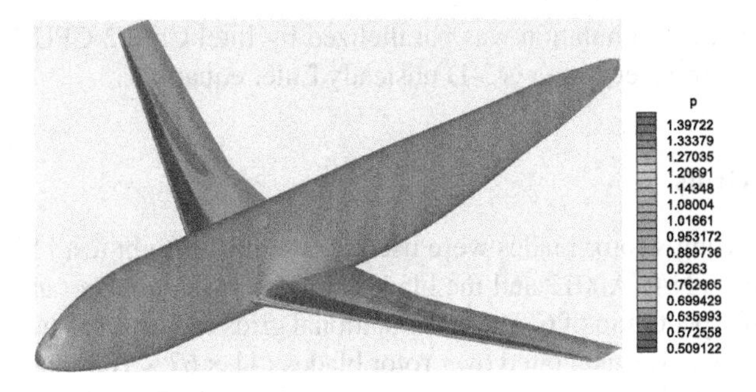

Fig. 1. DLR-F6 configuration and pressure distribution [3]

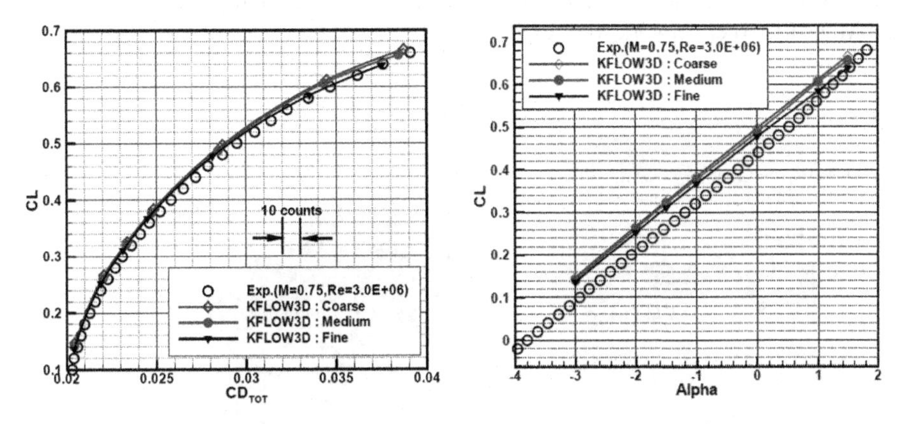

Fig. 2. (left) Drag Polar [3] and (right) CL vs. AOA curves [3]

3 Rotor Blades

The helicopter rotor blades were simulated in hovering flight and forwarding flight using the parallelized solver. To consider motions of rotor blades such as pitching and flapping, the overset grid technique wasapplied, and due to large grid size and computing time, the simulation was parallelized by Intel Core 2 CPU 2.13GHz 20 nodes. The governing equation is 3-D unsteady Euler equations.

3.1 Overset Grids

Caradonna & Tung's rotor blades were used for the hover flight test [4]. The airfoil of rotor blades is NACA0012 and the blades have no twist, no taper and 1.143m of length, and an aspect ratio of 6. The computational grids consist of 5 shown in Figure 2. The H-type grid was generated over rotor blades ($41 \times 67 \times 105$) and background grid ($89 \times 97 \times 97$). For the forward flight tests, two kinds of blades were considered. In the non-lift forward fight test, the Caradonna & Tung's rotor blades with an aspect ratio of 7 was used [5]. AH-1G rotor blades with a linear twist angle of $-10°$ were used in the forward flight with pitching and flapping motion [6]. The computational grids of AH-1G rotor blades also consist of 5 blocks that are the structured overset grid as Figure 3. The H-type grid also was generated over rotor blades ($65 \times 67 \times 133$) and background grid ($89 \times 97 \times 97$). The AH-1G rotor blades have an aspect ratio of 9.8. A Linux PC cluster with 20 nodes Intel Core 2 CPU 2.13GHz processors and a 100Mbps Ethernet network were used.

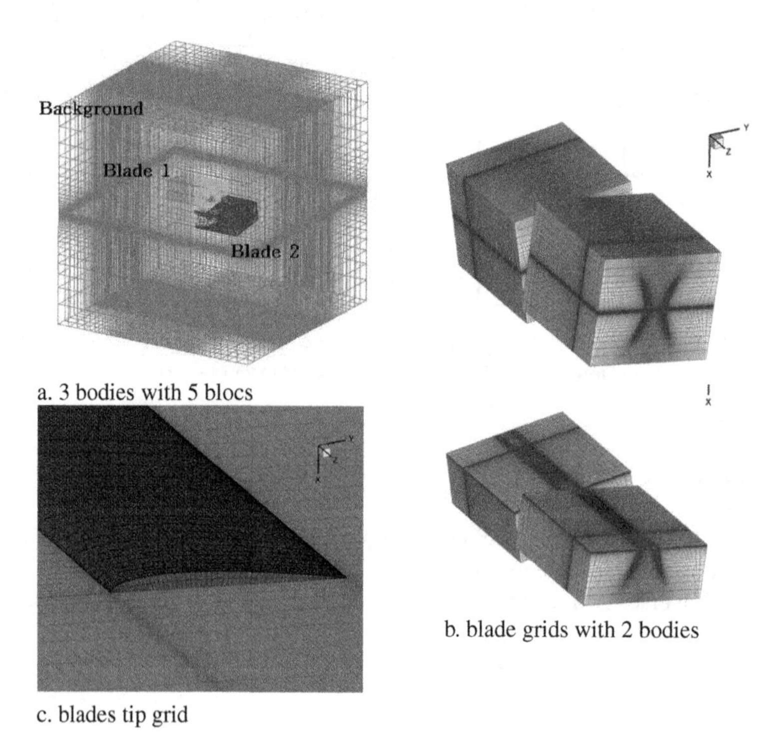

a. 3 bodies with 5 blocs

b. blade grids with 2 bodies

c. blades tip grid

Fig. 3. Overset grid

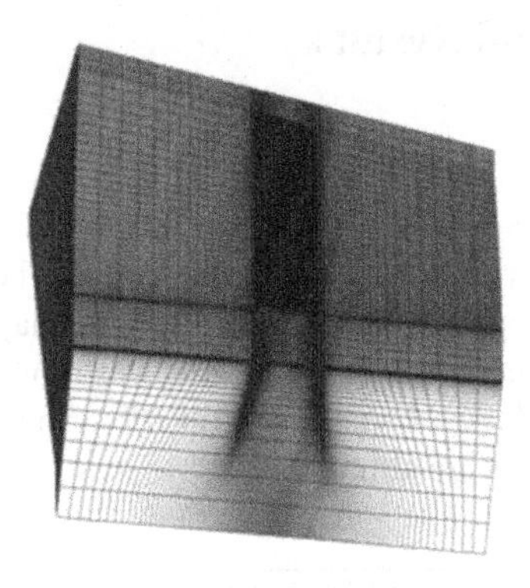

Fig. 4. AH-1G Rotor Blades

3.2 Numerical Analysis Method

The solver, KFLOW3D, has been developed for the solution of RANS (Reynolds Averaged Navier-Stokes Equations). In this study, the governing equations were unsteady Euler equations for the compressible inviscid flow discretized by a cell centered finite-volume scheme using a structured grid. The discretization of the inviscid fluxes has been realized with upwind scheme, 3rd -order MUSCL (Monotone Upstream Scheme for Conservation Laws) with minmod limiter using Roe-FDS (Flux Difference Splitting) with the Harten's entropy fix function [8][9][10]. The steady state is reached using the unsteady equations with the diagonalized ADI time integration scheme. A dual time stepping method is also used to increase the time accuracy of the scheme [11][12]. Riemann Invariant was applied to the boundary condition for convergence and stability.

3.3 Overset Grid Technique and Parallel Implementation

The overset grid method can be applied to any complex geometry which can be divided into some sub-grids having their own meshing. The independent grid system transfers information to each sub-grid using interpolation point by domain connectivity method. In this study, the domain connectivity method was used for modified overset grid technique for the sake of the robustness and accuracy of flow analysis by arranging interpolation points of overlapped grids away from the boundary of multi-sub-grid. A structured grid assembly is parallelized by using a static load balancing tied to the flow solver based on the grid size of a decomposed domain. The parallel algorithm is modified for a distributed memory system. To parallelize the grid assembly, message passing programming model based on the MPI library is implemented using the SPMD (Single Program Multiple Data) paradigm. The parallelized flow solver can be executed in each processor with the static load balancing by pre-processor.

4 Rotor Flight Test and Results

4.1 Hovering Flight Test

The collective pitch angle is 8° and tip Mach number is 0.439. In this test, the rotor blades were rotated in 6 revolutions and it takes about 6 hours of wall clock time. The thrust coefficient is shown in Table 2 and Fig 6 displays the sectional pressure distribution in hovering flight test case. The results were compared with experimental data at 0.5, 0.68, 0.8, and 0.96 of span location(r/R). The results of the numerical computation and experiments are almost same.

Thrust Coefficient (CT)	
Experiment	0.00459
Numerical result	0.00473

Table 2. Thrust coefficient in hovering flight

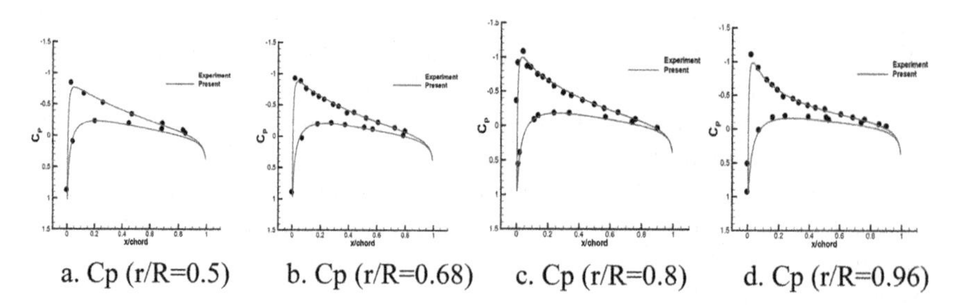

a. Cp (r/R=0.5) b. Cp (r/R=0.68) c. Cp (r/R=0.8) d. Cp (r/R=0.96)

Fig. 5. Pressure coefficient of rotor blades in hover flight

4.2 Non-lift Forwarding Flight Test

The collective pitch angle is 0°, tip Mach number is 0.8 and advancing ratio is 0.2. These test data were usually used for validation of forward flight by several researchers. In this test, the rotor blades were fixed at 0° collective pitch angle. That means the pitching and flapping motion were not considered. Also, the rotor blades were rotated with 6 revolutions and it takes about 6 hours of wall clock time. Figure 5 shows the pressure distribution of upper surface rotor blades in non-lift forwarding flight test case. The strong shock was occurred at 90 azimuth angle because tip Mach number is 0.8 and advancing ratio is 0.2. Figure 6 displays the results that were compared with experimental data on 30°, 60°, 90°, 120°, 150°, 180° azimuth angle at 0.8925 of span location(r/R). The reason why the pressure difference of upper and lower surface is zero is that the collective pitch angle is zero and there is no pitch and flapping motion and the strong shock also is shown at advancing side in Figure 6. The results of the numerical computation and experiments are almost same as seen in Figure 6.

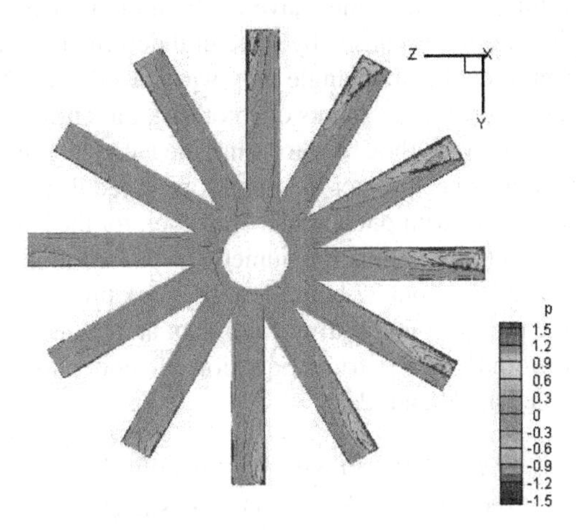

Fig. 6. Pressure coefficient of rotor blades in non-lift forward flight

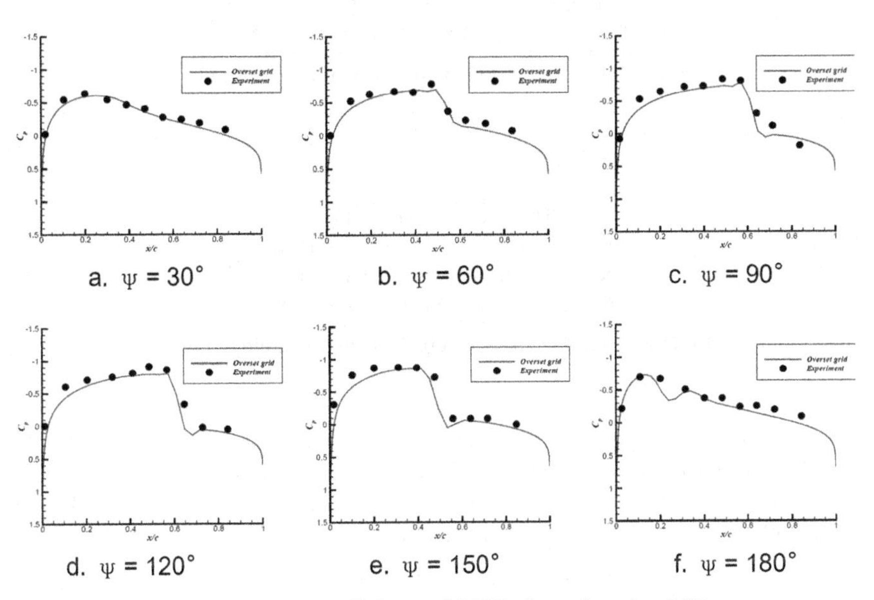

Fig. 7. Pressure coefficient at 0.8925 of span location (r/R)

4.3 Forwarding Flight Test

The forwarding flight test with pitching and flapping motion was conducted using AH-1G rotor blades. The rotor trim for pitching angle also was considered in forwarding test [6]. In this test, the blades have cyclic motions with the pitching and flapping. The flapping motion is caused by structural deformation such as the aeroelasiticity, and the trim of cyclic pitch angle only was considered. The rotor trim was conducted to meet zero periodic moments of rotor disk direction in reasonable averaged thrust. For the iteration method of rotor trim, the Newton-Raphson method was used [7]. The 5 revolutions need to get the proper averaged thrust and 3 revolutions also need to get the result of trim pitch angle. However, for the convergence of rotor trim, 3 iterations also need. Therefore, the numerical forward flight test can carry out after 30 revolutions. It takes about 32 hours of wall clock time.

Table 3 shows the pitching angles after rotor trim and table 4 displays the thrust coefficients with trim and without trim. Equation of motions of the pitching and flapping are represented at (1) and (2).

$$\theta(\psi) = \theta_0 + \theta_{1c}\cos(\psi) + \theta_{1s}\sin(\psi) \tag{1}$$

$$\beta(\psi) = \beta_0 + \beta_{1c}\cos(\psi) + \beta_{1s}\sin(\psi) \tag{2}$$

Degree	θ_0	θ_{1c}	θ_{1s}	β_{1c}	β_{1s}
Experiment[5]	6	1.7	-5.5	2.13	-0.15
Numerical result	6.11	1.32	-5.07	2.13	-0.15

Table 3. Pitching and Flapping angle after rotor trim

Thrust Coefficient (CT)	
Experiment	0.00464
Numerical result (Trim)	0.00464
Numerical result (No Trim)	0.00448

Table 4. Averaged Thrust Coefficients in forwarding flight

The results were compared with experimental data at 0.6, 0.75, 0.86, and 0.91 of span location(r/R) pictured in Figure 7, and the numerical results with rotor trim are almost same with experimental computation in Figure 7 and show the weak BVI effect at 90° azimuth angle that is on advancing side.

5 Conclusion

The hover and forward flight aerodynamic analyses were conducted using parallelized overset grid. For the high performance computation, 20 nodes of Intel Core 2

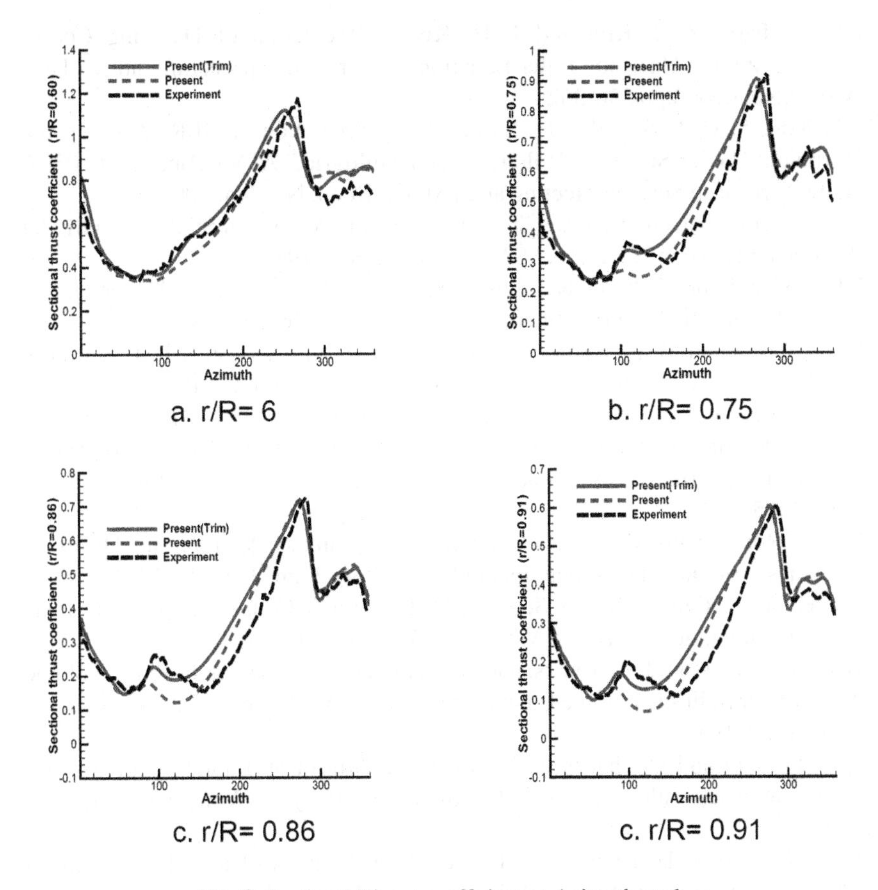

Fig. 8. Sectional Thrust coefficient as Azimuth angle

2.13GHz processors were used by applying the parallelized overset grid technique. The rotor trim of cyclic pitch angle also was considered in forward flight. The numerical results with rotor trim are much more similar to experimental data than the results without rotor trim.

Acknowledgement

This study has been supported by the KARI under KHP Dual-Use Component Development Program funded by the MKE

[1] Eugene Kim, J. H. Kwon and S. H. Park, " Parallel Performance Assessment of Moving Body Overset Grid Application on PC Cluster", pp59-66, Parallel Computational Fluid Dynamics-Parallel Computing and Its Applications 2007.

[2] S. H. Park, Y. S. Kim and J. H. Kwon, "Prediction of Damping Coefficients Using the Unsteady Euler Equations", Journal of Spacecraft and Rockets, Vol.40,No.3,2003,pp.356-362

[3] Y. Kim, S. H. Park and J. H. Kwon, "Drag Prediction of DLR-F6 Using the Turbulent Navier-Stokes Calculations with Multigrid," AIAA Paper 2004-0397, 42nd Aerospace Science Meeting and Exhibit, Reno, NV, Jan. 2004.

[4] F. X. Caradonna and C. Tung, "Experimental and Analytical Studies of a Model Helicopter Rotor in Hover," NASA TM-81232, Sep. 1981.

[5] F. X. Caradonna, F. X., G. H. Laub and C. Tung, "An Experimental Investigation of the Parallel Blade-Vortex Interaction," NASA TM-86005, 1984.

[6] J. L. Cross and M. E. Watts, "Tip Aerodynamics and Acoustics Test: A Report and Data Survey," NASA-RP-1179, NASA Ames Research Center, Dec. 1988.

[7] K. H. Chung, C. J. Hwang, Y. M. Park, W. J. Jeon and D. J. Lee, " Numerical Predictions of Rotorcraft Unsteady air-loadings and BVI noise by using a time-marching free-wake and acoustic analogy", 31th European Rotorcraft Forum, 2005.

[8] P. L. Roe, "Approximate Riemann Solver, Parameter Vectors and Differnce Schemes", Journal of Computational Physics, Vol 43, pp. 357-372, 1981.

[9] A. Harten, "High Resolution Schemes for Hyperbolic Conservation Laws," Journal of Computational Physics, Vol. 49, pp.357-393, 1983.

[10] W. K. Anderson, J. L. Tomas, and B. Van Leer, "Comparison of Finite Volume Flux Vector Splittings for the Euler Equations," AIAA Journal, Vol. 24, No. 9, pp. 1453-1460. 1986.

[11] T. Pulliam and D. chaussee, " A diagonal form of an implicit approximate-factorization algorithm", Journal of Computational Physics, Vol. 39, pp. 347-363, 1981.

[12] K. W. Cho, J. H. Kwon, and S. Lee., "Development of a Fully Systemized Chimera Methodology for Steady/Unsteady Problems", Journal of Aircraft, Vol. 36, No. 6, pp. 973-980, Nov. Dec. 1999

Large scale massively parallel computations with the block-structured elsA CFD software

M. Gazaix[1] *, S. Mazet[2], and M. Montagnac[3]

[1] Computational Fluid Dynamics and Aeroacoustics Department, ONERA, 92322 Chatillon Cedex, France
[2] Communication & Systèmes, Le Plessis Robinson, France
[3] Advanced Aerodynamics and Multiphysics, CERFACS, 42, Avenue Gaspard Coriolis. 31057 Toulouse Cedex 01, France

Abstract. We describe work performed inside the elsA (http://elsa.onera.fr) block-structured CFD software to analyse and improve parallel computation efficiency. Details of MPI implementation are discussed. Computational results are given for realistic industrial configurations ($28\ 10^6$ mesh points) on several massively parallel platforms up to 1024 processors.

1 Introduction

The elsA software is a widely used CFD code developed by Onera, Cerfacs and other research centers [1]. This software elsA is able to simulate a wide range of aerodynamic flows occuring in aircraft, turbomachinery or helicopter configurations. elsA is now an essential component of the design process of major industrial partners (Airbus, Safran, Eurocopter, MBDA, CEA).

elsA solves the compressible Navier-Stokes equations using a cell-centered, finite volume, block-structured formulation. Spatial discretization schemes include second order upwind or centred schemes; the semi-discrete equations are integrated in time, either by multistage Runge-Kutta schemes with implicit residual smoothing, or by backward Euler integration with implicit operator solved with robust LU relaxation methods. A very efficient multigrid technique can be selected to further accelerate convergence.

In order to cope with more and more geometrically complex configurations, in addition to matching block connectivity, highly flexible advanced techniques of multiblock structured meshes are available including non conforming block connectivity, hierarchical mesh refinement, and Chimera technique for overlapping meshes.

elsA is mostly written in C++, with intensive numerical computations delegated to Fortran subroutines. Users interact with the elsA kernel with a Python scripting interface.

* This work is partially funded within the framework of the Airbus FUSIM project.

D. Tromeur-Dervout (eds.), *Parallel Computational Fluid Dynamics 2008*,
Lecture Notes in Computational Science and Engineering 74,
DOI: 10.1007/978-3-642-14438-7_11,

elsA has been ported to most high-performance computing platforms, achieving good CPU efficiency on both scalar multi-processor computers and vector computers. To run in parallel, elsA uses a standard coarse-grained SPMD approach where each block is allocated to a processor, and communications between blocks are implemented with the MPI standard library. Several blocks can be allocated to the same processor. An internal load balancing tool with block splitting capability is available to allocate automatically blocks to processors.

As any other parallel scientific software, the benefits of parallelism are twofold:

- increase the amount of computer memory available, thus allowing to increase the number of mesh points;
- reduce the (wall clock) return time between user request and full job output.

Indeed, if there is no strong time constraint to get a job output, the best practice to optimise the computing platform throughput is to choose the minimum number of computing nodes providing the required memory. Conversely, if a time constraint is encountered such as a deadline in the design process, the number of processors is computed taking into account the estimated single processor cost and the parallel efficiency.

2 Detailed analysis of elsA parallel performance

Earlier versions of elsA exhibit a marked decrease of parallel efficiency when the number of processors increases. In the following subsections, the probable causes of this degradation are identified and corrections are described.

2.1 Load balancing

As the number of processors increases, it is well known that load balancing is harder to maintain (Amdahl's law). This is specially difficult to achieve with structured multiblock code, where the problem of finding the best "block splitting" can be challenging, particularly if the number of coarse grid levels has to be maintained. It must be acknowledged that partitioning an unstructured mesh is inherently simpler, where several public domain tools are available (Metis, Scotch, Chaco...). We have observed that many claims of poor parallel scaling are ill-founded and often vanish if care is exercised to obtain good enough load balancing. Note however that contrary to the unstructured case, memory constraints are much lower for block-structured splitting, so there is no need to parallelize the splitting algorithm itself to avoid memory bottlenecks.

A load balancing algorithm is available in elsA, closely following the strategy described by A. Ytterström [2] ("greedy" algorithm). Note that block splitting respects the multigrid constraint.

When block splitting does not occur, the load balancing algorithm is very simple. Blocks are first ordered by size (number of cells), and dispatched in a "zig-zag" way on the NP processors: each of the NP biggest blocks are assigned to a separate processor, the NP next blocks are each assigned to a separate processor, in reverse order of sizes, and so on until no blocks are left. In the test case studied, the following rule of thumb applies: load balancing degradates rapidly when the number of processors reaches one tenth of the number of blocks. For a test case of circa 350 blocks, the load balancing thus achieved degradates rapidly above 32 processors, and for a test case of circa 1000 blocks, the load balancing thus achieved degradates rapidly above circa 100 processors. To overcome this degradation, it is necessary to split the largest blocks, using the automatic splitting capability of elsA internal load balancing tool.

In addition, we have also tested the Metis tool to load balance a limited number of configurations [4]. We have observed that on a large configuration (1037 blocks, 128 processors) only pmetis gives a result; kmetis fails to give a distribution.

2.2 MPI message scheduling

Once the blocks are assigned to processors in a balanced way, one can classically setup a communication graph. Vertices in this graph are processors. An edge is added between two vertices for each boundary condition that links two structured blocks, resulting in a multi-graph.

During each communication phase, every edge of the communication graph gives rise to one message being passed between the two underlying processors.

One singularity of the elsA software is the use of send/receive/replace strategy for message passing. Specifically, this strategy is used between blocks that are connected with "coincident" meshes on either side of the interface. An important consequence of this is that the global ordering of communications is crucial to parallel scalability. The send/receive/replace strategy, while saving memory and buffer management, allows for little hardware data buffering since each MPI call (`MPI_Sendrecv_replace()`) must wait for all the requested data to be arrived in order to return.

The first implementation of MPI message exchanges between blocks did not pay attention to message ordering. Basically, the MPI messages ordering was determined by the ordering of the topology description in the Python input script.

This bottleneck can be illustrated on a simple four vertices graph as in figure 1. For the ordering on the left of the figure, communications take place in the following order:

1. First wave: edge 1, processor $P1$ and $P2$,
2. Second wave: edge 2, processor $P2$ and $P3$,
3. Third wave: edge 3, processor $P3$ and $P4$,
4. Fourth wave: edge 4, processor $P4$ and $P1$.

Thus processor $P4$ must wait for all communications between processors $P1$ and $P2$ and then between processors $P2$ and $P3$ to be over before it can start communicating with processor $P3$.

In the ordering on the right of the figure, communications take place as follows:

[4] Metis can not split structured blocks.

1. First Wave: edge 1 and 2, processor $P1$ and $P2$, processors $P3$ and $P4$,
2. Second Wave: edge 3 and 4, processor $P2$ and $P3$, processors $P4$ and $P1$.

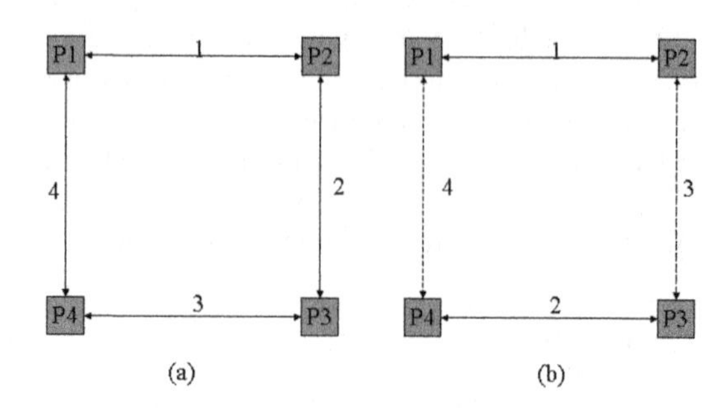

Fig. 1. Different global communication orderings for a simple graph

Each "communication wave" (edges 1 and 2, and edges 3 and 4 in figure 1(b)), is a set of independent edges of the graph (so-called "matching" in graph theory).

The generalization to bigger graphs is straightforward. Note G_0 the whole graph and put $N_0 = 0$. At step $i >= 1$, do the following:

- determine a set of N_i independent edges on the communication graph,
- number the edges (communications) in this set from $N_{i-1}+1$ to $N_{i-1}+N_i$,
- remove processed edges from graph G_i to obtain graph G_{i+1}
- continue until no edges are left.

As an example of efficiency improvement, the return time for a 128 processor computation was reduced by a factor of 2 when using the re-scheduling algorithm.

Remarks

Three remarks are in order.

It is known that the problem of finding the optimal set of independent edges is NP-complete. In the cases presented here, the greedy heuristic gives good results because of the sizes of the graphs involved (no more than a few hundred processors at the moment). It is our opinion that sophisticated heuristics would not significantly improve the graph coverage (number of connected nodes) for graphs of these sizes.

A basic improvement to the greedy heuristic is to order edges according to their weight. In this case, the weights of edges in one set are close, and every message in one communication "wave" should take approximately the same amount of time to travel. The weight of an edge is linked to the amount of communication incurred, i.e. the size of the interface meshes involved.

2.3 Test of different implementations of MPI messages

Use of `MPI_Sendrecv_replace()`

The original MPI parallel implementation uses point-to-point MPI communication routine (`MPI_Sendrecv_replace()`). As shown in section 2.2, it was necessary to develop a scheduler to reach high performances.

Concatenation of messages to reduce latency

In the current elsA version, each matching block connectivity is implemented with a call to `MPI_Sendrecv_replace()`. However, since several matching boundaries may induce communications with the same couple of processors, another implementation has been tested, using a single buffer for each couple of processors; during the send stage, the buffer concatenates (gather) all the messages to be exchanged; after the receive step, the buffer data are scattered to all the individual matching block connectivities. The reduction in the number of exchanged messages may lead to better efficiency, specially if the communication network used by MPI has a high latency. Up to now, we have not observed any improvement using message concatenation; however, the current implementation use unnecessary buffer copy to fill the concatenated exchanged buffers, and so is far from optimak; work is in progress to remove such unnecessary data copy.

Use of `MPI_Bsend()`

In all cases, use of `MPI_Bsend()` lead to strong decrease in parallel efficiency.

Replacing blocking with non-blocking MPI routines

Another idea was to develop a non-blocking communication scheme to get rid of the scheduler. The other major interest of this approach is to enable the overlapping of communication and computation. Nevertheless, the latter has not been achieved up to now.

In the first stage, all processors are told to get ready to receive MPI messages with a non-blocking receive instruction. The second stage consists in sending messages. As shown in table 1, no scheduler is needed anymore.

2.4 Influence of block splitting on numerical convergence

Since the implicit stage is done on a block basis, block splitting performed to achieve a good load balancing may reduce the numerical efficiency of the implicit algorithm. In practice, we have never observed any significant convergence degradation.

comm. scheme	greedy/sched.	greedy/no sched.	Metis/sched.	Metis/no sched
0	108.47	471.47	85.88	252.50
1	21.62	35.51	29.90	34.37
2	22.06	34.27	30.24	33.02
3	21.67	21.54	29.98	29.91

Table 1. 128 processors, CPU time spent to run one solver iteration on IBM BlueGene/L, 0 : `MPI_BSend()`, 1 : `MPI_Sendrecv_replace()`, 2 : `MPI_Sendrecv_replace()` `with message concatenation`, 3 : `MPI_ISend()` (asynchronous)

3 elsA parallel efficiency

In this section, we present the parallel speedup measured on several parallel platforms, using the same test case.

3.1 Test case description

We have chosen a realistic industrial configuration, typical of civil aircraft configurations computed with elsA. It is a 27.8 10^6 mesh points, with 1037 blocks. The numerical settings include a multigrid algorithm (3 levels), and Spalart (one equation) turbulence model. The following plots give the computed parallel speedup, with the following assumptions:

- Since the total CPU time is dominated by CPU time spent in the time loop, the initialization phase (Python script interpretation, mesh file reading,...) is not taken into account.
- Post-processing has been switched off, both during time loop and after loop termination.
- For large number of processors, the configuration has to be splitted; for example for 1024 processors, we end up with 1774 blocks. This splitting leads to a small memory and CPU increase.

3.2 IBM BlueGene/L

The minimum number of processors which allows to run the test case is 64. See Figure 2.(a).

3.3 BULL Novascale (Itanium Montecito 1.6GHz)

This platform is equipped with an Infiniband network. See Figure 2.(b).

3.4 SGI Altix XE (Xeon Woodcrest 3GHz dual core)

See Figure 2.(c).

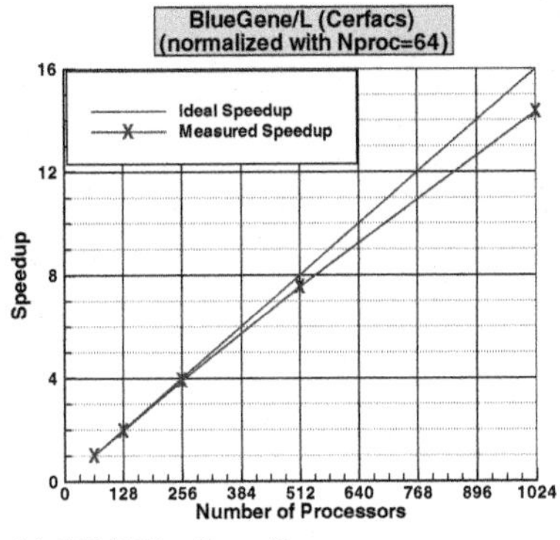

(a) IBM BlueGene/L.

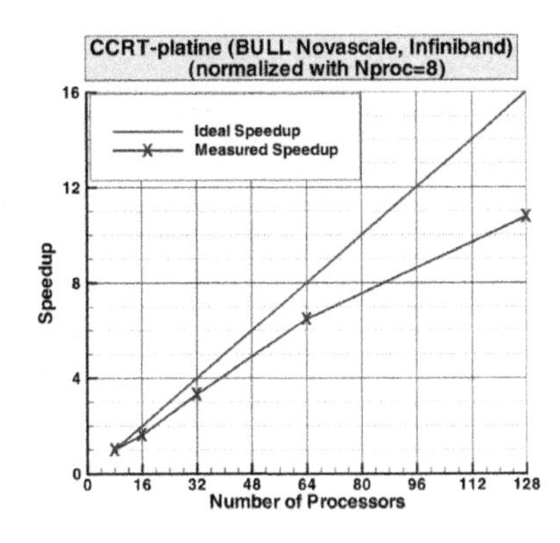

(b) BULL Novascale (Itanium Montecito 1.6GHz).

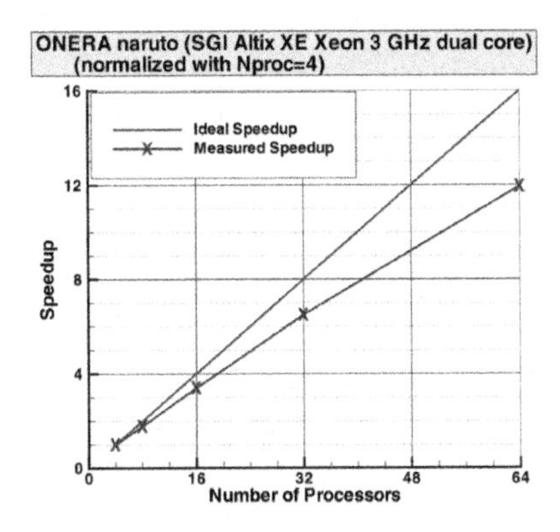

(c) SGI Altix XE (Xeon Woodcrest 3GHz dual core).

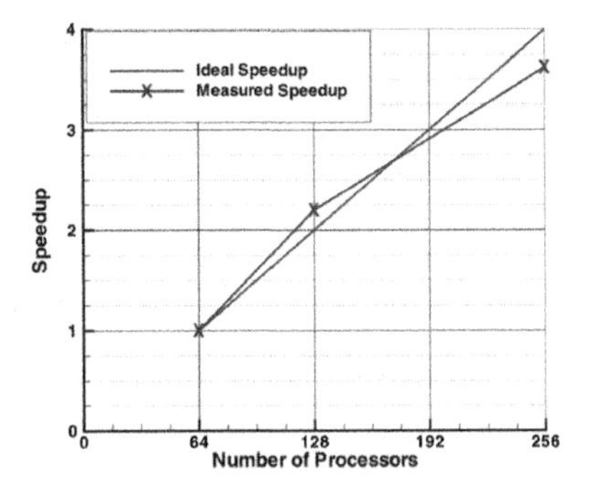

(d) (HP Cluster Opteron AMD 2.4 GHz).

Fig. 2. Parallel Speed-up of the Elsa code on different computing architectures

3.5 HP Cluster (Opteron AMD 2.4 GHz)

See Figure 2.(d).

[1] L. Cambier, J.P. Veuillot, Status of the elsA CFD software for Flow Simulation and Multidisciplinary Application, AIAA Paper 2008-664, Reno (January 2008).

[2] Anders Ytterström, A Tool for Partitioning Structured Multiblock Meshes for Parallel Computational Mechanics, International Journal of High Performance Computing Applications, Vol. 11, No. 4, 336-343 (1997)

Applications on Hybrid Unstructured Moving Grid Method for Three-Dimensional Compressible Flows

Hiroya Asakawa[1] Masashi Yamakawa[1] and Kenichi Matsuno[2]

[1] Division of Mechanical and System Engineering, Kyoto Institute of Technology, Matsugasaki, Sakyo-ku, Kyoto 606-8585, Japan yamakawa@kit.ac.jp
[2] Division of Mechanical and System Engineering, Kyoto Institute of Technology, Matsugasaki, Sakyo-ku, Kyoto 606-8585, Japan matsuno@kit.ac.jp

Abstract. For CFD problem with a complicated moving boundary, the unstructured moving-grid finite-volume method has been proposed and its ability recognized. However, the method was limited for applications using inviscid compressible flows. In this paper, the method is developed to apply to three-dimensional Navier-Stokes equations for viscous compressible flows. We formulate a control volume for prismatic element well adapted to unstructured mesh. Then, the method is applied to a flow around oscillating ONERA M6 airfoil at high Reynolds nmber. And the computation is executed in OpenMP parallel environment.

Key Words: Unstructured Grid, Compressible Flow, OpenMP Parallelization

1 Introduction

A numerical simulation of flows around moving complicated bodies is very interesting. To simulate such a flow field in the body-fitted coordinated system, great skewness of gridline may occur according to the motion of the body. In this case, it is necessary to consider a geometric conservation law[1] for moving grid. So, if the geometric conservation law is not satisfied, a numerical result will have error affecting movement of grid. For the issue, we have proposed the unstructured moving-grid finite-volume method[2]. A characteristic of the method is that it satisfies a conservation of flow parameter on moving grid, so it satisfies a geometric conservation law on such a moving grid in addition to satisfy a physical conservation law. The method adopts a control volume in a space-time unified domain on the unstructured grid system, then it is implicit and is solved iteratively at every time-step. However, the method was limited for applications using inviscid compressible flows on unstructured grid system which consist of tetrahedral meshes in three-dimensional grid system. However, generation of tetrahedral mesh for boundary layer is quite difficult. In those regions the large gradients occur in the direction normal to the surface, which requires mesh of very large aspect ratio. Thus, the prismatic mesh is superior in capturing the directionality of the flowfield over viscous regions. So, we adapt a hybrid grid system of a prismatic mesh for boundary layer regions and

D. Tromeur-Dervout (eds.), *Parallel Computational Fluid Dynamics 2008*,
Lecture Notes in Computational Science and Engineering 74,
DOI: 10.1007/978-3-642-14438-7_12,

tetrahedral mesh for other region. In this paper, there are two objectives. First, in a three-dimensional viscous flow, we propose the method that physical and geometric conservation laws are satisfied in moving grid system. Second, A numerical simulation of flow around a body is computed using moving grid method in OpenMP parallel environment. Then, it is applied to three-dimensional compressible flow in hybrid unstructured grid system.

2 Hybrid Unstructured Moving-Grid Finite Volume Method

2.1 Governing Equation

The Navier-Stokes equation in three-dimensional coordinate system is written in the conservation law form as follow:

$$\frac{\partial \mathbf{q}}{\partial t}+\frac{\partial \mathbf{E}}{\partial x}+\frac{\partial \mathbf{F}}{\partial y}+\frac{\partial \mathbf{G}}{\partial z}=\frac{1}{Re_\infty}\left(\frac{\partial \mathbf{E}_V}{\partial x}+\frac{\partial \mathbf{F}_V}{\partial y}+\frac{\partial \mathbf{G}_V}{\partial z}\right) \tag{1}$$

where, $\mathbf{q}$ is a vector of conservative variables,$\mathbf{E}$, $\mathbf{F}$ and $\mathbf{G}$ are inviscid fluxes, $\mathbf{E}_V$, $\mathbf{F}_V$ and $\mathbf{G}_V$ are viscous fluxes.

2.2 Numerical Algorithm

Then, the equation(1) is integrated for control volume of hybrid mesh as shown in Fig.1. In the case of three-dimensional system, the method is featured treatment of a control volume on the space-time unified domain (x, y, z, t), which is four-dimensional system in order that the method satisfies the geometric conservation laws. The present method is based on a cell-centered finite-volume method, thus we define flow variables at the center of a cell in unstructured mesh. So the cells are prism and tetrahedron in three-dimensional (x, y, z, t)-domain. When grid moves, the control volume becomes a complicated polyhedron in the (x, y, z, t)-domain.

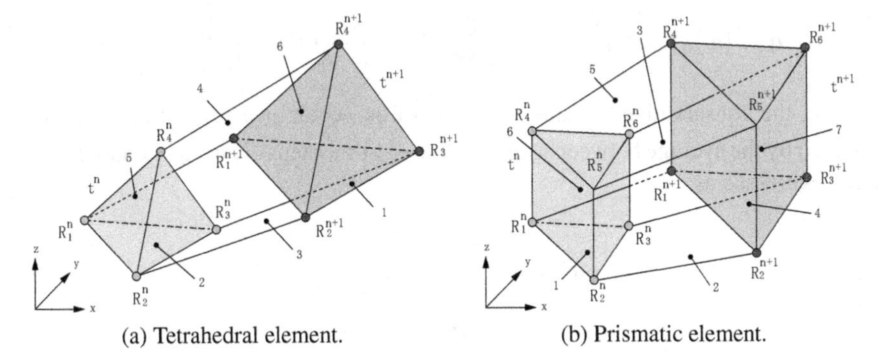

(a) Tetrahedral element. (b) Prismatic element.

Fig. 1. Control volume

For these control volume, the governing equation is integrated as follow:

$$\int_{\Omega}\left[\frac{\partial \mathbf{q}}{\partial t}+\frac{\partial \mathbf{E}}{\partial x}+\frac{\partial \mathbf{F}}{\partial y}+\frac{\partial \mathbf{G}}{\partial z}-\frac{1}{Re_{\infty}}\left(\frac{\partial \mathbf{E}_V}{\partial x}+\frac{\partial \mathbf{F}_V}{\partial y}+\frac{\partial \mathbf{G}_V}{\partial z}\right)\right]d\Omega=0. \quad (2)$$

The equation can be written as the equation

$$\int_{\Omega}\left[\left(\frac{\partial}{\partial x},\frac{\partial}{\partial y},\frac{\partial}{\partial z},\frac{\partial}{\partial t}\right)\cdot\left(\mathbf{E}-\frac{1}{Re_{\infty}}\mathbf{E}_V,\mathbf{F}-\frac{1}{Re_{\infty}}\mathbf{F}_V,\mathbf{G}-\frac{1}{Re_{\infty}}\mathbf{G}_V,\mathbf{q}\right)\right]d\Omega=0. \quad (3)$$

The equation can be represented using Gauss's Theorem:

$$\sum_{l=1}^{N_s+2}(\mathbf{q}\mathbf{n}_t+\Phi)_l dS_l=0 \quad (4)$$

where

$$\Phi = \mathbf{H}-\mathbf{H}_V, \quad (5)$$

$$\mathbf{H} = \mathbf{E}\mathbf{n}_x+\mathbf{F}\mathbf{n}_y+\mathbf{G}\mathbf{n}_z, \quad (6)$$

$$\mathbf{H}_V = \frac{1}{Re_{\infty}}(\mathbf{E}_V\mathbf{n}_x+\mathbf{F}_V\mathbf{n}_y+\mathbf{G}_V\mathbf{n}_z). \quad (7)$$

Here S is boundary of control volume surface. $\mathbf{n}_l=(\mathbf{n}_x,\mathbf{n}_y,\mathbf{n}_z,\mathbf{n}_t)_l$ $(l=1,2,\cdots,N_s+2)$ is the normal vector of control volume surface, and the length of the vector equals to the area of the surface. Then N_s is number of boundary, so in the case of prism $N_s=5$. $l=1,2,\cdots,N_s$ is presented as volume which is generated according to movement from n to $n+1$ time-step. $l=N_s+1,N_s+2$ is presented as cell itself at n time-step and $n+1$ time-step. The volume at n and $n+1$ time-step of the control volume are perpendicular to t-axis, and therefore they have only $\mathbf{n}_t$ component and correspond to the areas in the (x, y, z)-space at time t^{n+1} and t^n, respectively. Thus, Eq.(4) can be expressed as,

$$\mathbf{q}^{n+1}(n_t)_{N_s+2}+\mathbf{q}^{n}(n_t)_{N_s+1}+\sum_{l=1}^{N_s}\left[\mathbf{q}^{n+1/2}n_t+\boldsymbol{\Phi}^{n+1/2}\right]_l dS_l=0 \tag{8}$$

Here, the conservative variable vector and flux vector at $n+1/2$-time step are estimated by the average between n-time and (n+1)-time steps. Thus, $\mathbf{q}^{n+\frac{1}{2}}$ and $\boldsymbol{\Phi}^{n+\frac{1}{2}}$ can be expressed as,

$$\mathbf{q}^{n+\frac{1}{2}} = \frac{1}{2}(\mathbf{q}^{n+1}+\mathbf{q}^{n}), \tag{9}$$

$$\boldsymbol{\Phi}^{n+\frac{1}{2}} = \frac{1}{2}(\boldsymbol{\Phi}^{n+1}+\boldsymbol{\Phi}^{n}). \tag{10}$$

It is Crank-Nicolson type. However, Calculating flowfield at high Reynolds number, we use backward Euler type to get stability follow as:

$$\mathbf{q}^{n+\frac{1}{2}} = \mathbf{q}^{n+1}, \tag{11}$$

$$\boldsymbol{\Phi}^{n+\frac{1}{2}} = \boldsymbol{\Phi}^{n+1}. \tag{12}$$

The flux vectors are evaluated using the Roe flux difference splitting scheme [3] with MUSCL approach, as well as the Venkatakrishnan limiter [4]. The method uses the variable at $n+1$-time step, and thus the method is completely implicit. We introduce sub-iteration strategy with a pseudo-time approach [5] to solve the implicit algorithm. Now by defining that the operating function L(qn+1) as Eq.(13), the pseudo-time sub-iteration is represented as Eq.(14).

$$L\left(\mathbf{q}^{n+1,<m+1>}\right) = \frac{1}{\Delta t(n_t)_{N_s+2}}\Bigg[\mathbf{q}^{n+1,<m+1>}(n_t)_{N_s+2}+\mathbf{q}^{n}(n_t)_{N_s+1}$$

$$+\sum_{l=1}^{N_s}\left[(1-\theta)\left(\mathbf{q}^{n+1,<m+1>}n_t+\boldsymbol{\Phi}^{n+1,<m+1>}\right)+\theta\left(\mathbf{q}^{n}n_t+\boldsymbol{\Phi}^{n}\right)\right]_l dS_l\Bigg] \tag{13}$$

$$\frac{\partial \mathbf{q}}{\partial \tau} = -L\left(\mathbf{q}^{n+1,<m+1>}\right) \tag{14}$$

Here, the equation is Crank-Nicolson type at $\theta = 1/2$, and it is backward Euler type at $\theta = 0$ Where m is index for iteration, and τ is pseudo-time step. To solve Eq.(14), we adopted the LU-SGS implicit scheme [6] to advance $\delta\tau$ pseudo-time step. When m inner iteration is converged, we can get $n+1$-time step solution $\mathbf{q}^{n+1}$.

3 Application

3.1 Calculating condition

The hybrid unstructured moving-grid finite-volume method is applied to a flow around oscillating ONERA M6. The number of prism elements is 865,836 and the number of tetrahedron is 935,566, total is 1,801,402. An initial grid (whole appearance, cut view, around the wing) are shown as Fig2. The hybrid grid is generated by Marching step method [7]. Then the grid is moved according to a motion of the wing by Spling method [8].

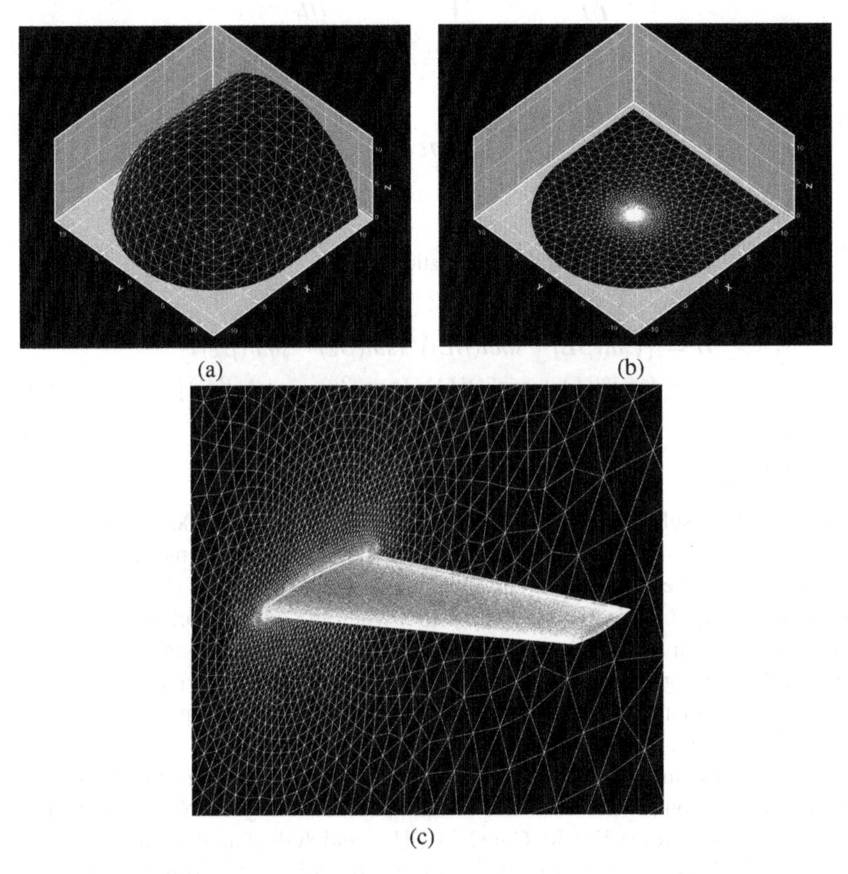

(a) (b) (c)

Fig. 2. Computational grid.

The flowfield is computed at $Re = 11.7x10^6$, $M_\infty = 0.84$, $\alpha = 3.06$. Initial condition is steady solution of this case. A computational region is shown as Fig.3. A Oscillation is defined as an amplitude at a wingtip by Eq.(15). Here, $\beta = 1.875/L$. Then, a maxium amplitude at a wingtip is 0.1L and A Frequency is 60Hz. In this case, we introduce Spalart-Allmaras one equation model [9] as turbulence model.

$$y(z,t) = A \cdot Hsin(\omega \cdot t) \tag{15}$$

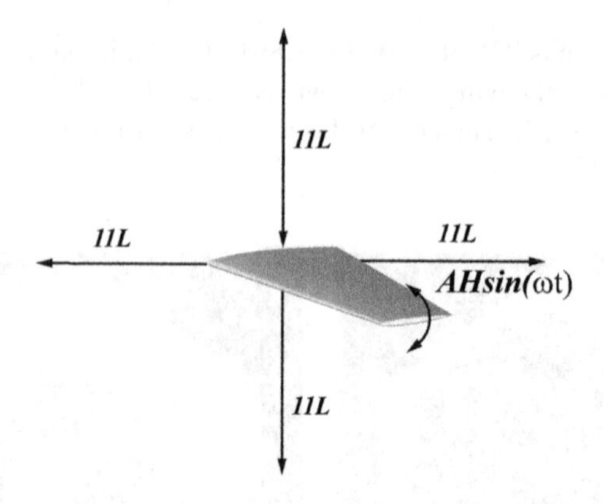

Fig. 3. Computational region.

$$H = \{(sin(\beta L) - sinh(\beta L)) \cdot (sin(\beta z) - sinh(\beta z)) + (cos(\beta L) + cosh(\beta L)) \cdot (cos(\beta z) - cosh(\beta z))\} \tag{16}$$

3.2 Result

Fig.4 shows the results of the flow field (pressure contours) at t = 0.000, 1.125, 2.375, 3.625, 4.625 and 5.875 respectively. We can see moving shocks on the wing surface. When the wing travels to upward, the shock generated on center of the wing surface moves to in front of wing. On the other hand, when the wing travels to downward, interaction between the shock generated in front of wing surface and generated in center of the wing generates clearly Lambda shock. This result confirmed that the method can calculate flow field and is promising to develop a three-dimensional viscous flow around moving body.

The parallel computation of the hybrid unstructured moving-grid finite-volume method is carried on for the flow field using OpenMP library on PC. The computer has two Intel processors (E6850: Core2 3.0GHz) and 8GB shared memory. The operating system is Windows Vista 64bit, fortran compiler is Intel Fortran 10.1. Then, a results of the estimating is shown in Table 1.

There was little number of the samples, however, we cannot get enough speed up in these environment. It is thought that we should examine a method to adopt technique such as division of the domain or MPI parallel environment.

Number of elements	1,300,000	1,800,000
Speed up	0.950	0.980

Table 1. Parallel computing performance

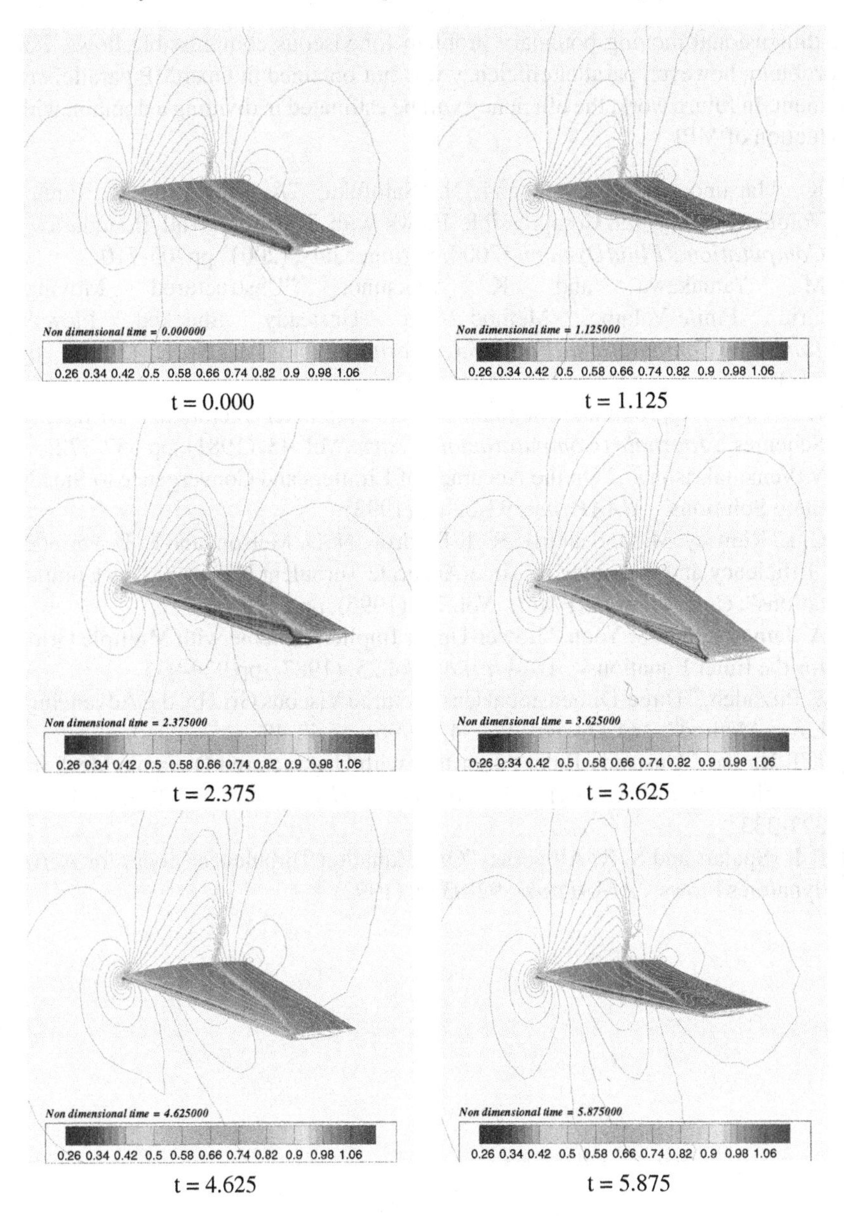

Fig. 4. Pressure contours

4 Conclusions

The unstructured moving-grid finite-volume method was formulated for three-dimensional viscous flows and developed in tetrahedral-prismatic hybrid grid system. Then, the result of a flow around oscillating wing showed applicable to three-dimensional moving boundary problem for viscous compressible flows. For this problem, however, parallel efficiency was not obtained in OpenMP parallel environment. In future work, the efficiency will be estimated in dividing a domain, with introduction of MPI.

[1] K. Matsuno, K. Mihara. and N. Satofuka, "A Moving-Mesh Finite-Volume Scheme for Compressible Flows with Traveling Wall Boundaries", *ComputationalFluidDynamics*2000, Springer July, (2001) pp705-710.
[2] M. Yamakawa and K. Matsuno, "Unstructured Moving-Grid Finite-Volume Method for Unsteady Shocked Flows", *JournalofComputationalFluidsEngineering*, Vol. 10, No.1, (2005), pp 24-30.
[3] P. L. Roe, "Approximate Riemann Solvers Parameter Vectors and Difference Schemes", *JournalofComputationalPhysics.*, Vol. 43, (1981), pp 357-372.
[4] V. Venkatakrishnan, "On the Accuracy of Limiters and Convergence to Steady State Solutions", *AIAAPaper*, 93-0880, (1993)
[5] C. L. Rumsey, M. D. Sanetrik, R. T. Biedron, N. D. Melson, and E. B. Parlette, "Efficiency and Accuracy of Time-Accurate Turbulent Navier-Stokes Computations", *ComputersandFluids*, Vol. 25, (1996), pp 217-236.
[6] A. Jameson, and S. Yoon, "Lower-Upper Implicit Scheme with Multiple Grids for the Euler Equations", *AIAAJornal*, Vol.25, (1987), pp 929-935
[7] S. Pirzadeh, "Three-Dimensional Unstructured Viscous Grid by the Advancing-Layer Method", *AIAAJornal*, Vol.34, (1996), pp 43-49
[8] J. T. Batine, "Unsteady Euler Algorithm with Unstructured Dynamic Mesh for Complex-Aircraft Aerodynamics Analysis", *AIAAJornal*, Vol.29, (1991), pp 327-333
[9] P. R. Spalart and S. R. Allmaras, "One-Equation Turbulence Model for Aerodynamics Flows", *AIAAPaper*, 92-0349, (1992)

Progressive Development of Moving-Grid Finite-Volume Method for Three-Dimensional Incompressible Flows

Shinichi ASAO[1], Sadanori ISHIHARA[1], Kenichi MATSUNO[2] and Masashi YAMAKAWA[2]

[1] Kyoto Institute of Technology, Matsugasaki, Sakyo-ku, Kyoto 606-8585, Japan
asao@fe.mech.kit.ac.jp
[2] Kyoto Institute of Technology, Matsugasaki, Sakyo-ku, Kyoto 606-8585, Japan
matsuno@kit.ac.jp

Abstract. In this paper, parallel computation of three-dimensional incompressible flows driven by moving multiple bodies is presented using a new moving embedded zonal grid method. Moving embedded zonal grid method is the method such that the embedded zonal grid can move freely in one direction in a main grid which covers the whole of the flow field. The feature of the method is to satisfy both geometric and physical conservation laws simultaneously. The method is applied to a flow driven by two cubes moving in the originally stationary fluid. The computation was performed using multi-core CPUs and the parallel algorithm has been implemented in the Poisson solver and the performance has been tested.

keyword: Incompressible flow, Moving-Grid Finite-Volume method, OpenMP

1 Introduction

Today, one of the interesting problems in Computational Fluid Dynamics is an unsteady flow and it is very important to calculate a moving boundary problem. Especially, in the case that body moves in the fluid is interesting on engineering. For example, these are a flow by stir, a flow around the turbine, and so on.

When we simulate such flow field, we encounter some problems to be overcome. One of the problems is on grid system. When the body moves in the flow field, a conventional single body-fitted grid system is hard to adjust the motion of body. One of the most popular methods for such a moving boundary problem is an overset grid method [1] where sub-grid placed around a moving body moves in main-grid. However the overset grid method does not satisfy conservation law rigorously in general due to unavoidable interpolation error. From the view point of accurate computation, the body-fitted single grid system is desirable, but is not difficult for the case that the body travels long distance in the flow field because of resultant highly skewed grid.

D. Tromeur-Dervout (eds.), *Parallel Computational Fluid Dynamics 2008*,
Lecture Notes in Computational Science and Engineering 74,
DOI: 10.1007/978-3-642-14438-7_13,

To overcome this problem, we have proposed a moving embedded-grid method [2] for two-dimensional space. The method patches a local grid generated around a body in a stationary "main grid" and shifts the embedded local grid in any direction through the main grid with eliminating front-side grids and adding back-side grid with satisfying both physical and geometric conservation laws. As a flow solver, the moving-grid finite-volume method [3] is modified and used in the moving embedded-grid method.

The purpose of this paper is two fold. One is to introduce the moving embedded zonal grid method as the first step in building up the three-dimensional moving embedded-grid method. In this method, the movement of embedded zonal grid is limited to one direction. The interactive flow field driven by the movement of the multiple cubes is demonstrated. The other is to implement a parallel procedure into present method and to investigate the parallel performance of the present method on the multi-core and/or parallel environment of the small laboratory.

2 Moving Embedded zonal grid method

2.1 Governing Equations

Governing equations are the continuity equation and the incompressible Navier-Stokes equations and are written in divergent form as,

$$\nabla \cdot q = 0 \tag{1}$$

$$\frac{\partial q}{\partial t} + \frac{\partial E}{\partial x} + \frac{\partial F}{\partial y} + \frac{\partial G}{\partial z} = 0 \tag{2}$$

where, q is velocity vector, E, F and G are flux vectors of x, y and z direction respectively. These flux vectors are composed of advection, pressure and viscous term.

2.2 Moving Grid Finite Volume Method

To assure the geometric conservation laws, we adopt a control volume in the space-time unified domain (x,y,z,t), which is four-dimensional in the case of three-dimensional flows. Now, Eq.(2) can be written in divergence form as,

$$\tilde{\nabla} \cdot \tilde{\mathscr{F}} = 0 \tag{3}$$

where,

$$\tilde{\nabla} = \left[\frac{\partial}{\partial x} \; \frac{\partial}{\partial y} \; \frac{\partial}{\partial z} \; \frac{\partial}{\partial t} \right]^T, \quad \tilde{\mathscr{F}} = \left[E \; F \; G \; q \right]^T \tag{4}$$

The present method is based on a cell-centered finite-volume method and, thus, the flow variables are defined at the center of the cell in the (x,y,z) space. The control

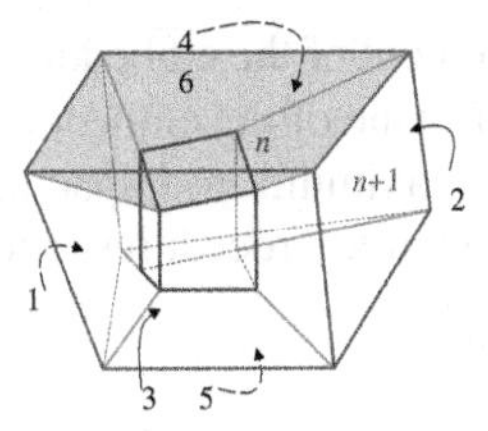

Fig. 1. Schematic view of control volume in (x,y,z,t) space-time unified domain.

volume becomes a four-dimensional polyhedron in the (x,y,z,t)-domain, as schematically illustrated in Figure 1.

We apply volume integration to Eq.(2) with respect to the control volume illustrated in Figure 1. With use of the Gauss theorem, Eq.(3) can be written in surface integral form as,

$$\int_{\Omega} \tilde{\nabla} \cdot \tilde{\mathscr{F}} \mathrm{d}\Omega = \oint_{S} \tilde{\mathscr{F}} \cdot \tilde{\tilde{n}} \mathrm{d}S = \sum_{l=1}^{8} \left(\tilde{\mathscr{F}} \cdot \tilde{n} \right)_l = 0 \tag{5}$$

Here, $\tilde{\tilde{n}}$ is a outward unit vector normal to the surface, S, of the polyhedron control volume Ω, and, $\tilde{n}_l = (n_x, n_y, n_z, n_t)_l$, ($l$ = 1,2,... 8) denotes the surface normal vector of control volume and its length equals to the boundary surface area in four-dimensional (x,y,z,t) space. The upper and bottom boundary of the control volume ($l = 7$ and 8) are perpendicular to t-axis, and, therefore they have only n_t component and its length is corresponding to the volume of the cell in the (x,y,z)-space at time t^n and t^{n+1} respectively. Thus, Eq.(5) can be expressed as,

$$q^{n+1}(n_t)_8 + q^n(n_t)_7 + \sum_{l=1}^{6} \left(\tilde{\mathscr{F}}^{n+1/2} \cdot \tilde{n} \right)_l = 0 \tag{6}$$

To solve Eq.(6), we apply the SMAC method[4]. Thus, Eq.(6) can be solved at three following stages. The equation to be solved at the first stage contains the unknown variables at $n+1$-time step in the flux terms. Thus the equation is iteratively solved using the LU-SGS method[5]. The equation to be solved at the second stage is the Poisson equation about the pressure correction. This equation is iteratively solved using the Bi-CGSTAB method[6]. The flux vectors are evaluated using the QUICK method. While the flux vectors of the pressure and viscous terms are evaluated in the central-difference-like manner.

2.3 Moving Embedded Zonal Grid Method

The present moving embedded zonal grid method is the method such that the zone grid in which bodies are included is embedded in the main grid and the embedded zonal grid moves in one direction freely with adding and/or removing gird plane in the main grid as illustrated in Figure 2. The front grid plane of the embedded zonal gird is eliminated from the main grid to avoid grid folding according to decrease of the grid spacing due to the movement of the embedded zonal grid, while a grid plane is added newly between the rear plane of the embedded zonal grid and the main grid in order to keep the allowable maximum grid spacing. Hence the present method includes essentially the two inevitable procedures: addition and elimination of grid plane.

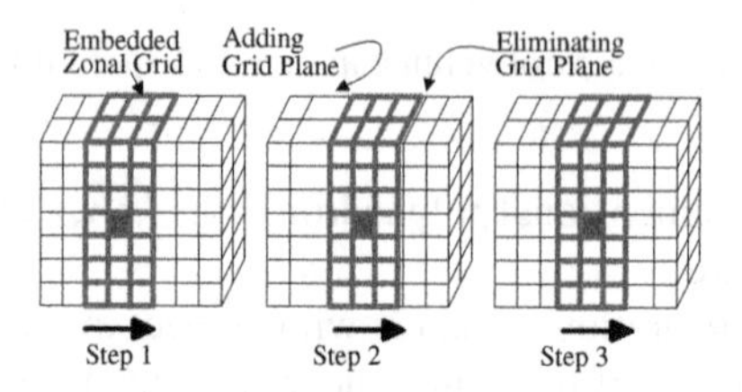

Fig. 2. Moving embedded zonal grid method.

The addition of the grid plane is performed as follows. Suppose that the grid cell I at time step n deforms as well as moves to new position, and separates into two grid cells I and I' at time step $n+1$, as illustrated in Figure 3(a). This means that the new grid plane i' is inserted or added between the grid plane $i-1$ and i. Under this situation, we have to treat two control volume, I and I', to get solutions at time step $n+1$. For control volume I, the Eq.(6) is applied and no any other extra procedure is necessary. For newly appearing control volume I', we need to modify Eq.(6) considering that the volume of cell I' at time n is zero. This means that $(n_t)_7 = 0$ for control volume I'. Thus the finite volume equation for I' becomes:

$$q^{n+1}(n_t)_{8I'} + \sum_{l=1}^{6} \left(\tilde{\mathscr{F}}^{n+1/2} \cdot \tilde{n} \right)_l = 0 \tag{7}$$

The elimination of the grid plane is accomplished through merging the two related control volumes. Suppose that the i'-th grid plane is eliminated at $(n+1)$-time step. Then we consider that the i'-th and i-th planes of cells at n-time step are merged at $(n+1)$-time step and only i-th plane of cells remains. In this situation, the control volume in space-time unified domain, (x,y,z,t), can be illustrated in Figure 3(b). Thus Eq.(8) replaces Eq.(6).

$$q^{n+1}(n_t)_{8I} + q^n(n_t)_{7I} + q^n(n_t)_{7I'} + \sum_{l=1}^{6} \left(\tilde{\mathscr{F}}^{n+1/2} \cdot \tilde{n} \right)_l = 0 \tag{8}$$

In the present method, the addition and elimination of grid plane are performed simultaneously when the grid spacing at the boundary between the main and embedded grids exceeds a user-specified tolerance.

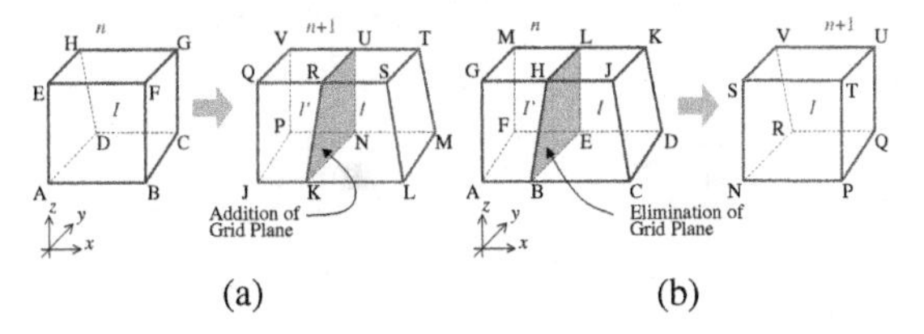

Fig. 3. Control volumes for (a) adding grid plane and (b) eliminating grid plane.

3 Numerical results

3.1 Validation 1: Geometric Conservation Laws

At first it is necessary to confirm that the method satisfies the "Geometric Conservation Law" (GCL condition)[7], which means the method can calculate an uniform flow on the moving mesh system without any error, when the embedded grid is moving in main grid with adding and eliminating grid planes. To verify that the method satisfies the GCL condition, we apply the present method to the uniform flow. After 100,000 time step, the error keeps the order of 10^{-17}, in a word, machine zero, which means the method satisfies GCL perfectly.

3.2 Validation 2: Computation of Incompressible Flow-Field Driven by Moving Cube in the Stationary Fluid

In order to check the present method, we have compared two physically equivalent flow fields. The one is the semi-steady flows driven by the cube with a constant speed in a stationary flow field, the other is the steady flows around the cube in an uniform flow. As the calculation condition, the Reynolds number is 40 in the both

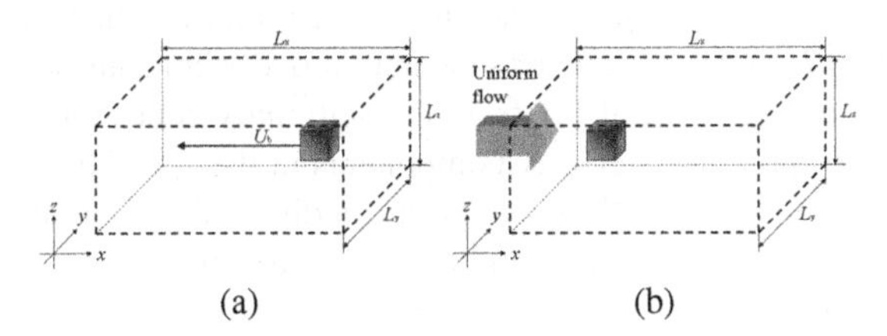

Fig. 4. (a) Moving cube with a constant speed in a stationary flow field. (b) The flows around the cube in a uniform flow.

cases. Figure 5 shows the vorticity contours. Since two physically equivalent flow fields are obtained, the reliability of code is confirmed.

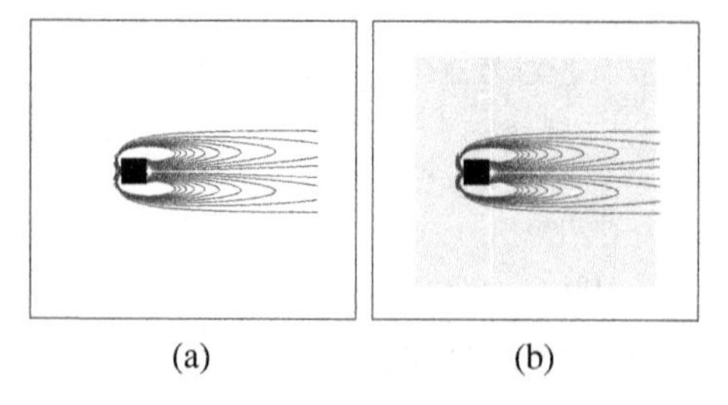

Fig. 5. Vorticity contours ((a) Using moving embedded zonal grid method, (b) Reference solution).

3.3 Computation of Incompressible Flow-Field Driven by Moving Twin Cubes in the Stationary Fluid

As an application of the present method, we simulate a flow-field driven by twin cubes moving a stationary fluid, as illustrated in Figure 6.

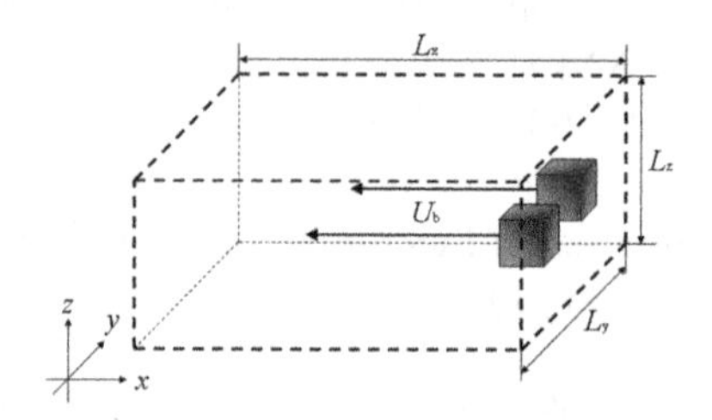

Fig. 6. The flows driven by multiple cubes.

The size of the whole computational domain is $L_x = 15.0L$, $L_y = 10.0L$, $L_z = 10.0L$, where L is the side length of the cube. The initial position of the cubes are $x = 12.5L$, $y = 4L, 6L$, and $z = 5L$. The twin cubes move at constant speed of $U_b = 1.0$ after the constant acceleration of 0.1 from the stationary condition. The initial stationary condition of pressure, velocity components in the x, y, z directions are given by $p = 1.0$, $u = v = w = 0.0$. As the calculation condition, the Reynolds number is 40 and time step size is $\Delta t = 0.01$. The number of the main grid is $151 \times 101 \times 101$ and the embedded zonal grid is $14 \times 101 \times 101$. Figure 7 shows the pressure distribution on the cubes as well as the vorticity contours at t = 5.0, 10.0 and 15.0.

The pressure in front of the cubes becomes higher as the speed of cube increases. We can capture the twin vortex which corresponds to Re = 40. The vortex made by the left body joins the vortex made by the right body at time $t = 15$. Therefore these are physically-meaningful flows, and it is confirmed that the moving embedded zonal grid method is useful for this application.

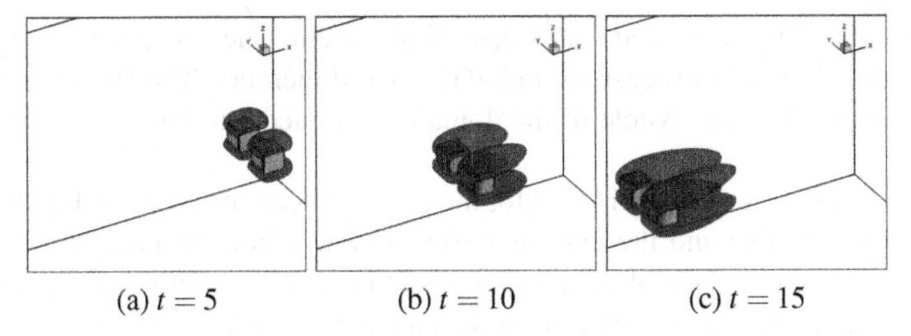

(a) $t = 5$ (b) $t = 10$ (c) $t = 15$

Fig. 7. Surface pressure distribution and vorticity contour.

4 Parallel implementation

The second stage of the SMAC method is the Poisson solver for the pressure correction, where the Bi-CGSTAB method is used in the present method. In spite of the efficiency of the Bi-CGSTAB method, this Poisson solver still consumers 90 % of whole of computer time. Thus we have implemented parallelization on this stage. The Bi-CGSTAB method includes the ILU decomposition. When we parallelize the ILU decomposition, we have to consider the data dependencies. We have applied the "Domino method" to the present method. The detail of the present "Domino method" is described as fallows.

(i) The calculation region is divided as illustrated in Figure 8(a).
(ii) The calculation region 1 of Figure 8(b) is calculated sequentially.
(iii) The calculation region 2,3 and 4 of Figure 8(c) is calculated in parallel after calculation of (ii) has finished.
(iv) The calculation region 5,6,7,8,9 and 10 of Figure 8(d) is calculated in parallel after calculation of (iii).
(v) The calculation region of Figure 8(e)∼(h) are calculated in a similar way.

Thus, the method can avoid the data dependences.

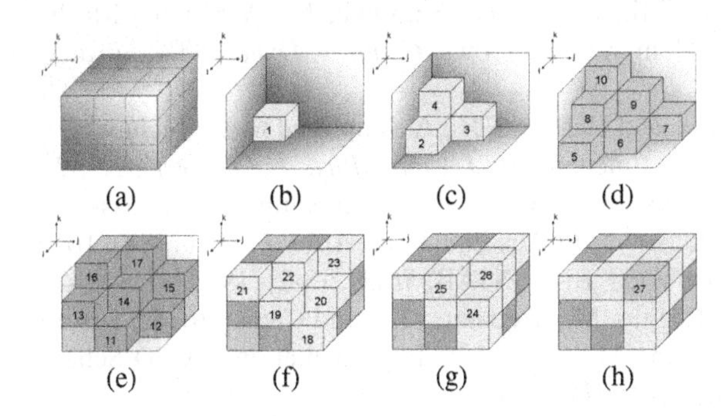

Fig. 8. Schematic diagrams of Domino method ($3 \times 3 \times 3$).

The parallel computation is carried out using OpenMP library on PC for a flow where two cubes move in the stationary fluid. As the operating system, we adopt

Fedora core 6, which is a 64bit Linux operating system. The computer has one Intel Core-two-duo 2.66GHz processors and 4GB shared memory. The Fortran compiler is Intel Fortran 10.0. The whole of the domain is divided into $64(=4\times4\times4)$ subdomains.

As a result, the "Speed up" is 1.19 for the two CPU-core processors. For this case, the result is miserable and the expected efficiency was not obtained in the present numerical experiment. One of the reasons might be that the number of processors is different from the division number in some calculation regions, and the parallelization efficiency has been down. With this result, we need further efforts to improve efficiency.

5 Concluding remarks

In this paper, the moving embedded zonal grid method has been introduced and applied to the flows driven by moving multiple bodies. As a result, physically-meaningful flows are obtained, and it is confirmed that the moving embedded zonal grid method is useful for the simulation of a flow-field driven by multiple bodies moving a stationary fluid. The second stage of the flow solver, the part of the Poisson solver, is parallelized using "Domino" method, and the expected speed up is not obtained. Further attempts have been continued for further parallel efficiency in the laboratory computer environment.

Acknowledgement. This work was supported by KAKENHI(18560169).

[1] Steger, J. L., Dougherty, F. C. and Benek, J. A., A Chimera Grid Scheme; Advances in Grid Generation, *ASME*, 5, (1983), 55-70.

[2] Asao, S. and Matsuno, K., A Moving Embedded-Grid Method for Simulation of Unsteady Incompressible Flows with Boundaries Traveling Long Distances, *JSME*, 74-737 B, (2008), 25-33.

[3] Matsuno, K., Mihara, K., and Satofuka, N., A Moving-Mesh Finite-Volume Scheme for Compressible Flows, *Comp. F. Dyna. 2000*, Springer, (2001), 705-710.

[4] Amsden, A.A. and Harlow, F.H., A Simplified MAC Technique for Incompressible Fluid Flow Calculations, *J. Comp. Phys*, 6, (1970b), 322-325.

[5] Yoon, S. and Jameson, A., Lower-Upper Symmetric-Gauss-Seidel Method for the Euler and Navier-Stokes Equations, *AIAA J.*, 26, (1988), 1025-1026.

[6] H. A. van der Vorst, Bi-CGSTAB, *SIAM*, 13-2, (1992), 631-644

[7] Zhang, H., Reggio, M., Trepanier, J.Y. and Camarero, R., Discrete Form of the GCL for Moving Meshes and Its Implimentation in CFD Schemes, *Computers and Fluids*, 22, (1993), 9-23.

Part IV

Boundary methods

Flow Computations Using Embedded Boundary Conditions on Block Structured Cartesian Grid

T. Kamatsuch[1], T. Fukushige[2] and K. Nakahashi[3]

[1] Fundamental Technology Research Center, Honda R&D Co. Ltd., Chuo 1-4-1,Wako 351-0193, Japan. toshihiro_kamatsuchi@n.f.rd.honda.co.jp
[2] Fundamental Technology Research Center, Honda R&D Co. Ltd., Chuo 1-4-1,Wako 351-0193, Japan. takahiro_fukushige@n.f.rd.honda.co.jp
[3] Department of Aerospace Engineering, Tohoku University, Aoba 6-6-01, Sendai 980-8579, Japan. naka@ad.mech.tohoku.ac.jp

1 Introduction

Computational Fluid Dynamics (CFD) has been increasingly relied on as an important tool for the aerodynamic design. However, mesh generation for 3D complex geometries is still the most time-consuming task in CFD. Recently, Cartesian grid methods have become popular because of their advantages of fast, robust, and automatic grid generation. Authors proposed a new approach based on a block-structured Cartesian grid approach, named Building-Cube Method (BCM, [5]). This method is aimed for large-scale, high-resolution computations of flows. In the original BCM, however, the wall boundary is defined by a staircase representation to keep the simplicity of the algorithm. Therefore, to resolve boundary layers, vast amounts of grid points are required especially for high-Reynolds number flows of practial applications. In this paper, a new embedded boundary treatment suitable for high-Reynolds viscous flows on BCM is proposed. The scalability performance on cluster computers is also investigated with this method.

2 Building-Cube Method

The BCM basically employs a uniformly-spaced Cartesian mesh because of the simplicity in the mesh generation, the numerical algorithm, and the post processing. A flow field is described as an assemblage of building blocks of cuboids, named "*Cube*", as shown in Fig. 1. The geometrical size of each cube is determined by adapting to the geometry and the local flow length by the omnitree refinement method. In each cube, a uniformly-spaced Cartesian mesh is generated as shown in Fig. 2. All cubes have the same mesh density so that the local computational resolution is determined by the cube size. The same mesh density within all cubes also keeps simple algorithm in the parallel computations.

D. Tromeur-Dervout (eds.), *Parallel Computational Fluid Dynamics 2008*,
Lecture Notes in Computational Science and Engineering 74,
DOI: 10.1007/978-3-642-14438-7_14,

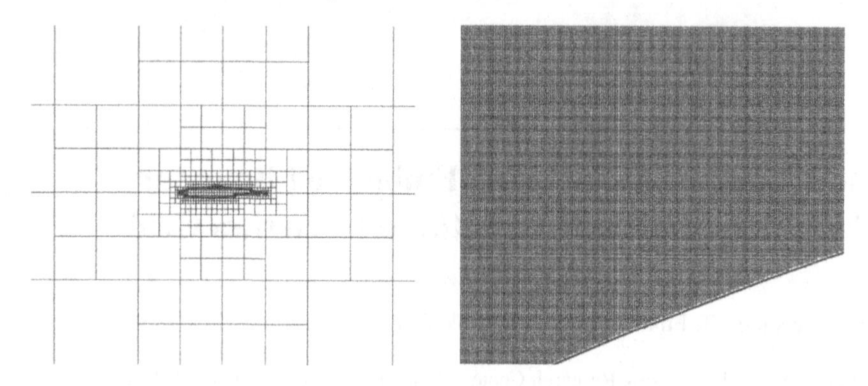

Fig. 1. Cube boundaries around RAE2822 airfoil

Fig. 2. Cartesian mesh inside the cube (256x256 cells)

3 Governing Equations

Non-dimensionalized Navier-Stokes equations governing compressible viscous flows can be written in the Cartesian coordinates $x_j (j = 1,2,3)$ as

$$\frac{\partial \mathbf{U}}{\partial t} + \frac{\partial \mathbf{F}_j}{\partial x_j} - \frac{1}{Re}\frac{\partial \mathbf{G}_j}{\partial x_j} = 0, \tag{1}$$

where $\mathbf{U} = (\rho, \rho u, \rho v, \rho w, \rho e)$ is the vector of conservative variables, ρ is the density, u, v, w, are the velocity components in the x_1, x_2, x_3 directions and e the specific total energy. The vector **F**(**U**) and **G**(**U**) represent the inviscid and viscous flux vectors respectively, and *Re* is the Reynolds number. Spalart and Allmaras' one equation model ([9]) for turbulent flow is used to evaluate eddy viscosity.

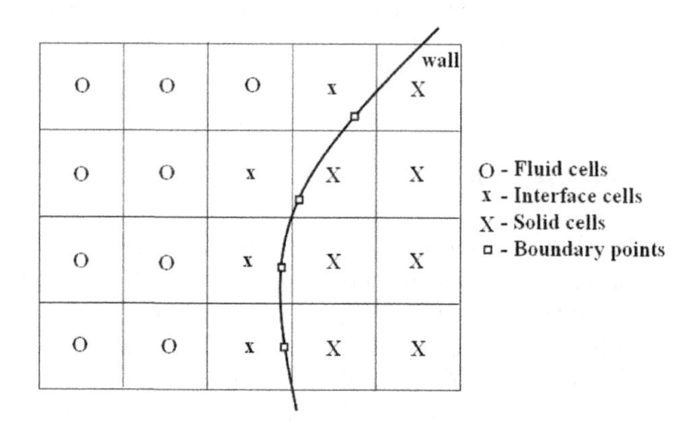

Fig. 3. Nodal assignment for boundary conditions

4 Numerical Method

4.1 Boundary Conditions

The boundary conditions are treated by the embedded boundary approach. This concept was first introduced for performing simulation of flow in a heart as the immersed boundary method ([7]). In this approach a body force is introduced at interface cell i to impose boundary conditions as shown in Fig. 3.
The original Navier-Stokes equations of Eq. (1) can then be written as follows:

$$\frac{\partial \mathbf{U}}{\partial t} = \mathbf{RHS}_i + \mathbf{S}_i. \tag{2}$$

$\mathbf{RHS}_i$ contains inviscid and viscous fluxes, and is given by

$$\mathbf{RHS}_i = -\left(\frac{\partial \mathbf{F}_j}{\partial x_j} - \frac{1}{Re}\frac{\partial \mathbf{G}_j}{\partial x_j}\right). \tag{3}$$

Body force $\mathbf{S}_i$ is defined by

$$\mathbf{S}_i = -\mathbf{RHS}_i + \mathbf{RHS}_{IB}. \tag{4}$$

$\mathbf{RHS}_{IB}$ is the components of the body force term applying boundary condition, and is given by

$$\mathbf{RHS}_i = -\left(\frac{\partial \mathbf{F}_j}{\partial x_j}\right)_{IB} + \frac{1}{Re}\left(\frac{\partial \mathbf{G}_j}{\partial x_j}\right)_{IB}, \tag{5}$$

where, $(\partial \mathbf{F}_j/\partial x_j)_{IB}$ and $(\partial \mathbf{G}_j/\partial x_j)_{IB}$ are the inviscid and viscous fluxes respectively, imposed by the immersed approach. For all interface cells, inviscid and viscous flux term of Eq. (5) is discretized with a gridless method ([3]). In this approach, however, gridless method was used as the boundary conditions of Euler computations on the Cartesian cells. In this work, these gridless approaches are extended to high-Reynolds number flow simulations with compressible Navier-Stokes equations.
The gridless method requires the notion of a "*cloud*" and its nodal shape functions. The spatial derivative of any particular variable f in the jth coordinate direction is finally written in terms of a derivative shape function as

$$\frac{\partial f}{\partial x_j} = \sum_{k\in C(i)} \frac{\partial \phi_{ik}}{\partial x_j} f_{ik}. \tag{6}$$

Here, $C(i)$ is the set of cloud points for a given point i. Let f_{ik} denote the value of functions f at the midpoint of the edge ik , where $k \in C(i)$. The inviscid flux of $\mathbf{RHS}_{IB}$ can then be estimated with the formula of Eq. (6) as

$$\left(\frac{\partial \mathbf{F}_j}{\partial x_j}\right)_{IB} = \sum_{k\in C(i)} \frac{\partial \phi_{ik}}{\partial x_j}(\mathbf{F}_j)_{ik} = \sum_{k\in C(i)} \mathbf{E}_{ik}. \tag{7}$$

The flux term $\mathbf{E}$ at the midpoint is expressed as

$$\mathbf{E} = \begin{bmatrix} \rho U \\ \rho u_i U + \frac{\partial \phi}{\partial x_i} p \\ U(\rho e + p) \end{bmatrix}, \tag{8}$$

where U is so-called contravariant velocity defined as

$$U = \frac{\partial \phi}{\partial x_j} u_j. \tag{9}$$

In the evaluation of the numerical fluxes at the midpoints, the approximate Riemann solver ([6]) is used, instead of directly introducing artificial damping terms. With this approach, no intrinsic measure of length is required and upwinding can be naturally introduced into the body force term for the embedded boundary approach. The viscous terms of the $\mathbf{RHS}_{IB}$ contain the following derivative

$$\frac{\partial}{\partial x}\left(\mu \frac{\partial u}{\partial x}\right). \tag{10}$$

The second derivative is estimated using Eq. (6) as

$$\left.\frac{\partial}{\partial x}\left(\mu \frac{\partial u}{\partial x}\right)\right|_i = \sum_{k \in C(i)} \frac{\partial \phi_{ik}}{\partial x}\left(\mu \frac{\partial u}{\partial x}\right). \tag{11}$$

For practical applications, the Reynolds number may become very large ($\sim 10^7$). In such high-Reynolds number flow, computation only with Cartesian mesh is still difficult by current computer capacity, since a very fine mesh is required to resolve viscous sublayers in the boundary regions. In the present work, the embedded method is extended for using a "*Subgrid*" to retain near-wall accuracy while reducing the number of overall Cartesian mesh points. A subgrid consists of new computational nodes, which is added independently to the Cartesian mesh as shown in Fig. 4. The inviscid and viscous fluxes of the subgrid are also discretized by using the gridless approach described above.

4.2 Solution Algorithm

The governing equations are solved based on a cell-centered, finite-volume scheme on Cartesian mesh. In this discretization, preconditioning system ([10]) is used in order to improve the efficiency for solving low Mach number flow problem with variable or constant density. The numerical flux is computed using an approximate Rieman solver of Harten-Lax-van Leer-Einfeldt-Wada (HLLEW, [6]). The primitive variables at the dual cell interfaces are evaluated by MUSCL approach ([11]) in order to achieve more than third-order accuracy. The lower/upper symmetric Gauss-Seidel (LU-SGS, [12]) implicit method is used for the time integration. In order to keep the second-order accuracy in time, a sub-iteration scheme ([4]) is employed.

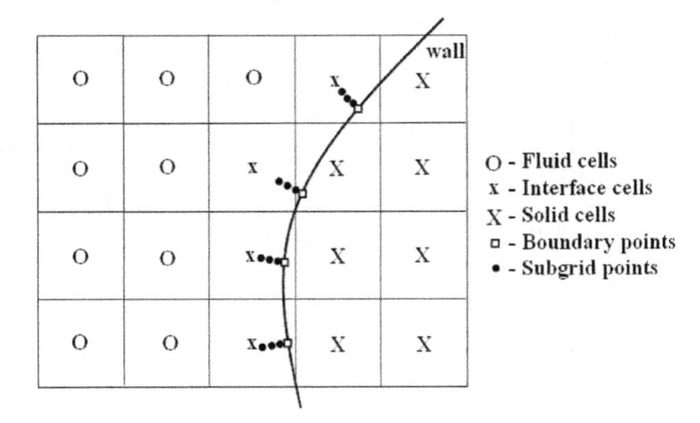

Fig. 4. Nodal assignment for boundary conditions with Subgrid

4.3 Parallelization

Because of the same mesh density of all cubes, the parallel computation of the present method is straightforward. It is easily implemented by distributing the cubes to the CPUs in a sequential manner as shown in Fig. 5. For the parallelization, the flow solver was modified using the MPI library for parallel computers. The domain decomposition technique is used for the parallelizing strategy. All cubes should be partitioned so that the leaf cubes are equally distributed to each processor, since only the leaf cubes are used for flow computation. In this work, Morton ordering ([8]), which is one of the Space-filling curves, was employed for domain decomposition because of a simple recursive algorithm. It was tested on the Opteron cluster with 64 CPUs. The scalability result of Fig. 6 is for flow computation using about 4 million Cartesian cells. With 64 CPUs, the speedup was about 54 times of the single CPU computation.

5 Numerical Results

Flow simulations around the Ahmed body ([1]) with the 35° slant-back configurations were performed. Computational mesh used here is shown in Figs. 7 and 8. The configurations for the computations have been specifically chosen to match Lienhart's experiment ([2]). The inflow velocity was 40 [m/s]. This results in a Reynolds number Re = 2.8 million (based on the body length of 1044 [mm]). For the front part, the profiles of the mean streamwise velocity on the symmetry plane are compared with the experimental data in Fig. 9. The flow upstream of the body is properly produced in the simulation. In Fig. 10 the profiles of the mean streamwise velocity at the rear body part on the symmetry plane are shown with the experimental results. The simulation results are in good agreement with experimental data.

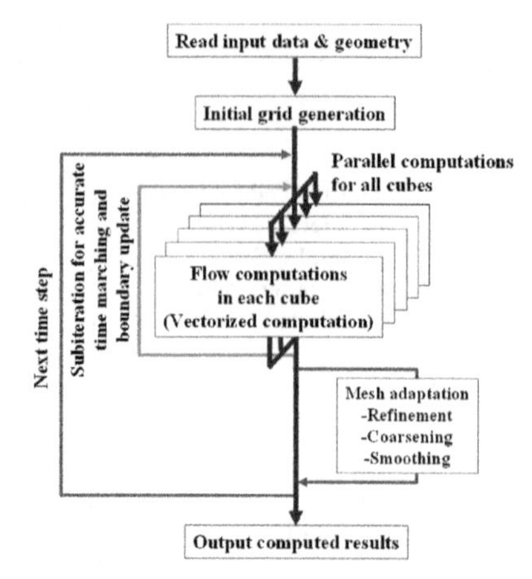

Fig. 5. Overall flow-solution procedure

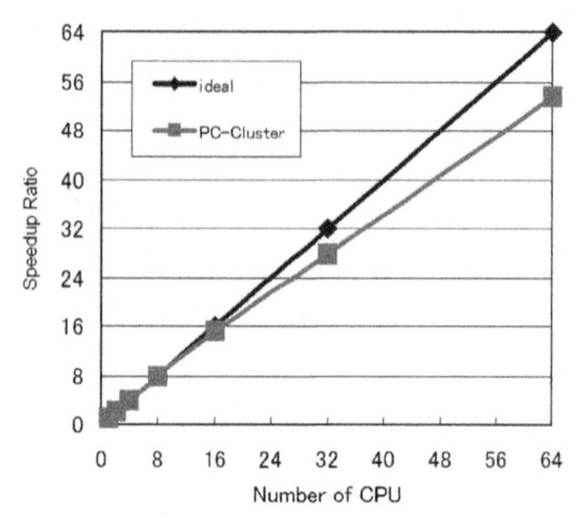

Fig. 6. Scalability test on the Opteron computer

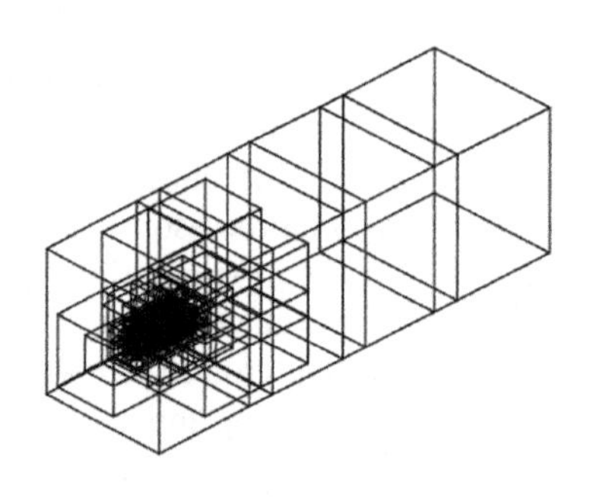

Fig. 7. Cube frames in the computational domain

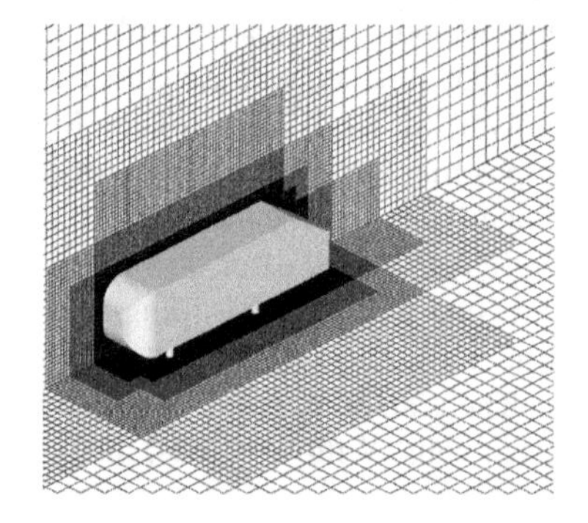

Fig. 8. Mesh cells around the body

6 Conclusion

A new embedded boundary treatment suitable for the Buildin-cube method was discussed for high-Reynolds number flow computations using parallel computers. In this approach, a body force terms were introduced only at boundary cells to impose wall boundary conditions. This method was extended for using a subgrid to retain near-wall accuracy while reducing the number of overall Cartesian cells. Because of the same mesh density of all cubes, the parallel computation of the present method was straightforward. All cubes were partitioned so that the leaf cubes were equally distributed to each processor. It was tested on the Opteron cluster and the scalability result showed good performance with present approach. The developed method was

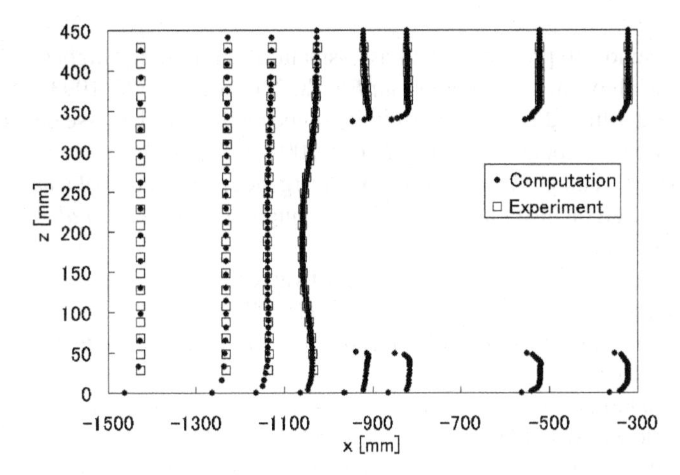

Fig. 9. Mean streamwise velocity profiles at the front body on the symmetry plane

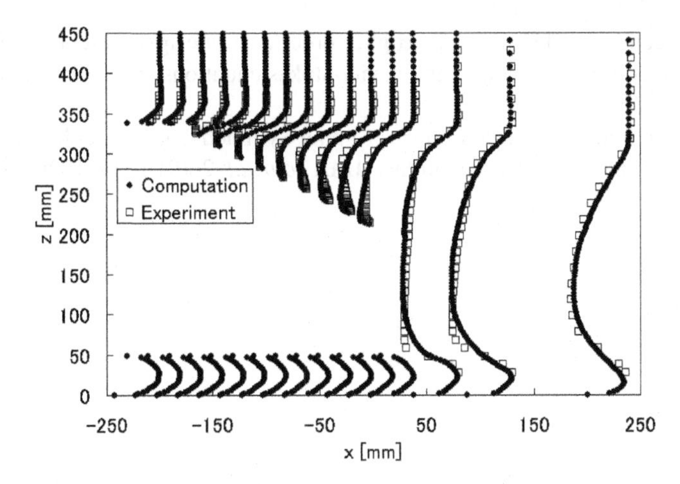

Fig. 10. Mean streamwise velocity profiles at the rear body and near wake on the symmetry plane

used to compute viscous flow past Ahmed body, and the simulation results were in good agreement with experimental data.

[1] S. R. Ahmed, G. Ramm, and G. Faltin. Some salient features of the time averaged ground vehicle wake. *SAE Paper*, 840300, 1984.

[2] H. Lienhart, C. Stoots, and S. Becker. Flow and turbulence structures in the wake of a simplified car model (ahmed model). *DGLR Fach Sym. der AG STAB*, 2000.

[3] H. Luo, J. D. Baum, and R. Lohner. A hybrid cartesian grid and gridless method for compressible flows. *Journal of Computational Physics*, 214(2):618–632,

2006.
[4] K. Matsuno. Improvement and assessment of an arbitrary-higher-order time-accurate algorithm. *Computer and Fluids*, 22(2/3):311–322, 1993.
[5] K. Nakahashi and L. S. Kim. Building-cube method for large-scale, high resolution flow computations. *AIAA Paper*, 2004-0423, 2004.
[6] S. Obayashi and G. P. Guruswamy. Convergence acceleration of a navier-stokes solver for efficient static aeroelastic computations. *AIAA Journal*, 33(6):1134–1141, 1995.
[7] C. S. Peskin. Flow patterns around heart valves: A numerical method. *Journal of Computational Physics*, 10(2):252–271, 1972.
[8] H. Samet. *The Design and Analysis of Spatial Data Structures*. Addison-Wesley, 1990.
[9] P. R. Spalart and S. R. Allmaras. A one-equation turbulence model for aerodynamic flows. *AIAA Paper*, 92-0439, 1992.
[10] J. M. Weiss and W. A. Smith. Preconditioning applied to variable and constant density time-accurate flows on unstructured meshes. *AIAA Journal*, 33(11):2050–2057, 1995.
[11] S. Yamamoto and H. Daiguji. Higher-order-accurate upwind schemes for solving the compressible euler and navier-stokes equations. *Computer and Fluids*, 22(2/3):259–270, 1993.
[12] S. Yoon and A. Jameson. Lower-upper symmetric-gauss-seidel method for euler and navier-stokes equations. *AIAA Journal*, 26(9):1025–1026, 1988.

Computation of Two-phase Flow in Flip-chip Packaging Using Level Set Method

Tomohisa Hashimoto[1], Keiichi Saito[2], Koji Morinishi[3], and Nobuyuki Satofuka[4]

[1] Department of Mechanical Engineering, Kinki University, 3-4-1 Kowakae, Higashi-Osaka, Osaka, 577-8502, Japan, hasimoto@mech.kindai.ac.jp
[2] Plamedia Corporation, Honcho, Nakano-ku, Tokyo, 164-0012, Japan, saito@plamedia.co.jp
[3] Department of Mechanical and System Engineering, Kyoto Institute of Technology, Matsugasaki, Sakyo-ku, Kyoto, 606-8585, Japan, morinisi@kit.ac.jp
[4] The University of Shiga Prefecture, 2500 Hassaka-cho, Hikone-shi, Shiga, 522-8533, Japan, satofuka.n@office.usp.ac.jp

In flip-chip packaging technology, the underfill encapsulation is one of the important processes to obtain a significant improvement in fatigue lifetime for the solder joints between IC chip and substrate. The advanced design of electronic devices aiming at the enhancement of the performance involves the increase of the number of solder bumps, smaller size of the IC chip and smaller gap height between IC chip and substrate. That leads to various problems caused by the flow behavior, such as voids in underfill and mis-placed IC chip. The numerical analysis is more and more strongly required for simulating the underfill flow behavior, including the condition of dispensing the underfill material on the substrate. In fact, it is desirable to predict the filling time, the final fillet shape formed around IC chip and the occurrence of air trap especially around the solder bump in the underfill process, considering the effect of contact angle, viscosity and surface tension of the underfill material for increasing the reliability of flip-chip packaging.

We developed a numerical method for simulating the underfill flow in flip-chip packaging, especially for designing the optimum condition of solder joint performance [6]. The two types of processes are presented for applying the underfill encapsulant to the gap between IC chip and substrate. One is conventional capillary flow type and the other is no-flow type. Both underfilling processes are illustrated in Fig. 1. In the capillary flow type, multiple processing steps are involved. The solder joints between IC chip and substrate are formed, which is called solder bump reflow, and then thermosetting epoxy resin is driven into the cavity by capillary action. After the resin is completely filled, the assembly is taken to an oven where the resin is cured. On the other hand, the no-flow type was invented to reduce the processing steps in the capillary flow type, which provides cost savings. The epoxy resin is directly dispensed on the substrate and is compressed by pre-heated IC chip. While the IC chip is mounted on the substrate, the solder joints are formed with curing of

D. Tromeur-Dervout (eds.), *Parallel Computational Fluid Dynamics 2008*,
Lecture Notes in Computational Science and Engineering 74,
DOI: 10.1007/978-3-642-14438-7_15,

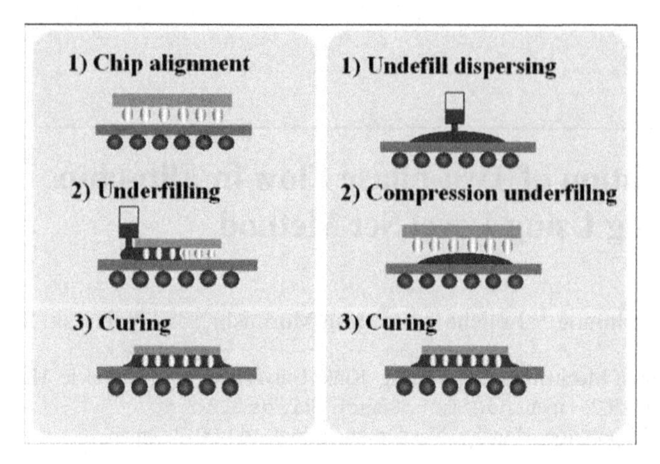

Fig. 1. The underfilling processes of conventional capillary flow type (left) and no-flow type (right)

the resin. We need to understand the flow behavior and filling time of underfill material for various solder bump patterns; solder bump diameter, bump pitch and gap height, taking into account the drag force acting on the solder bump. In the underfill flow analysis, the governing equations for three-dimensional incompressible flow are solved by using the finite difference method (FDM) incorporating the pseudo compressibility approach [7] on a non-uniform Cartesian grid. In the numerical method, a central difference scheme with artificial dissipation is used for the spatial discretization. The forward Euler explicit method is used as an iterative scheme in the pseudo time integration method. Our basic concept of numerical approach to the underfill encapsulation process can be found in [5, 4]. The level set method [8] is used as an interface capturing algorithm to represent the gas-liquid interface. The continuum surface force (CSF) model [1] is used for treating the surface tension. It is assumed that temperature distribution in the underfill material is uniform. The power-law model is adopted as a constitutive equation for treating the mould flow behavior of non-Newtonian fluid. The simulations especially in the no-flow type are carried out by coupling the Navier-Stokes equations and the equations of motion of IC chip.

For solving such a large scale problem, parallel computation is needed. In [2, 3], an efficient parallel computational algorithm has been studied for analyzing the compressible flow around a multi-element airfoil and incompressible flow around a moving circular cylinder using an overset grid method. In the implementation on a parallel computer, the domain decomposition method is used. The parallel code is applied to the underfill flow simulation.

In this paper, the model is presented for providing the underfill material, moving in a dispenser along the circumference of the IC chip.

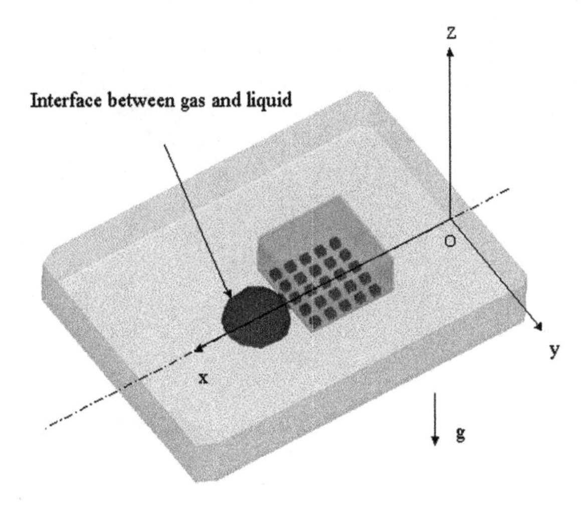

Fig. 2. The analytical model for the conventional capillary flow type

Table 1. Density, viscosity and surface tension

	density [kg/m^3]	viscosity [Pa · s]	surface tension [N/m]
gas	1.0	1.0×10^{-5}	–
liquid	1.0×10^2	1.5×10^{-2}	1.0×10^{-2}

1 Numerical Simulation of Capillary Flow Undefill

As one of the analytical models for the conventional capillary flow type, a semi-spherical liquid as shown in Fig. 2, or a sequence of spherical liquids dispensed on the substrate near the IC chip is driven by capillary action into the cavity with a gap height between IC chip and substrate. The solder bump in the gap is modeled as the rectangular cylinder. In this model, the surface tension is caused by prescribing the contact angle on the surface of IC chip and substrate. The effect of gravity force is included. The dimension of die is about 1.0mm $\times$ 1.0mm, and the gap height between IC chip and substrate is about 0.1mm. The array pattern of solder bump is 5×5. The bump diameter is 0.1mm and the bump pitch is 0.2mm. The number of grid points is about 37000. In the properties of gas and liquid, the density, viscosity and surface tension are shown in Table 1.

2 Numerical Results

In the case1, the solder bump between IC chip and substrate is neglected and the flow front profile and filling time are evaluated by changing the contact angle. The

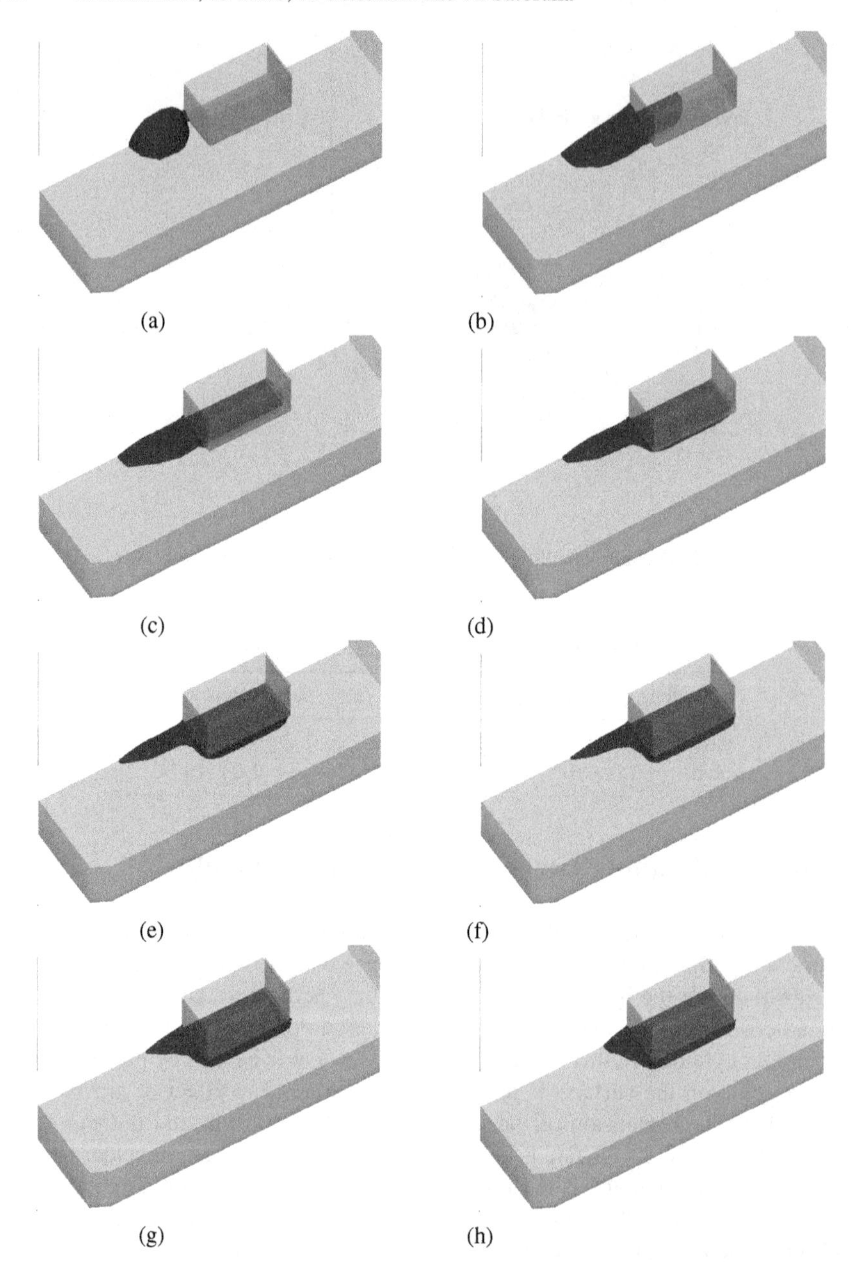

Fig. 3. The initial state of dispensing a semi-spherical liquid (a), the histories of propagating interface with increasing the time (b)-(g) and the final fillet shape (h) in the existence of no solder bumps (case1)

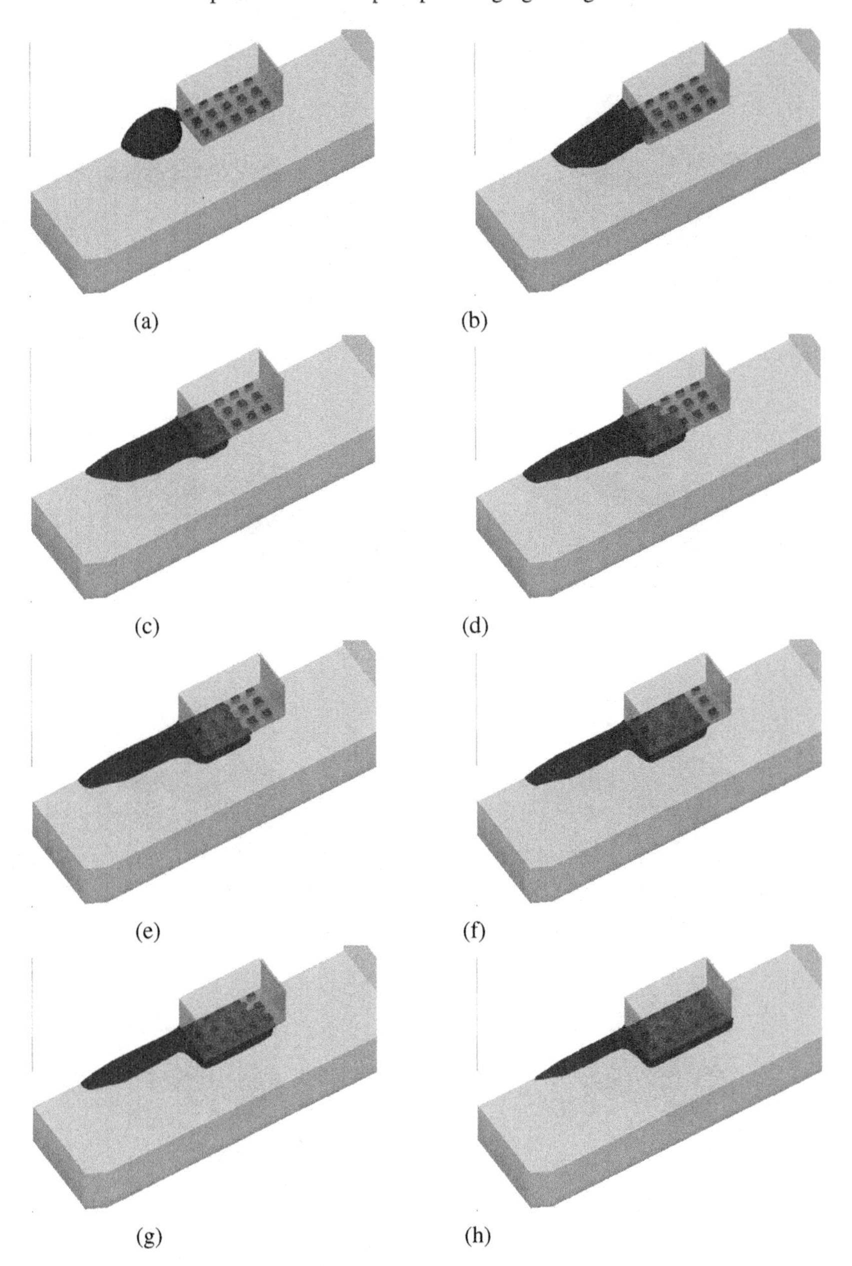

Fig. 4. The initial state of dispensing a semi-spherical liquid (a), the histories of propagating interface with increasing the time (b)-(g) and the final fillet shape (h) in the existence of solder bumps (case2)

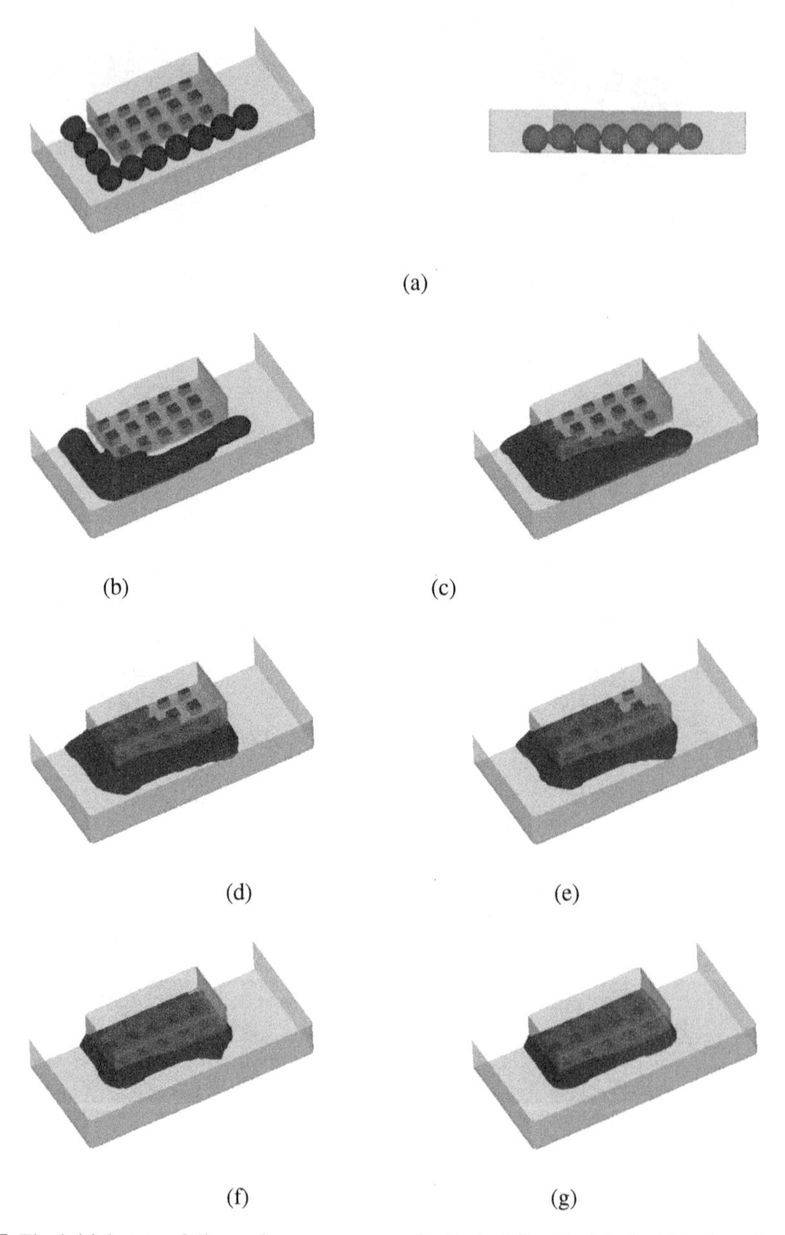

Fig. 5. The initial state of dispensing a sequence of spherical liquids (a), the histories of propagating interface with increasing the time (b)-(f) and the final fillet shape (g) in the existence of solder bumps (case3)

filling times of the contact angles of 30°, 45° and 60° are about 1.0s, 1.2s and 1.4s, respectively. The initial state, the histories of the propagating gas-liquid interface with increasing the time and the final fillet shape obtained for the contact angle 30° are shown in Fig. 3. It is observed that as the contact angle becomes larger, the filling time is longer due to decrease of the flow velocity, and that the final fillet shapes for the three contact angles are different, depending on the collapse of semi-spherical liquid and the spread especially in the plus direction of x-axis. It is found that the effect of capillary action on the flow behavior is one of the most important factors in predicting the filling time and the final fillet shape. It is also confirmed that the curve of flow front is a meniscus. For the next cases, the solder bump of rectangular cylinder between IC chip and substrate is considered, fixing the contact angle 30°. In the case2, the histories of the propagating gas-liquid interface with increasing the time and the final fillet are shown in Fig. 4. The filling time is about 3.5s. The spread of liquid in the plus direction of x-axis is larger than that of case1 due to the increase of flow resistance to the solder bumps. In the case3, the providing the underfill material, moving a dispenser along the circumference of the IC chip is modeled using a sequence of spherical liquids at the initial state as shown in Fig. 5(a). The center position and radius of the spherical liquids is determined by the flow rate of the underfill material at the gate of the dispenser moving at a speed. The height between center position of the spherical liquids and substrate increases for the time delay. The histories of the propagating interface with increasing the time and the final fillet shape are shown in Fig. 5(b)-(f) and (g), respectively. The filling is about 3.0s and the good final fillet shape is obtained.

3 Conclusions

In the model of providing the underfill material, moving in a dispenser along the circumference of the IC chip, it is considered that the use of a sequence of spherical liquids is very effective. In the future work, the numerical results should be confirmed by comparing with the available experimental data. The parallelization is essential for treating a large scale problem and reducing the computational time.

[1] J. U. Brackbill, D. B. Kothe, and C. Zemach. A continuum method for modeling surface tension. *J. Comput. Phys.*, 100:335–354, 1992.
[2] T. Hashimoto, K. Morinishi, and N. Satofuka. Parallel computation of multigrid method for overset grid. In *ICCFD3*, pages 167–174, 2002.
[3] T. Hashimoto, K. Morinishi, and N. Satofuka. Parallel computation of vortex-induced vibration of a circular cylinder using overset grid. In *Parallel Computational Fluid Dynamics*, pages 403–410, 2003.
[4] T. Hashimoto, K. Morinishi, and N. Satofuka. Numerical simulation for impact of elastic deformable body against rigid wall under fluid dynamic force. In *ICCFD3*, pages 375–380, 2004.
[5] T. Hashimoto, K. Morinishi, and N. Satofuka. Numerical simulation of unsteady flow around deformable elastic body under fluid dynamic force. In *ECCOMAS 2004*, number 364, pages 1–19, 2004.

[6] T. Hashimoto, S. Tanifuji, K. Morinishi, and N. Satofuka. Computers & fluids. *Numerical simulation of conventional capillary flow and no-flow underfill in flip-chip packaging*, 37(5):520–523, 2008.
[7] S. E. Rogers and D. Kwak. Upwind differencing scheme for the time-accurate incompressible navier-stokes equations. *AIAA Journal*, 28:253–262, 1990.
[8] M. Sussman, P. Smeraka, and S. Osher. A level set approach for computing solutions to incompressible two-phase flow. *J. Comput. Phys.*, 114:146–159, 1994.

A Parallel Immersed Boundary Method for Blood-like Suspension Flow Simulations

F. Pacull[1] and M. Garbey[2]

[1] Fluorem - Ecully, France - fpacull@fluorem.com
[2] Department of Computer Science - University of Houston - Houston, USA - garbey@cs.uh.edu

This paper presents a numerically efficient implementation of the Immersed Boundary Method (IBM), originally developed by [7] to simulate fluid/elastic-structure interactions. The fluid is assumed to be incompressible with uniform density, viscosity, while the immersed boundaries have fixed topologies with a linear elastic behavior. Based on the finite-difference method, a major numerical advantage of the IBM is the high level of uniformity of mesh and stencil, avoiding the critical interpolation processes of the cut-cell/direct methods. The difficulty of accurately simulating interaction phenomena involving moving complex geometries can be overcome by implementing large and parallel IBM computations on fine grids, as described in [1]. While this paper is restricted to a two-dimensional low-Reynolds-number flow, most of the concepts introduced here should apply to three-dimensional bio-flows. We describe here the decomposition techniques applied to the IBM, in order to decrease the computational time, in the context of the parallel Matlab toolbox of [3]. Finally, we apply the method to a blood-like suspension flow test-case.

1 The Fluid/elastic-structure interaction equations

1.1 The Context

Blood is a suspension flow: its primary purpose is transporting cells. The proportion of blood volume occupied by red blood cells is normally between forty and fifty percent. This implies the implementation of multi-scale models, taking into account these numerous small immersed bodies. However, we restrict ourselves only to a small-scale domain and present here a parallel simulation of a cavity flow with a significant volume of immersed elastic bodies. For simplicity reasons, further necessary reductions are also made on the fluid, assumed to be Newtonian, incompressible and inert, while blood is non-Newtonian, slightly compressible and with highly complex chemical properties.

As discussed by [8], various techniques have been proposed in the literature to treat moving boundary problems. The IBM, originally developed by C.S. Peskin,

D. Tromeur-Dervout (eds.), *Parallel Computational Fluid Dynamics 2008*,
Lecture Notes in Computational Science and Engineering 74,
DOI: 10.1007/978-3-642-14438-7_16,

combines respectively Eulerian and Lagrangian descriptions of flow and moving elastic boundaries, using discrete Dirac delta functions as integration kernels to switch from one description to the other. The incompressible Navier-Stokes and elasticity theories can be unified by the same set of equations to get a combined model of the interaction. There are numerous applications of the IBM in Bio-Engineering or in more general Computational Fluid Dynamics applications. We give, in the following subsection, a brief description of the method.

1.2 The Temporal discretization of the IBM

A complete and accurate introduction to the IBM can be found in [7]. We start with the incompressible Navier-Stokes equations of momentum and mass conservation:

$$\rho\left[\frac{\partial V}{\partial t}+(V.\nabla)V\right] = -\nabla P+\mu\Delta V+F \tag{1}$$

$$\nabla.V = 0 \tag{2}$$

The primitive variables are V and P, respectively the velocity and pressure of the fluid, which physical parameters are the uniform viscosity μ and the uniform density ρ. The fluid domain $\Omega \in \mathbb{R}^d$ $(d=2,3)$ is described by the Cartesian coordinate vector x. $\Gamma \subset \Omega$ is the immersed elastic boundary, which curvilinear dimension is m $(m<d)$. X is the Lagrangian position vector of Γ, expressed in the d-dimensional Cartesian referential. The Lagrangian vector f is the local elastic force density along Γ, also expressed in the Cartesian referential. The IBM requires the extrapolation of the Lagrangian vector f into the Eulerian vector field F, which is then plugged into Eq. (1). A distribution of Dirac delta functions δ is used for that purpose:

$$F(x,t)=\int_{\Gamma} f(s,t)\delta\left(x-X(s,t)\right)ds=\begin{cases} f(s,t) & \text{if } x=X(s,t) \\ 0 & \text{otherwise}\end{cases} \tag{3}$$

The motion of the immersed boundary should match the motion of the neighboring fluid particles thanks to a no-slip boundary condition. Eq. (4) approximates this latter condition using the Dirac delta function as an interpolating tool for V, from Ω to Γ:

$$\frac{\partial X(s,t)}{\partial t}=\int_{\Omega} V(x,t)\delta(x-X(s,t))dx=\begin{cases} V(X(s,t),t) & \text{if } x=X(s,t) \\ 0 & \text{otherwise}\end{cases} \tag{4}$$

The immersed boundary obeys Hooke's law of elasticity, i.e., the tension $\mathscr{T}$ of the immersed boundary is a linear function of the strain. For a one-dimensional boundary, we have:

$$\mathscr{T}(s,t)=\sigma\left|\frac{\partial X(s,t)}{\partial s}\right| \tag{5}$$

where σ is the boundary elasticity coefficient. The local elastic force density f is defined as:

$$f(s,t) = \frac{\partial\left(\mathscr{T}(s,t)\tau(s,t)\right)}{\partial s}, \quad \tau(s,t) = \frac{\partial X(s,t)/\partial s}{|\partial X(s,t)/\partial s|} \tag{6}$$

τ is the unit tangent vector to Γ. Finally, by plugging Eq. (5) into the set of Eqs.(6), we get:

$$f(s,t) = \sigma \frac{\partial^2 X(s,t)}{\partial s^2} \tag{7}$$

The practical implementation of the IBM of Peskin offers dozens of different possibilities regarding the choice of the temporal scheme, the space discretization, the discrete approximation of the Dirac function and so on. Overall, there is clearly a compromise between the stability of the scheme that suffers from sharp sources terms in the equations, for the pressure to be discontinuous, and accuracy that needs this numerical feature. We refer to the thesis of the first author [5] and its bibliography for an extensive comparison of possible implementations against standard benchmark problems.

For the temporal discretization of the fluid equations, we use the following fractional-step method (H is the convective term):

$$\rho\left[\frac{V^* - V^n}{\Delta t}\right] = -\nabla P^{n-\frac{1}{2}} + \frac{\mu}{2}\Delta\left(3V^n - V^{n-1}\right) - \frac{\rho}{2}\left[3H(V^n) - H(V^{n-1})\right] + \frac{3}{2}F^n - \frac{1}{2}F^{n-1} \tag{8}$$

$$\Delta\Pi = \frac{\rho}{\Delta t}\nabla . V^* \tag{9}$$

$$\rho\left[\frac{V^{n+1} - V^*}{\Delta t}\right] = -\nabla\Pi \tag{10}$$

$$P^{n+\frac{1}{2}} = P^{n-\frac{1}{2}} + \Pi \tag{11}$$

This is a second-order scheme, with only one system to solve per time step, Eq. (9). Similarly, the elastic body motion is discretized using the midpoint rule for the time derivative, the trapezoidal rule and an Adams-Bashforth scheme respectively for the fluid velocity and the boundary position extrapolated at time step $t^{n+\frac{1}{2}}$:

$$\frac{X^{n+1} - X^n}{\Delta t} = \int_\Omega \left(\frac{V^{n+1} + V^n}{2}\right)\delta\left(x - \frac{3}{2}X^n + \frac{1}{2}X^{n-1}\right)dx \tag{12}$$

Finally, the local elastic force density f^{n+1} is computed as a function of X^{n+1} and extrapolated onto Ω:

$$F^{n+1} = \int_\Gamma f^{n+1}\delta\left(x - X^{n+1}\right)ds \tag{13}$$

Spatially, we use a staggered grid with second-order discretization. The widths of the divergence operator and Laplace operator stencils are equal respectively to one space step and two space steps, leading to less smear out for the numerical approximation of the pressure front at the immersed boundary location. The discrete Dirac delta function used in this study is the traditional 2D function δ_h that has a $4h$ support (h being the uniform space step):

$$\delta_h(x_1,x_2) = \frac{1}{h}\phi\left(\frac{x_1}{h}\right)\frac{1}{h}\phi\left(\frac{x_2}{h}\right) \tag{14}$$

with:

$$\phi(r) = \begin{cases} \frac{1}{8}\left(3-2|r|+\sqrt{1+4|r|-4r^2}\right) & \text{if } |r| \leq 1 \\ \frac{1}{8}\left(5-2|r|-\sqrt{-7+12|r|-4r^2}\right) & \text{if } 1 < |r| \leq 2 \\ 0 \text{ if } |r| > 2 \end{cases} \tag{15}$$

2 The Parallel flow solver

Most of the computational cost of the IBM at each time-step corresponds to the pressure equation of the projection scheme Eq. (9). In the case of a closed driven cavity, this is a Poisson equation with homogeneous Neumann boundary conditions. Consequently, we used the analytic additive Aitken-Schwarz algorithm of [2], which is an excellent candidate to allow efficient distributed computing with high latency networks. We also refer to [1] for more details on the method. The Matlab parallel toolbox used to implement the solver is MatlabMPI, a set of Matlab scripts that implements a subset of MPI and allow any Matlab program to be run on a parallel computer. We now describe successively the pressure and IBM solvers.

2.1 The Pressure solver

The rectangular uniform mesh is decomposed into an unidirectional partition of overlapping strip domains. The method is a post-process of the standard Schwarz method with an Aitken-like acceleration of the sequences of interfaces produced with the block-wise Schwarz relaxation. We use a Fourier expansion to describe the solution on the interface and initially compute analytically each damping factor for each wave number. Thus, the Aitken acceleration for each wave number can be performed independently. The algorithm of this exact solver is summarized as follows:

- step 0: compute analytically each damping factor for each wave number
- step 1: solve the one-dimensional system corresponding to the zeroth Fourier mode and subtract this mode from the right-hand side (because of the homogeneous Neumann boundary conditions)
- step 2: perform one additive Schwarz iterate
- step 3: apply the generalized Aitken acceleration on the interfaces

  - 3.1: compute the expansion of the traces on the artificial interfaces from the Schwarz iterate
  - 3.2: apply the generalized Aitken acceleration separately to each wave coefficient, in order to get the exact solution on the artificial interfaces, expressed in the Fourier basis
  - 3.3: transfer back the interface values into the physical space
- step 4: compute the solution for each subdomain in parallel
- step 5: add the field corresponding to the zeroth Fourier mode

2.2 The IBM solver

The main steps of the parallel flow solver are:

- step 1: exchange interface data (from V^n and $P^{n-\frac{1}{2}}$) in order to compute locally V^*
- step 2: exchange interface data (from V^*) in order to compute locally the right-hand side of the pressure equation
- step 3: solve the pressure equation with the Aitken-Schwarz pressure solver described above
- step 4: compute V^{n+1} and $P^{n+\frac{1}{2}}$

Fig. 1 shows the speedup of the flow solver with respect to the number of processors and for three different domain sizes with our parallel implementation of the complete Navier-Stokes code. Performance speedup is based on the reference time provided by the code running with two subdomains on two processors. This speedup is therefore significantly better than what one obtains by comparing our parallel code with its sequential version. The speedup of the parallel code with two processors compared to the sequential code running on one processor is only 1.56. The overhead in the sequential code comes partially from the nature of the algorithm itself, which requires two subdomain solves, and partly from the fact that the parallel code has many more lines of code to be interpreted by Matlab. Another aspect of the parallelization is that we gain computational time in the decomposition of the operator process: this time is divided by the square of the number of subdomains since the complexity of the decomposition is proportional to the bandwidth of the subdomains. While the speedup obtained in Fig. 1 is not optimal, the MatlabMPI implementation still seems very attractive and fitted for solving relatively large size problems.

For the boundary treatment, the difficulty is that the discrete Dirac delta function has a support larger than the overlap. Each moving boundary point has a zone of influence corresponding to its support, which can be spread across two contiguous sub-domains, but which should only be taken into account once in the fluid/structure interaction. For example, in the no-slip boundary condition Eq. (4), we split the summation operator between the n_{sub} non-overlapping sub-domains $\{\Omega_i\}_{1\leq i\leq n_{sub}}$:

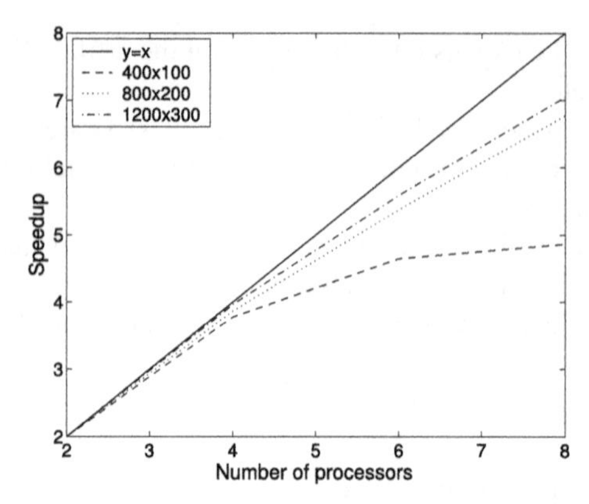

Fig. 1. Speedup of the flow solver with respect to the number of processors and for three different domain sizes

$$\frac{\partial X(s,t)}{\partial t} = \int_{\Omega} V(x,t)\delta(x - X(s,t))dx$$
$$= \sum_{i=1}^{n_{sub}} \int_{\Omega_i} V(x,t)\delta(x - X(s,t))dx \qquad (16)$$

This means that one process needs to know the position the boundary points located inside but also in the neighborhood of its subdomains Ω_i. The width of this neighborhood is depending on the width of the δ support. This is also true for the local force density extrapolation Eq. (3).

3 The Test-case

The geometry is a square cavity $\Omega = [0,1]^2$ in which the upper wall is sliding and with a fixed cylinder of radius 0.1 and center $(0.5, 0.75)$ implemented using the direct forcing method of [4]. The physical parameters are $\mu = \rho = 1$, $\sigma = 100$. The upper wall velocity is $u(x,1) = 40x(1-x)$. We have included 35 bubbles of radius 0.05, which represent around 31% of the fluid domain. Initially, the fluid is at rest, then the moving bodies are pushed between the cylinder and the sliding wall, as shown on Fig. 2. To prevent the moving bubbles from sticking to the cylinder, which may happen when the immersed boundary force terms of both kinds overlap, we found that the fixed boundary force term in the momentum equation should be slightly spread on, in our case on the nodes neighboring the cylindric obstacle. Contacts generate numerical perturbations and thus require a minimum level of discratization accuracy. In order to better visualize the dynamic of the simulation in Fig. 2, we added a cross inside each bubble, these crosses are not actual elastic boundaries but visual markers.

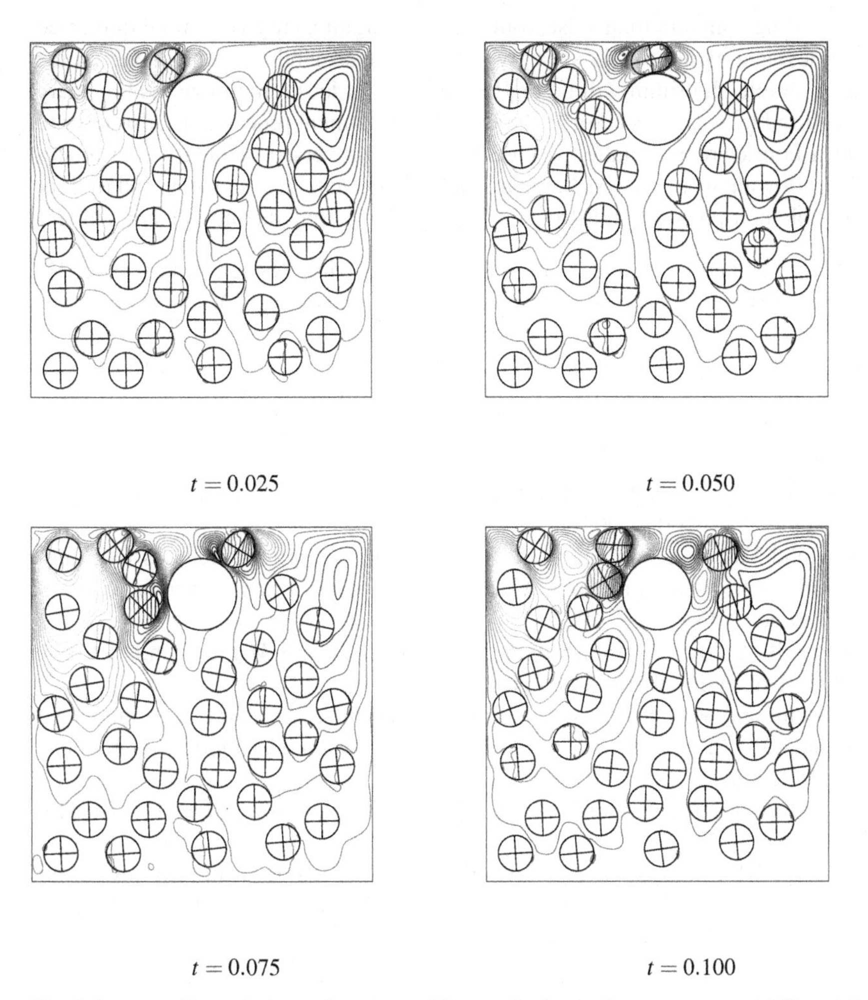

Fig. 2. Immersed boundaries and contour of the verctical velocity component at different times

4 Conclusion

We presented an efficient implementation of the IBM, based on the Aitken-Schwarz algorithm. Although strong model reductions have to be made toward blood flow simulations, it is of particular interest that these techniques, which are relatively simple to implement, give access to such complex fluid flow problems, taking full advantage of parallel computers. Regarding our particular test-case, we noted that the domain partition needs to be strictly non-overlapping for the boundary treatment. The total number of immersed boundaries being reasonable in the present case, each process could manage to gather all the Lagrangian boundary data at each time step, which are the curvilinear position and force density vectors X and s. However, the corresponding computational cost would increase significantly with more immersed boundaries, and would finally strongly penalize the scalability of the solver. In the future, a simple algorithm will be implemented, in order to minimize the boundary data required by each processor in the boundary treatment process. We also refer to [6] for a Fourier representation of the discretized immersed boundary coordinates vectors, allowing us to use less boundary points, thus gaining efficiency, and to filter the possible non-physical high-frequency oscillations along the moving boundaries.

[1] M. Garbey and F. Pacull. A Versatile incompressible Navier-Stokes solver for blood flow application. *International Journal for Numerical Methods in Fluids*, 54(5):473–496, 2007.

[2] M. Garbey and D. Tromeur-Dervout. Aitken-Schwarz method on Cartesian grids. In N. Debit, M. Garbey, R. Hoppe, J. Périaux, and D. Keyes, editors, *Proc. Int. Conf. on Domain Decomposition Methods DD13*, pages 53–65. CIMNE, 2002.

[3] J. Kepner and S. Ahalt. MatlabMPI. *Journal of parallel and distributed computing*, 64(8):997–1005, 2004.

[4] J. Mohd-Yusof. Combined immersed boundaries/b-splines methods for simulations of flows in complex geometries. *Ctr annual research briefs, Stanford University, NASA Ames*, 1997.

[5] F. Pacull. *A Numerical Study of the Immersed Boundary Method*. Ph.D. Dissertation, University of Houston, 2006.

[6] F. Pacull and M. Garbey. A numerically efficient scheme for elastic immersed boundaries. In *(to appear) Proc. Int. Conf. on Domain Decomposition Methods DD18*, 2008.

[7] C. S. Peskin. The Immersed boundary method. *Acta Numerica*, 11:479–517, 2002.

[8] W. Shyy, M. Francois, H.-S. Udaykumar, N. N'dri, and R. Tran-Son-Tay. Moving boundaries in micro-scale biofluid dynamics. *Applied Mechanics Reviews*, 54(5):405–454, September 2001.

Part V

High Order methods

3D Spectral Parallel Multi-Domain computing for natural convection flows

S. Xin[1], J. Chergui[2], and P. Le Quéré[2]

[1] Cethil, UMR 5008, INSA-Lyon, UCBL, 9, Rue de la Physique, 69621 Villeurbanne, France
[2] L.I.M.S.I.-C.N.R.S., BP 133, 91403 Orsay Cedex, France

1 Introduction

Natural convection in cavities is a well-known configuration of CFD community. Recently there is increasing interest in numerical study of 3D natural convection at high Rayleigh number [3, 5, 7, 8] and the corresponding direct numerical simulations of transitional flow regime are approaching the limiting case of single-processor computing even on a computer of NEC-SX8. Parallel computing is therefore needed for studying higher Rayleigh number and turbulent flow regime.

Spectral Chebyshev collocation method [1, 2] has been widely used for investigating natural convection in cavities and suits especially the flow regime of separated boundary layers because of the special distribution of the collocation points (Gauss points). But it suits less the stretched geometries and the corresponding multi-cellular flow structures [10]. Even in a square cavity when the instabilities of the vertical boundary layer move more and more upstream with increasing Rayleigh number, the distribution of Gauss points is less suitable: for example, a grid transformation technique has been used for a 2D square cavity at a Rayleigh number of 10^{10} [6, 9]. In these cases, multi-domain or domain decomposition can be applied in order to overcome the drawbacks of Gauss points. In the sense of parallel computing, parallel multi-domain is needed to overcome the drawbacks related to the collocation points.

The present work is thus realized under the two-fold motivations. Natural convection in a cubic differentially heated air-filled cavity is chosen as an example to illustrate the motivations and numerical results of parallel computing are presented.

2 Basic ideas

Multi-domain is first associated with Schur complement or influence matrix technique in order to handle the problem on the interfaces between the subdomains. Take a 1D Helmholtz equation $\left(\frac{d^2}{dx^2} - \lambda\right) f = s$ for example, the initial domain is divided

D. Tromeur-Dervout (eds.), *Parallel Computational Fluid Dynamics 2008*,
Lecture Notes in Computational Science and Engineering 74,
DOI: 10.1007/978-3-642-14438-7_17, © Springer-Verlag Berlin Heidelberg 2010

into subdomains and both f and its first derivative should be continuous on the interfaces. Spectral collocation method leads to a block-structured global matrix which are coupled through the interface conditions. To take advantage of the block structure, it is usual to impose Dirichlet conditions (continuous f) on the interfaces and forget the condition of continuous df/dx: different blocks can be solved independently on different processors. The condition of continuous df/dx can be recovered by using the Schur complement which is constructed and inverted in a preliminary stage as there exists a linear relationship between the function and its first derivative on the interfaces. The well-known two iteration procedure is the following: in the first iteration the independent blocks are solved with guessed interface values to obtain the guessed subdomain solutions, one calculates on the interfaces the df/dx differences of the guessed subdomain solutions and the corrections to the guessed interface values by using the inverse of Schur complement; in the second iteration the independent blocks are solved with the corrected interface values and the final subdomain solutions satisfy the conditions of continuous f and df/dx on the interfaces.

For parallel computing, one independent block is given to one processor and one processor has either a guessed or a final solution of one subdomain. Communications between processors are needed to construct Schur complement and calculate the df/dx differences on the interfaces: for example, one processor gives the df/dx value of the guessed solution on the left interface to its left neighboring processor and this is a MPI_Sendrecv communication. In this way, one processor has either only one column of the Schur complement or only one value of df/dx difference (except for the last one which may have nothing if the model problem is not periodic). Although this is not an interesting case for one 1D problem, it becomes interesting for a set of 1D problems: the first processor has the first column of all the Schur complements (or all the df/dx differences on the first interface), the second processor has the second column of all the Schur complements (or all the df/dx differences on the second interface), etc. In this particular situation, an MPI_Alltoall communication will allow each processor to possess a certain number of Schur complements or the df/dx differences on all the interfaces for a certain number of 1D problems. Concerning the interface problems, one processor will invert a certain number of Schur complements in order to obtain the corrections to the guessed interface values. Another MPI_Alltoall communication will allow each processor to possess the corrections to the guessed values on the right interface. Each processor should give then the corrections to its right neighboring processor so that the final subdomain solution can be obtained using the corrected interface values.

It is important to note that for the 1D problems the corresponding Schur complements are tridiagonal. Therefore in practice the Schur complements are not inverted directly because the TDMA (Tri-Diagonal Matrix Algorithm) is more suitable and cost effective.

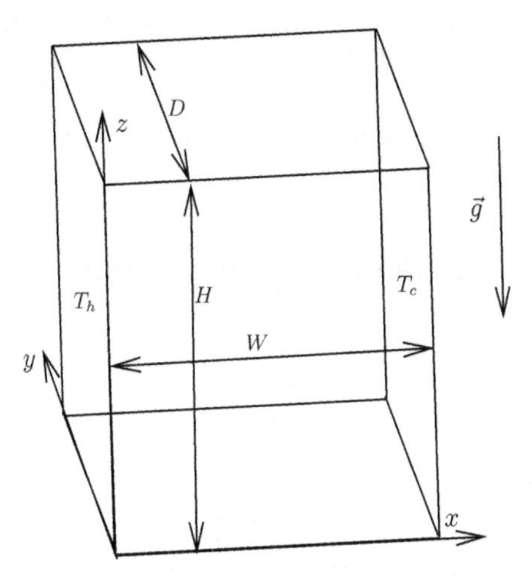

Fig. 1. Configuration of a differentially heated cavity. The vertical walls at $x = 0$ and W are submitted to a temperature difference $\Delta T = T_h - T_c$ and other walls are adiabatic.

3 Physical problem and mathematical equations

In order to apply the above ideas, we restricted ourselves to the cases of 1D multi-domain, *i.e.* the 3D computational domain is decomposed only in one of the three spatial directions, and homogeneous condition type on the same boundary.

We consider natural convection in a cubic differentially heated air-filled cavity (Figure 1): no slip condition is applied on the cavity walls, two faced vertical walls are submitted to a temperature difference $\Delta T = T_h - T_c$ and other walls are adiabatic. When using cavity height H (also width and depth) as reference length, $\kappa Ra^{1/2}/H$ as reference velocity and reduced temperature $\Theta = (T - T_0)/(T_h - T_c)$ where $Ra = (g\beta \Delta T H^3)/(\nu\kappa)$ is Rayleigh number and $T_0 = (T_c + T_f)/2$, the governing incompressible Navier-Stokes equations under the Boussinesq assumption are written in dimensionless form as follows:

$$
\begin{aligned}
&\frac{\partial u}{\partial x} + \frac{\partial v}{\partial y} + \frac{\partial w}{\partial z} = 0 \\
&\frac{\partial u}{\partial t} + V.\nabla u = -\frac{\partial p}{\partial x} + \frac{Pr}{Ra^{1/2}}\nabla^2 u \\
&\frac{\partial v}{\partial t} + V.\nabla v = -\frac{\partial p}{\partial y} + \frac{Pr}{Ra^{1/2}}\nabla^2 v \\
&\frac{\partial w}{\partial t} + V.\nabla w = -\frac{\partial p}{\partial z} + \frac{Pr}{Ra^{1/2}}\nabla^2 w + Pr\Theta \\
&\frac{\partial \Theta}{\partial t} + V.\nabla \Theta = \frac{1}{Ra^{1/2}}\nabla^2 \Theta
\end{aligned}
\tag{1}
$$

where $(x,y,z) \in [0,1] \times [0,1] \times [0,1]$, $V = (u,v,w)$, $V.\nabla = u\frac{\partial}{\partial x} + v\frac{\partial}{\partial y} + w\frac{\partial}{\partial z}$, $\nabla^2 = \frac{\partial^2}{\partial x^2} + \frac{\partial^2}{\partial y^2} + \frac{\partial^2}{\partial z^2}$ and $Pr = \nu/\kappa$ is Prandtl number.

4 Numerical schemes and parallel algorithm

Equations (1) are discretized in time by the well-known second-order Euler-backward Adams-Bashforth scheme. This leads essentially to Helmholtz equations apart from the velocity-pressure coupling. When using any variant of projection method [4], one obtains a pseudo-Laplacian operator for pressure and has to solve a pseudo-Poisson equation (a special Helmholtz equation). In this way, Equations (1) are reduced to Helmholtz equations after the time discretization. The parallel algorithm is thus illustrated through a model Helmholtz equation.

Given f, one of u, v, w, p and Θ, and $(\nabla^2 - \lambda)f = s$. The domain in one of x, y and z directions can be divided into subdomains and for convenience x direction is chosen for illustration. Each subdomain is discretized by spectral methods and the continuity of f and its first derivative in x is required on the interfaces. Due to the assumption of homogeneous condition type on the same boundary, all the subdoamins have the same continuous and discret operators in both y and z directions. For the model equation, it is the operators of the second-derivatives modified by the corresponding boundary conditions. In practice, these discret operators are full matrices and they are diagonalizable and diagonalized when using spectral collocation methods to solve the model equation. In the eigen-spaces of the discret operators, the model equation is written as follows: $\left(\frac{\partial^2}{\partial x^2} - \lambda + \lambda_j + \lambda_k\right) f_{jk} = s_{jk}$ where λ_j and λ_k are respectively the eigen-values of the discret operators in y and z direction. One thus has a set of 1D problems and the basic ideas described above can be applied to the model equation and to the equations (1). For each variable, instead of a Schur complement in the physical space, one has a set of small-block Schur complements in the eigen-spaces: the Schur complement in the physical space is diagonalized in the eigen-spaces into tridiagonal blocks. The tridiagonal blocks are divided into groups, distributed each group to one processor through MPI_Alltoall communications and solved by the well-known TDMA.

It is important to note that when spectral collocation methods are used to solve a 3D Helmholtz equation the resolution method is usually the total diagonalisation. Diagonalising the Schur complement into blocks in the eigen-spaces is the natural and convenient way to do it. Although the MPI_Alltoall and MPI_Sendrecv communications used to construct and distribute the small-block Schur complements may be expensive, in the context of time-stepping or time-marching, the construction and inversion of small-block Schur complements are done only once and they represent only a preprocessing step before the time loop. In fact, for each variable, one time-step needs one neighbor-to-neighbor (MPI_Sendrecv) communication of $\frac{\partial f}{\partial x}$

on the interfaces (in order to obtain the $\frac{\partial f}{\partial x}$ differences), one MPI_Alltoall communication of $\frac{\partial f}{\partial x}$ differences on the interfaces, one MPI_Alltoall communication of the corrections of f on the interfaces (after each processor solves the corresponding small-block Schur complement problems) and one final neighbor-to-neighbor (MPI_Sendrecv) communication of the f corrections. As can be seen later, the present parallel algorithm is very efficient and cost effective.

5 Code behavior and numerical results

A numerical code in F90 has been developed under MPI (Message Passing Interface). In order to understand the parallel performance of the new code, parallel computation with increasing subdomains (or processors) for different spatial resolutions has been performed on a small cluster of Opteron 250 with Gigabit switch network. The tests concern natural convection in differentially heated cavities: the domain in z direction has been decomposed into subdomains, the spatial resolution in x and y directions is kept the same, *i.e.* 50×50 and three spatial resolutions in z direction, 20, 30 and 50, have been tested. The Walltime per processor (subdomain) shown in Figure 2 remains approximately constant. This behavior is confirmed by computations performed with more processors on the IBM Power4 and Power4+ of IDRIS: unfortunately it was not possible to choose the type of nodes and the two types of nodes used prevented us from providing similar curves as in Figure 2. Nevertheless, it is to note that different tests performed with 30 and 60 processors showed that the communication time is less than 15 percent of the total Walltime.

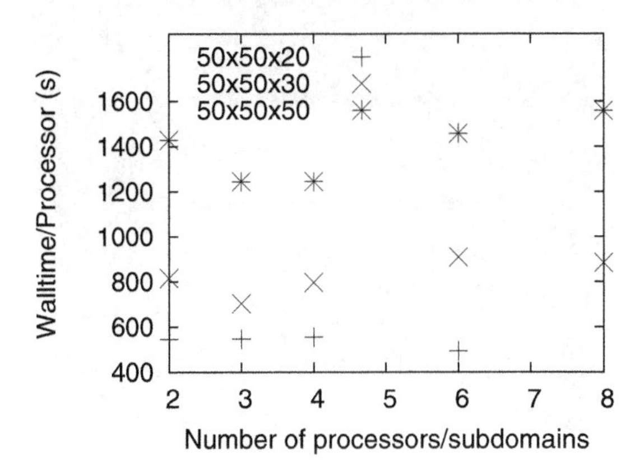

Fig. 2. Walltime per subdomain (or processor) for different spatial resolution. The domain in z direction is divided into subdomains.

The parallel code has also been used for benchmarking: numerical simulations have been done for a cubic differentially heated cavity at $Ra = 10^6$ [5]. The number of subdomains ranges from 2 to 6 and the spatial direction in which the domain is decomposed into subdomains is either x or y direction. The benchmark solutions obtained with 3 subdomains (or processors) are listed in Table 1: reasonable agreement is observed. Figure 3 displays the corresponding isotherms obtained at $Ra = 10^6$ with 4 subdomains (or processors).

NP=3	DD=2	DD=1	Reference [5]
Nu_0	8.6401	8.6399	8.6407
u_{max}	.12634	.12695	.12697
v_{max}	.02544	.02550	.02556
w_{max}	.23412	.23032	.23672

Table 1. Benchmark solutions at $Ra = 10^6$ obtained by parallel computations with 3 subdomains (or processors) (NP=3). DD=2 means that the y direction is decomposed and the corresponding subdomain resolution is $50 \times 20 \times 50$. DD=1 means that the x direction is decomposed and the corresponding subdomain resolution is $20 \times 50 \times 50$. Nu_0 is averaged Nusselt number on the hot wall ($x = 0$), u_{max}, v_{max} and w_{max} are respectively the maxima of velocity components.

Fig. 3. Isotherms at $Ra = 10^6$. The domain in y direction is divided into 4 subdomains.

Previous studies [5, 3] showed that the onset of time-dependent flows occurs at approximately $Ra = 3.2 \times 10^7$ in a cubic differentially heated air-filled cavity. The instable modes break the reflection symmetry in y direction, their spatial distribution is mainly located about $y = 0.5$. Parallel computations have been realized for $Ra = 3.5 \times 10^7$ using 4 subdomains (or processors), the decomposed direction is y direction and a subdomain resolution of $80 \times 20 \times 80$. As the first two instable modes have almost the same critical Rayleigh number, symmetry constraining has been applied: either the 2D centro-symmetry in (x,z) or the full 3D centro-symmetry has been imposed. Figure 4 depicts the numerical results of x component of vorticity field: instable structures about $y = 0.5$ are clearly shown and the results are similar to those presented in [3]

Fig. 4. Instantaneous x component of vorticity obtained at $Ra = 3.5 \times 10^6$. The front is the cold wall at $x = 1$. On the left, the 2D centro-symmetry in (x,z) is imposed while on the right the full 3D centro-symmetry has been forced.

6 Summary

A 3D spectral parallel multi-domain code has been developed using Schur complement for interface problems and validated with known results of natural convection in a differentially heated cavity in the literature.

It solves the unsteady incompressible Navier-Stokes equations and is based on parallel algorithm for solving Helmholtz equations. In fact, any time scheme reduces the N-S equations to Helmholtz equations and any variant of projection method leads to a pseudo-Poisson equation for pressure (a particular Helmholtz equation).

In order to have efficient parallel algorithm, Schur complement of the interface problems is not treated in physical space but in the eigen-spaces: it is diagonalized into small diagonal blocks and this reduces efficiently Walltime related to the interface problems. Tests performed on IBM Power4 and Power4+ show that the communication time remains less than 15 percent of the total Walltime even when 60 processors are used. The developed parallel code has good scalability and will be used for doing DNS of 3D turbulent natural convection flows in cavities.

Acknowledgement: The present work has been conducted under LIMSI research project ASP. Computations have been performed at IDRIS under research project 70326.

[1] C. Bernardi and Y. Maday. Approximations spectrales de problèmes aux limites elliptiques. In Collection Mathématiques & Applications. Springer Verlag, Berlin, 1992.

[2] C. Canuto, M.Y. Hussaini, A. Quarteroni, and T.A. Zang. Spectral methods in fluid dynamics. Springer Verlag, New York, 1988.

[3] G. de Gassowski, S. Xin, and O. Daube. Bifurcations et solutions multiples en cavité 3D différentiellement chauffée. C. R. Acad. Sci., Ser. II b, 331:705–711, 2003.

[4] J.-L. Guermond and L. Quartapelle. On the approximation of the unsteady Navier-Stokes equations by finite element projection methods. Numer. Math., 80(5):207–238, 1998.

[5] G. Labrosse, E. Tric, H. Khallouf, and M. Betrouni. A direct (peudo-spectral) solver of the 2D/3D Stokes problem : Transition to unsteadiness of natural-convection flow in a differentially heated cubical cavity. Num. Heat Trans. B, 31:261–276, 1997.

[6] P. Le Quéré. A modified Chebyshev collocation algorithm for directsimulation of 2D turbulent convection in differentially heated cavities. Finite Element in Ana. & Design, 16:271–283, 1994.

[7] S.H. Peng and L. Davidson. Large eddy simulation for turbulent buoyant flow in a confined cavity. Int. J. Heat & Fluid Flow, 22:323–331, 2001.

[8] J. Salat, S. Xin, P. Joubert, A. Sergent, F. Penot, and P. Le Quéré. Experimental and numerical investigation of turbulent natural convection in a large air-filled cavity. Int. J. Heat & Fluid Flow, 25(5):824–932, 2004.

[9] S. Xin and P. Le Quéré. Numerical simulations of two-dimensional turbulent natural convection in differentially heated cavities of aspect ratios 1 and 4. In Eurotherm ERCOFTAC Workshop on Direct and Large Eddy Simulation, pages 423–434, Univ. of Surrey, UK, 1994.
[10] S. Xin, X. Nicolas, and P. Le Quéré. Stability analyses of longitudinal rolls of Poiseuille Rayleigh-Bénard flows in air-filled channels of finite transversal extension. Num. Heat Trans. A, 50(5):467–490, 2006.

3D time accurate CFD simulations of a centrifugal compressor

Andreas Lucius[1] and Gunther Brenner[1]

Institute of Applied Mechanics, Department of Fluid Mechanics,
Clausthal University of Technology, Adolph-Roemer Str. 2a,
38678 Clausthal Zellerfeld, Germany

andreas.lucius@tu-clausthal.de
gunther.brenner@tu-clausthal.de

In this paper results of unsteady CFD computations of a centrifugal compressor are presented and compared to steady state solutions. Two cases are examinated, the first case is a rotor with spiral casing close to the design point. This simulations shows the unsteady rotor stator interaction. In the second case the same rotor followed by a vaneless diffuser without spiral casing is simulated at part load. The aim is to show the unsteady rotating stall effect without disturbance coming from the geometry. A frequency analysis for both cases is conducted. As expected the blade passing frequency and its multiples are dominating in the calculation with casing. For the rotating stall typical unsteady vortices are observed. The unsteady pressure fluctuations are a magnitude higher in comparison to the calculation at design point.

1 Introduction

CFD simulations are today state of the art in the design and optimization process for turbomachinery applications. Progress in development of numerical techniques and the increasing computational power make it possible to run 3D Navier Stokes codes on small workstations. For standard problems in the turbomachinery design process - like optimization in terms of effciency for the design point - this is sufficient. For a typical calculation only a part of the impeller e.g. the flow around one or two blades is modelled, with periodic boundary conditions applied in the circumferential direction. This approach is valid only for axially symmetric geometries. For this reason it is not applicable, if a volute is used as a collector for the fluid. A complete analysis of machine performance is only possible, if the full 360 Rotor coupled with the stator geometry is considered. In addition to the need of including the whole machine with rotor and stator domains, transient effects appear in the flow. This is especially the case for operation in off-design conditions. In such an operational point the flow

D. Tromeur-Dervout (eds.), *Parallel Computational Fluid Dynamics 2008*,
Lecture Notes in Computational Science and Engineering 74,
DOI: 10.1007/978-3-642-14438-7_18, © Springer-Verlag Berlin Heidelberg 2010

conditions do not match the volute design resulting in an asymmetric pressure distribution. For this reason the rotating blades are loaded with differing back pressure during one rotation, resulting in a strongly unsteady flow in the impeller itself. Thus a more detailed transient analysis for off-design operation is needed, which requires substantially more computational ressources.
In the past several simplified approaches have been attempted to account for this effect. Hillewaert has simulated the transient rotor stator interaction with an inviscid unsteady flow solver [5]. Hagelstein et al. [2] used an unsteady Euler code using a correction term accounting for viscous effects. Their simulation showed in general a good agreement with measurements, but at low mass flow seperation was not predicted. In the work of Treutz [7] a transient Navier Stokes solver was applied to simulate the unsteady effects in a water pump. He pointed out the differences between steady state solutions and time accurate results. In his work the whole characteristic curve was calculated in one simulation via reduction of massflow in time. Allthough grid dependence was pointed out to be small, the number of grid points is very low. (160,000 CVs for the whole pump including volute were used). Turunen-Saaresti [6] also used an unsteady Navier Stokes solver for the calculations. He used a quite coarse grid neglecting the tip clearance.
For the present paper different approaches for coupling the rotating impeller and the stationary volute are tested and compared against transient calculations. Computational time and effort were evaluated, in order to reach accurate solutions economically. All calculations were performed on the HLRN High performance computer in Hannover, using the general purpose CFD code ANSYS CFX 10.0 with up to 16 processors in parallel. In the end results are compared with performance measurements obtained for an industrial compressor. Furthermore a transient simulation of the rotating stall phenomenon was conducted. For this simulation the impeller without volute was calculated, since stall is a part load problem induced by leading edge separation. The results show the typical travelling vortices and strong pressure fluctuations, which may lead to resonance problems.

2 Transient effects in turbomachinery

There are several sources of unsteadiness in turbomachines ranging from turbulent fluctuations at very small time and length scales up to rotating stall and surge where the length scales are of the dimension of the machine [3]. The turbulent fluctuations are usually modelled, since the direct simulation of turbulence is not possible for high Re number flows. In addition a resolution of turbulence is often not of interest, but its influence on the mean flow.
For this investigation the unsteady effects resulting from rotor stator interactions are of main interest. One source of instability is due to the interaction of the rotor wake with the volute tongue appearing at timescales depending on the number of blades and the rotational speed. On the other hand a perfect axially symmetric pressure field cannot be expected. In the volute tongue region a distortion of the pressure field is usually observed. In addition to that, a volute can only be designed for one

operational point. If the machine operates in off-design conditions, the pressure distortion will increase, resulting in a highly unsteady back pressure for the blading. In part load an additional unsteady phenomenon appears, which is referred to as rotating stall [1]. With decreasing flow rate the inlet velocity triangle changes, resulting in separation on the suction side of the blades. This separation leads to a reduced flow through the stalled channel, influencing the adjacent flow channels. The stall cells travel through the blading at a fraction of machine rotational speed, introducing another source of instability. The simulation of this effect is another scope of this work.
As was shown before, unsteady effects appear at many time and length scales. Near the design point steady state calculations may be sufficient, but in off-design conditions transient effects cannot be neglected. Apart from accurately predicting the flow through the compressor, transient results are interesting with respect to Fluid Structure Interaction [4] or noise prediction. With further development in these fields unsteady CFD calculations in turbomachines will become more common.

3 Numerical Procedure

For all CFD calculations the finite volume code ANSYS CFX 10.0 was used. The code solves the Reynolds Averaged Navier Stokes equations, turbulence was modelled using the SST k-ω model. In all simulations transition was neglected. The wall boundary layers were calculated using scalable wall functions. As the grid resolution plays an important role, a study of grid convergence was first conducted. For the blading a block structured grid was generated, using a J-grid topology. This topology avoids highly skewed elements at the leading edge of the impeller. The smallest angles were of 34 degrees, on blade surfaces a high orthogonality with near 90 angles was achieved. The blade tip clearance was gridded with an own O-grid topology with minimum 8 nodes in the gap. The grid contains a stationary inlet domain, one blade passage of the rotor and a vaneless diffuser (Fig. 1a). The inlet boundary is placed 3 tube diameters upstream from the leading edge, the outlet is placed at 1.6 times the rotor outlet diameter. Table 1 shows the results obtained for single passage calculations with periodic boundary conditions. These calculations took up to 5 hours on a standard PC in serial mode.In terms of pressure rise convergence was reached already for the coarse grid. The isentropic pressure coefficient is defined as

$$\Psi_s = \frac{2\Delta h_s}{u_2^2} \tag{1}$$

where Δh_s is the isentropic change in static enthalpy and u_2 is the rotational velocity at the trailing edge. The maximum y^+ is found at a very small amount of cells at the trailing edge of the rotor blade. As you can see from the computed tip clearance mass flow, the resolution in the tip gap influences the results. The coarse grid had 8 nodes in the tip gap whereas 10 respectively 14 nodes were used in the two other cases. Thus at least 10 nodes are needed to obtain accurate tip clearance flow. Due to the small differences in pressure rise and the high number of nodes for the full machine,

the coarse grid was chosen or further calculations.
The full machine model contains 3 domains (Fig. 1b): the inlet domain, the rotor containing 13 flow channels, and the volute. Due to the complexity of the geometry a single structured grid could not be applied at the volute tongue region. Here separate structured meshes were generated and merged in the flow solver. The rest of the volute and the cone diffuser were meshed with an own block structured grid. The total number of nodes was 2.7 million for this case. The computations were done on the HLRN 1 system in Hannover (Norddeutscher Verbund für Hoch und Höchstleistungsrechnen). The complex consists of 16 p690 compute servers with 32 processors each. Until now up to 16 processors were used in parallel. MPICH was applied as parallelization algorithm. The time step was chosen in order to resolve one revolution with 64 respectively 128 time steps, which corresponds to approximately 5 (10) timesteps per blade passage.

Table 1. Grid convergence

Nr. of nodes	max y^+	mean y^+	Ψ_s	relative tip clearance mass flow [%]
143,000	358	31	1.06	15.9
251,000	115	18.9	1.05	16.9
405,000	45.2	9.1	1.05	16.8

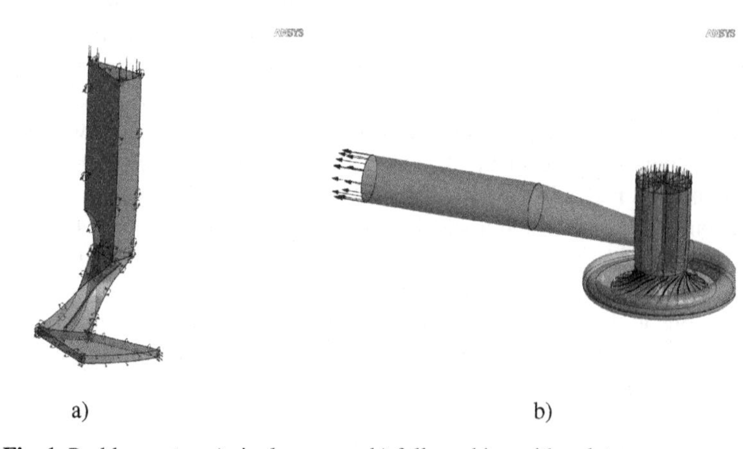

Fig. 1. Problem setup a) single passage b) full machine with volute

4 Results

4.1 Compressor with volute

For coupling stationary and rotating domains 3 methods exist. For steady state simulations the interface can be modelled either as a mixing plane or as a frozen rotor. A mixing plane averages the circumferential profiles of all variables and transfers only the mean values to the stator domain. The frozen rotor interface directly transfers the circumferential profiles. A steady state solution with frozen rotor interfaces depends on the actual position of the rotor. For this reason several calculations are usefull in order to obtain mean values. The most accurate approach is to couple the rotor and the stator in a transient simulation. First of all steady state simulations were conducted. The operational point was chosen near the design point at measured rotational speed which was about 60 % of the design speed. First a mixing plane interface was applied. The rotor wakes disappear in the vaneless diffuser due to averaging at the interface. The pressure coefficient was calculated to 0.99, the measured value was 1.11. For the frozen rotor interface, only a small change in the pressure rise was observed. In order to see an influence of the actual rotational position, the blades were rotated in several steps 25 % of one pitch (27.3 °), but the influence is less than 1 % between the different cases (Table 2).

The transient calculation was done using 128 time steps per revolution. The time averaged pressure coefficient was calculated to 1.05, but no real periodic state could be found within 5 rotations of the impeller. This may be a result of a separation appearing in the cone diffuser. The deviation of about 10 % for the steady state calculations is quite high. For a steady state calculation the computational time was about 17 h, the transient simulation took about 70h for one revolution. Both were done with 8 CPUs in parallel. An analysis of the pressure signal shows a change of the transient behaviour after 3 revolutions, where a superposition of low frequency fluctuations is observed. A FFT analysis of the time dependent pressure signal shows, that the blade passing frequency is dominating, especially near the rotor. Similar low frequency fluctuations can be found at both locations (Fig. 2).

Table 2. Comparison of simulated results

	stage	0°	6.9°	13.9°	20.8°	transient	measurement
Ψ_s	0.994	1.002	0.999	0.997	1.003	1.05	1.11

4.2 Compressor without volute

The rotating stall phenomenon was simulated at part load at 65% of mass flow compared to the point with maximum efficiency. In contrast to the results shown before, the compressor volute was not included in these calculations to avoid disturbance coming from the asymetric geometry. The fluid domain is a full circle setup of

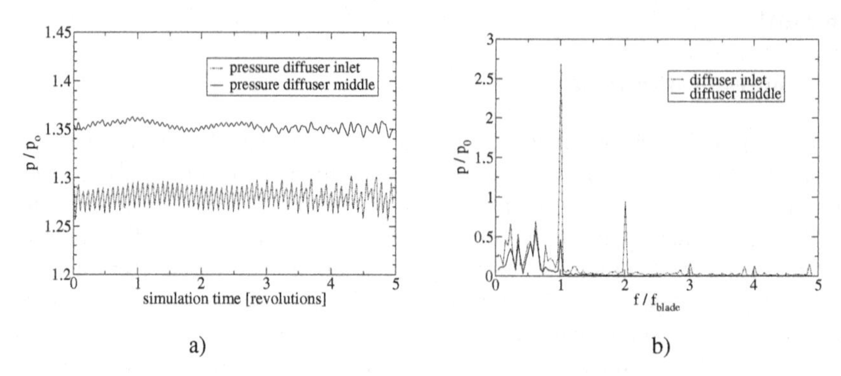

Fig. 2. Simulated pressure at two locations, a) variation in time b) FFT transformed signal

the single passage calculation conducted for the convergence study. The operational point was chosen according to the low limit of the performance curve determined from steady state single passage calculations. In Fig. 3 one can see the stall cells displayed as areas with negative streamwise velocity. The stalled regions travel with a negative rotational velocity through the machine seen from the rotating frame of reference.

The pressure history shows a noticeable difference when changing from 64 to 128

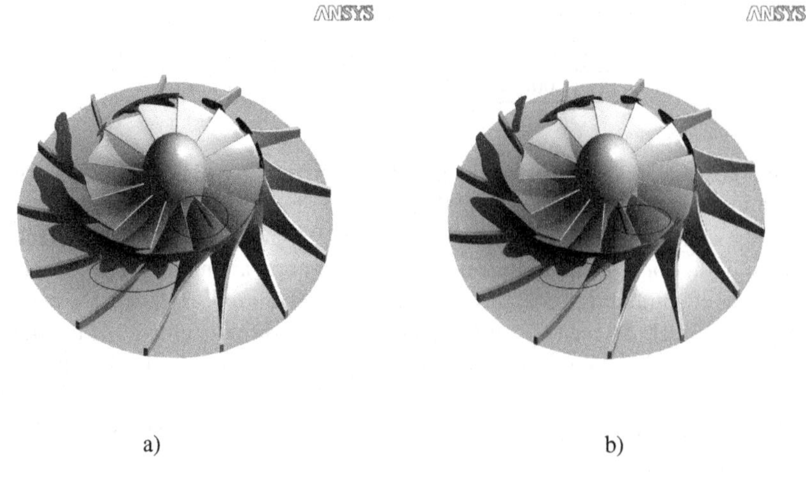

a) b)

Fig. 3. Travelling rotating stall cells displayed as negative streamwise velocity, a) and b) two different time steps

time steps per revolution (Fig. 4a). For this reason only the last revolutions with the smaller time step were analyzed in more detail. We observed strong fluctuations in pressure, but the signal seems random with no visible periodic behaviour. For this case the FFT analysis shows, that high amplitude fluctuations at low frequencies are

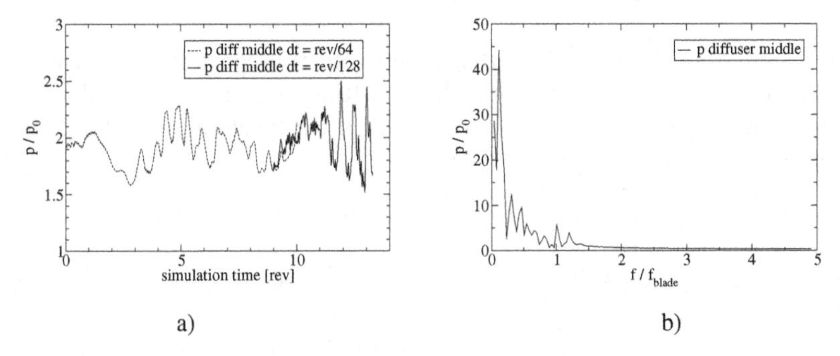

Fig. 4. a) Pressure history during rotating stall b) FFT transformed signal

dominating, while the blade frequency is of minor importance (Fig. 4b). This is expected since rotating stall usually appears at frequencies lower than rotational speed. In order to obtain an accurate value of the stall frequency, a simulation time of at least 8 or more revolutions is needed.

5 Summary

As expected, a transient simulation for the rotor with volute gives a better results compared to steady state simulations in terms of pressure rise. The relatively high difference in comparison to the measurements may be explained with the apperance of a large separated area in the outlet diffuser following the volute. This separation may not be captured correctly with the used turbulence model. The difference between the 4 frozen rotor positions is neglegible, a stage interface showed nearly the same result. The computation time for the transient calculation was about 20 times compared to a single steady state solution, but gives superior results. Such an increase in effort is usually not nessecary to obtain accurate results near the design point, but it is important if the transient behaviour is of interest. In case of the rotating stall effect we were able to simulate the typical behaviour of stalled cells travelling through the rotor and to determine low frequency high amplitude pressure fluctuations. The simulation time was too short to determine an accurate value for the stall frequency. The calculation of transient phenomena like rotating stall requires transient CFD. Since pressure fluctuates in time, blade vibrations may appear leading to damage of the blading. CFD is able to calculate the frequencies of pressure fluctuations and their amplitudes, which enables to identify possible resonance. This is especially true for off - design conditions were other than the blade passing frequency are dominating excitation mechanisms.

6 Acknowledgements

The research project was funded through the EFRE (European Regional Development Fund). The authors also thank Mr. H.J. Ring and Mr. J. Adomako (Piller Industrieventilatoren GmbH, Mohringen) for their contributions and providing the measurement data.

[1] Lakshminarayana B. *Fluid Dynamics and Heat Transfer of Turbomachinery*. John Wiley and Sons Inc, 1996.
[2] Hagelstein D., Hillewaert K., Van den Braembussche R.A., Engada A., Keiper R., and Rautenberg M. Experimental and numerical investigatation of the flow in a centrifugal compressor volute. *ASME Journal of Turbomachinery*, 122:22–31, 2000.
[3] Greitzer E.M. An introduction to unsteady flow in turbomachines. In *Thermodynamics and fluid mechanics of turbomachinery*, volume 2. Martinus Nijhoff Publishers, 1985.
[4] Dickmann H.P., Secall Wimmel T., Szwedopwicz J., Filsinger D., and Roduner C.H. Unsteady flow in a turbocharger centrifugal compressor 3d-cfd-simulation and numerical and experimental analysis of impeller blade vibration. *ASME Paper*, (GT2005-68235).
[5] Hillewaert K. and Van den Braembussche R.A. Numerical simulation of impeller-volute interaction in centrifugal compressors. *ASME Journal of Turbomachinery*, 121:603–608, 1999.
[6] Turunen-Saaresti T. *Computational and Experimental Analysis of Flow Field in the Diffusers of Centrifugal Compressors*. PhD thesis, Lappeeranta University of Technology, 2004.
[7] G. Treutz. *Numerische Simulation der instationären Strömung in einer Kreiselpumpe*. PhD thesis, Darmstadt University of Technology, 2002.

Part VI

Parallel Algorithms and Solvers

Multicolor SOR Method with Consecutive Memory Access Implementation in a Shared and Distributed Memory Parallel Environment

Kenji Ono[1,2] and Yasuhiro Kawashima[3]

[1] RIKEN, 2-1 Hirosawa, Wako, 351-0198, Japan keno@riken.jp
[2] Division of Human Mechanical Systems and Design, Faculty and Graduate School of Engineering, Hokkaido University, N13, W8, Kita-ku, Sapporo, 060-8628, Japan
[3] Fujitsu Nagano Systems Engineering, 1415 Midori-cho, Tsuruga, Nagano, 380-0813, Japan kawashima.yasuh@jp.fujitsu.com

1 Introduction

Elliptic partial difference equations like Poisson's equation are used in many fields of application. However, the coefficient matrix of the derived algebraic equation is large and sparse, and so its inversion is expensive. Various iterative methods are used to solve such a sparse matrix system. Although there have been many studies on solving the large sparse matrix system [1, 2, 3, 4, 5, 6, 7], there have been few reports on the implementation and performance of the iterative method with multicolor ordering. In this paper, a novel implementation technique to enhance the performance of the 2-colored SOR method is proposed, which eliminates the recursion for the standard 7-point stencil on the Cartesian grid in three dimensions. The performance of the multicolor SOR method is investigated on both a shared memory vector/parallel computer and a symmetric multiprocessor machine in a distributed memory environment.

2 Multicolor Ordering with Sorting of Memory Layout

2.1 Issues and Concept of Improvement

A linear system can be solved by various iterative methods such as the Jacobi, SOR, and conjugate gradient methods. The Jacobi method offers high execution speed and simple implementation in terms of vectorization and parallelization, but is slow to converge. Meanwhile, although the SOR method converges faster than the Jacobi method, its performance is low due to the forward recursion of the iterative procedure, which disturbs full vectorization of a loop. A hyper-plane or a multicolor ordering is often introduced to eliminate the recursion of the SOR iteration.

D. Tromeur-Dervout (eds.), *Parallel Computational Fluid Dynamics 2008*,
Lecture Notes in Computational Science and Engineering 74,
DOI: 10.1007/978-3-642-14438-7_19, © Springer-Verlag Berlin Heidelberg 2010

Two colors are sufficient to eliminate the dependency of the recursion for the 7-point stencil, which is generated from the Laplace operator. To take account of the vectorization and/or the parallelization, the minimum number of colors is preferable, because both the vector length and granularity of the parallelization increase as the number of colors decreases. Furthermore, fewer color minimizes the number of times of synchronization, thus boosting parallel performance.

There is an important performance issue in code implementation. Here, let us consider the discretized Laplace equation (1) of 7-point stencil on a three-dimensional Cartesian mesh.

$$p_{i,j,k} = p_{i-1,j,k} + p_{i+1,j,k} + p_{i,j-1,k} + p_{i,j+1,k} + p_{i,j,k-1} + p_{i,j,k+1} \quad (1)$$

It is common to describe a multicolor SOR code by a stride in a do-loop in *List 1*, which yields stride memory access (SMA). However, SMA prevents efficient usage of cache and vector pipelines. On the other hand, the natural ordering realizes consecutive memory access (CMA), and so CMA is expected to offer better performance than SMA. Therefore, a novel technique that combines the SOR method and CMA implementation is proposed.

List 1 Pseudo code for 2-colored ordering SOR with stride memory access.

```
do c=0,1
  do k=2,nx-1
  do j=2,nx-1
  do i=2+mod(k+j+c,2), nx-1, 2
    s0= p(i+1,j  ,k  ) + p(i-1,j  ,k  ) &
      + p(i  ,j+1,k  ) + p(i  ,j-1,k  ) &
      + p(i  ,j  ,k+1) + p(i  ,j  ,k-1)
    ss=(s0-p(i,j,k))
    p(i,j,k)=p(i,j,k)+omg*ss
    er = er + ss*ss
  end do
  end do
  end do
end do
```

Note: Variables nx, c and er represent a dimension size, color and error, respectively. Array p indicates some potential.

2.2 Implementation of 2-Colored Ordering SOR with Consecutive Memory Access

The 2-colored ordering as shown in Fig. 1 is well known. First, a computational domain of size $imax \times jmax \times kmax$ is divided by a $2 \times 2 \times 2$ sub-domain that forms a unit block in three dimensions. Then, the computational domain is consists of $NI \times NJ \times NK$ blocks, where $NI = imax/2+1$ for index i. Therefore, the dimension size is limited to an even number in each direction. The shape of the dimension can be described by P(*NI***NJ***NK*, 4, 2). The first element in the dimension represents the serialized one-dimensional address of the block. The second and third elements represent the position index inside a block and color, which are indicated in Fig. 1.

Next, the relationship between variables is described for each position in the unit block. For example, the code of equation (1) can be written by *List 2* for variable P(i, 1, 1) in Fig. 1.

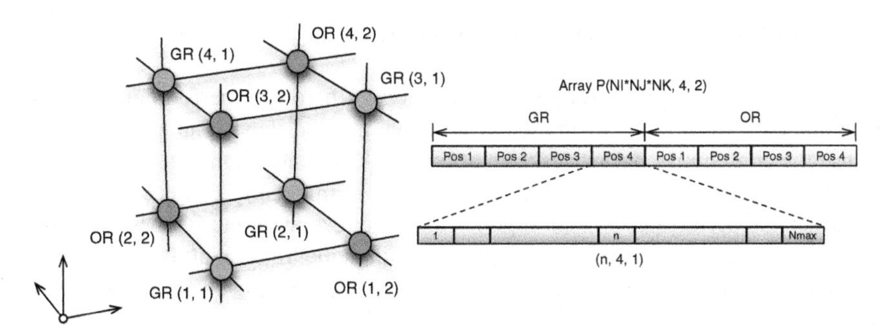

Fig. 1. 2-colored ordering with consecutive memory access. Indexes in parenthesis in the left figure are position (Pos = 1, 2, 3, 4) and color (OR = 1, GR = 2), respectively. Index N in the right figure has four continuous variables corresponding to inside a unit cube in the left figure.

List 2 Pseudo code of CMA for P(i, 1, 1).

```
do i=ist,ied
  s0= p(i    , 1, 2) + p(i-1  , 1, 2)*f2 &
    + p(i    , 2, 2) + p(i-ni , 2, 2)*f4 &
    + p(i    , 3, 2) + p(i-nij, 3, 2)*f6
  ss= (s0-p(i,1,1))
  p(i,1,1)=p(i,1,1)+omg*ss
  er = er + ss*ss
end do
```

Note: Variables *ist* and *ied* represent the start and end address of a block. A neighbor index beyond the own block is taken from a neighbor block in consideration of periodicity. The load balance becomes optimal because loop i is parallelized and vectorized simultaneously.

2.3 Performance Evaluation

We consider the Laplace equation $\nabla^2 \varphi = 0$ on the Cartesian grid system to investigate the convergence rate and the performance for iterative methods. Details of problem setting are given in [7]. A norm to determine the convergence is defined by:

$$\sqrt{\sum_{i,j,k} \left(\left| \varphi^{m+1} - \varphi^{m} \right| \right)_{i,j,k}}, \tag{2}$$

where m indicates the number of iterations. The computational space is defined by a unit cube $\Omega = [0, 1]^3$, which is divided by $NX \times NX \times NX$, where NX is dimension size.

3 Results and Discussion

3.1 Performance on a Shared Memory Vector Computer

Table 1 shows measured performance obtained by a hardware performance monitor on one CPU of a Cray Y-MP8/664. The Jacobi method shows high performance because the consecutive data access can utilize the high bandwidth of a vector computer. Although the natural ordering SOR method is inherently consecutive memory access, the performance is not high because the recursion interference prevents the code from vectorizing. When a programmer implements the multicolor SOR by stride memory access, the memory-bound occurs at runtime even in a vectorized code.

Therefore, it is important to employ a strategy for the division and sorting of the variable array such that the processor can access the memory consecutively. The measured MFLOPS score of multicolor CMA code indicates that it is as fast as the Jacobi code which can utilize the full bandwidth of the vector computer as shown in Table 1. It was confirmed that the expected performance was obtained on the vector computer.

Fig. 2 shows the measured convergence history on the vector machine. In this test case, the size of dimension NX is 33. It is observed that the convergence rate of the Jacobi method is the worst among them, and one of the 2-colored SOR-SMA and the 2-colored SOR-CMA are identical. Therefore, the multicolor SOR method by the CMA implementation is the best candidate from the aspect of the both performance and convergence rate.

Table 1. Performance (MFLOPS) on a shared memory vector machine.

Jacobi	SOR	Multicolor SMA	Multicolor CMA
236	60	150	232

Next, the parallel efficiency on a Cray C90, which is a similar vector/parallel machine, is shown in Fig. 3. The Jacobi method is written by the triple loop of indices k, j, i. The outer loop k is parallelized and the inner loop i is vectorized. In this case, according to the number of CPUs, an imbalance of the load may reduce the performance. This phenomenon is of particular note when the residue number of outer loop is small compared to the number of CPUs. For the same reason, the efficiency of the SMA multicolor SOR method becomes the worst among the methods because of the inefficient stride memory access. On the other hand, since both the Jacobi and CMA

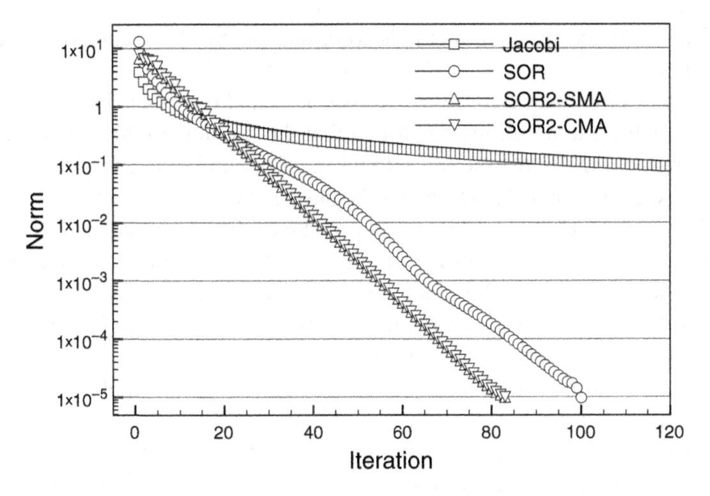

Fig. 2. Convergence history for several iterative methods against norm defined by Euation (2).

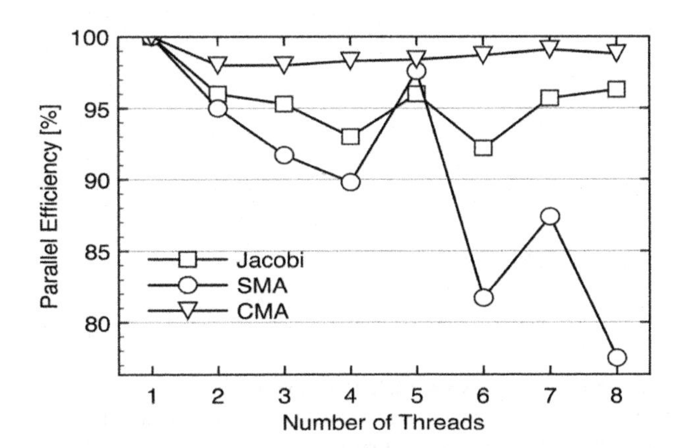

Fig. 3. Efficiency of thread parallelization on a Cray C90.

code are consecutive memory access, the performance is better than the SMA code. Especially, the CMA code shows excellent parallel efficiency because the inner loop of the array for each color is serialized and hence the load balance becomes optimal. It was found that the CMA multicolor SOR code is as fast as the Jacobi code and shows excellent parallel efficiency on a shared memory vector computer.

3.2 Performance on a Symmetric Multiprocessor Computer

The proposed CMA implementation of the SOR method on a SMP machine is examined. The code was parallelized by OpenMP directives. The test environment was a Xeon 5360 (4 cores x dual CPUs, 3GHz) with Intel compiler 10.1 on a Mac OSX

10.5.4. The evaluated iterative methods were the Jacobi, the 2-colored SOR-SMA, and SOR-CMA.

Fig. 4 shows the measured timing for in-cache and out-of-cache sizes and number of threads for one iteration. It is found that the 2-colored CMA code shows the best performance among the methods. In terms of parallel efficiency, the CMA code does not show good parallel performance because the CMA code is already saturated beyond two threads. However, the execution time of the CMA code is the shortest.

The CMA code have the reason for high performance on the scalar machine. That is the relationship between the bandwidth and the amount of load/store. In terms of bandwidth, the number of arrays to be loaded and stored is important for the performance. The number of loads/stores of the SOR group is two. Meanwhile, the Jacobi need an additional work array. The CMA also needs an small additional work array for sorting, but this is trivial. In fact, the feature of fewer amount of load/store and fast memory access allows high performance. This aspect has a larger influence for the SMP machine than the shared memory vector machine because of the limited bandwidth.

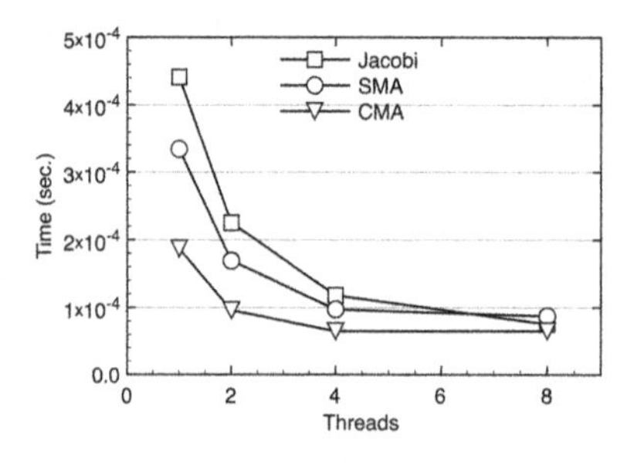

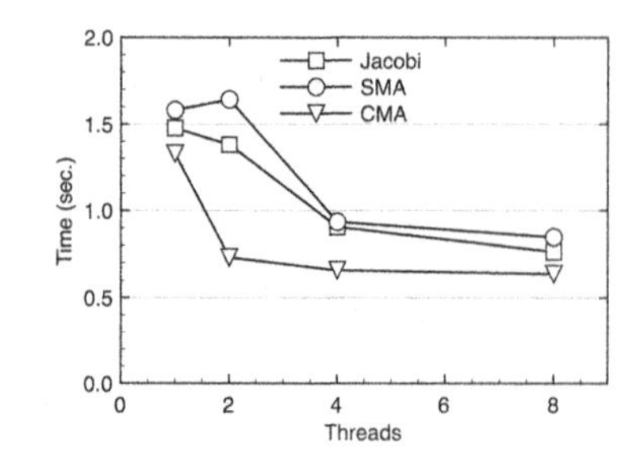

Fig. 4. Timing for in-cache size (NX = 33, left) and out-of-cache size (NX = 513) against number of threads. Vertical and horizontal axes indicate time and number of threads, respectively. In all cases, the CMA code shows the best performance.

3.3 Performance in a Distributed Memory Parallel Environment

The performance was investigated in a distributed environment. The OpenMPI library was used for the parallel computation with the domain decomposition method. The computer resource used was the Quest system in RIKEN, which consists of 1024 nodes (2048 cores) of a Core2Duo cluster (L7400, 1.5GHz, L2 4GB). The interconnect is GbE and 2Gbps for each node. Fig. 5 shows the measured performance for

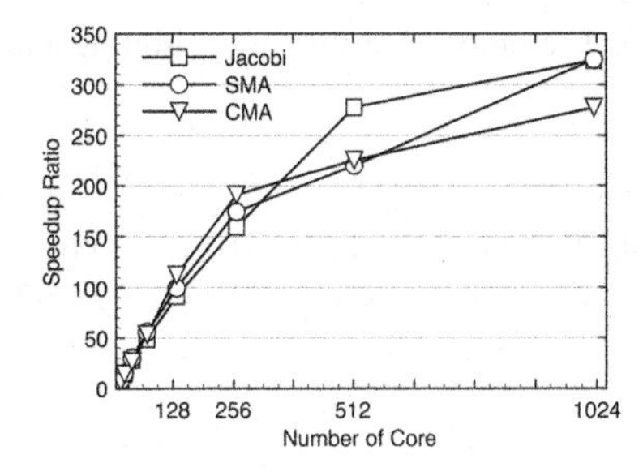

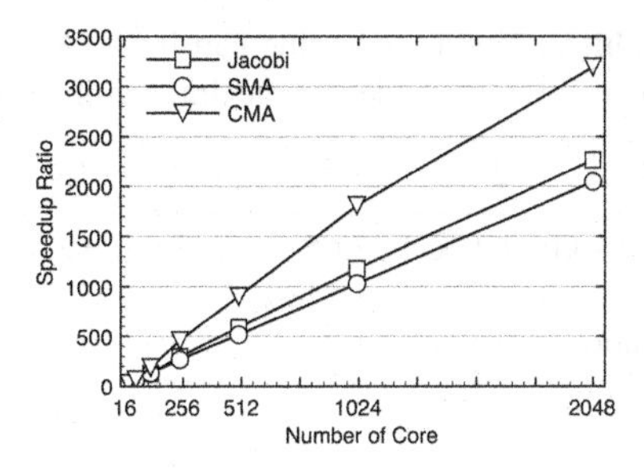

Fig. 5. Comparison of performance on Core2Duo cluster. Intel Compiler 10.1, option O3 was used. The left figure shows the scalability based on elapsed time. The right figure shows the scalability based on only calculation time. CMA indicates super-linear behavior and the other methods show good scalability.

the fixed size of $416 \times 416 \times 416$. This is a strong scaling test, which is obtained for the case of one core per node.

Scalability based on measured elapsed time indicates that all methods show good scalability up to 256 nodes and saturated behavior above 256 nodes. However, regarding scalability based on the timing for only the calculation part as shown in the right panel of Fig. 5, all methods show very good scalability, and CMA in particular is super-linear. The is because beyond 512 nodes, the computation size for each color array becomes less than $26 \times 26 \times 26$, which fits into cache.

One issue is the communication part. Table 2 shows the communication cost per iteration. It is found that almost all of the time cost is occupied by communication over 256 nodes and that the CMA code is faster below 256 nodes. As mentioned above, beyond 512 nodes, all data are in-cache in CMA code, so the communication cost becomes relatively large. At present, since the communication part was implemented by a brute-force method, there is still room for improvement. In this environment, the measured results reveal that the stencil-based calculation is almost saturated for this size of problem.

In the CMA implementation, we need to sort the array once depending on the ordering before and after the iteration. Table 3 indicates the cost of sorting. Three arrays and one array need to be sorted before and after iteration, respectively. The timing data showed a relative cost ratio for pre-sorting to post-sorting of almost 3:1. This sorting cost is equivalent to the cost for one or two iterations.

Table 2. Cost of communication part for one iteration (percent).

Node	2	4	8	16	32	64	128	256	512	1024
Jacobi	1.5	2.7	6.3	8.8	13.3	21.0	29.9	46.1	52.8	72.5
2-colored SMA	1.2	2.5	4.4	7.9	9.0	14.9	24.8	34.7	57.3	68.4
2-colored CMA	2.1	3.4	6.4	9.4	14.0	21.3	40.7	58.2	75.0	84.7

Table 3. Timing of sorting part in CMA code (sec). The ratio is the sum of pre-sorting and post-sort time divided by the time for one iteration.

Node	2	4	8	16	32	64	128	256	512
pre-sort	6.96e-1	3.65e-1	1.79e-1	9.08e-2	4.69e-2	2.38e-2	1.78e-2	6.07e-3	3.06e-3
post-sort	2.21e-1	1.13e-1	5.71e-2	2.89e-2	1.47e-2	7.77e-3	9.47e-3	2.04e-3	1.05e-3
Ratio	1.4	1.2	1.2	1.3	1.2	1.3	2.3	1.1	0.7

4 Conclusions

A new implementation of the multicolor SOR was proposed including the variable sorting to achieve consecutive memory access. Its effect was confirmed on a shared memory vector/parallel and SMP machines, and it was found that the multicolor SOR method with consecutive memory access demonstrates excellent performance on both types of machine. The CMA multicolor SOR method retains almost the same convergence rate of the original SOR method and its execution speed is much faster than the original SOR on a vector computer. On the other hand, the parallel performance on a vector/parallel machine was greatly improved by one-dimensional serialization of the array, which leads to excellent load balance between cores. On an SMP machine, the CMA code showed the best performance among the test codes. In addition, good scalability was confirmed up to 256 nodes using a medium size of $416 \times 416 \times 416$ for all tested methods in a distributed memory parallel environment. Especially, for the strong scaling test case, the CMA code showed super-linear behavior for a small number of CPUs compared with the other methods. Thus, if the size of the problem becomes bigger as the number of CPUs increases, i.e., the weak scaling case, then scalability seems to be promising.

Acknowledgement

This research was supported by the Research and Development on Next-Generation Integrated Simulation of Living Matter, which is part of the Development and Use of the Next-Generation Supercomputer Project of the Ministry of Education, Culture, Sports, Science and Technology (MEXT).

[1] L.M. Adams and H.F. Jordan. Is sor color-blind? *SIAM J. Sci. Stat. Comput.*, 7(2):1507–1520, apr 1986.

[2] S. Doi and A. Lichnewsky. Some parallel and vector implementations of preconditioned iterative methods on cary-2. *International Journal of high Speed Computing*, 2(2):143–179, 1990.

[3] S. Fujino, M. Mori, and T. Takeuchi. Performance of hyperplane ordering on vector computers. *J. Comp. and Appl. Math.*, 38:125–136, 1991.

[4] R.J. Leveque and L.N. Trefethen. Fourier analysis of the sor iteration. ICASE Report 86-63, NASA, 1986.

[5] E.L. Poole and J.M. Ortega. Multicolor iccg methods for vector computers. *SIAM J. Numer. Anal.*, 24(6), 1987.
[6] R.S. Varga. *Matrix Iterative Analysis*. Springer, second edition, 2000.
[7] M. Yokokawa. Vector-parallel processing of the successive overrelaxation method. JAERI-M 88-017, Japan Atomic Energy Research Institute, 1988.

Proper Orthogonal Decomposition In Decoupling Large Dynamical Systems

Toan Pham[1], Damien Tromeur-Dervout[1]

Université de Lyon, Université Lyon 1, CDCSP/Institute Camille Jordan
UMR5208-U.Lyon1-ECL-INSA-CNRS, F-69622 Villeurbanne Cedex, France
tpham@cdcsp.univ-lyon1.fr,
Damien.Tromeur-Dervout@cdcsp.univ-lyon1.fr

Abstract. We investigate the proper orthogonal decomposition (POD) as a powerfull tool in decoupling dynamical systems suitable for parallel computing. POD method is well known to be useful method for model reduction applied to dynamical system having slow and fast dynamics. It is based on snapshot of previous time iterate solutions that allows to generate a low dimension space for the approximation of the solution.Here we focus on the parallelism potential with decoupling the dynamical system into subsystems spread between processors. The non local to the processor sub-systems are approximated by POD leading to have a number of unknowns smaller than the original system on each processor. We provide a mathematical analysis to obtain a criterion on the error behavior in using POD for decoupling dynamical systems. Therefore, we use this result to verify when the reduced model is still appropriated for the system in order to update the basis. Several examples show the efficient gain in term of computational effort of the present method.
Keywords: POD, reduced-order modelling, dynamical systems, parallel computing.

1 Introduction

Proper orthogonal decomposition (POD) is known as an application of the singular value decomposition (SVD) to the approximation of general dynamical systems. Usually, it is used to separate the low and fast dynamics of a dynamical system. In this paper, we focus on it use to solve large dynamical systems on a parallel system leading to a robust approach based on a priori criterion to communicate updated data between processors.

Related works on decoupling large dynamical systems on parallel architecture are based on extrapolation techniques. In [4], the decoupling is done by extrapolating components involved on the sub-system but not managed by the current processor. This technique has been extended to asynchronous adaptative time steps [11]. In [13] or in [12], some multirate integration formulae are applied for solving large dynamical systems in parallel. Nevertheless, the major drawbacks of using these methods based on extrapolation technique are first the only feedback comes from an

D. Tromeur-Dervout (eds.), *Parallel Computational Fluid Dynamics 2008*,
Lecture Notes in Computational Science and Engineering 74,
DOI: 10.1007/978-3-642-14438-7_20,

a posteriori error estimate between the extrapolated in time data and the received corresponding data from other systems. If the error is to high, the integration must restart from a previous checking point with changing the extrapolation formulae reducing the order approximation of the extrapolation or the delay in the data to compute the extrapolation. Second, for stiff non linear dynamical system that must be solved by implicit time integration method, only the components managed by the processor are involved in the Jacobian matrix leading to less influence of the other components of the system in the computing.
Our approach consists in the decoupling of the system through the model reduction method that provides an answer to the two drawbacks mentioned above. First, we can derive an a priori error estimate, to know when the data from other sub systems must be updated. Second, we still incorporate the influence of the other sub-systems (extra-systems) in the non linear solve of the current sub-systems (intra-system). Indeed, extra-systems solved in reduced form are very computational attractive. When applying model reduction to dynamical systems, the snapshots of previous time steps are used to compute the POD basis vectors to yield an optimal representation of the data in sense of optimal least squares approximation. Combined with Galerkin projection method, we can generate a lower dimensional model of the systems. We extend this notion of model reduction to decouple the system into smaller subsystems. Each subsystem is solved separately on parallel computer and a model reduction of the other subsystem is performed.
Basic error estimation for reduced order models has been proposed in [14] - the original problem is linearized in the neighborhood of the initial time. Other result can be found in [9], where bounds for errors resulting from reduced models has been computed. In [7], we found some results on the influence of perturbations in the original system on the quality of the approximation given by the reduced model. Here our goal is to find a criterion to know when we need to update the POD basis to compute the extra-subsystems with no introducing error in the intra-subsystem.

The plan of this paper is as follows. In section 2 we recall some properties of the POD, while in section 3.2 we give an analysis of the error on the global system when subsytems are approximated by POD. Then we derive an a priori estimated to update the POD basis that appoximated the subsytems. Section 4 gives the decoupling algorithm and its parallel implementation. Numerical tests and performances are summarized in section 5 before the conclusion in section 6.

2 Proper Orthogonal Decomposition and Model reduction

2.1 Singular Value Decomposition

Property 0.1. For any matrix $A \in \mathbb{C}^{m\times n}$, there exist two unitary matrices $U \in \mathbb{C}^{m\times m}$ and $V \in \mathbb{C}^{n\times n}$ such as:

$$U^T AV = \Sigma = \operatorname{diag}(\sigma_1,\ldots,\sigma_m) \tag{1}$$

$\sigma_1 \geq \cdots \geq \sigma_m \geq 0$. This decomposition is called the SVD of the matrix A, $\sigma_i = \sqrt{\lambda_i(A^*A)}$ are called singular values of A where λ_i is the ith eigenvalue.

A can be written as : $A = \sum_{i=1}^{m} \sigma_i u_i v_i^*$

Remark 0.1. Given a matrix $A \in \mathbb{C}^{n \times m}$, the solution $\hat{X}$ of

$$\min_{X \in \mathbb{C}^{n \times m}, rank(X) \leq k < rank(A)} \|A - X\|_2 = \sigma_{k+1}(A) \tag{2}$$

is obtained by truncating the SVD to the k first singular modes: $\hat{X} = \sigma_1 u_1 v_1^* + \sigma_2 u_2 v_2^* + \ldots \sigma_k u_k v_k^*$ (c.f [1] or [8]).

2.2 POD in model reduction

POD method is widely used in generating a lower dimensional model for dynamical systems. We present here a procedure called Garlerkin projection. Once found the approximating subspace to the data set of the system, we are seeking for the corresponding vector field on the subspace that represents the reduced order model.
Given a system of dimension n:

$$x(t)' = \mathbf{f}(t, x(t)) \tag{3}$$

$$x(t_0) = x_0 \tag{4}$$

for $t \in [t_0, t_f]$, $x, x_0 \in \mathbb{R}^n$ and $\mathbf{f} : \mathbb{R}^n \times \mathbb{R} \to \mathbb{R}^n$. Collect the solutions of (3)-(4) at m time points and form the matrix $\mathbb{X} = [x(t_1), \ldots, x(t_m)]$. POD consists of finding the matrix $P \in \mathbb{R}^{n \times n}$ defining the projection onto a subspace $\mathcal{S} \in \mathbb{R}^n$ minimizing the total square distances $\|\mathbb{X} - P\mathbb{X}\|^2$. Using eq. (2), the singular decomposition of the matrix of snapshots $\mathbb{X} = U \Sigma V^T$ gives a such minimizer as $P = U_k U_k^T \in \mathbb{R}^{n \times n}$ where U_k correspond to the k columns of U associated to singular values $\sigma_1 \geq \ldots \geq \sigma_k$. U_k spanes as a natural orthogonal basis the space $\mathcal{S}$. The reduced model to the system (3) is constructed by projecting onto $\mathcal{S}$ the vector field $f(s,t)$ at each point $s \in \mathcal{S}$. If ξ are the subspace coordinates of s, then the reduced model is:

$$\xi'(t) = U_k^T \mathbf{f}(U_k \xi(t), t) \tag{5}$$

where $\mathbf{x}(t) \approx U_k \xi(t)$, $\xi(t) \in \mathbb{R}^k$.

2.3 POD for decoupling systems

In this part we present how POD can be useful in decoupling system for parallel purpose. Given an ODE system of state variable $x(t) = (x_1(t), x_2(t))$:

$$x(t)' = \mathbf{f}(t, x(t)) = \begin{pmatrix} f(t, x_1, x_2) \\ g(t, x_1, x_2) \end{pmatrix}, \; x(0) = x_0 \tag{6}$$

We take some snapshots of the whole system by solving globally the system. Let say q steps are taken for initialization:

$$\mathbb{X} = (\mathbf{x}_1, \mathbf{x}_2, \ldots, \mathbf{x}_q) \tag{7}$$

from which we extract $\mathbb{X} = \begin{pmatrix} \mathbb{X}_1 \\ \mathbb{X}_2 \end{pmatrix} = \begin{pmatrix} x_{1,1} \dots x_{1,q} \\ x_{2,1} \dots x_{2,q} \end{pmatrix}$. The SVD will be computed separately for each subsystem $\mathbb{X}_1 = U_1 S_1 V_1^T$ and $\mathbb{X}_2 = U_2 S_2 V_2^T$. Let y_i denote the full variable x_i $(i = 1,2)$, and α_i $(i = 1,2)$ reduced variables representing the main part of the full variables x_i, $i = 2,1$. We can separate the original system into two independent sub-systems:

$$(S1) \quad \begin{pmatrix} y_1' \\ \alpha_2' \end{pmatrix} = \begin{pmatrix} f(t, y_1, U_2\alpha_2) \\ U_2^T g(t, y_1, U_2\alpha_2) \end{pmatrix} \tag{8}$$

$$(S2) \quad \begin{pmatrix} \alpha_1' \\ y_2' \end{pmatrix} = \begin{pmatrix} U_1^T f(t, U_1\alpha_1, y_2) \\ g(t, U_1\alpha_1, y_2) \end{pmatrix} \tag{9}$$

In the last equations, the systems $(S1)$ respectively $(S2)$ can be solved alone as long as the reduced variables α_2 (resp. α_1) is still representing correctly the dynamics of the other variable y_2 (resp. y_1) of the current sub-system. If the last condition fails, we need to update with the new POD computed from the other system. In the next section, we derive a posteriori criterion to know when the updates of the POD must occur.

3 Analysis of POD in solving dynamical system

Several studies of the error estimate for a reduced model was introduced. In [7], the error is estimated by introducing the adjoint forward model and using the statistical sample estimation method. In [14], an algorithm for computing errors for reduction methods is presented. This approach consists of linearizing the original problem around the initial time, and the numerical error estimation is given.
We propose to establish the error formulation for our method. At first we look at the linear case for the simple model reduction, then the non-linear problems will be treated. Then this study will be extended for the parallel decoupling algorithm to validate our approach.

3.1 Analysis of the reduced model

Given an dynamical system as described in (3)-(4) with the reduced model as in (5):

$$\xi'(t) = U^T A U \xi(t) \tag{10}$$

and consider $e(t) = y(t) - U\xi(t)$. By liearlizing to the 1-st order (see [7]):

$$e'(t) = (UU^T - I) f(t, x(t)) + J_f(t, y(t)) e(t) + \mathcal{O}(\|e\|^2)$$

where J_f is the Jacobian matrix of the function f.

Remark 0.2. Starting with the initial condition $e(0) = 0$, the solution vanishes in the neighborhood of 0. If UU^T represents the projection onto the original space, then $e(t)$ remains bounded as long as $Ax(t)$ belong to the kernel of the projection defined by $\mathbf{P} = UU^T$. This is an important criterion to verify when the reduced model is still appropriated to the full model.

3.2 Analysis of POD in decoupling systems

Consider at the system (6). We apply POD method to have two separated systems (8) and (9) and we are looking for the error estimates of the decoupled solution. Consider at first the system (8), denote its solution by $(\hat{y}_1, \hat{y}_2)^T = (y_1, U_2\alpha_2)^T$. Then we introduce the error as $e_1 = \hat{y}_1 - x_1$ and $e_2 = \hat{y}_2 - x_2 = U_2\alpha_2 - y_2$. Linearizing the right hand side to the first order gives up to:

$$
\begin{aligned}
e_1' &= J_f(y_1)e_1 + J_f(y_2)e_2 + \mathcal{O}(\|\mathbf{x}-\mathbf{y}\|^2) \\
e_2' &= (U_2U_2^T - I)g(x_1,x_2,t) + U_2U_2^T[J_g(y_1)e_1 + J_g(y_2)e_2] + \mathcal{O}(\|\mathbf{x}-\mathbf{y}\|^2)
\end{aligned}
$$

where J is the jacobian matrix. We drop remaining terms in last equation to obtain:

$$\mathbf{e}'(t) = \Phi\mathbf{e}(t) + \Theta(t) \tag{11}$$

where $\mathbf{e}(t) = (e_1(t), e_2(t))^T$, $\Phi = \begin{pmatrix} J_f(y_1) & J_f(y_2) \\ U_2U_2^T J_g(y_1) & U_2U_2^T J_g(y_2), \end{pmatrix}$ and $\Theta(t) = \begin{pmatrix} 0_{n_1} \\ (U_2U_2^T - I)g \end{pmatrix}$ where 0_{n_1} is a zero vector of length n_1.

Remark 0.3. From (11), assumed no error in building the SVD, i.e. $\mathbf{e}(t_0) = 0$, and the error is bounded when $\Theta(t)$ "small enough". This can be viewed as:

$$y_2' = g \in Ker(U_2U_2^T - I) \tag{12}$$

Indeed, the right hand side of the equation results in an error estimation for the reduced system and then the decoupled solution (e_1 is function of e_2). If $(\mathscr{U}_2 - \mathscr{I})y_2'(t) = \varepsilon$ (a "small" vector), then $\mathbf{e}(t) = \varepsilon t$ and if T is finite, the error term remains bounded.

In a technical point of view, as long as we compute the decoupled numerical solution, one needs to verify that $x_2' = g$ belongs to the kernel of the projection. Otherwise, if the last condition fails, we need to update the basis.

4 Parallel implementation and algorithm

We decouple the system (6) into two separated sets of states. We propose the algorithm that follows:

Algorithm 1 Parallel algorithm with POD

```
Initialization of the POD basis from time iterate snapshots
Send the first components of the POD to construct the reduced system
if RECEIVE the new basis then
    Update to the reduced sub-system
end if
Run the two models
if Check the orthogonality of y_i's to the projection as in (12) succeed then
    Go to the next step
else
    Reconstruction of the new basis and go to the next step
end if
```

This algorithm is suitable for parallel computer. The point is, once the orthogonal basis is no longer representative to the system, we can either *restart a whole new basis* or *incrementally shifting POD basis*. The restarting technique just drop the old POD and recompute the POD from newer data set. In further implementation, we are considering to explore the technique that allows to reuse the main part of the POD and shifting only for some newer values c.f. [6] and [10]. Let consider a n dimensional system. The complexity of an implicit numerical method is mainly due to function and jacobian evaluations and the complexity of inverting the mass matrix within Newton (or Newton-liked) iterations.

The main advantage of decoupled POD technique is in the fact that the decoupled system lies on a reduced dimensional space. Even if the gain in function evaluations are the same in both cases ($f(x)$ and $U'f(U\xi)$ have the same complexity), but the reduced order equation takes advantage in jacobian evaluation: since the reduced system is n_1, n_2 dimensional($n_1 + n_2 < n$), thus all numerical Jacobian evaluations of the required for Newton iteration is greatly reduced (by $0\left(\frac{n_1^2+n_2^2}{n^2}\right)$ factor). Beside, one of the most advantage of decoupled method results from solving the linear system.

The method requires also the evaluation of the POD basis. As in [5], let say our data required for constructing the POD lie in $\mathbb{R}^{n\times m}(n \gg m$- by construction we chose only a "small" step number compared to the dimension of the system), then we can obtain the POD with the cost of $6mn^2 + 20n^3$-which is relatively small since we choose m in general much smaller than n. Therefore, on parallel computers, computing the POD can be done by separated processor.

5 Numerical tests

All the SVD are applied for q steps, the SVD is truncated to the first k singular values. In general, the $n-k$ last singular values are dropped since they represent noise modes. k is taken such that:$\sigma_l < 1\cdot e^{-6}\sum_{i=1}^{m}\sigma_i,\ \forall l > k$. Once subsystem received the orthonormal basis, its goes for the next r steps, and send those q for the other to reevaluate the SVD. Then we define a cycle i as q SVD steps, r steps for independent solvers. And we loop for p cycles to the end of the simulation.

We apply our decoupled POD method for splitting the ODE system into subsystems. Via MPI interface, the new basis is communicated for each subsystems by sending messages. We measure the the performance of our new method by comparing it with a single solver execution. The quality of the decoupled solution is also considered.

Diurnal Kinetic Advection-Diffusion problem

The problem consists of a pair of kinetics-advection-diffusion partial differential equations. The PDE can be written as:

$$\frac{\partial c^i}{\partial t} = K_h \frac{\partial^2 c^i}{\partial x^2} + V\frac{\partial c^i}{\partial x} + \frac{\partial}{\partial y}K_v(y)\frac{\partial c^i}{\partial y} + R^i(c^1, c^2, t) \quad (i = 1, 2) \tag{13}$$

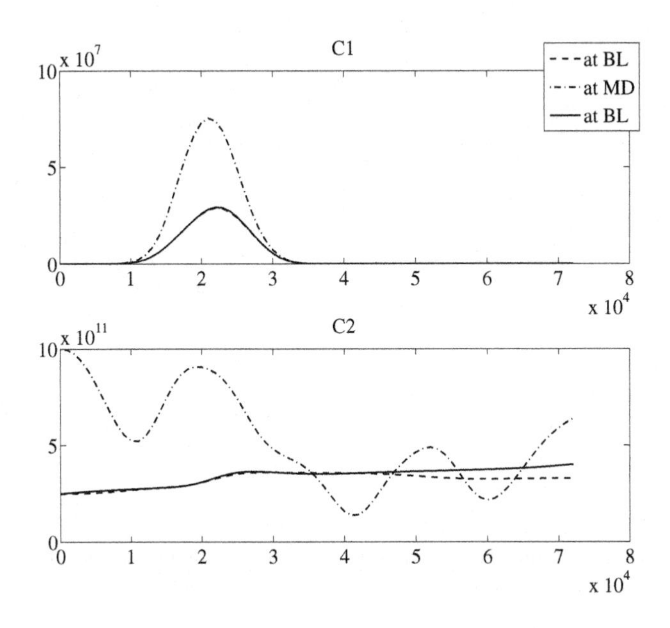

Fig. 1. Reference solution in time at some sample points: BL bottom left, MD: middle, TR: top right on the space grid

Table 1. Compare a multi-subsystems version vs single solver version: $e(t_{end}) = \|u - y\|_2 / \|y\|_2$

N procs	Single	2		3		
system	S	$S1$	$S2$	$S1$	$S2$	$S3$
Size	800	420	420	307	307	306
Fevals	2651	3957	5455	4102	6175	6281
LUs	771	1152	1336	1053	1627	1614
Elapsed Times (s)	667.51	200.5	200.5	31.41	31.46	31.45
$e(t_{end})$	0.0	$1.05 \cdot 10^{-2}$		$1.81 \cdot 10^{-2}$		

where $t \in [0, 7.26 \cdot 10^4]$ and

$$R^1(c^1, c^2, t) = -q_1 c^1 c^3 - q_2 c^1 c^2 + 2q_3(t) c^3 + q_4(t) c^2, \tag{14}$$

$$R^2(c^1, c^2, t) = q_1 c^1 c^3 - q_2 c^1 c^2 - q_4(t) c^2 \tag{15}$$

$0 \le x \le 20$, $30 \le y \le 50$ (in km).The various constants and parameters are: $K_h = 4.0 \cdot 10^{-6}$,$V = 10^{-3}$, $K_v = 10^{-8} exp(y/5)$, $q_1 = 1.63 \cdot 10^{-16}$, $q_2 = 4.66 \cdot 10^{-16}$, $c^3 = 3.7 \cdot 10^{16}$ and

$$q_i(t) = \begin{Bmatrix} exp(-a_i / sin\omega t), & \text{for } sin\omega t > 0 \\ 0, & \text{for } sin\omega t \le 0 \end{Bmatrix} \quad (i = 3, 4)$$

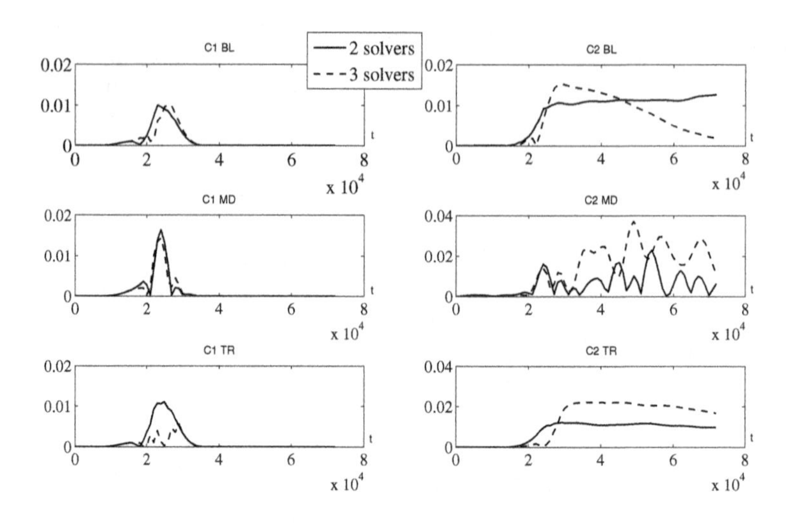

Fig. 2. Relative error in time compared to the reference solution at individual sample points: 2-solver and 3-solver solution compared to the reference solution

where $\omega = \pi/43200$, $a_3 = 22.62$, $a_4 = 7.601$. Homogeneous Neumann boundary conditions are imposed on each boundary. We discretize spatially with standard finite difference on a 20×20 mesh, giving and ODE system of size 800.

For a constant step size version, we use $q = 30$ time steps for constructing the POD. We exchange the basis every $r = 90$ time steps. The time step is constant at 10 (s). The solver used is based on GEARs method, c.f. [2] for more detail on the CVODE solver.

The solution at some sample points are ploted in the figure 1 and the relative error can be showed in the figure 2. The individual relative error compared to the reference solution is not larger than 2%.

The numerical properties of our solver are reported in the table 1. Some statistics as Size of the systems, number of function evaluations (*Fevals*), number of LU decomposition (*LUs*) done for solving linear system, elapsed time and the norm of the error at the end of the simulation.

Our method results in high performance code (great speed up according to the higher number of proccessors).

Beside, we do not ommit here that since we use the DENSE linear solver, the LU decomposition is therefore applied for dense full matrix. And the reduced models have drastic advantage to make it reduced LU decomposition compared to the full model. That explain the high speed up obtained in this case. In general, we expect to have a little lower performance when using sparse (or band) solver.

6 Conclusions and future work

Our first implementation of the method show a promising method for decoupling dynamical systems. The method works well with stiff problem and has significant performance. Especially for large dynamical systems provided from CFD and other field of applications (c.f. [3]), reduced order method can drastically have reasonable approximated solutions with lower cost.

We analysed the method and established its error formulation, that allows to have an a priori estimate on the validity of the POD basis to represent for a sub-system the influence of the other sub-systems.

We are currently considering the new approach of shifting the POD for better performance, as well as optimizing function and jacobian evaluation for the reduced system.

[1] Moody T. Chu, Robert E. Funderlic, and Gene H. Golub. A rank-one reduction formula and its applications to matrix factorizations. *SIAM Rev.*, 37(4):512–530, 1995.

[2] S.D. Cohen and A.C. Hindmarsh. CVODE, a stiff/nonstiff ODE solver in C. *Computers in Physics*, 10(2):138–143, 1996.

[3] F. Fang, CC Pain, IM Navon, MD Piggott, GJ Gorman, P. Allison, and AJH Goddard. Reduced order modelling of an adaptive mesh ocean model. *International Journal for Numerical Methods in Fluids, sub-judice*, 2007.

[4] M. Garbey and D. Tromeur-Dervout. A parallel adaptive coupling algorithm for systems of differential equations. *J. Comput. Phys.*, 161(2):401–427, 2000.

[5] Gene H. Golub and Charles F. Van Loan. *Matrix computations*, volume 3 of *Johns Hopkins Series in the Mathematical Sciences*. Johns Hopkins University Press, Baltimore, MD, second edition, 1989.

[6] Ming Gu and Stanley C. Eisenstat. Downdating the singular value decomposition. *SIAM J. Matrix Anal. Appl.*, 16(3):793–810, 1995.

[7] Chris Homescu, Linda R. Petzold, and Radu Serban. Error estimation for reduced-order models of dynamical systems. *SIAM Rev.*, 49(2):277–299, 2007.

[8] Lawrence Hubert, Jacqueline Meulman, and Willem Heiser. Two purposes for matrix factorization: a historical appraisal. *SIAM Rev.*, 42(1):68–82 (electronic), 2000.

[9] K. Kunisch and S. Volkwein. Galerkin proper orthogonal decomposition methods for a general equation in fluid dynamics. *SIAM J. Numer. Anal.*, 40(2):492–515 (electronic), 2002.

[10] Marc Moonen, Paul Van Dooren, and Filiep Vanpoucke. On the *QR* algorithm and updating the SVD and the *URV* decomposition in parallel. *Linear Algebra Appl.*, 188/189:549–568, 1993.

[11] T. Pham and F. Oudin-Dardun. $c(p,q,j)$ scheme with adaptive time step and asynchronous communications. volume to appear of *Lecture Note in Computational Science and Engineering, Parallel CFD 2007*. Lecture OPTNote in Computational Science and Engineering, 2009.

[12] Stig Skelboe. Accuracy of decoupled implicit integration formulas. *SIAM J. Sci. Comput.*, 21(6):2206–2224 (electronic), 2000.
[13] Stig Skelboe. Adaptive partitioning techniques for ordinary differential equations. *BIT*, 46(3):617–629, 2006.
[14] Senol Utku, Jose L. M. Clemente, and Moktar Salama. Errors in reduction methods. *Comput. & Structures*, 21(6):1153–1157, 1985.

Performance Analysis of the Parallel Aitken-Additive Schwarz Waveform Relaxation Method on Distributed Environment

Hatem Ltaief[1] and Marc Garbey[2] *

[1] University of Tennessee, Department of Electrical Engineering and Computer Science, 1508 Middle Drive, Knoxville TN 37996, USA ltaief@eecs.utk.edu
[2] University of Houston, Department of Computer Science, 501 PGH Hall, Houston TX 77204, USA garbey@cs.uh.edu

1 Introduction

Grids provide an incredible amount of resources spread geographically to scientists and researchers, but the poor performance of the interconnection network is often a limiting factor for most parallel applications based on partial differential equations where intensive communications are required. Furthermore, the evolution towards multicore architecture has further exacerbated the problem and a strong endeavor is emerging from the scientific community to effectively exploit parallelism at an unprecedented scale. Indeed, standard softwares and algorithms need to be rethought and redesigned to be able to benefit from the power that new generations of multicore processors offer (hundreds of thousands of nodes). The Aitken-Additive Schwarz Waveform Relaxation (AASWR) for parabolic problems [9] is the analogue of the Aitken-like acceleration method of the Additive Schwarz algorithm (AS) for elliptic problems [1]. We have generalized the original algorithm of [9] to a grid-efficient Parallel AASWR version (PAASWR). The first results were successfully presented at *PARCFD'07* [7] on a single cluster computing. The Parallel Non Blocking Windowed approach was selected to be the most efficient in solving a million of unknowns. The main contribution of the paper is to present the performance analysis of the method in grid computing environment. The fundamental concept of the AASWR method is to postprocess the sequence of interfaces generated by the domain decomposition solver. Many researchers have studied the optimization of the transmission conditions for domain decomposition techniques using other approaches - see for example [6, 2, 4, 10, 3] and their references. Ideally a numerical solver should satisfy load balancing, scalability, robustness and fault tolerance to face the grid challenge. We will show how the PAASWR method for parabolic problems is able to achieve those features. The paper is organized as follows. Section 2 recalls the fundamental

* Research reported here was partially supported by Award 0305405 from the National Science Foundation.

D. Tromeur-Dervout (eds.), *Parallel Computational Fluid Dynamics 2008*,
Lecture Notes in Computational Science and Engineering 74,
DOI: 10.1007/978-3-642-14438-7_21,

steps of the three dimensional space and time PAASWR algorithm. Section 3 describes the parallel implementation on the grid. Section 4 presents a comparison of the experimental results performed on a single parallel machine and on distributed grid computing. Finally, section 5 summarizes the study and defines the future work plan.

2 The PAASWR Algorithm

2.1 Definition of the Problem and its Discretization

The problem definition is the following Initial Boundary Value Problem (IBVP):

$$\frac{\partial u}{\partial t} = L[u] + f(x,y,z,t), (x,y,z,t) \in \Omega = (0,1)^3 \times (0,T) \quad (1)$$

$$u(x,y,z,0) = u_o(x,y,z), \quad (2)$$

completed by Dirichlet boundary conditions. L is a separable second order linear elliptic operator. We assume that the problem is well posed and has a unique solution. The domain $\Omega = (0,1)^3$ is decomposed into q overlapping strips $\Omega_i = (X_l^i, X_r^i) \times (0,1)^2$, $i = 1..q$ with $X_l^2 < X_r^1 < X_l^3 < X_r^2, ..., X_l^q < X_{q-1}^r$. We write the discretized problem as follows

$$\frac{U^{n+1} - U^n}{dt} = D_{xx}[U^{n+1}] + D_{yy}[U^{n+1}] + D_{zz}[U^{n+1}] \quad (3)$$
$$+ f(X,Y,Z,t^{n+1}),\ n = 0, \ldots, M-1,$$

with appropriate boundary conditions corresponding. To simplify the presentation, let us deal with homogeneous Dirichlet boundary conditions. We introduce the following expansion for the discrete solution using the sine base functions:

$$U^n(X,Y,Z,t) = \Sigma_{j=1}^{M_y} \Sigma_{k=1}^{M_z} \Lambda_{j,k}^n(X,t)\, sin(jy)\, sin(kz)\,, \quad (4)$$

$$u_o(X,Y,Z) = \Sigma_{j=1}^{M_y} \Sigma_{k=1}^{M_z} \lambda_{j,k}(X) sin(jy)\, sin(kz) \quad (5)$$

$$\text{and } f(X,Y,Z,t^n) = \Sigma_{j=1}^{M_y} \Sigma_{k=1}^{M_z} f_{j,k}^n(X,t^n). \quad (6)$$

M_y and M_z are the number of modes in the y and z direction. Then, by plugging into the discrete solution from (4-6) to (3), we end up with the following independent one dimensional $M_y \times M_z$ problems based on the Helmotz operator:

$$\frac{\Lambda_{j,k}^{n+1} - \Lambda_{j,k}^n}{dt} = D_{xx}[\Lambda_{j,k}^{n+1}] - (\mu_j + \mu_k)\Lambda_{j,k}^{n+1} + f_{j,k}(X,t^{n+1}), \quad (7)$$
$$n = 0, \ldots, M-1,$$
$$\Lambda_{j,k}^0 = \lambda_{j,k}(X). \quad (8)$$

μ_j and μ_k are respectively the eigenvalues of D_{yy} and D_{zz}. This algorithm generates a sequence of vectors $W^n = (\Lambda^n_{2,l}, \Lambda^n_{1,r}, \Lambda^n_{3,l}, \Lambda^n_{2,r}, \ldots, \Lambda^n_{q,l})$ corresponding to the boundary values on the set

$$\mathscr{S} = (X_l^2, X_r^1, X_l^3, X_r^2, \ldots, X_l^q, X_r^{q-1}) \times (t^1, ..., t^M)$$

of the $\Lambda_{j,k}$ for each iterate n. The trace transfer operator is decomposed into $M_y \times M_z$ independent trace transfer operators as well and is defined as follows:

$$W^n_{j,k} - W^\infty_{j,k} \rightarrow W^{n+1}_{j,k} - W^\infty_{j,k}.$$

Let $P_{j,k}$ be the matrix of this linear operator. $P_{j,k}$ has a bloc diagonal structure and can be computed prior to the PAASWR and once for all. In the next section, we describe the general algorithm.

2.2 General Algorithm

The general PAASWR algorithm can then be summarized in three steps:

1. Compute the 1st iterate of ASWR for the 3D parabolic problem (1).
2. Expand the trace of the solution in the eigenvectors basis and solve the linear problem component wise (Aitken-like acceleration)

$$\begin{aligned} (Id - P_{j,k}) W^\infty_{j,k} &= W^1_{j,k} - P_{j,k}\, W^0_{j,k}, \\ \forall j \in \{1, \ldots, M_y\},\ \forall k &\in \{1, \ldots, M_z\}. \end{aligned} \tag{9}$$

   Assemble the boundary conditions $W^\infty = \sum_{j=1}^{M_y} \sum_{k=1}^{M_z} W^\infty_{j,k}\, sin(jy)\, sin(kz)$.
3. Compute the 2nd iterate using the exact boundary value W^∞.

The computation of each subdomain in the first and third steps can be processed by any linear solvers of choice (Multigrid, Krylov, etc.). In the application, we use a combination of Fourier transform and LU decomposition. More details on the above steps and their proof can be found in [9]. In the following section, we will discuss the parallel implementation for the grid.

3 Parallel Implementation extended to the Grid

3.1 The Concept of Subdomain Solver and Interface Solver

From the different steps explained in section 2.2, we determine two process groups: the Subdomain Solver (SS) processes and the Interface Solver (IS) processes. The SS processes work on the parallel computation of the solution of the IBVP in each subdomain $\Omega_i \times (0,T)$, $i = 1 \ldots q$ with q the number of subdomains (Step 1 and 3 in 2.2) The IS processes execute the Aitken-like acceleration and solve the interface problem (Step 2 in 2.2). Not only are they part of the main computation, they also checkpoint *indirectly* the application. Compared to [5] with the spare process concept, we do not add any overheads since it is part of the PAASWR algorithm to save in main memory the interface subdomain solutions. This algorithm is, therefore, called *naturally* fault tolerant. Figure 1 draws this framework. We have one dimensional processor topology implemented in each process group. There are no local neighborhood communications within the groups. The communications are only established between the SS and the IS groups. We do not need as many IS processes as SS processes since the main time-consuming task (the subdomain computation) is performed by the SS processes. In the next section, we describe a new Parallel Non

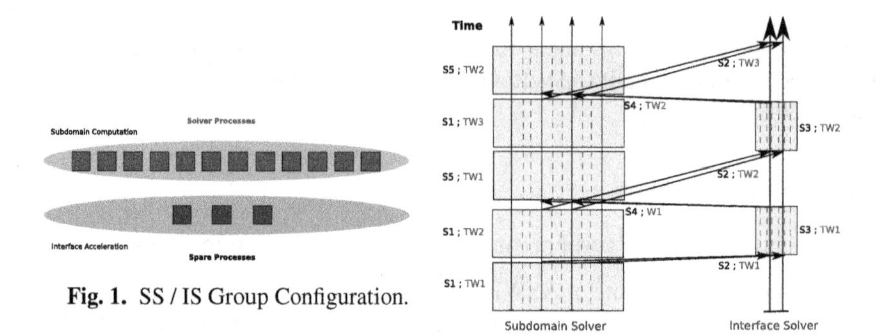

Fig. 1. SS / IS Group Configuration.

Fig. 2. The Parallel Non Blocking Windowed Approach.

Blocking Windowed Version revisited for the grid which are built upon this structure of groups.

3.2 The Parallel Non Blocking Windowed Version revisited

The Parallel Non Blocking Windowed approach was introduced in [7]. Its goal was to allocate more IS processes to handle first the simultaneous communications and second, to parallelize the interface problem resolution. Besides the parallelization in space coming from the domain decomposition in the SS group, this version appends a second level of parallelization in the IS group for the interface problem resolution. That was possible thanks to the nature of the interface problem. Indeed, all eigenvector components are given by the solution of a set of $M_y \times M_z$ parabolic problems completely **decoupled**. One can then distribute straightaway these problems on a set of IS processes, for example, by allocating to each one $M_y/nsp \times M_z$ problems, nsp being the number of IS processes. Figure 2 presents the new framework. This method is very efficient and benefits the best from the SS / IS concept. It is also very challenging to set it up, since many ongoing communications have to be cautiously managed. Figure 3 highlights the critical sections where communications are involved. For simplicity's sake, we did not represent all the communication channels between the two process groups. For instance, in Figure 3(a), the SS process **0** distributes equally its subdomain interface to **all** IS processes. This operation corresponds actually to the collective communication *MPI_SCATTER* function where the current SS process and all the processes from the IS group are involved. The same communication scheme is repeated for each SS process. Collective operations are blocking and may dramatically slow down overall performance on a parallel machine, let alone on a grid. Therefore, we have instead developed our own non blocking communication schemes which offer more opportunities to overlap communication by computation. Conversely, in Figure 3(b), the IS process **0** sends back the new computed interface solutions to **all** SS processes. This operation corresponds in fact to the collective communication *MPI_GATHER* function. The same communication scheme is repeated for each IS process. Again, for performance purposes we did not take this direction and preferred instead to implement it in a non blocking manner. Further, the TW size (*Mopt*) must be estimated empirically to get optimal

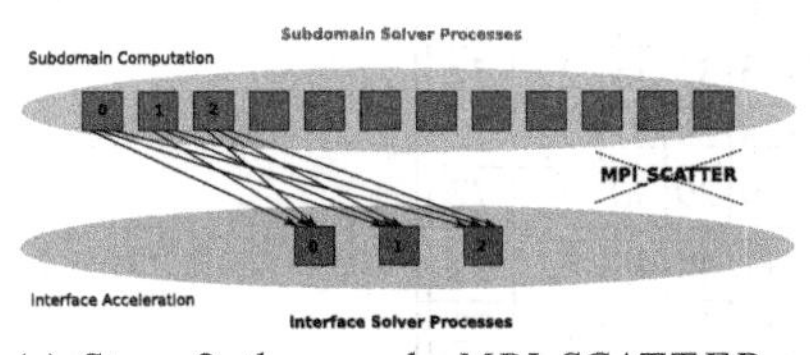

(a) Stage 2: the pseudo *MPI_SCATTER*.

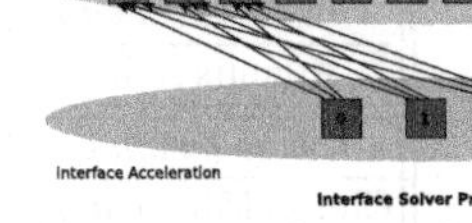

(b) Stage 4: the pseudo *MPI_GATHER*.

Fig. 3. Communication Patterns.

performance. Indeed, we need to find a balance between the space data size and the number of time steps per TW to better minimize the idle process time between the SS and IS groups. Another methodology would be to build a statistical model with some parameters (number of mesh points in each direction and number of time steps per TW) to theoretically determine the *Mopt*. A similar approach for load balancing has been implemented in [8]. In the next section, we present a comparison of the experimental results performed on a single parallel machine and on distributed grid computing.

4 Performance Results on Distributed Environment

The first tests have been performed on a single parallel system to select the best approach to be used later for the grid. At the same time, we determined empirically the ideal *Mopt* to reduce the waiting time. Each SS process is in charge of solving one subdomain and the ratio of SS / IS processes is $2:1$. On a single parallel machine, the Parallel Non Blocking Windowed approach appeared to be the most efficient method and was capable of solving 100^3 unknown problems with 25 time steps in 4.56 seconds with a window size of 3. The network performance may not be affected by dealing with sendings and receivings of small messages so often (each 3 time steps). The network of the parallel system seems to handle frequent and small messages (TW=3) better than occasional and large messages (TW=6). Then, we experiment with the selected Parallel Non Blocking Windowed algorithm on the grid and see whether a TW size of three time steps is also applicable for distributed computing environments. We are distributing the whole domain among three heterogeneous sites: the Itanium 2 cluster ($1.3GHz$) *Atlantis* in Houston (U.S.A.), the Xeon EM64T cluster ($3.2GHz$) *Cacau* in Stuttgart (Germany) and the Itanium 2 cluster ($1.6GHz$) *Cluster150* in Moscow (Russia). The application runs for 25 or 26 time steps depending on the TW size. Table 1 presents the data repartition on each host per subdomain that satisfies load balancing. Three different global sizes are presented: small ($90 \times 72 \times 72$), medium ($120 \times 72 \times 72$), and large ($170 \times 72 \times 72$). Cacau obtains the largest data allocation and seems to be the fastest machine. To lower the communication overhead between remote sites due to the slow network interconnect, a filter has been implemented to only send half of the modes in each direction (Y and Z) of the interface solutions, sufficient to still keep the spectral approximation better than the finite difference approximation. Therefore, from table 1, the actual total number of modes in the two directions sent is 36×36. We could actually even decrease the ratio of SS / IS processes since the interface problems solved by the IS processes are now smaller. Figures 4(a), 4(b) and 4(c) represent the execution time in seconds with the different data grid sizes depending on the TW size. We show the overall performance when dis/enabling communications between the SS and IS groups. One can notice that the higher the TW size, the lower the communication time and thus, the better the execution time. It takes 16.3 seconds to solve 24 subdomains on the grid with a total number of 7 million unknowns and 25 time steps. Figure 4(d) presents the scalability with the optimal TW size of 6 time steps. The elapsed time stays roughly identical when doubling the number of SS processes and keeping the same local subdomain size.

Mopt	**3**	**4**	**5**	**6**	N_y	N_z
Cacau *small*	**32**	**33**	**32**	**33**	72	72
Cluster150 *small*	30	28	29	27	72	72
Atlantis *small*	28	29	29	30	72	72
Cacau *medium*	**46**	**47**	**47**	**49**	72	72
Cluster150 *medium*	44	41	42	38	72	72
Atlantis *medium*	40	42	41	43	72	72
Cacau *large*	**60**	**62**	**62**	**65**	72	72
Cluster150 *large*	59	54	55	50	72	72
Atlantis *large*	51	54	53	55	72	72

Table 1. Local Subdomain Grid Sizes in X space direction.

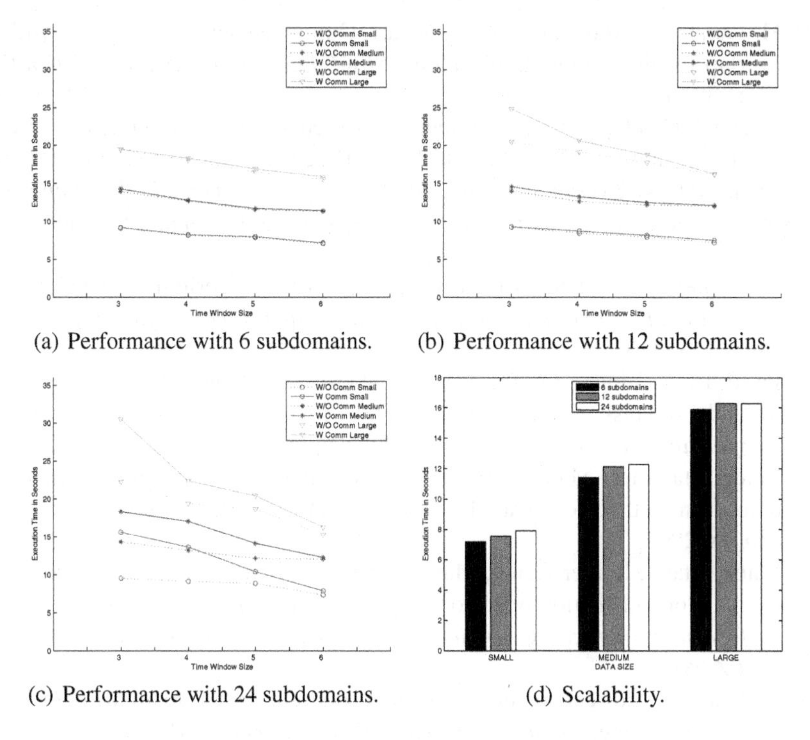

(a) Performance with 6 subdomains. (b) Performance with 12 subdomains.

(c) Performance with 24 subdomains. (d) Scalability.

Fig. 4. PAASWR Execution Time in Seconds.

5 Conclusion

In this paper, we have described how PAASWR can achieve load balancing, scalability, robustness and fault tolerance under grid environments. The Parallel Non Blocking Windowed methodology is the most efficient compared to the other approaches. Indeed, the identification of five successive stages in the general algorithm permits applying a pipelining strategy and therefore, takes advantage of the SS / IS process concept. Also, the parallelization of the interface problem resolution makes PAASWR scalable as the number of subdomains increases. Furthermore, PAASWR is *naturally* fault tolerant, and in case of failures can restart the computation from the interface solutions located in the IS process main memory. The application will then terminate as if no failures occurred.

[1] Nicolas Barberou, Marc Garbey, Matthias Hess, Michael M. Resch, Tuomo Rossi, Jari Toivanen, and Damien Tromeur-Dervout. Efficient metacomputing of elliptic linear and non-linear problems. *J. Parallel Distrib. Comput.*, 63(5):564–577, 2003.

[2] Bruno Després. Décomposition de domaine et problème de Helmholtz. *C.R. Acad. Sci. Paris*, 1(6):313–316, 1990.

[3] M. Gander, F. Magoules, and F. Nataf. Optimized schwarz methods without overlap for the helmholtz equation. *SIAM Journal on Scientific Computing*, 24(1):38–60, 2002.
[4] Martin J. Gander, Laurence Halpern, and Frédéric Nataf. Optimized Schwarz methods. In Tony Chan, Takashi Kako, Hideo Kawarada, and Olivier Pironneau, editors, *Twelfth International Conference on Domain Decomposition Methods, Chiba, Japan*, pages 15–28, Bergen, 2001. Domain Decomposition Press.
[5] Marc Garbey and Hatem Ltaief. On a fault tolerant algorithm for a parallel cfd application. *Parallel Computational Fluid Dynamics, Theory and Applications*, pages 133–140, 2006.
[6] Pierre Louis Lions. On the Schwarz alternating method. II. In Tony Chan, Roland Glowinski, Jacques Périaux, and Olof Widlund, editors, *Domain Decomposition Methods*, pages 47–70, Philadelphia, PA, USA, 1989. SIAM.
[7] Hatem Ltaief and Marc Garbey. A parallel aitken-additive schwarz waveform relaxation method for parabolic problems. *Parallel Computational Fluid Dynamics*, 2007.
[8] Hatem Ltaief, Rainer Keller, Marc Garbey, and Michael Resch. A grid solver for reaction-convection-diffusion operators. *Submitted to the Journal of High Performance Computing Applications (University of Houston Pre-Print UH-CS-07-08).*
[9] M. Garbey. A direct solver for the heat equation with domain decomposition in space and time. *Domain Decomposition in Science and Engineering XVII, Editor Ulrich Langer et Al, Lecture Notes in Computational Science and Engineering No 60, Springer*, pages 501–508, 2007.
[10] V. Martin. An optimized schwarz waveform relaxation method for unsteady convection-diffusion equation. *Applied Num. Math.*, 52(4):401–428, 2005.

Aitken-Schwarz Acceleration not based on the mesh for CFD

D. Tromeur-Dervout[1]

Université de Lyon, Université Lyon 1, CNRS, Institut Camille Jordan, 43 blvd du 11 novembre 1918, F-69622 Villeurbanne-Cedex, France
dtromeur@cdcsp.univ-lyon1.fr

1 Introduction

The generalized Schwarz alternating method (GSAM) was introduced by [4]. Its purely linear convergence in the case of linear operators suggests that the convergent sequence of trace solutions at the artificial interfaces can be accelerated by the well known process of Aitken convergence acceleration. This is the basis of the Aitken-Schwarz method proposed by [1]. In [3] authors extend the Aitken acceleration method to nonuniform meshes, by developing a new original method to compute the Non Uniform Discrete Fourier Transform (NUDFT) based on the function values at the nonuniform points. Nevertheless, the acceleration used was based on the mesh, as it requires an orthogonal basis related to this mesh to decompose the iterated solution at the artificial interfaces. In this paper, we develop a technique that have the same benefits as the Fourier transform: an orthogonal basis to represent the iterated solution and a decrease of the coefficient of the solution related to this basis. This technique creates a robust framework for the adaptive acceleration of the Schwarz method, by using an approximation of the error operator at artificial interfaces based on *a posteriori* estimate of the modes behavior.

The structure of this paper is as follows. Section 2 introduces the pure linear convergence of the GSAM. Then section 3 extends the Aitken acceleration of the convergence technique to the vectorial case with the help of Singular Value Decomposition (SVD). Then the robustness of the method has been validated on Darcy flow problem containing permeability coefficients with considerable contrast in 4.

2 Aitken-Schwarz Method for Linear Operators

Consider $\Omega = \Omega_1 \cup \Omega_2$ where $\Omega_i, i = \{1,2\}$ are two overlapping subdomains. Let $\Gamma_i = (\partial\Omega_i \backslash \partial\Omega) \cap \Omega_{mod(i+1,2)}, i = \{1,2\}$ the artificial interfaces of the DDM. The Generalized Schwarz Additive Method algorithm [4] to solve the linear problem $L(x)u(x) = f(x), \forall x \in \Omega$, $u(x) = g(x), \forall x \partial\Omega$ writes:

D. Tromeur-Dervout (eds.), *Parallel Computational Fluid Dynamics 2008*,
Lecture Notes in Computational Science and Engineering 74,
DOI: 10.1007/978-3-642-14438-7_22,

Algorithm 2 General Schwarz Alternating method

Require: Λ_i's some operators, λ_i's constants
1: Starting $n=0$ and arbitrary conditions u_2^0 compatible with boundary conditions on Γ_2
2: **while** $(||u_{1|\Gamma_1}^{2n+1} - u_{1|\Gamma_1}^{2n-1}|\Gamma_1||_\infty > \varepsilon)$ **do**
3: Solve

$$L(x)u_1^{2n+1}(x) = f(x),\ \forall x \in \Omega_1, \tag{1}$$
$$u_1^{2n+1}(x) = g(x),\ \forall x \in \partial\Omega_1 \backslash \Gamma_1, \tag{2}$$
$$\Lambda_1 u_1^{2n+1} + \lambda_1 \frac{\partial u_1^{2n+1}(x)}{\partial n_1} = \Lambda_1 u_2^{2n} + \lambda_1 \frac{\partial u_2^{2n}(x)}{\partial n_1},\ \forall x \in \Gamma_1. \tag{3}$$

4: Solve

$$L(x)u_2^{2n+2}(x) = f(x),\ \forall x \in \Omega_2, \tag{4}$$
$$u_2^{2n+2}(x) = g(x),\ \forall x \in \partial\Omega_2 \backslash \Gamma_2, \tag{5}$$
$$\Lambda_2 u_2^{2n+2} + \lambda_2 \frac{\partial u_2^{2n+2}(x)}{\partial n_2} = \Lambda_2 u_1^{2n+1} + \lambda_2 \frac{\partial u_1^{2n+1}(x)}{\partial n_2},\ \forall x \in \Gamma_2. \tag{6}$$

5: **end while**

For example $(\Lambda_1 = I, \lambda_1 = 0, \Lambda_2 = 0, \lambda_2 = 1)$ gives the Schwarz Neumann-Dirichlet Algorithm. In [4] it is shown that if $\lambda_1 = 1$ and Λ_1 is the Dirichlet to Neumann mapping operator (DtN) at Γ_1 associated to the homogeneous PDE in Ω_2 with homogeneous boundary condition on $\partial\Omega_2 \cap \partial\Omega$ then GSAM converges in two steps. Let be $\Omega = \Omega_1 \cup \Omega_2$, $\Omega_{12} = \Omega_1 \cap \Omega_2$, $\Omega_{ii} = \Omega_i \backslash \Omega_{12}$. Then $e_i^n = u - u_i^n$ in Ω_i satisfies :

$$(\Lambda_1 + \lambda_1 S_1) R_1 e_1^{2n+1} = (\Lambda_1 - \lambda_1 S_{11}) R_{22} P_2 R_2^* R_2 e_2^{2n} \tag{7}$$
$$(\Lambda_2 + \lambda_2 S_2) R_2 e_2^{2n+2} = (\Lambda_2 - \lambda_2 S_{22}) R_{11} P_1 R_1^* R_1 e_1^{2n+1} \tag{8}$$

with the projection $P_i : H^1(\Omega_i) \rightarrow H^1(\Omega_{ii})$, S_i (respectively S_{ii}) the DtN mapping operator in Γ_i associated to the problem in Ω_i (respectively $\Gamma_{mod(i,2)+1}$ associated to the problem in Ω_{ii}), the trace operators $R_i : H^1(\Omega_i) \rightarrow H^{1/2}(\Gamma_i)$, and $R_{ii} : H^1(\Omega_{ii}) \rightarrow H^{1/2}(\Gamma_{mod(i,2)+1})$ R_i^* to be the left inverse operator of R_i, i.e.,such that $R_i^* :\ R_i^* R_i = I$, $\forall g \in H^{1/2}(\Gamma_i)$, $L(x) R_i^* g = 0, R_i^* g = g\ on \Gamma_i, R_i^* g = 0\ on\ \partial\Omega_i \backslash \Gamma_i$. Equations (7) and (8) can be written as:

$$\begin{pmatrix} R_1 e_1^{2n+1} \\ R_2 e_2^{2n+2} \end{pmatrix} = \begin{pmatrix} \mathbb{P}_1 \mathbb{P}_2 & 0 \\ 0 & \mathbb{P}_2 \mathbb{P}_1 \end{pmatrix} \begin{pmatrix} R_1 e_1^{2n-1} \\ R_2 e_2^{2n} \end{pmatrix} \stackrel{def}{=} \mathbb{P} \begin{pmatrix} R_1 e_1^{2n-1} \\ R_2 e_2^{2n} \end{pmatrix} \tag{9}$$
$$\text{with}\ \mathbb{P}_i = (\Lambda_i + \lambda_i S_i)^{-1} (\Lambda_i - \lambda_i S_{ii}) R_{jj} P_j R_j^* \tag{10}$$

Intensive works in the domain decomposition community try to approximate the Steklov-Poincaré operators $\lambda_i S_{ii}$ by the operator Λ_i see for example [4]. The idea of the Aitken-Schwarz methodology is to try to build exactly or approximately the operator $\mathbb{P}$ (or $\mathbb{P}_2\mathbb{P}_1$) instead of $\lambda_i S_{ii}$. We have then more flexibility to approximate the operator $\mathbb{P}$ and we can derive it explicitly for some separable problem on simple geometries for which the matrix $\mathbb{P}$ is diagonal and the acceleration can be applied on each component [2]. Then the solution at artificial interface is obtained by the Aitken formula: $R_1 u^\infty = (I-\mathbb{P})^{-1}(R_1 u^3 - \mathbb{P} R_1 u^1)$. The discretizing of the operators does not change the nature of the pure linear convergence of the GSAM.

3 The Aitken acceleration for a sequence of vectors

This section focus on the Aitken acceleration to a sequence of vectors that extends quite naturally the scalar case with the help of the SVD. Let us define first the extension of the pure linear convergence for the vectorial case:

Definition 0.1. *A sequence of vectors* $(u^i)_{i\in N}$ *defined on* $\mathbb{R}^n$ *converges purely linearly towards a limit* u^∞ *if:* $u^{i+1} - u^i = \mathbb{P}(u^i - u^{i-1})$, $\forall i > 1$, *where* $\mathbb{P}$ *is a* $\mathbb{R}^{n\times n}$ *non singular matrix independent of* i*.*

Let the relaxed Jacobi $\mathscr{G}$ be the iterative method to solve a linear system $Au = b$ whith $u, b \in \mathbb{R}^n$ and $A \in \mathbb{R}^{n\times n}$ is decomposed in $A = D - E - F$ where D is the main diagonal , $-E$ (resp. $-F$) the strict lower (resp. upper) triangular part of A. $u^{i+1} = \mathscr{G}(u^i)$ writes $u^{i+1} = \omega D^{-1}(b + (E+F)u^i)) + (1-\omega)u^i$ for $\omega \in]0,2[$. Consequently the error satisfies $u^{i+1} - u^\infty = (I - \omega D^{-1}A)(u^i - u^\infty)$ and thus $\mathbb{P} = (I - \omega D^{-1}A)$ independently of i. Then Aitken acceleration gives: $u^\infty = (I-\mathbb{P})^{-1}(u^1 - \mathbb{P}u^0)$ with $(I-\mathbb{P})^{-1} = \frac{1}{\omega}A^{-1}D$, and $(u^1 - \mathbb{P}u^0) = (u^1 - (I-\omega D^{-1}A)u^0) = \omega D^{-1} b$. Consequently the Aitken acceleration technique consists to solve the problem directly. The interest to apply this acceleration will not be obvious if we cannot save some computing in the building of $\mathbb{P}$ and $(I-\mathbb{P})^{-1}$. The following algorithm of the vectorial Aitken acceleration based on the sequence of vectors written in their original canonical basis of $\mathbb{R}^n$ named physical space writes:

Algorithm 3 Vectorial Aitken acceleration in the physical space

Require: $\mathscr{G} : \mathbb{R}^n \to \mathbb{R}^n$ an iterative method having a pure linear convergence
Require: $(u^i)_{1\le i\le n+1}$, $n+1$ successive iterates of $\mathscr{G}$ starting from an arbitrary initial guess u^0

1: Form $E^i = u^{i+1} - u^i$, $0 \le i \le n$
2: **if** $\left[E^{n-1}, \ldots, E^0\right]$ is invertible **then**
3: $\quad \mathbb{P} = \left[E^n, \ldots, E^1\right]\left[E^{n-1}, \ldots, E^0\right]^{-1}$
4: $\quad u^\infty = (\mathbf{I_n} - \mathbb{P})^{-1}(u^{n+1} - \mathbb{P}u^n)$
5: **end if**

This algorithm is limited to a sequence of small size vectors because it needs $n+1$ of iterates related to the vector size n. The main difficulty is to invert the matrix $\left[E^{n-1}, \ldots, E^0\right]$ which can be closed to singular. The objective is then to save as mush computing as possible with the SVD.

3.1 The Singular Value Decomposition

A SVD of a real $n \times m$ $(n > m)$ matrix A is the factorization $A = \mathbb{U}\Sigma\mathbb{V}^*$, where $\mathbb{U} = [U_1, \ldots, U_m]$ is an $n \times m$ matrix with orthonormal columns, Σ is an $m \times m$ nonnegative diagonal matrix with $\Sigma_{ii} = \sigma_i$, $1 \leq i \leq m$ and the $m \times m$ matrix $\mathbb{V} = [V_1, \ldots, V_m]$ is orthogonal. The left $\mathbb{U}$ and right $\mathbb{V}$ singular vectors are the eigenvectors of AA^* and A^*A respectively. It readily follows that $Av_i = \sigma_i u_i$, $1 \leq i \leq m$. Assume that the $\sigma_i, 1 \leq i \leq m$ are ordered in decreasing order and there exits r such that $\sigma_r > 0$ while $\sigma_r + 1 = 0$. Then A can be decomposed in a dyadic decomposition:

$$A = \sigma_1 U_1 V_1^* + \sigma_2 U_2 V_2^* + \ldots + \sigma_r U_r V_r^*. \tag{11}$$

This means that SVD produces an orthonormal basis for representing the data series in a certain least squares optimal sense as follows:

Theorem 0.1. *$X_* = \sigma_1 U_1 V_1^* + \sigma_2 U_2 V_2^* + \ldots + \sigma_k U_k V_k^*$ is a non unique minimizer of the problem $\min_{X, rankX=k} ||A - X||_2$ reaching the value $\sigma_{k+1}(A)$.*

Consider the matrix $A, B \in \mathbb{R}^n$, the Fan inequalities write $\sigma_{r+s+1}(A+B) \leq \sigma_{r+1}(A) + \sigma_{s+1}(B)$ with $r, s \geq 0$, $r+s+1 \leq n$. Considering the perturbation matrix E such that $||E|| = O(\varepsilon)$, then $|\sigma_i(A+E) - \sigma_i(A)| \leq \sigma_1(E) = ||E||_2$, $\forall i$. This good properties allow us to search the acceleration of the convergence of the sequence of vectors in the basis linked to its SVD.

Proposition 0.1. *Let $(u^i)_{1 \leq i \leq m}$ m successive iterates satisfying the pure linear convergence property: $u^i - u^\infty = \mathbb{P}(u^{i-1} - u^\infty)$. Then there exists an orthogonal basis $\mathbb{U} = \left[U^1, U^2, \ldots, U^m\right]$ of a subset of $\mathbb{R}^n$ such that $u^i = \Sigma_{k=1}^m \alpha_k^i U^k, \forall i \in \{1, ..., m\}$ with a decrease of α_k^i with respect to k. Moreover, $(\alpha_1^\infty, \ldots, \alpha_m^\infty)^t \stackrel{def}{=} \mathbb{U}^* u^\infty$, the limit of the sequence of vectors in the space generated by $\mathbb{U}$ satistfies:*

$$(\alpha_1^{j+1} - \alpha_1^\infty, \ldots, \alpha_m^{j+1} - \alpha_m^\infty)^t = \hat{P}(\alpha_1^j - \alpha_1^\infty, \ldots, \alpha_m^j - \alpha_m^\infty)^t, j = 1, \ldots, m. \tag{12}$$

where $\hat{P} \stackrel{def}{=} \mathbb{U}^* P \mathbb{U}$.

proof: by the theorem 0.1 there exist a SVD decomposition of $\left[u^1, \ldots, u^m\right] = \mathbb{U}\Sigma\mathbb{V}$ and we can identify α_k^i as $\sigma_k V_{ik}^*$. The orthonormal property of $\mathbb{V}$ associated to the decrease of σ_k with increasing k lead to a decrease of α_k^i with respect to k. Taking the pure linear convergence of u^i in the matrix form, and applying $\mathbb{U}^*$ leads to:

$$\mathbb{U}^*(u^i - u^\infty) = \mathbb{U}^* P \mathbb{U}\mathbb{U}^*(u^{i-1} - u^\infty) \tag{13}$$

$$(\alpha_1^i - \alpha_1^\infty, \ldots, \alpha_m^i - \alpha_m^\infty)^t = \hat{P}(\alpha_1^{i-1} - \alpha_1^\infty, \ldots, \alpha_m^j - \alpha_m^\infty)^t \tag{14}$$

We can then derive the following algorithm:

Proposition 0.2. *Algorithm 4 converges to the limit u^∞.*

Algorithm 4 Vectorial Aitken acceleration in the SVD space with inverting

Require: $\mathscr{G} : \mathbb{R}^n \to \mathbb{R}^n$ an iterative method having a pure linear convergence
Require: $(u^i)_{1\leq i\leq m+2}$, $m+2$ successive iterates of $\mathscr{G}$ starting from an arbitrary initial guess u^0

1: Form the SVD decomposition of $\mathbf{Y} = \left[u^{m+2}, \ldots, u^1\right] = \mathbf{USV}'$
2: set l the index such that $l = \max_{1\leq i\leq m+1} \{\mathbf{S}(i,i) > tol\}$, $\{$ex.:$tol = 10^{-12}.\}$
3: set $\hat{\mathbf{Y}}_{1:l,1:l+2} = \mathbf{S}_{1:l,1:l}\mathbf{V}^t_{1:l,m-l:m+2}$
4: set $\hat{\mathbf{E}}_{1:l,1:l+1} = \hat{\mathbf{Y}}_{1:l,2:l+2} - \hat{\mathbf{Y}}_{1:l,1:l+1}$
5: **if** $\hat{\mathbf{E}}_{1:l,1:l}$ is invertible **then**
6: $\quad\hat{\mathbf{P}} = \hat{\mathbf{E}}_{1:l,2:l+1}\,\hat{\mathbf{E}}^{-1}_{1:l,1:l}$
7: $\quad\hat{y}^{\infty}_{1:l,1} = (\mathbf{I}_l - \hat{\mathbf{P}})^{-1}\,(\hat{Y}_{1:l,l+1} - \hat{\mathbf{P}}\,\hat{Y}_{1:l,l})$ {Aitken Formula}
8: $\quad u^{\infty} = \mathbf{U}_{:,1:l}\,\hat{y}^{\infty}_{1:l,1}$
9: **end if**

Proof: as the sequence of vector u^i converges to a limit u^∞ then we can write $\Xi = \left[u^1, \ldots, u^m\right] = \left[u^\infty, \ldots, u^\infty\right] + E$ where E is a $n \times m$ matrix with decreasing coefficients with respect to the columns. The SVD of $\Xi^\infty = [u^\infty, \ldots, u^\infty]$ leads to have $U^1 = u^\infty$ and $\sigma_i(\Xi^\infty) = 0$, $i \geq 2$. The fan inequalities lead to have $\sigma_i(\Xi) \leq \sigma_1(E) = ||E||_2, i \geq 2$. Consequently, the algorithm 4 decreases the number of non zero singular values at each loop iterate.

In Algorithm 4 the building of $\mathbb{P}$ needs the inversion of the matrix $\hat{\mathbf{E}}^{-1}_{1:l,1:l}$ which can contain very small singular values even if we selected those greater than a tolerance. A more robust algorithm can be obtained without inverting $\hat{\mathbf{E}}^{-1}_{1:l,1:l}$. It consists on building $\mathbb{P}$ with applying the iterative method $\mathscr{G}$ to the selected columns of $\mathbb{U}$ that appears in Algorithm 4. Then $\hat{\mathbb{P}} = \mathbb{U}^*_{1:n,1:l}\mathscr{G}(\mathbb{U}_{1:n,1:l})$.

Two main parallel techniques to compute the SVD have been proposed [5] [6]. One consists in two phases: the first phase transforms the matrix in a bidiagonal matrix and then the second phase computes the singular values of this bidiagonal matrix based on a rank one modification.A quasi optimal implementation of a parallel divide and conquer algorithm using cyclic reduction was performed in [5]. The second approach consists in using the Jacobi method which is based on transformation with Given's rotations to compute the eigenvalues of a symmetric matrix. In [6] authors obtained comparable computational efficiency to the bidiagonalisation approach with more accuracy. Bidiagonal Divide and Conquer (BDC) algorithm computes all the singular values of a $N \times N$ matrix in $O(N^2)$ time and all the singular values and singular vectors in $O(N^3)$ time. By using the fast multipole method in [7], BDC can be accelerated to compute all the singular values in $O(Nlog_2N)$ time and all the singular values and singular vectors in $O(N^2)$.

4 Numerical results on Darcy equation

The 2D Darcy equation writes: $\nabla.(K(x,y)\nabla u) = f$, on$\Omega$, $u = 0$, on $\partial\Omega$

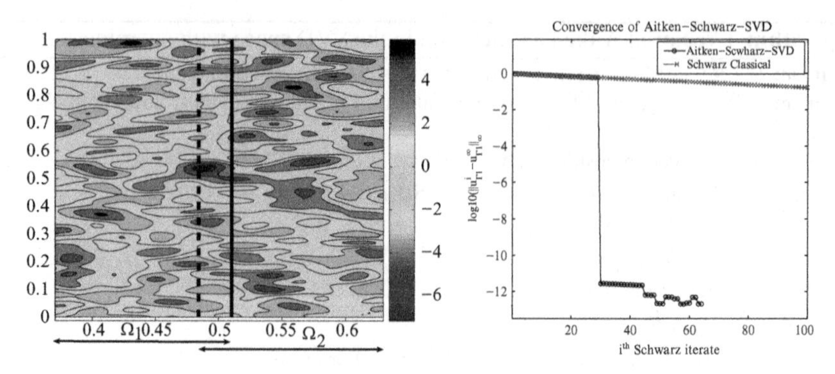

Fig. 1. Schwarz DDM accelerated by the Aitken SVD procedure: (left) random distribution of K along the interfaces , (right) the convergence of the Aitken-Schwarz

In the case of 2D medium, the computational domain is a regular grid on which a random hydraulic conductivity field K is generated. K follows a stationary log-normal probability distribution $Y = ln(K)$, which is defined by a mean m_Y and a covariance function $C_Y(x,y) = \sigma_Y exp(-[(\frac{x}{\lambda_x})^2 + (\frac{y}{\lambda_y})^2]^{\frac{1}{2}})$ where σ_Y is the variance of the log hydraulic conductivity and λ_x and λ_y) are the directional correlation length scales in each direction. The porous medium is assumed to be isotropic. To generate the random hydraulic field, a spectral simulation based on the FFT method is used . For sake of simplicity, we take the same value λ for λ_x and λ_y.

The σ and λ parameters have an impact on the stiffness of the linear system to be solved. The range of σ^2 is usually from 2 to 6 and the λ goes from 2 to 10. σ plays on the amplitude of the permeability K, for $\sigma^2 = 4$ the K varies in mean from $10^{-7.28}$ to $10^{7.68}$. λ represents the length scale for the change of K, smaller is λ greater is the probability that the K vary strongly from cell to cell.

The domain is discretized by a regular grid of stepsizes (h_x, h_y) and the operator is discretized by second order finite differences leading to a five points stencil where corresponding coefficients in the matrix are taken as the harmonic mean of permeability in the neighboring cells. The diagonal term is thus the negative sum of the four other terms.

We apply the Aitken-Schwarz method with overlapping GSAM algorithm with $\lambda_1 = \lambda_2 = 0$ and $\Lambda_1 = \Lambda_2 = \mathbb{I}$ on the domain $\Omega = [0,1]^2$ split in Ω_1 and Ω_2. The overlap is chosen to as $5h_x$ in order to have a small convergence. The solution is set to be $16*y*(y-1)*x*(x-1)*\exp(-40*((x-1/2)^2+(y-1/2)^2))$. The number of points in each direction is $n_x = 200$ and $n_y = 100$ Figure 1 left part gives the random distribution of the permeability K on the domain with $\lambda = 5$ and $\sigma^2 = 4$. As it can be seen, the permeability fluctuates strongly along the two interfaces. The permeability varies from $10^{-7.28} to 10^{7.68}$. Figure 1 right shows the convergence of the Aitken-Schwarz with the acceleration based on the SVD. We computed 30 Schwarz iterations to generate the SVD of the iterated solution trace on Γ_1. Then only 16 modes of this SVD are used in the acceleration process for the first acceleration. Based on this information we perform only 16 iterations for the second acceleration.

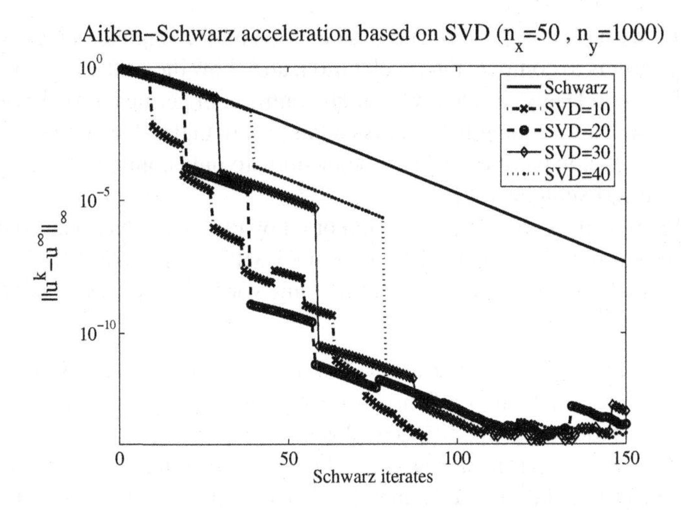

Fig. 2. Convergence of Aitken-Schwarz with respect to the number of iterates in the SVD

For the second acceleration, only four modes are needed and then the solution enter in the numerical noise area. The size of the vectors in the sequence is reduced due to the Schwarz domain decomposition, compared to the entire domain. Figure 2 gives the convergence of the Aitken Schwarz with the SVD applied recursively on the sequence of 10, 20, 30, and 40 iterates of Schwarz. The $\lambda = 5$ and $\sigma^2 = 5$, the number of discretizing points on the interface is set to $n_y = 1000$, the number of points n_x in each subdomain is set to 50. The overlap is again $5h_x$ cells. Consequently the Schwarz overlap is greater leading to a better convergence rate of the algorithm. The Aitken Schwarz exhibits quite good convergence even with a limit set of singular vectors; The convergence is reached in nearly the same amount of global Schwarz iterates for the 4 cases. The Algorithm 4 was used.

5 Conclusions

We present the Aitken-Schwarz methodology which is linked to the pure linear convergence of the Schwarz method when applied to linear operators. In case of a problem with separable operators, we can find some basis linked to the operator or to the mesh in order to expand the solution in this basis. Then the scalar Aitken acceleration can be applied to each mode. Morerover, when the operator is no more separable and/or the mesh is no more regular, the decoupling of modes of the solution is no more available. We then propose to accelerate the sequence of vectors generated by Schwarz with the Aitken acceleration in a matrix form. For this we proceed to the SVD decomposition of the sequence of vectors which has the property to generate an orthogonal set of vectors and a dyadic decomposition of the sequence of vectors in this basis. We obtain then a method for the acceleration that is mesh independent

and for which we have an a posteriori estimate based on the singular values. We propose two algorithms to compute the acceleration, and show their efficiency on Jacobi method and also on Darcy problem with high contrasts in permeability. Large scale computations on Darcy equation with this technique are under development. We are also looking to develop parallel SVD decomposition to only compute the basis associated to the largest singular values.
Acknowledgement:this work has been supported by the french National Agency of Research through the projects ANR-07-CIS7-004 MICAS and ANR-07-TLOG-06 LIBRAERO, the Région Rhône-Alpes in the framework of the project CHP of cluster ISLES

[1] M. Garbey and D. Tromeur-Dervout, *Two level domain decomposition for Multiclusters*, 12th Int. Conf. on Domain Decomposition Methods DD12, T. Chan & Al editors, ddm.org, pp. 325–339, 2001.
[2] M. Garbey and D. Tromeur-Dervout, *On some Aitken like acceleration of the Schwarz method.*, Internat. J. Numer. Methods Fluids 40(12), 1493–1513, 2002.
[3] A. Frullone, D. Tromeur-Dervout, *A new formulation of NUDFT applied to Aitken-Schwarz DDM on nonuniform meshes*, Parallel Computational Fluid Dynamics 2005, 493–500, 2006.
[4] B. Engquist, H. K. Zhao, *Absorbing boundary conditions for domain decomposition*, Appl. Numer. Math. 27 (4), 341–365, 1998.
[5] I. Bar-On, M. Leoncini, *Reliable parallel solution of bidiagonal systems* Numer. Math., 90, 415–440, 2002.
[6] Z. Drmac, K. Veselic, *New fast and accurate Jacobi SVD algorithm. I.* SIAM J. Matrix Anal. Appl. 29(4), 1322–1342, 2008.
[7] J. Carrier, L. Greengard, V. Rokhlin, *A fast adaptive multipole algorithm for particle simulations.* SIAM J. Sci. Statist. Comput., 9(4),669–686, 1988.

From extruded-2D to fully-3D geometries for DNS: a Multigrid-based extension of the Poisson solver

A. Gorobets, F. X. Trias, M. Soria, C. D. Pérez-Segarra, and A. Oliva

Centre Tecnològic de Transferència de Calor (CTTC)
ETSEIAT, c/ Colom 11, 08222 Terrassa, Spain
E-mail: cttc@cttc.upc.edu, web page: http://www.cttc.upc.edu

Abstract. Direct numerical simulation (DNS) of incompressible flows is an essential tool for improving the understanding of the physics of turbulence and for the development of better turbulence models. The Poisson equation, the main bottleneck from a parallel point of view, usually also limits its applicability for complex geometries. In this context, efficient and scalable Poisson solvers on fully-3D geometries are of high interest.

In our previous work, a scalable algorithm for Poisson equation was proposed. It performed well on both small clusters with poor network performance and supercomputers using efficiently up to a thousand of CPUs. This algorithm named Krylov-Schur-Fourier Decomposition (KSFD) can be used for problems in parallelepipedic 3D domains with structured meshes and obstacles can be placed inside the flow. However, since a FFT decomposition is applied in one direction, mesh is restricted to be uniform and obstacles to be 2D shapes extruded along this direction.

The present work is devoted to extend the previous KSFD algorithm to eliminate these limitations. The extension is based on a two-level Multigrid (MG) method that uses KSFD as a solver for second level. The algorithm is applied for a DNS of a turbulent flow in a channel with wall-mounted cube. Illustrative results at $Re_\tau = 590$ (based on the cube height and the bulk velocity $Re_h = 7235$) are shown.

Keywords: parallel 3D Poisson solver; Schur complement method; FFT; Multigrid; Preconditioned Conjugate Gradient; Wall-mounted cube; DNS;

1 Introduction

Direct numerical simulation (DNS) has become an important area of contemporary fluid dynamics, because its interest for improving the understanding of the physics of turbulence and because it is an essential tool for the development of better turbulence models. In this context, high resolution DNS results at of relatively complex geometries and configurations are of extreme importance for further progress. The main idea behind this is to assess the validity of turbulence models in more realistic configurations, understand their limitations and finally improve them. Therefore,

D. Tromeur-Dervout (eds.), *Parallel Computational Fluid Dynamics 2008*,
Lecture Notes in Computational Science and Engineering 74,
DOI: 10.1007/978-3-642-14438-7_23,

this is really a crucial issue since turbulence modelling ultimately becomes an essential tool for engineering applications. In this context, the availability of efficient and scalable Poisson solvers for fully-3D geometries is of extreme importance.

For simplicity, we restrict ourselves to parallelepipedic geometries. In such configurations, the complexity of DNS increases dramatically with the number of arbitrarily meshed directions. This is mainly due to the fact that the discrete Laplacian operator cannot be analytically diagonalized in such directions. Consequently, most of the DNS simulations have been restricted to flows with at least one homogeneous direction. In this context, efficient and scalable Poisson solvers applicable to solve fully-3D flows are of high interest. Immersed boundary (IB) methods has become an alternative to circumvent the Poisson solver problem. However, IB methods suffer strong grid limitations and can introduce significant non-physical effects.

In the previous version of our DNS code (see [4] and [1]) the Poisson solver was based on combination of FFT and Conjugate Gradient (CG) method (see [2], for instance) preconditioned with a direct Schur Decomposition (DSD) method (see [3], for example). The Fourier decomposition is used to uncouple the original 3D Poisson equation into a set of independent 2D planes. Then, each 2D problem in solved using a CG method preconditioned by a DSD algorithm. To do that, each plane is decomposed into *blocks* and each of them in solved with the DSD solver. However, the use of the FFT has the following restrictions:

- Mesh must be uniform in the direction where FFT is applied.
- Obstacles geometry is restricted to be extruded-2D through such direction.

In what follows such geometry will be denoted as limited-3D geometry as it is an extrusion of a 2D geometry with a constant mesh step. Therefore, the number of arbitrarily meshed directions can not be more than two.

The main goal of the present work has been to extend the KSFD solver to be able to solve fully-3D cases with non-uniform mesh in all three spatial directions (that allows to resolve all boundary layers) and arbitrary obstacles. The fairly good scalability of the original method should be preserved with a reasonable efficiency. Finally most promising results were obtained when combining a two-level Multigrid (MG) with the KSFD method.

2 Governing equations and numerical method for DNS

The non-dimensional incompressible Navier-Stokes equations in a parallelepipedic domain $\Omega = (0,L_x) \times (0,L_y) \times (0,L_z) \subset \mathbb{R}^3$ in primitive variables are considered

$$\frac{\partial \mathbf{u}}{\partial t} + (\mathbf{u} \cdot \nabla)\mathbf{u} = \frac{1}{Re}\Delta \mathbf{u} - \nabla p \; ; \qquad \nabla \cdot \mathbf{u} = 0 \tag{1}$$

where Re is the non-dimensional Reynolds number.

Equations (1) are discretized on a staggered grid in space by symmetry-preserving schemes by [7]. For the temporal discretization, a fully explicit dynamic second-order one-leg scheme is used for both convective and diffusive terms. Finally, to solve the pressure-velocity coupling a classical fractional step projection method is used. Further details about the time-integration method can be found in [6, 5].

3 On the extension of the KSFD algorithm for fully-3D problems

In this work, we investigate the feasibility of using the limited-3D solver, *i.e* KSFD solver, as an approximation of the fully-3D case. Initially, two different strategies have been explored:

- CG-KSFD: to use KSFD solver as preconditioner for a Krylov CG method.
- MG-KSFD: a two-level Multigrid (MG) approach using KSFD as a second-level solver. Then, a CG with a local band-LU preconditioner is used as smoother.

Original fully-3D system to be solved and limited-3D system are respectively denoted

$$A^{3D}x^{3D} = b^{3D} \tag{2a}$$

$$Ax = b \tag{2b}$$

where A^{3D} and A are both symmetric positive definite matrices. Both algorithms are briefly outlined in the next subsections. CG-KSFD approach is not considered further since it was substantially outperformed by MG-KSFD.

3.1 CG-KSFD approach:

Algorithm on i-th iteration:

1. Call to preconditioner: $M(A, z_i^{3D}, r_i^{3D})$
2. Using z_i^{3D} obtain new x_{i+1}^{3D} by means of CG algorithm inner operations.

where r_i^{3D} is the residual of (2a) on i-th iteration and A is matrix of the limited-3D system (2b).

Algorithm of preconditioner $M(A, z_i^{3D}, r_i^{3D})$:

1. Transfer r_i^{3D} by means of operator Q to limited-3D mesh: $r_i = Qr_i^{3D}$.
2. Solve $Az_i = r_i$ using KSFD algorithm.
3. Transfer z_i by means of operator P back to fully-3D case: $z_i^{3D} = Pz_i$

The choice of operators P and Q will be further considered.

3.2 MG-KSFD approach:

A two-level MG method is proposed. Second-level of MG becomes a limited-3D case that is then solved using the KSFD algorithm.

Algorithm on i-th iteration:

1. Smoother: Obtain approximate solution x_i^{3D} of (2a) using CG with local preconditioner. It does not demand any data exchange.
2. Calculate residual r_i^{3D} of system (2a).
3. Transform residual to second MG level $r_i = Qr_i^{3D}$
4. Solve error equation $Az_i = r_i$ on second level using KSFD algorithm.
5. Transform error from the limited-3D second level to fully-3D: $z_i^{3D} = Pz_i$
6. Correct $x_{i+1}^{3D} = x_i^{3D} + z_i^{3D}$

Here matrices Q and P represent restriction and prolongation operators.
The idea behind of the MG-KSFD method is the following: KSFD algorithm solving limited-3D case efficiently eliminates low-frequencies of error and smoother provides fast convergence on higher frequencies.

4 Solution of the second MG level

Solution of the second MG level is provided by KSFD solver (see [4] and [1]). For the sake of simplicity, we have kept the number of nodes on the second MG level equal to the first level. In this way, no additional data numeration and allocation is required. Two different remapping approaches have been tested:

- A first-order conservative interpolation. Conservative interpolation preserves the integral of the quantity:
$$\sum_{i=1}^{N} \Omega_i x_i = \sum_{i=1}^{N} \tilde{\Omega}_i \tilde{x}_i \qquad (3)$$
where Ω_i and $\tilde{\Omega}_i$ are control volumes of first and second level respectively, x_i and $\tilde{x}_i$ are values from the first and second level. This conservative inter-level transfer provides stability of the iterative process.
- No remapping. As the numbers of nodes are equal, vectors are taken without transformations (Q and P are identity matrixes)

Preliminary tests performed using first-order conservative interpolation (3) showed stable convergence but no substantial gain compared with second (no remapping) approach was observed. Higher-order accurate interpolation remapping schemes may improve the convergence. However, the main goal of the current work has been to demonstrate the robustness of the MG-KSFG algorithm without pretending to achieve the optimal performance yet. Hence, further research on more accurate restriction/prolongator operators remain as an interesting area for future research.

5 Motivation: DNS of a wall-mounted cube

Flow around a wall-mounted cube in a channel flow has been chosen as a first demonstrative DNS application for the MG-KSFD solver. Despite the simple geometry of the flow around parallelepipedic 3D bluff obstacles several aspects strongly limit their DNS simulations:

- Region around obstacle demands higher resolution. Thus, arbitrary meshing is desirable in order to achieve the resolution requirements without wasting a huge amount of computational resources in order regions.
- Additionally, the presence of a 3D bluff obstacle itself does not allow to apply directly an FFT-based method (unless some IB approach is used).

In the last two decades this configuration has been experimental and numerically studied. Results showed that this flow is mainly characterized by the appearance of a horseshoe-type vortex at the upstream face, an arc-shaped vortex in the wake of the cube, flow separation at the top and side faces of the cube and vortex shedding. However, most of the numerical studies have been performed using RANS and LES modelling techniques while accurate DNS simulations are quite scarce and limited to very low Reynolds number (the reader is referred to the review by [8] and the references therein). Moreover, since this flow configuration is used for benchmarking purposes to validate turbulence models and numerical methods the availability of DNS results at relatively high Reynolds numbers is of extreme importance.

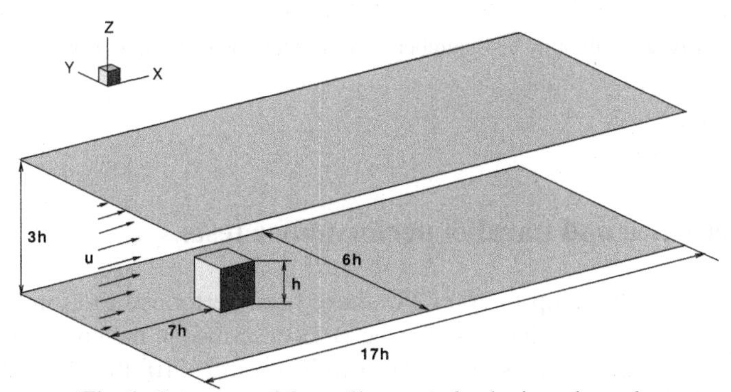

Fig. 1. Geometry of the wall-mounted cube in a channel.

The geometry of the wall-mounted cube in a channel is displayed in figure 1. The computational domain is $17h \times 6h \times 3h$ in the streamwise, spanwise and normal to the channel wall directions where h is the cube height. The upstream face of the cube is located at $7h$ from the inlet. For the sake of clarity[1] the following analytical profile has been prescribed at the inlet

[1] Influence of inflow boundary conditions have been studied. Transient fully-developed channel flow, channel flow averaged profile and analytic profile (4) have been considered. It was found that for cube locations far enough from the inlet ($\gtrsim 5h$, for the range of Re_τ-number

$$U^+ = U/u_\tau = min\left(y^+, k\ln y^+ + B\right) \tag{4a}$$
$$V^+ = W^+ \ = 0 \tag{4b}$$

where $y^+ = (y/H)Re_\tau$, $u_\tau = Re_\tau \nu/H$, $k = 0.25$ and $B = 5.0$. H is the channel half-height (in our case $H = 1.5h$). Convective boundary conditions are in imposed at the outflow. Global mass conservation is forced through a minor[2] correction of the outflow conditions. Non-slip boundary conditions are imposed at the channel and at the obstacle surfaces. Periodic boundary conditions are imposed at the y-direction.

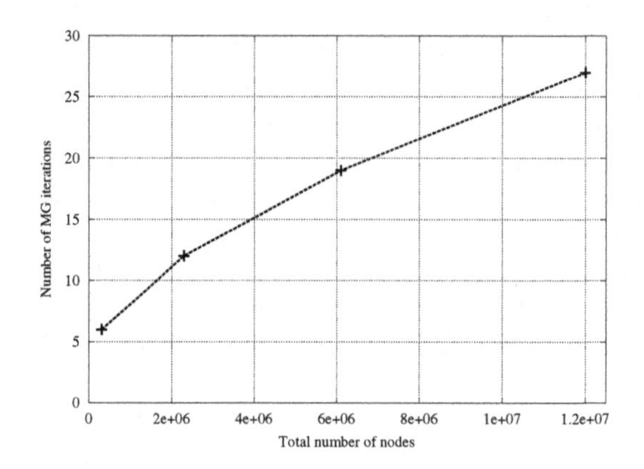

Fig. 2. Scalability test: number of MG iterations with mesh growth

6 Convergence and parallel performance tests

Several scalability tests up to 200 CPUs have been performed to estimate performance of the MG-KSFD algorithm for the DNS described in the previous section[3]. Averaged number of iterations for meshes varying from 2×10^5 to 1.2×10^7 nodes is displayed in figure 2. Re_τ-number was initially set to 425 for these tests. The solver has following configuration is the following:

- Residual tolerance for MG solver is set to 10^{-5}.

---

studied) no significant differents are observed. Therefore, since the analytical profile is the simplest and easiest-to-reproduce we adopted this inlet boundary conditions.

[2] In practice, several orders of magnitude lower than velocity values.

[3] Total number of nodes has been distributed according to spatial resolution requirements for the selected DNS problem. The criteria used here is the same used by [5]. Further details about the mesh spacing functions and geometry are beyond the scope of the present work.

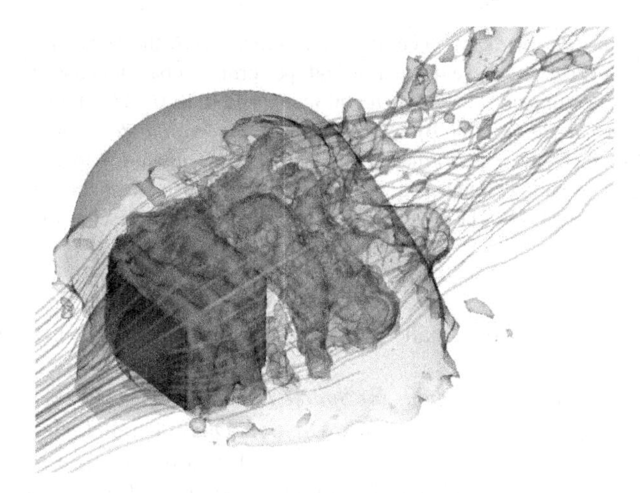

Fig. 3. Surface mounted cube in a channel flow at $Re_\tau = 590$ (based on the cube height and the bulk velocity $Re_h = 7235$). Pressure iso-surfaces with streamlines.

- Smoother consists on 15 preconditioned CG iterations. Jacobi (diagonal scaling) is used as preconditioner.
- Restriction/prolongation operators are identity matrix (no remapping).
- Residual tolerance for second MG level is set to 10^{-2}.

We observe that the number of MG iterations is increased about 4.5 times while the number of nodes is increased by a factor of 60. However, it must be noted that there are several parameters that can be tuned, in particular the number of smoother iterations, in order to improve the overall performance. For instance, preliminary DNS series for $Re_\tau = 590$ (based on the cube height and the bulk velocity $Re_h = 7235$) the number of MG iterations for a mesh of 24×10^6 nodes was around $6 \sim 7$ after further solver tuning (number of smoother iterations equal to 25). A illustrative snapshot of this DNS simulation in displayed in figure 3.

7 Conclusions and future research

A scalable algorithm for solving the Poisson equation arising from fully-3D problems has been proposed and tested. It is based on a two-level MG using KSFD as a second-level solver. Preliminary scalability tests show that the MG-KSFD algorithm performs good enough to carry out DNS simulations on meshes up to about $30 \sim 40$ millions of nodes using second-order discretization. The algorithm have been successfully applied for a DNS of a turbulent flow in a channel with a wall-mounted cube at $Re_\tau = 590$. Further research on more accurate restriction/prolongator operators and the influence of solver parameters remain as interesting areas for future research.

Acknowledgement. This work has been financially supported by the *Ministerio de Educación y Ciencia*, Spain, (Project: "Development of high performance parallel codes for the optimal design of thermal equipments". Contract/grant number ENE2007-67185) and a postdoctoral fellowship *Beatriu de Pinós* (2006 BP-A 10075) by the *Generalitat de Catalunya*.

Calculations have been performed on the IBM MareNostrum supercomputer at the Barcelona Supercomputing Center and on the MVS 100000 supercomputer at the Joint Supercomputer Center of RAS. The authors thankfully acknowledge these institutions.

[1] A. Gorobets, F. X. Trias, M. Soria, and A. Oliva. A scalable Krylov-Schur-Fourier Decomposition for the efficient solution of high-order Poisson equation on parallel systems from small clusters to supercomputers. *Computers and Fluids*, under revision, 2008.
[2] Yousef Saad. *Iterative Methods for Sparse Linear Systems*. PWS, 1996.
[3] M. Soria, C. D. Pérez-Segarra, and A. Oliva. A Direct Parallel Algorithm for the Efficient Solution of the Pressure-Correction Equation of Incompressible Flow Problems Using Loosely Coupled Computers. *Numerical Heat Transfer, Part B*, 41:117–138, 2002.
[4] F. X. Trias, A. Gorobets, M. Soria, and A. Oliva. DNS of Turbulent Natural Convection Flows on the MareNostrum Supercomputer. In *Parallel Computational Fluid Dynamics*, Antalya, Turkey, May 2007. Elsevier.
[5] F. X. Trias, M. Soria, A. Oliva, and C. D. Pérez-Segarra. Direct numerical simulations of two- and three-dimensional turbulent natural convection flows in a differentially heated cavity of aspect ratio 4. *Journal of Fluid Mechanics*, 586:259–293, 2007.
[6] F. X. Trias, M. Soria, C. D. Pérez-Segarra, and A. Oliva. A Direct Schur-Fourier Decomposition for the Efficient Solution of High-Order Poisson Equations on Loosely Coupled Parallel Computers. *Numerical Linear Algebra with Applications*, 13:303–326, 2006.
[7] R. W. C. P. Verstappen and A. E. P. Veldman. Symmetry-Preserving Discretization of Turbulent Flow. *Journal of Computational Physics*, 187:343–368, May 2003.
[8] A. Yakhot, T. Anor, H. Liu, and N. Nikitin. Direct numerical simulation of turbulent flow around a wall-mounted cube: spatio-temporal evolution of large-scale vortices. *Journal of Fluid Mechanics*, 566:1–9, 2006.

Parallel direct Poisson solver for DNS of complex turbulent flows using Unstructured Meshes

R.Borrell[1,2], O.Lehmkuhl[1,2], F.X.Trias[1], M.Soria[1], and A.Oliva[1]

[1] Centre Tecnològic de Transferència de Calor, Technical University of Catalonia, ETSEIAT, C/Colom 11, 08222 Terrassa, Spain cttc@cttc.upc.edu
[2] Termo Fluids, S.L., Mag Colet, 8, 08204 Sabadell (Barcelona), Spain

Abstract. In this paper a parallel direct Poisson solver for DNS simulation of turbulent flows statistically homogeneous in one spatial direction is presented. It is based on a Fourier diagonalization and a Schur decomposition on the spanwise and streamwise directions respectively. Numerical experiments carried out in order to test the robustness and efficiency of the algorithm are presented. This solver is being used for a DNS of a turbulent flow around a circular cylinder at $Re = 1 \times 10^4$, the size of the required mesh is about 104 M elements and the discrete Poisson equation derived is solved in less than one second of CPU time using 720 CPUs of Marenostrum supercomputer.

Keywords: parallel Poisson solver, Schur decomposition, FFT, DNS, unstructured meshes.

1 Introduction

Direct Numerical Simulation (DNS) of turbulent flows are rarely used for 'real applications'. This is because the size of the required mesh and the time step are proportional to $Re^{9/4}$ and $Re^{1/2}$ respectively. However, DNS are of high interest for the study of the physics of turbulent flows because the numerical results are obtained without modelling any term of the Navier-Stokes equations. Besides, DNS has become very important for the improvement and validation of new turbulence models.

In the numerical algorithm used for DNS, the resolution of a Poisson equation, which arises from the incompressibility constraint and has to be solved at least once at each time step, is usually the main bottleneck in terms of RAM memory and CPU time requirements. In this context, efficient and scalable algorithms for the solution of the Poisson equation are of high interest.

A parallel Schur-Fourier Decomposition algorithm for the solution of discrete Poisson equation on extruded unstructured meshes is proposed in the present work. This method has been used before to carry out DNS for a differentially heated cavity at high Rayleigh numbers using *Cartesian* grids [5]. The goal of the present work is to extend it to unstructured meshes.

D. Tromeur-Dervout (eds.), *Parallel Computational Fluid Dynamics 2008*,
Lecture Notes in Computational Science and Engineering 74,
DOI: 10.1007/978-3-642-14438-7_24, © Springer-Verlag Berlin Heidelberg 2010

2 Numerical method for DNS

2.1 Spatial discretization of Navier-Stokes equations

In this paper, turbulent flows that are statistically homogeneous in one spatial direction are considered. These flows can be handled very well using periodic boundary conditions in that direction. The absence of boundary layers, together with the homogeneity of the flow, yields to use extruded unstructured meshes with uniform step on the spanwise direction. The advantage of unstructured type, compared with body-fitted structured or Cartesian grids with cut cells, is that any computational domain can be easily dealt with. On the other hand, the refinement around internal bodies is local and does not generate unnecessary cells in other parts of the domain. However, from a computational point of view, the lack of structure yields to a more complex data management.

The finite volume discretization of the Navier-Stokes and continuity equations in an arbitrary mesh can be written as

$$\rho\Omega\frac{du_c}{dt}+\mathsf{C}(u_c)\,u_c+\mathsf{D}u_c+\Omega\mathsf{G}p_c=0_c \tag{1}$$

$$\mathsf{M}u_c=0_c \tag{2}$$

where $u_c\in\mathbb{R}^{3n}$ and $p_c\in\mathbb{R}^{n}$ are the velocity vectors and pressure, respectively. The matrix $\Omega\in\mathbb{R}^{3n\times 3n}$ is a diagonal matrix of velocity cell control volumes. The matrices $\mathsf{C}(u_c)$, $\mathsf{D}\in\mathbb{R}^{3n\times 3n}$ are the convective and diffusive operators, respectively. And finally, $\mathsf{G}\in\mathbb{R}^{3n\times n}$ represents the discrete gradient operator, and the matrix $\mathsf{M}\in\mathbb{R}^{n\times 3n}$ is the divergence operator.

The conservative nature of the Navier-Stokes equations is intimately tied with the symmetries of their differential operators. In this paper a symmetry-preserving discretization of the differential operators is used. This means that the discrete operators conserve the symmetry properties of its continuous counterparts. For further details about the symmetry-preserving discretization, the reader is referred to [6].

2.2 Time integration

The temporal discretization is carried out using a central difference scheme for the time derivative term, a fully explicit second-order one-leg scheme for $\mathsf{R}(u_c):=-\mathsf{C}(u_c)\,u_c-\mathsf{D}u_c$ ([6]), and a first-order backward Euler scheme for the pressure-gradient term. Incompressibility constraint is treated implicitly. Thus, the fully-discretized Navier-Stokes equations are obtained:

$$\mathsf{M}u_c^{n+1}=0 \tag{3}$$

$$\frac{(\beta+1/2)\,u_c^{n+1}-2\beta u_c^{n}+(\beta-1/2)\,u_c^{n-1}}{\Delta t}=\mathsf{R}\left((1+\beta)\,u_c^{n}-\beta u_c^{n-1}\right)-\mathsf{G}_c p_c^{n+1} \tag{4}$$

where the parameter β is computed each time-step to adapt the linear stability domain of the time-integration scheme to the instantaneous flow conditions in order to use the maximum Δt possible. For further details about the time-integration method, the reader is referred to [5].

To solve the velocity-pressure coupling, a classical fractional step projection is used. In this method, a discrete Poisson equation has to be solved at each time step. The matrix of this system $\mathsf{L} := \mathsf{M}\Omega_c^{-1}\mathsf{M}^*$ remains constant if the mesh does not change.

3 Poisson solver

As a result of the discretization, the Poisson equation obtained is:

$$\mathsf{L}x = b \tag{5}$$

where $\mathsf{L} \in \mathbb{R}^{n\times n}$ is a Laplacian operator, symmetric and positive definite. The periodicity and uniformity of the mesh on the spanwise direction allow to solve the Poisson equation by means of a Fourier diagonalization method. As a result of this, the problem is decoupled into a set of $2D$ systems reducing dramatically the RAM memory requirements and the arithmetical complexity of the algorithm.

The 2D systems are solved by means of a Schur Complement based decomposition method. In general, this method can be very fast compared with an iterative solver but some disadvantages are the RAM memory requirements and the preprocessing time. However, on this particular case the systems considered are 2D, and the memory resources are still reasonable with very large problems (see section 4). On the other hand, the pre-process is carried out only once (L is constant). Typically, for DNS applications the number of time steps is $10^5 \sim 10^6$, thus the computational cost (per time-step) of the pre-processing stage can be neglected.

The parallelization is done in two directions, on the spanwise direction dividing the set of frequencies, and on the streamwise direction by means of the Schur Complement based decomposition method.

3.1 Fourier diagonalization

When the mesh is uniform and periodic in one direction, the 1D restrictions of the Poisson equation in that direction are circulant matrices. On the other hand, the 2D restrictions of the Poisson equation are the same for all the planes. With this conditions, a Fourier diagonalization method can be used and the initial system is decomposed into a set of 2D mutually independent systems. The idea is that the system has a diagonal block structure in a spectral space. Thus the block associated to each frequency can be solved independently from the others. An important point of these methods is that the change of basis from the physical space to the spectral space and vice versa, can be accomplished with an FFT algorithm that has $O(n\log n)$.

The linear system associated to each frequency is of the form

$$(L_{2D} + D_i)\tilde{x}_i = \tilde{b}_i \qquad i \in \{1, ..., n_{per}\} \tag{6}$$

where $L_{2D} \in \mathbb{R}^{n_{2D} \times n_{2D}}$ is the 2D restriction of the Poisson equation replacing the diagonal elements by zeros. $D_i \in \mathbb{R}^{n_{2D} \times n_{2D}}$ are the diagonal matrices given by the spanwise eigenvalues. And $\tilde{x}_i, \tilde{b}_i \in \mathbb{C}^{n_{2D}}$ are the components of the i'th frequency, in the spectral space, of the vectors x and b. As the matrix of system 6 is real, the imaginary and real parts of $\tilde{x}_i$ are solved simultaneously by means of a matricial system.

Regarding the parallelization, the set of 2D systems is partitioned and each subset solved by a group of processors. The parallel solver used for the 2D systems is described in the next section. To apply the FFT to the spanwise sub-vectors it is necessary to have the values of all of their components. As a consequence, if the set of frequencies is partitioned, before applying FFT it is necessary to perform an *all_to_all* communication between processors in the same spanwise line. This is the main limitation for the parallelization in this direction.

3.2 Schur Complement Decomposition

After Fourier diagonalization is carried out, a set of independent 2D systems of equations is obtained. These systems are of the form

$$Ax = b \tag{7}$$

where $A \in \mathbb{R}^{n_{2D} \times n_{2D}}$ is symmetric and positive definite.

The main idea of Schur complement methods is to decompose the initial system of equations into a number of *internal* systems and one *interface* system. This decomposition verifies that the unknowns of different *internal* systems are not directly coupled but indirectly by the *interface*. As a consequence, the internal equations are separated from the distributed system and solved by each processor independently.

As a first step of the Schur Complement algorithm, a set of independent subsets of unknowns is evaluated $\mathscr{D} := \{\mathscr{U}_1, ..., \mathscr{U}_p\}$, this means that two unknowns of different subsets are not directly coupled by the system (7). The elements of these subsets are here named *internal* unknowns. If $\mathscr{D}$ was a partition of the total unknowns set $\mathscr{Y}$, the system would be diagonalisable by blocks. However, in general, this is not possible and it is necessary a subset of unknowns, named *interface* unknowns $\mathscr{S}$, that decouples the different inner subsets.

There are different options to select the internal and interface subsets. One option is to determine the interface as the local unknowns that are coupled with unknowns of other processors. This option is convenient because $\mathscr{S}$ can be determined locally by each processor, but its size ($|\mathscr{S}|$) is not minimal.

In this paper, in order to reduce the complexity of the interface system, its size is minimised. To determine $\mathscr{S}$, firstly a partition $\{\mathscr{Y}_1, ..., \mathscr{Y}_p\}$ of $\mathscr{Y}$ is accomplished using METIS software, [3]. Then, if two unknowns of different processors are coupled, only the unknown of the processor with higher rang is fixed as interface, the other is maintained as internal. After this process, each processor has a subset of interface and internal unknowns ($\mathscr{Y}_i = \mathscr{S}_i \sqcup \mathscr{U}_i$), and the internal sets are mutually uncoupled. However, the sizes of $\mathscr{S}_1,...,\mathscr{S}_p$ can be very different. Thus, it has been developed an algorithm to balance their load. In the main step of this algorithm, an unknown $k \in \mathscr{S}_i$ is moved to the internal set $\mathscr{U}_i$ and, at the same time, the unknowns of other processors coupled with k are moved to the interface. For further details the reader is referred to [1].

Therefore, if the unknowns are labelled with the order $\mathscr{U}_1,...,\mathscr{U}_p,\mathscr{S}_1,...,\mathscr{S}_p$ the system matrix has the following block structure:

$$\begin{bmatrix} A_{\mathscr{U}_1} & 0 & \cdots & 0 & F_{\mathscr{U}_1} \\ 0 & A_{\mathscr{U}_2} & \cdots & 0 & F_{\mathscr{U}_2} \\ \vdots & & & \vdots & \\ 0 & 0 & \cdots & A_{\mathscr{U}_p} & F_{\mathscr{U}_p} \\ E_{\mathscr{U}_1} & E_{\mathscr{U}_2} & \cdots & E_{\mathscr{U}_p} & A_{\mathscr{S}} \end{bmatrix} \begin{bmatrix} x_{\mathscr{U}_1} \\ x_{\mathscr{U}_2} \\ \vdots \\ x_{\mathscr{U}_p} \\ x_{\mathscr{S}} \end{bmatrix} = \begin{bmatrix} b_{\mathscr{U}_1} \\ b_{\mathscr{U}_2} \\ \vdots \\ b_{\mathscr{U}_p} \\ b_{\mathscr{S}} \end{bmatrix} \tag{8}$$

where $A_{\mathscr{U}_i}$ are the linear dependences between unknowns in $\mathscr{U}_i$. $A_{\mathscr{S}}$ the linear dependences in $\mathscr{S}$. $F_i \in \mathscr{U}_i \times \mathscr{S}$ the linear dependences between internal and interface unknowns and $E_{\mathscr{U}_i} = F^T_{\mathscr{U}_i}$. Gaussian elimination is applied to (8) and the *Schur Complement* matrix is derived:

$$C = A_{\mathscr{S}} - \sum_{i=1}^{p} E_{\mathscr{U}_i} A^{-1}_{\mathscr{U}_i} F_{\mathscr{U}_i} \tag{9}$$

In the solution stage the internal systems are solved twice: firstly to obtain the new r.h.s. for the interface system $b'_{\mathscr{S}} = b_{\mathscr{S}} - \sum_{i=1}^{p} E_{\mathscr{U}_i} A^{-1}_{\mathscr{U}_i} b_{\mathscr{U}_i}$, and secondly, after the Schur complement system $Cx_{\mathscr{S}} = b'_{\mathscr{S}}$ is solved, to obtain the internal unknowns $A_{\mathscr{U}_i} x_{\mathscr{U}_i} = b_{\mathscr{U}_i} - F_{\mathscr{U}_i} x_{\mathscr{S}}$.

The different algorithms based on a Schur decomposition depend on the solvers used for the interface and internal systems. In this paper a sparse LU factorization, [2], is used for the internal systems. On the other hand, a complete explicit evaluation of the inverse of the Schur Complement matrix is performed. The two methods are direct solvers, thus the solver derived from its composition is also direct. In general, for a system of dimension m, the explicit evaluation of the inverse is cumbersome in terms of RAM memory requirements $O(m^2)$. In addition, in terms of CPU time, is very costly as the system has to be solved m times to get the inverse matrix. However, when the systems decomposed are *two-dimensional*, the size of $\mathscr{S}$ is much smaller than m, consequently these inconvenients are not important. Once the inverse of $A_{\mathscr{S}}$ is evaluated the parallelization is straightforward.

After Fourier diagonalization, the matrices of all the independent 2D systems have the same sparcity pattern (see section 3.1), this implies that the communication topologies are the same. This situation is taken into account and the communications episodes of all the 2D solvers are done together. This procedure is specially beneficial in networks that have a high latency.

4 Numerical Experiments and DNS applications

The discretization method and the Poisson solver described were used to carry out a DNS of a flow past a circular cylinder at a Reynolds number of 3900. The domain considered was a rectangular box of dimensions $[-4D, 20D], [-4D, 4D], [0, \pi D]$ with a circular cylinder of diameter D centered in $(0,0,0)$. At the inflow $(u, v, w) = (1,0,0)$ was prescribed. Periodic boundary conditions in the spanwise direction were fixed and, for the rest of the boundaries, pressure based conditions were used (outflow boundaries). The validation was carried out comparing with experimental data from [4]. The mesh used for the discretization was a 6M element mesh partitioned into 46 CPU. After this validation a DNS of this flow at $Re = 1 \times 10^4$ is being studied. The mesh used for the discretization has 104 M elements, and it is composed by 96 planes of 1.08M nodes each one. In the figure (1) a snapshot of the vorticity isosurfaces of periodic component is shown.

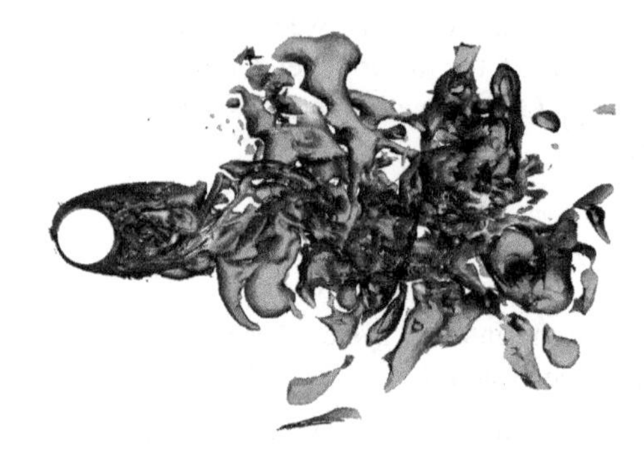

Fig. 1. Snapshot of the vorticity isosurfaces of periodic component for flow past circular cylinder at $Re = 1 \times 10^4$.

The figure 2 shows the results of the parallel performance tests for the Poisson solver. All these tests have been carried out in Marenostrum supercomputer. At the moment when the results were obtained, it was basicly composed of 10240 IBM Power PC 970MP processors at 2.3GHz interconnected with a Myrinet network.

In the left part of figure 2, a *strong scalability* test for the Direct Schur complement method is shown. The mesh considered is the 2D mesh used for the cylinder at $Re = 1 \times 10^4$. The results show that the speed up is ideal until 40 CPU. The degradation of the speedup is caused by the growth of the interface with the number of CPU.

In the right part of figure 2, the *weak scalability* of the parallelization in the spanwise direction is considered. The first point in the graphic is the CPU time required for the solution of the Poisson equation discretized in a mesh of 16 planes with 1.08M nodes each one (17.3M elements). The number of CPUs used is 120, and the time spend is 0.36 seconds. To solve this system no parallelization in the spanwise direction is carried out. The second point in the figure corresponds to a mesh of 32 planes

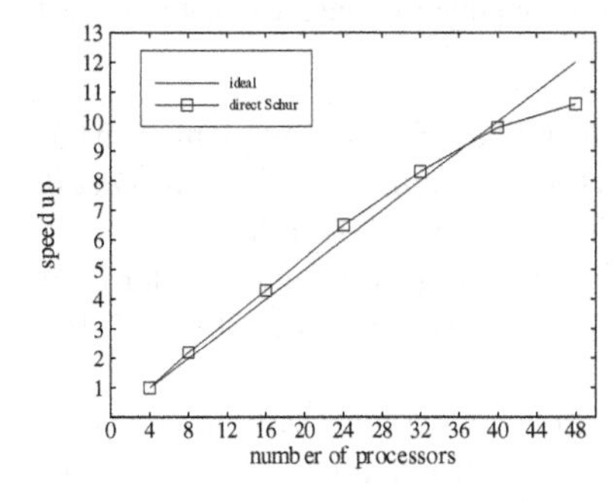

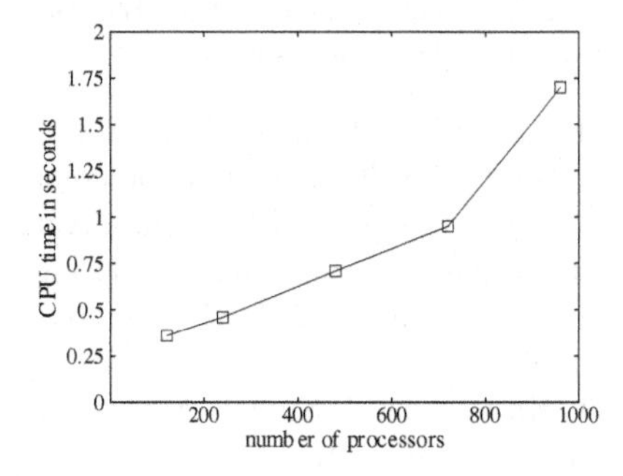

Fig. 2. *Left*: Strong *speedup* for direct Schur solver on a 1.08M mesh. *Right*: Scalability in spanwise direction, constant load per CPU is 1.4×10^5 unknowns.

(two blocks of 16 planes) parallelized with 240 CPUs. The successive points in the figure were obtained adding two blocks of 16 planes and 240 CPUs, to the mesh and the set of processors respectively. For example, the 4th point corresponds to a system of 104M unknowns, 720 CPUs, and the CPU time required to solve it was 0.95 seconds.The main reason of the *speedup* degradation is the *all_to_all* communication necessary to transport data from physical to spectral space and vice versa. It can be observed that up to 720 processors the *scalability* is good, the CPU time grows approximately 50% slower than the problem size. Nevertheless for more than 6 blocks the degradation grows faster.

5 Concluding remarks

A Poisson solver together with the basic ideas of the non-structured discretization of Navier-Stokes equations, for DNS simulation of flows statistically homogeneous in one spatial direction, are presented in this paper. This solver is based on a Fourier diagonalization and a Schur decomposition on the spanwise and streamwise directions respectively. It is being used for the simulation of flow past circular cylinder at $Re = 1 \times 10^4$. The numerical experiments, carried out with the discretization mesh, show an ideal *strong scalability* up to 40 CPUs of the streamwise parallelization. A good performance of the *weak scalability* on the spanwise direction up to 720 CPU is observed. However, the main benefit of this solver is that it can be very fast. For example, the resolution of the Poisson equation discretized in a 104M element mesh used for DNS of cylinder at $Re = 1 \times 10^4$, takes 0.95 seconds of CPU time with 720 CPUs of Marenostrum supercomputer.

Acknowledgement. This work has been financially supported by Termofluids S.L and by the Ministerio de Educacin y Ciencia, Spain, (Project: 'Development of high performance parallel codes for the optimal design of thermal equipments'. Contract grant number ENE2007-67185). Calculations have been performed on the JFF cluster at the CTTC, and on the IBM

MareNostrum supercomputer at the Barcelona Supercomputing Center. The authors thankfully acknowledge these institutions.

[1] R. Borrell, O. Lehmkuhl, M. Soria, and A. Oliva. Schur Complement Methods for the solution of Poisson equation with unstructured meshes. In *Parallel Computational Fluid Dynamics*, Antayla, Turkey, May 2007. Elsevier.
[2] T. A. Davis. A column pre-ordering strategy for the unsymmetric-pattern multifrontal method. *ACM Transactions on Mathematical Software*, 30, no. 2:165–195, 2004.
[3] G.Karypis and V.Kumar. MeTIS: A software package for partitioning unstructured graphs, partitioning meshes and computing fill-reducing ordering of sparse matrixes. Technical report, 1998.
[4] A. G. Kravchenko and P. Moin. Numerical studies of flow over a circular cylinder at $Re_D = 3900$. *Physic of Fluids*, 12:403–417, 2000.
[5] F. X. Trias, M. Soria, A. Oliva, and C. D. Pérez-Segarra. Direct numerical simulations of two- and three-dimensional turbulent natural convection flows in a differentially heated cavity of aspect ratio 4. *Journal of Fluid Mechanics*, 586:259–293, 2007.
[6] R. W. C. P. Verstappen and A.E.P. Veldman. Symmetry-Preserving Discretisation for Direct Numerical Simulation of Turbulence. *Lecture Notes in Physics*, 529, 2003.

A numerical scheme for the computation of phase transition in compressible multiphase flows

Vincent Perrier[1,2]

Université de Lyon 1
Institut Camille Jordan and CDCSP
43 Boulevard du 11 Novembre 1918
69622 Villeurbanne Cedex
perrier@math.univ-lyon1.fr

Abstract. This paper is devoted to the computation of compressible multiphase flows involving phase transition. The compressible model is the system of Euler, without viscosity. Our aim is to simulate such a system, and for that, it is mandatory to understand well the Riemann problem with such an equation of state. We then propose a 2nd order numerical scheme, which is validated and proved to be accurate on one dimensional cases. Last, a 2D version of the code is proposed.

Introduction

In this paper, we are interested in the simulation of phase transition in compressible flows. The model is the inviscid compressible Euler system

$$\begin{cases} \partial_t \rho + \operatorname{div}(\rho \mathbf{u}) = 0 \\ \partial_t(\rho \mathbf{u}) + \operatorname{div}(\rho \mathbf{u} \otimes \mathbf{u} + P\mathbf{I}) = 0 \\ \partial_t(\rho E) + \operatorname{div}((\rho E + P)\mathbf{u}) = 0 \end{cases}$$

where ρ is the density, $\mathbf{u}$ the velocity, P the pressure. E is the total energy

$$E = \frac{|\mathbf{u}|^2}{2} + \varepsilon$$

where ε is the specific internal energy. To close the system, an equation of state is necessary, that links the different thermodynamic parameters, for example $\varepsilon = \varepsilon(P,\rho)$.

The simulation of phase transition is difficult for (at least) two reasons

1. We hope that the system is hyperbolic (this is true provided $\left(\frac{\partial P}{\partial \rho}\right)_s$ is positive). Numerical approximation of hyperbolic systems is difficult, because of the non uniqueness of the weak solution.

D. Tromeur-Dervout (eds.), *Parallel Computational Fluid Dynamics 2008*,
Lecture Notes in Computational Science and Engineering 74,
DOI: 10.1007/978-3-642-14438-7_25,

2. The very model of equation of state for modelling phase transition is still an open question. The most widely spread model is the van-der-Waals model of [7]. In this case, system is not hyperbolic in the whole phase space, so that higher order terms are required for recovering the well-posedness of the Cauchy problem. In this paper, we are interested in a different model for which no regularization is needed.

This paper is organized as follows: in Section 1, we briefly record the model we use. In Section 2, we explain how to solve the Riemann problem with such a model of equation of state. Then in Section 3, we give a second order numerical scheme for simulating phase transition. In Section 4, we validate the scheme and give a two dimensions test.

1 Thermodynamic model

We denote by a subscript l all that refers to the liquid, and by a subscript g all that is linked to the gas. A subscript i will be used when the equation holds for both of the phases. We suppose that each of the phase has its own equation of state. The total specific energy ε, specific entropy s, and specific volume τ ($\tau = 1/\rho$) are equal to

$$\begin{aligned} \varepsilon &= y_l\varepsilon_l + (1-y_l)\varepsilon_g \\ s &= y_l s_l + (1-y_l)s_g \\ \tau &= y_l\tau_l + (1-y_l)\tau_g \end{aligned} \tag{1}$$

where y_l is the mass fraction of the liquid. In order to reduce the number of unknowns, we choose the most stable mixture state, which is the one that optimizes the total entropy with fixed total specific volume and energy. As proved in [6], the most stable state is

- either a pure liquid or a pure gas,
- or a mixture of both of the phases, with equality of pressure, temperature and energy.

To simplify, we suppose from now on that both of the phases are described by a perfect gas equation of state:

$$\varepsilon_i(P_i,\tau_i) = \frac{P_i\tau_i}{\Gamma_i}$$

where Γ_i is the Grüneisen constant. As explained in [6], this model is not able to account well for physics, but it nevertheless has the same mathematical characteristics as the realistic one. Moreover, all the computations can be led explicitly. All the details can be found in [6]. If we denote by

$$\alpha = \exp(1)\left(\frac{\Gamma_2^{\Gamma_2}}{\Gamma_1^{\Gamma_1}}\right)^{\dfrac{1}{\Gamma_1-\Gamma_2}},$$

by $\tau_i = \Gamma_i/\alpha$, and if we suppose that $\Gamma_l < \Gamma_g$ then the situation is as follows.

- for $\tau < \tau_l$, the most stable state is the liquid,
- for $\tau > \tau_g$, the most stable state is the gas,
- for $\tau_l < \tau < \tau_g$, the most stable state is a mixture at thermodynamic equilibrium; it follows the equation of state $\varepsilon = P/\alpha$.

We note that the equation of state is continuous on $\tau = \tau_i$, but it has two differents derivative on the left and on the right. This means that the sound velocity is discontinuous. This will induce problems in solving the Riemann problem, and this is the issue of the next section.

2 Solution of the Riemann problem

Solving the Riemann problem is mandatory for building a numerical scheme based on a Godunov' method. The solution is well known when the equation of state has a continuous derivative, and when the isentropes are convex, see [2]. Solving the Riemann problem for the Euler equations relies on computing the simple waves (see [2]) for the waves $u \pm c$, and then intersecting them in the (P,u) plane. Note that the computation of the velocity never deals with any problem, it is computed with the Riemann invariants or the Rankine-Hugoniot relations depending on the regularity of the wave. That is why we concentrate on the computation of the thermodynamic parameters along the wave curves.

As we said in the previous section, the derivatives of the equation of state are not continuous along $\tau = \tau_i$. Problems might occur when the wave curves cross $\tau = \tau_i$, which will be called in the following "phase transition". We then have to use the Liu criterion (see [4]) and other references ([8]) to build the wave curves. In the following, we detail how to compute them, depending on the thermodynamic state of the initial point.

2.1 The initial point is a gas

If a gaseous state undergoes an undercompressive wave, it begins by an isentrope, which means that τ increases. Therefore, it never meets any curve $\tau = \tau_i$, so that no phase transition occurs.

If a gaseous state undergoes a compressive wave, it begins by a shock, so that the line $\tau = \tau_g$ might be crossed. This is typically the situation described in [4]: the shock may be split into two shocks: a first one that leads to a saturated liquid (i.e. with $\tau = \tau_l$), followed by another shock.

2.2 The initial point is a mixture

If a mixture point undergoes a rarefaction wave, it may cross the line $\tau = \tau_g$. On this point, the characteristics of the left and right states are crossing, so that the wave curve cannot be composed only of an isentrope. Following [8], the rarefaction wave is a composite wave, which can be composed of

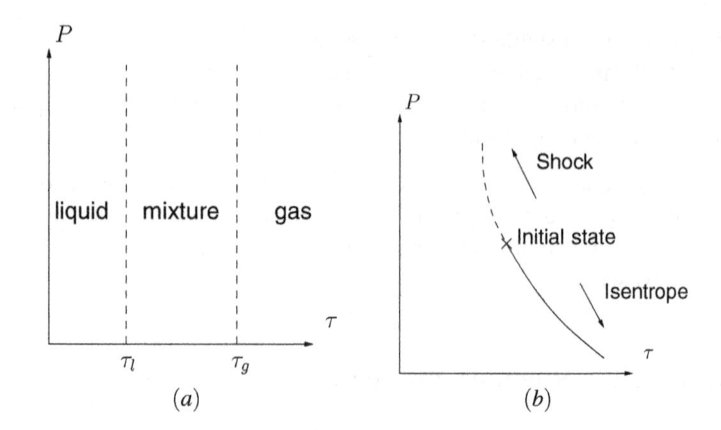

Fig. 1. (a): The thermodynamic plane is divided into three zones in which either a pure phase (liquid or gas) is stable, or a mixture is stable. On the lines $\tau = \tau_i$, the equation of state is continuous, but cannot be differentiated. (b): A wave curve in the thermodynamic plane (τ, P) is a decreasing curve. Under the initial state, the wave curve is an isentrope, whereas above, the wave curve is a shock.

- a mixture isentrope,
- or a mixture isentrope, followed by an undercompressive discontinuity
- or a mixture isentrope, followed by an undercompressive discontinuity, followed by a gaseous isentrope,
- or an undercompressive discontinuity
- or an undercompressive discontinuity, followed by a gaseous isentrope.

If a mixture point undergoes a shock, the wave curve may cross the line $\tau = \tau_l$. Nevertheless, as the Hugoniot curve remains convex, the shock cannot be split (see [4]).

2.3 The initial point is a liquid

If a liquid point undergoes a shock, τ decreases, so that it does not cross any curve $\tau = \tau_i$. Therefore, no phase transition occurs.

If a liquid point undergoes a rarefaction wave, it may cross the line $\tau = \tau_l$. On this point, characteristics do not cross, so that the wave curve can be continued by a mixture rarefaction wave, which may be a composite wave, as seen in the previous subsection.

3 Numerical scheme

3.1 Numerical scheme

Simulation of multiphase flows is a difficult task; the most advanced algorithms rely on the level set method, see [5]. Nevertheless, this algorithm is not conservative, and therefore is not accurate for capturing shocks.

Another way to simulate multiphase flows if the Volume Of Fluid (VOF) method. It nevertheless deals with many geometrical problems, and is also not very developed in the compressible framework.

Moreover, these two algorithms are well suited for the simulation of material interfaces, in which the interface moves at the velocity $\mathbf{u}$, whereas in the phase transition context, they move at a sonic or subsonic velocity.

The numerical scheme we use is based on [3]. Nevertheless, in this last reference, the way to deal with mixture was left unclear. One of the problems in simulation of multiphase flows is that the strong disparities in the nonlinearity of the equation of state induces pressure oscillations, see [1]. In the phase transition context, the nonlinearity of the equation of state strongly changes from one to the other phase, but also from one phase to the mixture zone. That is why we use a *three phase algorithm*, for which the mixture at thermodynamic equilibrium is considered as a third phase.

3.2 Second order extension

The second order accuracy is achieved by a MUSCL-Hancock strategy: first, the variables are interpolated and then limited in each cell. This limitation is done with a Van Albada limiter, and also takes into account the thermodynamic stability of the states. More precisely,

- either the color function α is such that $0 < \alpha < 1$, and then the color function is interpolated, the phase thermodynamic parameters and velocity are not interpolated.
- or the color function α is equal to 0 or 1, and the phase thermodynamic parameters and velocity are interpolated.

Then a time predictor-corrector scheme is used, with a special integration formula for dealing with nonconservative terms.

4 Numerical results

All the computations are made with $\Gamma_l = 0.9$ and $\Gamma_g = 0.2$. With this choice of Grüneisen coefficients, the limit of saturation of the phases are equal to $\rho_l = 0.566$ and $\rho_g = 2.544$. We first validate the code with Riemann problems in dimension 1. We then give a 2 dimensional test.

4.1 Validation test 1: split rarefaction wave

In the first test, the left state is composed of a mixture at thermodynamic equilibrium. Its density is equal to 1, and its pressure is equal to 5×10^5 Pa. The right state is composed of a gas, with density 0.1 and pressure 10^5 Pa. In the two sides, the velocity is equal to 0. Results obtained are shown on Figure 2, and perfectly match with the analytical solution. The pressure ratio is such that it will induce a rarefaction wave in the mixture zone, which will be split because of the phase transition.

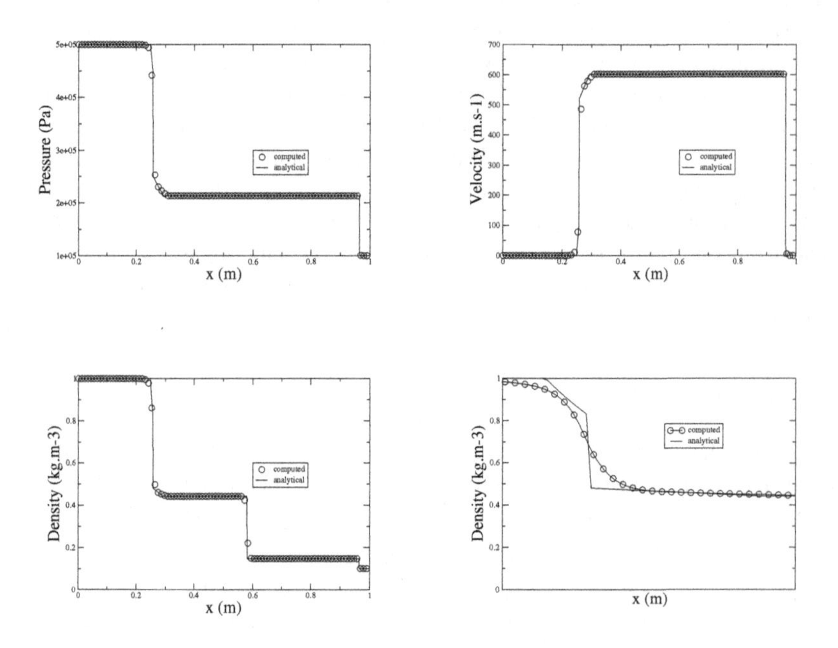

Fig. 2. Comparison of the numerical solution obtained and the analytical one for a split rarefaction wave for density, pressure and velocity. The last figure is a zoom on the split rarefaction wave.

4.2 Validation test 2: Liu solution of a split shock

In this test, the left state is a liquid, with velocity $100 \mathrm{m.s}^{-1}$, and the right state is a gas, with velocity $-100 \mathrm{m.s}^{-1}$. This induces a liquefaction shock, that is split, see [4]. Results and comment are on Figure 3.

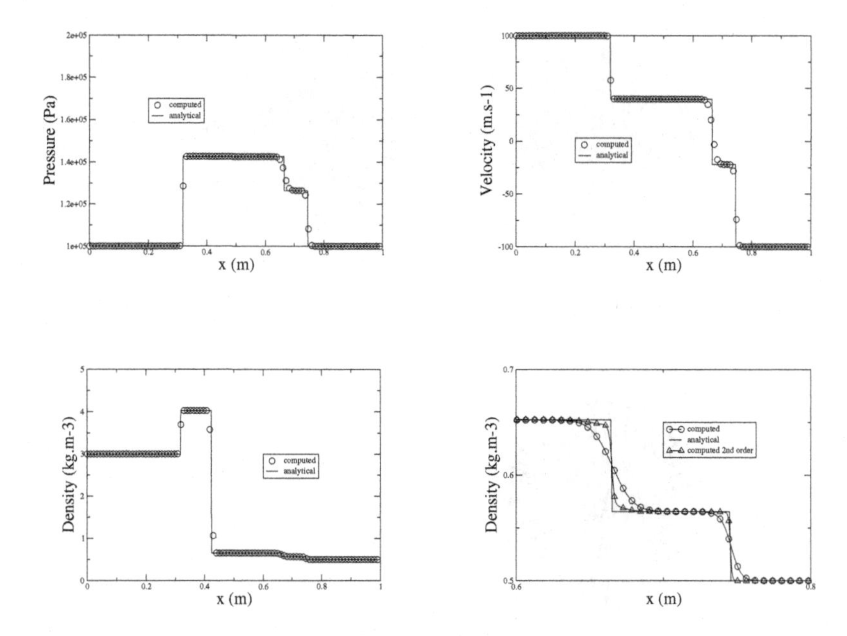

Fig. 3. Comparison of the numerical solution obtained and the analytical one for a shock splitting, for density, pressure and velocity. The last figure is a zoom on the split shock, in which we compare the first and second order computed solutions.

4.3 2 dimensional test

In this test, we consider a metastable phase transition. For details on the modelling and the way to solve the Riemann problem, see [6]. Results and comments are shown on Figure 4.

5 Conclusion

In this article, a numerical scheme for simulating phase transition was developed and validated. Originality of this method relies on that it is fully variational; its finite volume formulation gives a good potentiality for being extended to higher order with the Discontinuous Galerkin method, which is a compact method; the time integration is led explicitly. Therefore it has a strong potentiality for being easily used on parallel environment.

[1] Rémi Abgrall. How to prevent pressure oscillations in multicomponent flow calculations: a quasi-conservative approach. *J. Comput. Phys.*, 125(1):150–160, 1996.

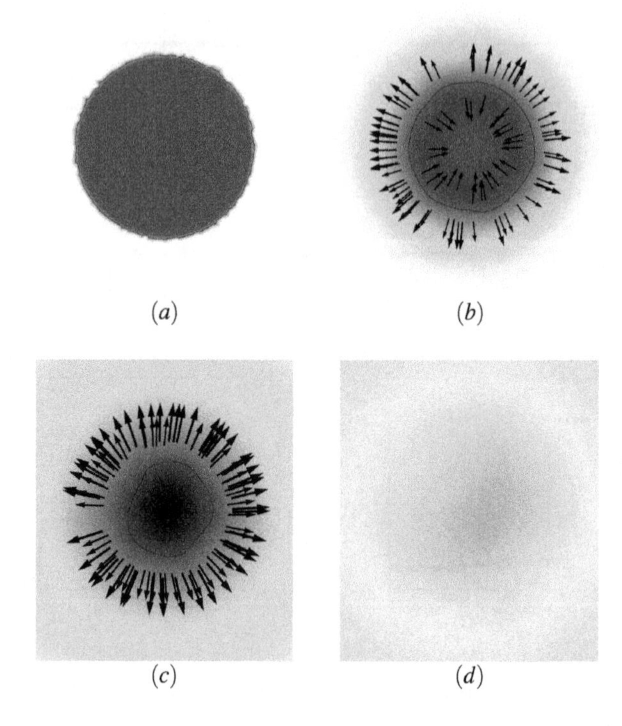

Fig. 4. On this four figures, the density is in grey-scale: the higher is the darker; the velocity is shown with arrows, and the bubble is represented with an iso of the volume fraction. At the beginning, a bubble of liquid is inside a gas (a). The liquid is metastable, which induces a phase transition front, and sonic waves: a shock is emitted inside (b), then focus in the center of the bubble (c). Last, the bubble disappears (d), and behaves as a single phase problem which goes at rest.

[2] Edwige Godlewski and Pierre-Arnaud Raviart. *Numerical approximation of hyperbolic systems of conservation laws*, volume 118 of *Applied Mathematical Sciences*. Springer-Verlag, New York, 1996.

[3] Olivier Le Métayer, Jacques Massoni, and Richard Saurel. Modelling evaporation fronts with reactive Riemann solvers. *J. Comput. Phys.*, 205(2):567–610, 2005.

[4] Tai Ping Liu. The Riemann problem for general systems of conservation laws. *J. Differential Equations*, 18:218–234, 1975.

[5] Stanley Osher and Ronald Fedkiw. *Level set methods and dynamic implicit surfaces*, volume 153 of *Applied Mathematical Sciences*. Springer-Verlag, New York, 2003.

[6] Vincent Perrier. The Chapman-Jouguet closure for the Riemann problem with vaporization. *SIAM J. Appl. Math.*, 68(5):1333–1359, 2008.

[7] Marshall Slemrod. Dynamic phase transitions in a van der Waals fluid. *J. Differential Equations*, 52(1):1–23, 1984.
[8] Burton Wendroff. The Riemann problem for materials with nonconvex equations of state. II. General flow. *J. Math. Anal. Appl.*, 38:640–658, 1972.

Part VII

Lattice Boltzman and SPH Methods

Lattice Boltzmann Simulations of Slip Flow of Non-Newtonian Fluids in Microchannels

Ramesh K. Agarwal[1] and Lee Chusak[1]

Mechanical, Aerospace and Structural Engineering Department, Washington University in St. Louis, MO 63130 email: rka@wustl.edu

Abstract. This paper considers the application of Lattice Boltzmann Method (LBM) to non-Newtonian flow in micro-fluidic devices. To set ideas, we first consider the pressure driven gaseous slip flow with small rarefaction through a long micro-channel and formulate the problem in LB framework. The non-Newtonian fluids are characterized by the non-linear stress-strain constitutive models formulated by Casson, Carreau & Yasuda, Herschel, and Cross, and the well known power law model. The formulation of the LBM for slip flow of non-Newtonian flow is presented. For planar constant area micro-channel for power law fluid, it is possible to obtain an analytical solution for both no-slip and slip flow. For other non-Newtonian fluid models, LBM results are compared with the numerical solutions obtained by using the commercial software FLUENT. The LBM results agree well with the analytical solutions and the numerical solutions. Small differences in the results are noticed using the different models characterizing the non-Newtonian flow.
Keywords: Lattice Boltzmann Method, Non-Newtonian Fluid Flows

1 Introduction

Historically originating from the seminal work of Frisch, Hasslacher, and Pomeau [1] in 1986 on lattice gas automata (LGA), the lattice Boltzmann method (LBM) has recently developed into an alternative and very promising numerical scheme for simulating fluid flows [2]. The lattice Boltzmann algorithms are simple, fast and very suitable for parallel computing. It is also easy to incorporate complicated boundary conditions for computing flows on complex geometries. The algorithms have been successfully applied to compute flows modeled by the incompressible Navier-Stokes equations including reactive and multiphase flows. Attempts have also been made to include the turbulence models in LBM.

Unlike the conventional numerical methods which directly discretize the continuum equations of fluid dynamics on a finite-difference, finite-volume or finite-element mesh, the LBM derives its basis from the kinetic theory which models the microscopic behavior of gases. The fundamental idea behind LBM is to construct the

D. Tromeur-Dervout (eds.), *Parallel Computational Fluid Dynamics 2008*,
Lecture Notes in Computational Science and Engineering 74,
DOI: 10.1007/978-3-642-14438-7_26,

simplified kinetic models that capture the essential physics of microscopic behavior so that the macroscopic flow properties (calculated from the microscopic quantities) obey the desired continuum equations of fluid dynamics. Thus LBM is based on the particle dynamics governed by a simplified model of the Boltzmann equation, the simplification is usually to the nonlinear collision integral. In 1992, a major simplification to the original LBM was achieved by Chen et al. [3] and Qian et al. [4] by employing a single relaxation time approximation due to Bhatnagar, Gross and Krook (BGK) to the collision operator in the lattice Boltzmann equation. In this lattice BGK (LBGK) model, one solves the evolution equations of the distribution functions of fictitious fluid particles colliding and moving synchronously on a symmetric lattice. The symmetric lattice space is a result of the discretization of the particle velocity space and the condition for synchronous motions. That is, the discretizations of time and particle phase space are coherently coupled together. This makes the evolution of lattice Boltzmann equation very simple; it consists of only two steps: collision and advection. Furthermore, the advection operator in phase space (velocity space) is linear in contrast to the nonlinear convection terms in the macroscopic continuum equations of fluid dynamics. Thus, this simple linear advection operator in LBM combined with the simplified BGK collision operator results in the recovery of nonlinear macroscopic convection. It has been shown by Qian et al. [4] among others, using multiple scale expansion that the local equilibrium particle distribution function obtained from the BGK-Boltzmann equation can recover the Navier-Stokes equations and the incompressible Navier-Stokes equations can be obtained in the nearly incompressible limit of LBGK method.

Thus, there are three essential ingredients in the development of a lattice Boltzmann method for a single physics or multi-physics fluid flow problem which are needed to be completely specified: (1) a discrete lattice on which the fluid particles reside, (2) a set of discrete velocities e_i to represent particle advection from one node of the lattice to its nearest neighbor, and (3) a set of rules for the redistribution of particles on a node to mimic collision processes in the fluid, which are provided by the distribution functions f_i of these particles; the evolution of distribution functions in time (for a discrete time step Δt) is obtained by solving the LBGK equation. The LBGK equation for f_i requires the knowledge of the equilibrium distribution function $f_i^{(0)}$. The discrete velocities e_i are determined so that the macroscopic density and momentum satisfy the constraints $\rho = \sum_i f_i$ and $\rho u = \sum_i f_i e_i$ respectively, where u is the macroscopic-averaged fluid velocity. Therefore, the determination of appropriate equilibrium particle distribution function for a given fluid flow problem is essential for solving the problem by LBM.

Accurate treatment of boundary conditions (B.C.) is very important in any numerical scheme. In LBM, the standard treatment of no-slip boundary condition at a solid wall is bounce back boundary condition which is second-order accurate on a flat wall. On a curved wall, the bounce back B.C. becomes first-order accurate and most treatments proposed in the literature for increasing the accuracy to secondorder require that the particle distribution function be handled with given macroscopic quantities. For slip flow on the wall, the specular B.C. is normally employed. For slip flow in microgeometries, the correct formulation of slip boundary conditions is very important for obtaining accurate results with both the conventional finite-volume continuum solvers and the LBGK solvers. For flows in continuum-transition regime at moderate Knudsen numbers, the correct treatment of temperature-jump and slip-velocity boundary conditions becomes even more important for accurate calculation of the flow field. The correct formulation of slip boundary conditions remains a major research issue in all flow solvers. Beginning with the formulation of Maxwell and Smoulchowski [5], significant advances have been made in recent years to improve on their seminal work. For isothermal LBGK solvers, the earliest paper of Niu, Doolan and Chen [6] employed the modified bounce-back type boundary condition by assuming slip velocity known from the data. Lim et al. [7] successfully employed the specular B.C. at low Knudsen numbers. Tang, Tao and He [8] have recommended the combined bounce-back and specular B.C. Several researchers have proposed kinetic B.C. to account for the particles and solid surface interactions in a more realistic manner to include the diffuse-scattering [9]. In this paper, we consider the application of LBM to flow in micro-fluidic devices, which requires special consideration because of the variation in Knudsen number as the fluid moves along these devices driven by pressure or acceleration. We first consider the pressure driven gaseous slip flow with small rarefaction through a long micro-channel and formulate the problem in LB framework. We follow the approach by Lim et al. [7]. The accuracy of the LB solution is checked by comparing it with analytical solution with slip boundary condition and the numerical solutions of Navier-Stokes and augmented Burnett equations without and with slip boundary condition. For planar microchannel flow, this work has been reported earlier by Agarwal [10]. Here we report the results for slip flow in a lid-driven microcavity. Next, we consider the flow of non-Newtonian fluids characterized by the non-linear stress-strain constitutive models formulated by Casson, Carreau & Yasuda, Herschel, and Cross, and the well known power law model. All these models are described in the various books [11, 12]. The LB formulation for non- Newtonian flow is described. It is similar to the approach described in papers by Gabbanelli et al. [13] and Ashrafizaadeh and Bakhshaei [14] for no-slip flow. For slip flow calculations described in this paper, the approach of Lim et al. [7] is included in the formulation. For planar constant area micro-channel for power law fluid, it is possible to obtain an analytical solution for both no-slip and slip-flow. For other non-Newtonian fluid models, LBM results are compared with the numerical solutions obtained by using the commercial software FLUENT. The LBM results agree well with the analytical solutions and the numerical solutions. Small differences in the results are noticed using different models characterizing the non- Newtonian flow.

2 Brief review of basic theory of laticce Boltzmann method

We briefly describe here the basic equations for the simplest and most widely used form of LBM, known as the Lattice-BGK (LBGK) method. For simplicity, we consider a square lattice in 2D, as shown in Figure 1, with unit spacing on which each node has eight nearest neighbors connected by eight links. Particles can only reside on the nodes and move to their nearest neighbors along the links in unit time. There are two types of moving particles: the particles that move along the axis with speed $|e_i| = 1, i = 1,2,3,4$ and the particles that move along the diagonals with speed $|e_i| = \sqrt{2}, i = 5,6,7,8$. Also, there are rest particles with speed zero at each node. The occupation of these three types of particles is described by the single particle distribution function f_i where the subscript i indicates the velocity direction. The distribution function f_i is the probability of finding a particle i at node x at time t with velocity e_i. We assume that the particle distribution function f_i evolves in time according to the LBGK equation:

$$f_i(x+e_i\Delta t, t+\Delta t) = f_i(x,t) + \Omega_i \Delta t, \text{ where } \Omega_i = -\frac{1}{\tau}[f_i(x,t) - f - i^{(0)}(x,t)], \quad (1)$$

which is a discretized form of the discrete BGK equation

$$\frac{\partial f_i}{\partial t} + e_i.\nabla f_i = -\frac{1}{\tau}[f_i - f - i^{(0)}]. \quad (2)$$

In equations (1) and (2), f_i^0 is the equilibrium particle distribution function and τ is the single relaxation time which controls the rate of approach to equilibrium. The hydrodynamic equilibrium particle distribution function (derived from the Maxwellian) is given by:

$$f_i^{(0)} = \rho w_i [1 + \frac{u.e_i}{c_S^2} + \frac{uu.(e_i.e_i - c_S^2 I)}{2c_S^4}] \quad (3)$$

where ρ is the fluid density, u is the flow speed, I is the identity tensor, and c_S is the lattice sound speed defined by the condition:

$$c_S^2 I = \sum_i w_i e_i e_i. \quad (4)$$

In equation (4), w_i are a set of directional weights normalized to unity. These weights are given as $w_0 = 4/9, w_1 = w_2 = w_3 = w_4 = 1/9, and w_5 = w_6 = w_7 = w_8 = 1/36$. The local equilibria obey the following conservation relations:

$$\rho = \sum_i f_i, \rho u = \sum_i f_i e_i, \text{ and } \sum_i f_i e_i e_i = \rho [uu + c_S^2 I] \quad (5)$$

In the limit of long wavelengths, where a particles mean free-path sets the scale; the fluid density and velocity satisfy the Navier-Stokes equations for a quasi-incompressible fluid. The macroscopic fluids equations can be derived using the Chapman-Enskog expansion as:

$$\frac{\partial \rho}{\partial t} + \nabla.(\rho u) = 0, \quad (6)$$

$$\frac{\partial \rho u}{\partial t} + \nabla.(\rho uu) = -\nabla P + \nabla.[\mu(\nabla u)_S + \lambda(\nabla.u)I]. \quad (7)$$

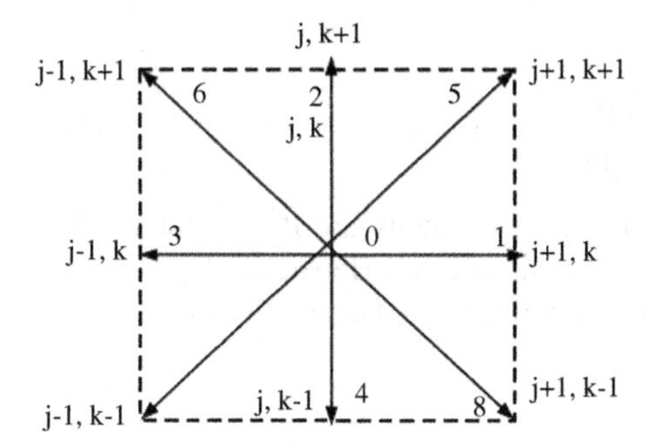

Fig. 1. Nine-Speed Square Lattice.

where P is the fluid pressure, $(\nabla u)_S$ is the symmeterized strain tensor, μ is the dynamic viscosity, and λ is the bulk viscosity. According to the definition of the pressure P, the LBGK fluid obeys an ideal equation of state. Using the standard linear transport theory, with careful handling of the artifacts the lattice introduces, the dynamic and bulk viscosity coefficients become :

$$\mu = c_S^2(\tau - \frac{\Delta t}{2})\rho, \text{ and } \lambda = (1 - 2c_S^2)\rho[1 - \frac{\Delta t}{2}] \tag{8}$$

In equation (7), $P = \rho.c_S^2$. The derivation of LBGK equation (1) assumes that the particles velocities are much smaller than the sound speed and the flow is isothermal, thus the flow field is quasi- incompressible. For computing the LBGK solution, a uniform lattice with equally spaced points is created with square cells. The relaxation time t is calculated from equation (8). The flow field is initialized by assuming a distribution of density and velocity field. The initial values of the distribution function (as equilibrium distribution function $f_i^{(0)}$ at $t = 0$) are then determined on the lattice from equations (5). The updating of the particle distribution functions fi at subsequent time steps is done as described in equation (1). The procedure is repeated until the convergence of the distribution function is obtained. The macroscopic variables are then calculated from equations (5). In equations (1) (5), i represents summation over all lattice points.

3 Non-newtonian fluid models

In literature [11, 12], a number of constitutive models have been proposed to describe the behavior of shear-thinning and shear-thickening non-Newtonian fluids. These models are briefly described below. Let $T = \rho[uu + c_S^2] = \sum_i f_i e_i e_i, T_v = \mu(S)S, S = (\frac{\partial u_i}{\partial x_j} + \frac{\partial u_j}{\partial x_i})$, and $\dot{\gamma} = \sqrt{\frac{1}{2}S : S}$ then in LB method for non-Newtonian fluid, we employ the following relations:

$$S = \frac{1}{2\tau c_S^2} T_v, \text{ or } S = \frac{1}{2\tau c_S^2} \sum_i (f_i - f_i^{(0)}) e_i e_i \tag{9}$$

The relaxation time is given by:

$$\frac{\tau - \tau_\infty}{\tau_0 - \tau_\infty} = \frac{\mu - \mu_\infty}{\mu_0 - \mu_\infty} = F(\dot{\gamma}(\tau)) \tag{10}$$

The various non-Newtonian models employed in the study are Power law model, Casson model, Carreau- Yasuda model, Cross model and Hershel-Bulkley model [11, 12]. The details of these models are not described here but are given in [11, 12].

4 Results

4.1 Newtonian and Non-Newtonian Flow in a Microcavity

These computations were performed to validate the LB method for non-Newtonian fluids. Numerical solutions in a lid-driven square cavity were computed for Newtonian (n =1.0) and shear thinning Power law fluid (n = 0.5) using both the LBGK method and FLUENT. Identical solutions were obtained on the same uniform grid. Figures 2 and 3 show the solutions for normalized u- velocity along the y-axis at the center of the cavity and normalized v-velocity along the x-axis at the center of the cavity. These graphs show the difference between the Newtonian and non-Newtonian solutions as the Reynolds number increases.

4.2 No-Slip and Slip Flow of Non-Newtonian Fluids in a Microchannel

Figure 4 shows the LB solutions for various non-Newtonian fluid models. LB solutions compare well with the analytical solutions for power law fluid and with FLUENT solutions for other non-Newtonian models. Figure 5 shows the LB solutions for a power law fluid which are in excellent agreement with the analytical solutions. In these calculations, Pin/Pout = 2.28; Umax is the maximum velocity at the centerline of the channel at the exit.

Figures 6, 7 and 8 show the LB solutions for slip flow of a non-Newtonian fluid in a microchannel using the power law shear-thickening model, power law shear thinning model and Carreau-Yasuda shear-thinning model respectively. It should be noted that these solutions are in excellent agreement with the FLUENT solutions. The flow conditions are Pin/Pout = 2.28, Knin = 0.088, Knout = 0.2; Ubar is the average velocity.

5 Parallelization

The parallelization of the LB code is straightforward. Computations were performed on a 126processor SGI Origin 2000 parallel supercomputer. In the calculations, only 16 processors were used. SGI Origin 2000 is a cache-coherent non-uniform access multiprocessor architecture. The memory is physically distributed among the nodes

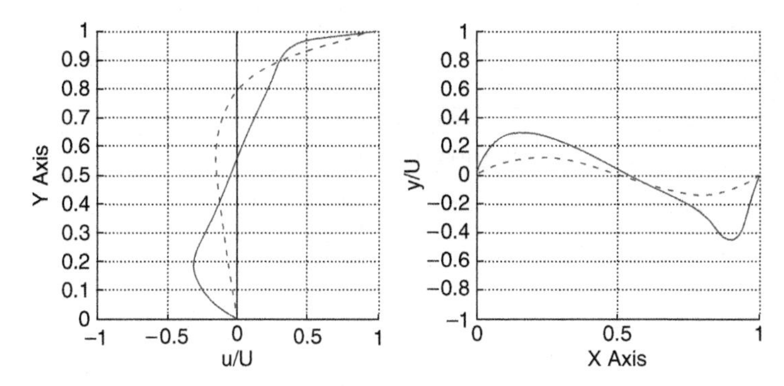

Fig. 2. Identical LB and Fluent solutions for flow in a lid-driven square cavity; Re = 1000, 128x128 uniform grid, solid line: Newtonian fluid (n = 1), dashed line: Power law fluid (n = 0.5) .

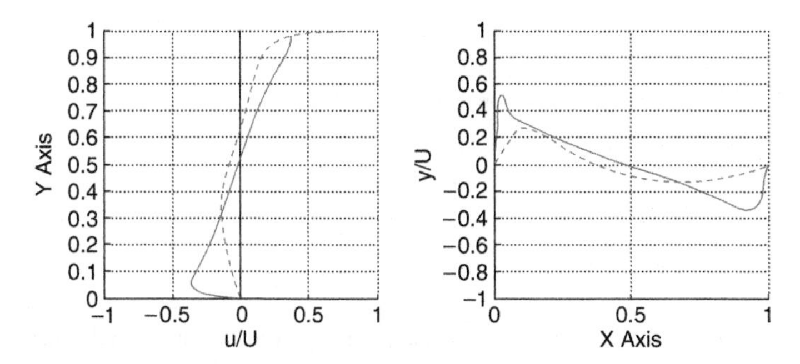

Fig. 3. Identical LB and Fluent solutions for flow in a lid-driven square cavity; Re = 10000, 512x512 uniform grids,__Newtonian fluid (n = 1).

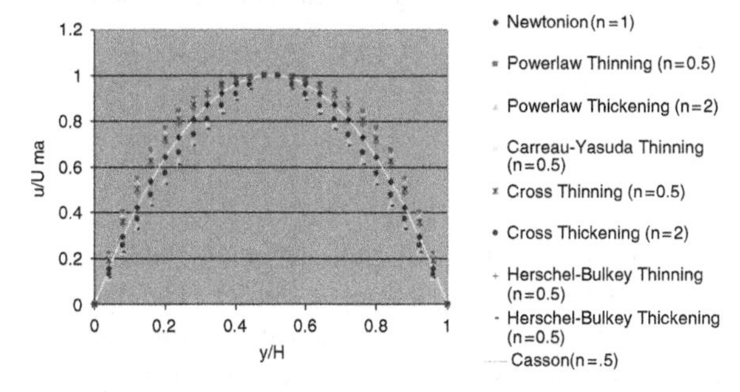

Fig. 4. Velocity profiles at the exit of the microchannel with no slip boundary condition.

but is globally accessible to all processors through interconnection network. The distribution of memory among processors ensures that memory latency is achieved. Parallelization was achieved by MPI. For the 512x512 grid calculation, 95.2% speedup

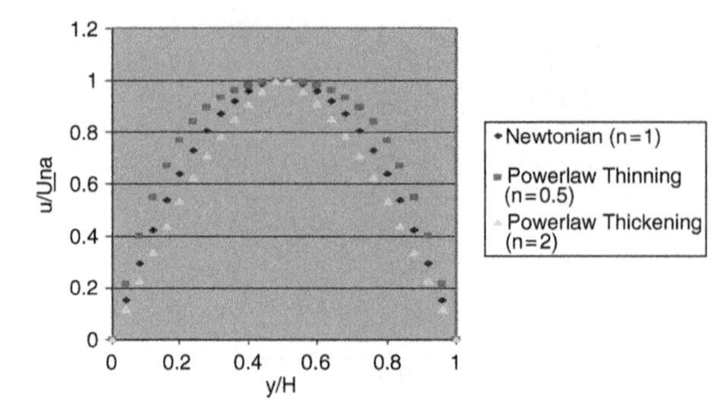

Fig. 5. Velocity profiles at the exit of the microchannel with no slip boundary condition.

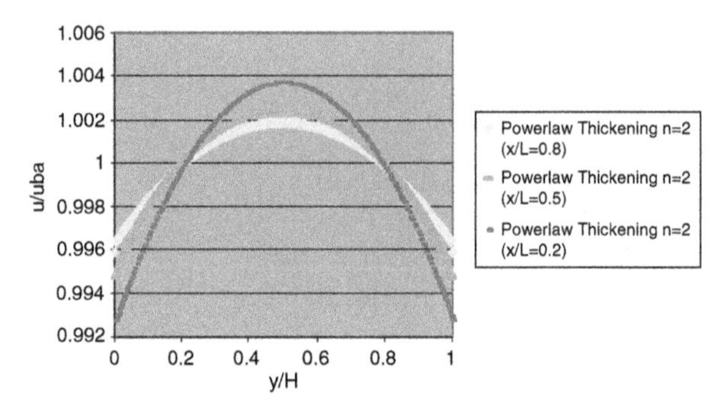

Fig. 6. LB Velocity profiles at the exit of the microchannel for power law shear-thickening fluid .

efficiency was achieved on 16 processors. On coarser girds, the speedup efficiency ranged between 91 to 93%.

6 Conclusions

A Lattice-Boltzmann method has been developed for computing Non-Newtonian slip flow of shearthinning and shear-thickening fluids in microgeometries. The method has been validated by comparing the LB solutions with analytical solutions (where available) and the numerical solutions of the Navier-Stokes equations using FLUENT.

[1] U. Frisch, B. Hasslacher, and Y. Pomeau, Lattice-Gas Automata for the Navier-Stokes Equations, Phys. Rev. Lett., Vol. 56, p. 1505 (1986).
[2] S. Chen and G.D. Doolen, Lattice Boltzmann Method for Fluid Flows, Annu. Rev. Fluid Mech., Vol. 30, pp. 329-364 (1998).

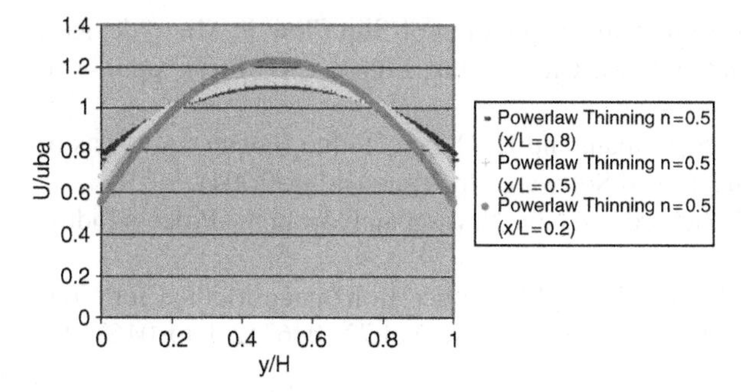

Fig. 7. LB Velocity profiles at the exit of the microchannel for power law shear-thinning fluid.

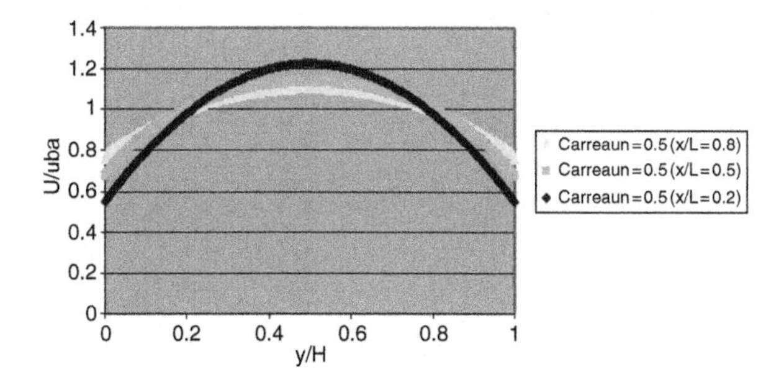

Fig. 8. LB Velocity profiles at the exit of the microchannel for Carreau-Yasuda shear-thinning fluid.

[3] H. Chen, S. Chen, and W.H. Matthaeus, Recovery of the Navier-Stokes Equations Using a Lattice- Gas Boltzmann Method, Phys. Rev. A, Vol. 45, pp. R 5339-5342 (1992).

[4] Y.H. Qian, D. DHumieres, and P. Lallemand, Lattice BGK models for Navier-Stokes Equations, Europhys. Lett., Vol. 17, pp. 479-484 (1992).

[5] J.C. Maxwell, On Stresses in Rarefied Gases Arising from Inequalities of Temperature, Phil. Trans. Royal Soc., London, Vol. 170, pp. 231-256 (1878).

[6] X. Niu, G. Doolen, S. Chen, Lattice Boltzmann Simulations of Fluid Flows in MEMS, J. Stat. Phys., Vol. 107, pp. 279-289 (2002).

[7] C.Y. Lim, C. Shu, X.D. Niu, Y.T. Chew, Application of Lattice Boltzmann Method to Simulate Microflows, Phys. Fluids., Vol. 107, pp. 2299-2308 (2002).

[8] G.H. Tang, W.Q. Tao, Y.L. He, Lattice Boltzmann Method for Gaseous Microflows Using Kinetic Theory Boundary Conditions, Phys. Fluids., Vol. 17, pp. 05101-1 to 058101-4 (2005).

[9] F. Toschi and S. Succi, Lattice Boltzmann Method at Finite Knudsen Numbers, Europhys. Lett., Vol. 69, pp.549-555 (2005).

[10] R.K. Agarwal, Lattice-Boltzmann Simulation of Slip Flow in Micro-devices, MEMS Engineering Handbook, M. Gad-El-Hak, Editor, CRC Press, pp. 8-1 to 8-15 (2005).

[11] H-C Huang, Z-H Li, A.S. Usmani and K. Ozbay, Finite Element Analysis of Non-Newtonian flow: Theory and Software, Springer-Verlag (2004).

[12] R. P. Chhabra and J. F. Richardson, Non-Newtonian Flow in the Process Industries, Elsevier, 1999.

[13] S. Gabbanelli, G. Drazer and J. Koplik, Lattice Boltzmann Method for Non-Newtonian (Power Law) Fluids, Phys. Rev. E, Vol.72, 046312-1 to 046312-7 (2005)

[14] M. Ashrafizaadeh and H. Bakhshaei, A Comparison of Non-Newtonian Models for Lattice- Boltzmann Blood Flow Simulations, Elsevier Science (2007).

Multiple Relaxation Time Lattice Boltzmann simulation of binary droplet collisions

Ernesto Monaco[1], Kai H. Luo[2] and Gunther Brenner[1]

[1] Institute of Applied Mechanics, Clausthal University,Germany [ernesto.monaco, gunther.brenner]@tu-clausthal.de

[2] Energy Technology Group, School of Engineering Sciences, Southampton University, United Kingdom K.H.Luo@soton.ac.uk

The Lattice Boltzmann method is employed to simulate binary droplet collisions. The Shan-Chen multiphase model, improved in the equation of state and in the incorporation of the body force, is integrated into the Multiple Relaxation Time scheme. Qualitative comparisons with the experiments show very good agreement.

1 Introduction

Droplet collisions are encountered in natural phenomena and in many industrial processes like spray applications. The parameters characterizing this phenomenon are the densities and viscosities of liquid and gaseous phases, respectively ρ_l, ρ_g, μ_l and μ_g, the surface tension σ, the droplets radii R_1 and R_2, their relative speed $\mathbf{U}$ and finally their displacement in the direction normal to $\mathbf{U}$. These quantities are reported in Fig. 1. The whole process is therefore described by five nondimensional quantities, namely the Weber and Reynolds numbers, respectively given by $We = \rho_L U^2(R_1+R_2)/\sigma$ and $Re = \rho_L U(R_1+R_2)/\mu_L$, the size and viscosity ratios, and finally the impact factor $B = \chi/(R_1+R_2)$. Experimental studies like [10, 1, 12]

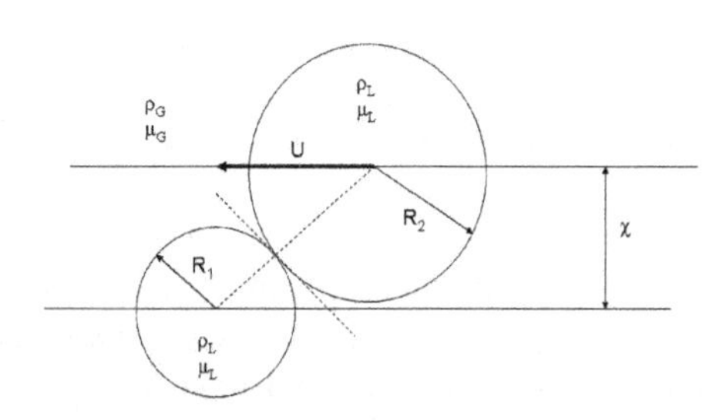

Fig. 1. Physical and geometrical parameters characterizing binary droplet collision.

D. Tromeur-Dervout (eds.), *Parallel Computational Fluid Dynamics 2008*,
Lecture Notes in Computational Science and Engineering 74,
DOI: 10.1007/978-3-642-14438-7_27, © Springer-Verlag Berlin Heidelberg 2010

allowed the identification of five possible collisional regimes: adopting the notation of [12], it is possible to observe "coalescence" (Regime I), "bouncing" (II), "coalescence with major deformation" (III), "head-on separation" (IV) and "off-centre separation" (V). These regimes can be identified by transition curves on $B-We$ plane. Coalescence occurs when, at very small We, the gas is pushed out of the gap between the approaching droplets till the thickness of this gap is reduced to the order of molecular interactions. If the minimum thickness is higher than this value, the droplets will bounce. Regime III occurs at intermediate We, when the initial kinetic energy is sufficient only to cause extensive deformation to the coalesced droplet . At high We number it is possible to observe Regimes IV or V,depending on B. The coalesced droplet can experience either disruption, that produces again two droplets,or fragmentation, producing catastrophic break-up into many small droplets.
Droplet collisions represent a challenging case for numerical simulations, because of the necessity of following the evolution of variable interfaces. As pointed out in [12], the amount of gas absorbed in the liquid surface layer during the collision is negligible, so mass conservation represents another issue for a numerical scheme. Two- and three-dimensional simulations were presented for instance in [7, 8], employing respectively the volume of fluid (VOF) and the front-tracking schemes ; in [9], remarkable agreement with the experiments of [12] by using the level-set method is observed. All these methods require some artificial technique for interface tracking discretizing the Navier-Stokes equation.
The Lattice Boltzmann method (LB) [3] is attracting more and more attention in multiphase flow simulations. Its kinetic nature allows the incorporation of interface models without suffering from the limitations in terms of length and time scales typical of Molecular Dynamics simulations. Besides, the method is simple to code and highly parallelizable, and does not present nonlinear convective terms like in classical CFD schemes.
In this study, the LB method is adopted to simulate binary droplet collisions at different We and B values. The goal is to reproduce the five possible collision regimes. To do that, the Shan-Chen (SC) [15] multiphase model improved by the inclusion of more effective equations of state (EOS) [18] and by the "Exact Difference Method" (EDM) [6] is used. To deal with low-viscosity flows, this model is integrated in the framework of a Multiple-Relaxation-Time scheme [4].

2 The Lattice Boltzmann Method for single and multiphase flows

The Boltzmann equation is the fundamental equation of kinetic theory. It expresses the rate of change of the "single particle probability distribution function" (PDF), which represents the probability of finding, at time t, a particle in position $\mathbf{x}$ with speed ξ, because of particle streaming and collisions. The LB equation can be obtained by discretizing the phase space of the Boltzmann equation over a lattice defined by a set of b finite speeds $\mathbf{e}_i$ [5]:

$$f_i(\mathbf{x}+\mathbf{e}_i\Delta t, t+\Delta t) - f_i(\mathbf{x},t) = -\Lambda_{ij}\left(f_j(\mathbf{x},t) - f_j^{eq}(\mathbf{x},t)\right), \quad i,j = 0,..,b-1 \quad (1)$$

The left hand side of (1) represents the streaming operator, while the right hand side represents the collision operator, which preserves mass and momentum and is completely local, making LB particularly suited for massively parallel computations. The collision matrix Λ is diagonal, with all its elements equal to the collision frequency $1/\tau$: (1) is thus called single-relaxation-time LB (SRT-LB). The collision relaxes the f_i to its equilibrium value f_i^{eq}, which depends on local macroscopic density and speed [17]. By means of a Chapman-Enskog expansion, the incompressible NS equation is recovered, in the limit of low Mach number. To achieve that, the lattice and the f_i^{eq} must be carefully chosen [3]. Density, momentum and energy are obtained as discrete momenta of the $\mathbf{f}^{eq}$, while pressure and kinematic viscosity are respectively given by $p = \rho c_s^2 \delta_{ij}$ and $\nu = \frac{2\tau-1}{6} c\Delta x$. The constant $c_s = \frac{\Delta x}{\Delta t}\sqrt{3}$ is termed "lattice speed of sound". Pressure is therefore obtained by an ideal gas equation of state (EOS).
To model non-ideal gas effects, the SC model introduces a forcing term derived as discrete gradient of an interparticle potential, $\psi(\rho)$, termed "effective mass":

$$\mathbf{F}(\mathbf{x},t) = -c_s^2 \psi \sum_{j=0}^{k-1} G_b w_k \psi(\mathbf{x}+\mathbf{e}_i\Delta t)\mathbf{e}_i. \quad (2)$$

The discrete gradient can be computed employing different numbers k of nodes and corresponding weights w_k [14]. In the original SC formulation, $\psi = \rho_0(1 - \rho/\rho_0)$. The general form of the EOS associated with the SC model is $p = \rho c_s^2 + \frac{1}{2} G_b c_s^2 \psi^2$.

The role of temperature is played by the "coupling constant" G_b. It is possible to show that this EOS exhibits a region of the $P-\rho$ curve at temperatures below the critical value (represented by a critical value of G_b) in which each pressure corresponds to three different densities, one for the liquid phase, one for the gas phase and the last in the region in which $\partial p/\partial \rho < 0$: the existence of this unphysical region allows to maintain self-generated sharp interfaces.
The force modelled by (2) modifies the local equilibrium velocity in f_i^{eq} depending on τ according to $\mathbf{U}^{eq} = \mathbf{U} + \mathbf{F}\tau/\rho$. This method is first order accurate in $\Delta\mathbf{U}$. A more effective way of including body forces in LB is the "Effective Difference Method" (EDM) presented in [6]: it consists in adding at right hand side of (1), the term $\Delta f_i^{eq} = \left(f_i^{eq}(\mathbf{U}+\Delta\mathbf{U},t) - f_i^{eq}(\mathbf{U},t)\right)$, where $\Delta\mathbf{U} = F\Delta t/\rho$.

The SC model as described so far cannot handle density ratios over $O(10)$. To address this limitation, the form of ψ proposed in [18] is adopted: this form can be used to incorporate any EOS (van der Waals, Peng-Robinson or others) into the LB scheme.

The SRT-LBM is very simple, but is known to fail when dealing with low viscosity flows. The Multiple-Relaxation-Times LB (MRT-LB) [4] overcomes this limitation by performing the collision not in f_i space as done so far, but in a moment space m_i determined by a linear transformation driven by the matrix T. See [4] for more details. The MRT-LB equation integrated by the EDM scheme reads as:

$$f_i(\mathbf{x}+\mathbf{e}_i\Delta t, t+\Delta t) - f_i(\mathbf{x},t) = -T_{ij}^{-1} S_{jk} \Delta m_k + \Delta f_i^{eq}, \tag{3}$$

In (3) the matrix $\mathsf{S} = \mathsf{T}^{-1}\Lambda\mathsf{T}$ is diagonal and its elements are the relaxation frequencies of the different hydrodynamic moments, while $\Delta m_k = m_k - m_k^{eq}$. Streaming and boundary conditions are performed in f_i space as before. The strategy proposed in [4] is adopted to code (3) with limited increase of computational cost.

3 Results

In the rest of the paper the $D3Q19$ model is used; the relaxation times are equal to 1 except $s_9 = s_{11} = s_{13} - s_{16} = 1/\tau$ [4]. The EOS used to compute ψ is the Carnahan-Starling [2]:

$$p^* = \rho RT \frac{1 - b\rho/4 + (b\rho/4)^2 - (b\rho/4)^3}{(1 - b\rho/4)^3} - a\rho^2, \tag{4}$$

with $a = 1, b = 4$ and $R = 1$.Finally, in (2), the discrete gradient operator has $6th$ order isotropy [14]
To illustrate the capabilities of the scheme described above, two preliminary tests have been conducted. The first test consists in comparing the coexistence curve obtained by numerical simulations the theoretical curve predicted by the Maxwell equal-area construction. The result is shown in Fig. 2, from which it is possible to see an excellent agreement. The maximum density ratio achieved was about 1340, as [18]. The second test consists in taking one spherical droplet in a fully periodical domain and in evaluating the minimum viscosity and the maximum density ratio achievable. The grid is $100 \times 100 \times 100$ and the initial radius is $30\Delta x$. At density ratio of 100 the minimum viscosity achieved was $1/500$.

In the following, LB simulations of binary droplet collisions are presented and qualitatively compared with corresponding cases of [12]. In all cases the density ratio is 100 and the droplets radii are $40\Delta x$. Fig. 3 corresponds to case a in [12], ($Re = 14.8, We = 0.2$ and $B = 0.2$, Regime I). The grid is $141 \times 121 \times 121$. The coalesced drop takes a cylindrical shape, then tends to reach a spherical configuration to minimize surface energy.

Regime III is illustrated in Fig. 4 (a-b), corresponding respectively to cases f ($Re = 210.8, We = 32.8$ and $B = 0.08$) and k ($Re = 327.7, We = 70.8$ and $B = 0.25$). The expected different behavior under different B is clearly reproduced. In the first case, the tangential component of the impact inertia is almost zero, and a thin disk is initially formed, that later contracts and forms a cylinder and then a dumbell, without any rupture of the film. In the second case, the higher tangential component of the impact inertia is expected to cause an extensive rotational motion, and that is precisely what the simulation shows. In both cases the grid is $251 \times 141 \times 141$ and the simulation took about 4 hours on 16 processors.

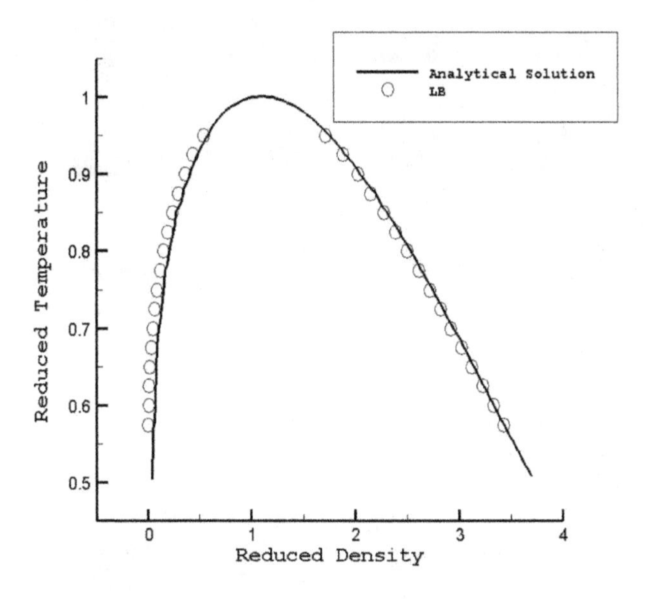

Fig. 2. Comparison between coexistence curve obtained from simulation and theoretical one relative to the CS-EOS.

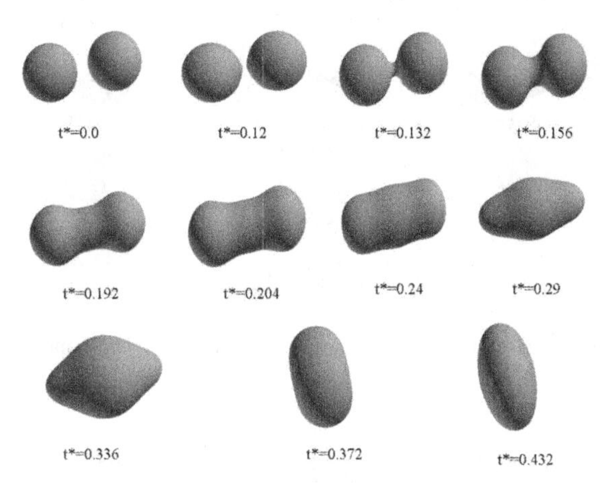

Fig. 3. Simulation of case a from [12]. The nondimensional time is computed as $t^* = tU/2R$

Fig. 5 is concerned with case g (Regime IV) ($Re = 228, We = 37.2$ and $B = 0.01$): the evolution is similar to case f previosly seen, but this time initial kinetic energy is enough to overtcome the surface tension force and split the coalesced drop. Again, good agreement is found with the experiments.

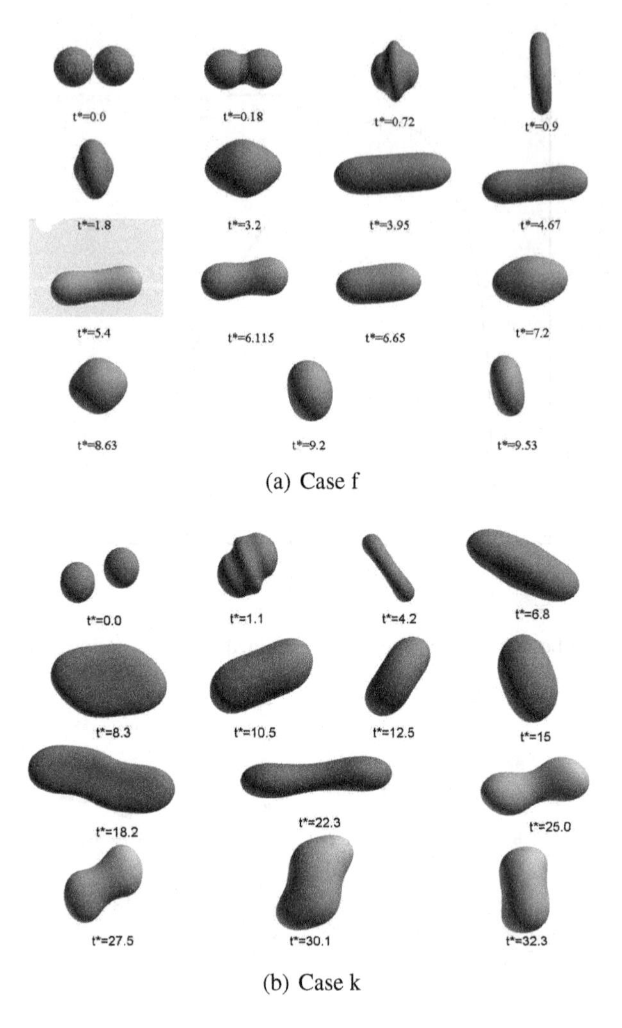

(a) Case f

(b) Case k

Fig. 4. Simulations of cases relative to Regime III from [12]. The nondimensional time is $t^* = tU/2R$

A further increase of B causes a reduction of the contact region: if We is high enough, like in case m ($Re = 302.8, We = 60.1$ and $B = 0.55$), Regime V is observed: the LB simulation in Fig. 6 clearly reproduces the formation of a thin neck linking two main globes and its subsequent rupture by means of the "end-pinching" mechanism, producing three droplets.

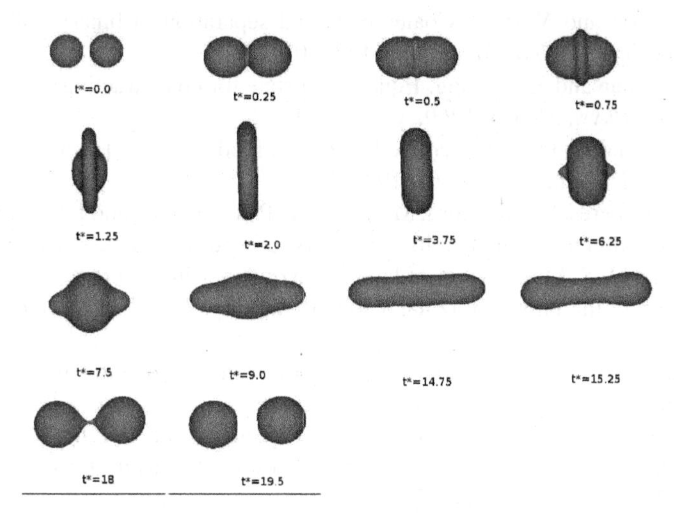

Fig. 5. Simulation of case g from [12]. The nondimensional time is $t^* = tU/2R$

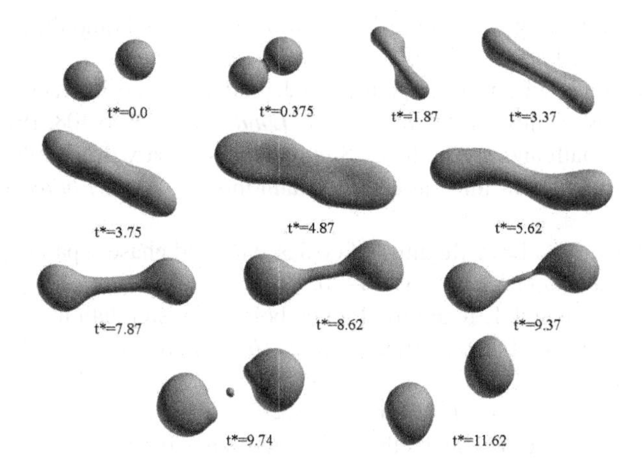

Fig. 6. Simulation of case m from [12]. The nondimensional time is $t^* = tU/2R$

4 Discussion and Conclusions

The LB formulation described in this study allows stable simulations of all the collision regimes except bouncing (not observed) at an higher density ratio than what reported in other similar works [16, 13, 11]. Besides, the SC model does not require two different sets of PDFs like the models employed in the aforementioned works. Density ratios similar to the experiments have been reached for static droplet tests, but collisions at high *We* are often unstable, due to the high spurious currents at interfaces. Their reduction, together with comparisons with analytical models, will be the subject of future study.

[1] N. Ashgriz and Y. Poo. Coalescence and separation in binary collisions of liquid drops. *J. Fluid. Mech.*, 221:183–204, 1990.
[2] N. Carnahan and K. Starling. Equation of state for non attracting rigid spheres. *J. Chem. Phys.*, 51:635, 1969.
[3] S. Chen and G. Doolen. Lattice boltzmann model for fluid dynamics. *Annu. Rev. Fluid Mech.*, 30:329–364, 1998.
[4] D. D'Humieres, I. Guinzburg, M. Krafczyk, P. Lallemand, and L.S. Luo. Multiple relaxation time lattice boltzmann method in 3d. *ICASE*, 20:329–364, 2002.
[5] X. He and L.S. Luo. Theory of lattice boltzmann method: From the boltzmann equation to the lattice boltzmann equation. *Annu. Rev. Fluid Mech.*, 30:329–364, 1997.
[6] A. Kupershtokh and D. Medvedev. Lattice boltzmann equation method in electrodynamic problems. *J. Electrostatics*, 2006.
[7] B. Lafaurie, C. Nardone, R. Scardovelli, S. Zaleski, and G. Zanetti. Modeling merging and fragmentation in multiphase flows with surfer. *J. Comput. Phys.*, 113:134, 1994.
[8] Y.J. Jan M. Nobari and G. Tryggvason. Head-on collision of drops-a numerical investigation. *Phys. Fluids*, 29, 1996.
[9] Y. Pan and K. Suga. Numerical simulations of binary liquid droplet collision. *Phys. Fluids*, 17, 082105, 2005.
[10] S.G. Jennings P.R. Brazier-Smith and J. Latham. The interaction of falling water drops: coalescence. *Proc. R. Soc. Lond. A*, 326:393–408, 1972.
[11] K.N. Premnath and J. Abraham. Simulations of binary drop collisions with a multiple-relaxation-time lattice-boltzmann model. *Phys. Fluids*, 17, 122105, 2005.
[12] J. Qian and C.K. Law. Regimes of coalescence and phase separation in droplet collision. *J. Fluid. Mech.*, 331:59–80, 1997.
[13] B. Sakakibara and T. Inamuro. Lattice boltzmann simulation of collision dynamics of unequal-size droplets. *Int. J. Heat Mass Transfer*, 51:3207–3216, 2008.
[14] M. Sbaraglia, R. Benzi, L. Biferale, S. Succi, K. Sugiyama, and F. Toschi. Generalized lattice boltzmann method with multirange pseudopotential. *Phys. Rev. E*, 2007.
[15] X.W. Shan and H.D. Chen. Lattice boltzmann model for simulating flows with multiple phases and components. *Phys. Rev. E*, 47:1815, 1993.
[16] F. Ogino T. Inamuro, S. Tajima. Lattice boltzmann simulation of droplet collision dynamics. *Int. J. Heat Mass Transfer*, 47, 2004.
[17] D. d'Humieres Y.H. Quian and P. Lallemand. Lattice bgk models for the navier-stokes equation. *Europhys. Lett.*, 17:479, 1992.
[18] P. Yuan and L. Schaefer. Equations of state in a lattice boltzmann model. *Phys. Fluids*, 18, 042101, 2006.

High-Performance Computing and Smoothed Particle Hydrodynamics

C. Moulinec[1], R. Issa[2], D. Latino[3], P. Vezolle[4], D.R. Emerson[1] and X.J. Gu[1]

[1] STFC Daresbury Laboratory, WA4 4AD, UK charles.moulinec@stfc.ac.uk
[2] EDF R&D, National Hydraulics and Environment Laboratory, and Saint-Venant Laboratory for Hydraulics, 6, quai Watier, 78400 Chatou, F reza.issa@edf.fr
[3] IBM Middle East, Dubai Internet City, P.O. Box 27242, Dubai, UAE dlatino@ae.ibm.com
[4] IBM France, Rue de la Vieille Poste, BP 1021, F-34006, Montpellier, F vezolle@fr.ibm.com

Abstract

Smoothed Particle Hydrodynamics is a gridless numerical method that can be used to simulate highly complex flows with one or more free surfaces. In the context of engineering applications, very few 3-D simulations have been carried out due to prohibitive computational cost since the number of particles required in 3-D is usually too large to be handled by a single processor. In this work, an improved version of a parallel 3-D SPH code, Spartacus-3D, is presented. Modifications to the code, which include a localisation of all the previously global arrays combined with a switch from global communications to local ones where possible, lead to a more efficient parallel code and allow for a substantial increase in the number of particles in a simulation.

1 Introduction

The Smoothed Particle Hydrodynamics (SPH) method has been successfully applied in fluid mechanics to simulate strongly distorted free surface flows in situations where classical Eulerian Computational Fluid Dynamics (CFD) approaches would fail because of grid steepness [2] [3] [4]. However, 3-D SPH simulations remain costly as a large number of particles are necessary in order to accurately simulate such flows and because of the nature of classical SPH, which requires a homogeneous initial particle distribution for incompressible flows. A parallel 3-D SPH code, namely Spartacus-3D, was developed in 2006 based on the MPI library [5], but most of the arrays were global. This approach limited the total number of particles that could be handled and the memory required by processors was largely over-estimated. Moreover, global communications were compulsary.

D. Tromeur-Dervout (eds.), *Parallel Computational Fluid Dynamics 2008*,
Lecture Notes in Computational Science and Engineering 74,
DOI: 10.1007/978-3-642-14438-7_28,

The Spartacus-2D code was developed in 1999 at EDF R&D, mainly for modelling coastal and environmental applications, such as spillways, dam breaking, and breaking waves. Based on this 2-D version, a 3-D version was first developed [1] in 2004 and initially parallelised in 2006 [5]. In this paper, details concerning the latest version of the code, where the focus has been on array localisation, will be presented. In Section 2, the methodology is explained and Section 3 introduces the equations in SPH framework. Section 4 discusses parallelisation and Section 5 highlights the new results and we conclude with some final remarks Section 6.

2 Methodology

The pseudo-compressible Navier-Stokes equations written in Lagrangian form for Newtonian fluids and incompressible, laminar flows augmented by the position equation read:

$$\frac{D\mathbf{u}}{Dt} = -\frac{1}{\rho}\nabla p + \nabla\cdot(\nu\nabla\mathbf{u}) + \mathbf{F^e}, \quad \frac{D\rho}{Dt} = \rho\nabla\cdot\mathbf{u}, \quad \frac{D\mathbf{r}}{Dt} = \mathbf{u}. \tag{1}$$

where $D\cdot/Dt$ is a Lagrangian derivative, $\mathbf{u}$ the velocity vector, t the time, ρ the density, p the pressure, ν the kinematic viscosity, and $\mathbf{F^e}$ an external force. In addition, '∇' and '$\nabla\cdot$' are the gradient and divergence operators, respectively. This system is closed by the following equation of state:

$$p = \frac{\rho_0 c_0^2}{\gamma}\left[\left(\frac{\rho}{\rho_0}\right)^{\gamma} - 1\right] \tag{2}$$

where ρ_0 is the reference density, c_0 a numerical speed of sound, and γ a constant coefficient. Equations 1 are discretised explicitly in time (first order) and the SPH approach is used to perform the spatial discretisation.

3 Equations in SPH framework

The SPH continuity equation and momentum equation are given by:

$$\frac{D\rho_a}{Dt} = \rho_a \sum_b \frac{m_b}{\rho_b}\mathbf{u}_{ab}\cdot\nabla_a w_{ab}, \tag{3}$$

$$\frac{D\mathbf{u}_a}{Dt} = -\sum_b m_b \frac{p_a + p_b}{\rho_a\rho_b}\nabla_a w_{ab} + 16\sum_b \frac{\nu}{\rho_a+\rho_b}\frac{\mathbf{u}_{ab}\cdot\mathbf{r}_{ab}}{r_{ab}^2}\nabla_a w_{ab} + \mathbf{F}^e_a, \tag{4}$$

where the subscripts a and b, respectively, represent the particle which the operators are calculated for, and its neighbours contained in the kernel compact support, $\mathbf{u}_{ab} = \mathbf{u}_a - \mathbf{u}_b$, $w_{ab} = w_h(r_{ab})$, $r_{ab} = |\mathbf{r}_{ab}| = |\mathbf{r}_a - \mathbf{r}_b|$. w_h is the kernel as a function of the smoothing length, $\nabla_a w_{ab}$ is the gradient of the kernel with respect to a, m is the particle mass, and $\mathbf{F}_a^e$ is an external force applied to particle a. Note that the pressure gradient expression selected here is anti-symmetric with respect to the subscripts a and b and the viscous term is symmetric. This allows a reduction in computational cost because, when calculating the SPH operators, b's contribution can always be deduced from a's.

4 Parallelisation

Equations 3 and 4 show that SPH operators are expressed as differences or sums of contribution from particles a and b, with a sum on b. Due to the particle motion and the fact that neighbouring particles (neighbouring particles of particle a are particles which belong to the compact support, whose centre is the position of a) at a given time step might not be neighbouring particles at the next time step, the search for particles b has to be performed at each temporal iteration. The CPU time would normally scale as N^3 (in 3-D) where N denotes the total number of particles, if the search for neighbours would be carried out over the whole set of particles. Since spline kernels have a compact support, each particle a is only linked to its closest neighbours b for which $r_{ab} < h_t$, where h_t is the kernel compact support size. A coarse Cartesian grid made of homogeneous cubic cells embeds the physical domain to speed-up the search for the links.

Profiling the serial version of Spartacus-3D shows that the search for particle a's links is the most expensive part of a temporal iteration, consuming up to 60% of the CPU time. When running in parallel, another section that takes a lot of time is the rebuilding of the list of particles by re-indexing them at each temporal iteration, which is done in order to minimise the communications between processors in the equation resolution stage. Some problems identified in the previous parallel version of the code came from the declaration of gobal arrays for the main variables (velocity, density, pressure, viscosity,...) and from the global communications required to build SPH operators. The way to circumvent most of these problems is described in the following, with a special focus on the attempt to use arrays whose size varies per processor, depending on the number of particles of a given processor augmented by the number of particles located on other processsors, which play a role in building the SPH operators.

In the new parallel version of Spartacus-3D, four main steps are carried out per temporal iteration. At the $(n+1)^{th}$ iteration, for instance it yields:

- Step 1: Generation of the new particle list,
- Step 2: Search for particle links,
- Step 3: Link particle re-indexing,
- Step 4: Resolution of the equations.

Steps 1 to 4 are detailed in the following sections.

4.1 Step 1. Generation of the new particle list

If the total number of processors is NP the general idea is to approximately assign N/NP particles a per processor to ensure load-balancing, while reducing the number of processors containing the neighbours b of the N/NP particles a. In the present version of the code, there is no weight depending on the computing effort required per particle, even though different operations are carried out for wall (respectively dummy) particles and fluid ones.

- The Cartesian grid is generated and the cubes denoted by C_i $(i = 1, NC)$ are linerarly ordered, where NC is the whole number of cubes. The loop is carried out over NC. This is parallelised.
- Denoting the number of particles per cube C_i as NPC_i, a temporary array $DISPTB_i$ is built from the NPC_is as the accumulated number of particles contained in the cubes already listed until cube of index 'i', *i.e.* $DISPTB_i = \sum_{j=1}^{i} NPC_j$. The loop is carried out over NC.
- NP blocks BLO_l are built from $DISPTB_i$, with approximately N/NP particles per block. The loop is over NC.
- The next step consists of detecting which particles (whose index is defined in the list corresponding to iteration n) located on a given processor $PROC_k$ belong to a given block BLO_l as defined in the previous item. At the end of this operation, each processor has access to the number of its particles located on block BLO_l, and to the local list of its own particles per block. Both arrays are not complete as the loop is performed over the particles present on each processor $PROC_k$ at iteration n, and not over the total number of particles N.
- The full local list assembled on the master node is then broadcast to the other processors. Note that this array is still global in this version of the code.
- The global list is built up from the local list.

4.2 Step 2. Search for particle links

Links between particles a and their neighbouring particles b used in the calculation of the operators are locally constructed by processors, with the help of the cube structure. The next step differs from the previous version of the code, in the sense that arrays containing the main variables are now locally defined rather than globally. For each particle located on a given processor, the neighbouring particles (which may not be on the same processor) are gathered in extra arrays, and the rank of those processors stored. If the first loop of Step 2 goes over particle a located on processor $PROC_k$ as before, the second one goes on the neighbouring cubes of the cube containing particles a, taking into account the particles not located on the processor where particle a is.

4.3 Step 3. Link particle re-indexing

After the search for the links, the particle links are re-indexed in order to limit the communications between processors in the calculation phase. This means that the array containing the list of neighbouring particles has to be updated.

4.4 Step 4. Resolution of the equations

The equations are solved explicitly. The operators are built per processor, with loops going over the list of particles on $PROC_k$ at iteration $n+1$. The neighbouring particles, which may not be located on the processor of concern, are collected in an external array and used to calculate the various operators. As a result, only local communication is necessary to build these operators.

5 Results

5.1 Description of the Machines

Simulations have been performed on Blue Gene/L, BlueGene/P and IBM AIX systems. Both single rack Blue Gene systems contain 1,024 chips, with 2 processor cores per chip in the L system and 4 processor cores per chip in the P system, giving a total of 2,048 cores and 4,096 cores, respectively. Memory is provided at 512 Mbytes per core in both systems. The basic processor in the L system is the Power 440 running at 700 MHz, whilst the P system uses a processor from Power 450 family running at 850 MHz. The HPCx system uses an IBM eServer 575 nodes for the compute and IBM eServer 575 nodes for login and disk I/O. Each eServer node contains 16 1.5 GHz 2Gb memory POWER5 processors. The main HPCx service provides 160 nodes for compute jobs for users, giving a total of 2,560 processors [6].

5.2 Accuracy

The first test is carried out to validate this laminar parallel version against the analytical Poiseuille solution. The Reynolds number based on the bulk velocity is equal to 100. The run is performed on 128 cores of BG/P. Figure 1 (Left) shows that Spartacus-3D's output compares well to the analytical solution.

5.3 Performance of the Code

Due to the update of the particle positions at each time step, each iteration consists of both a new partitioning to ensure load-balancing and the fluid dynamics calculation itself. For this approach, two types of performance are analyzed showing the total CPU time per iteration per processor, and the CPU time for the dynamics per iteration per processor, to be able to compare with any other CFD Eulerian software dealing with non-moving grids.
The results of the previous parallel version of the code are shown in Fig. 1 (Right) for $355,000$ particles. This was carried out on a cluster of thin nodes, each of them having 4Gb memory per processor. The code scales well up to 8 processors, but from 16 processors it can be seen that the generalized use of global communications has an important impact on the performance.

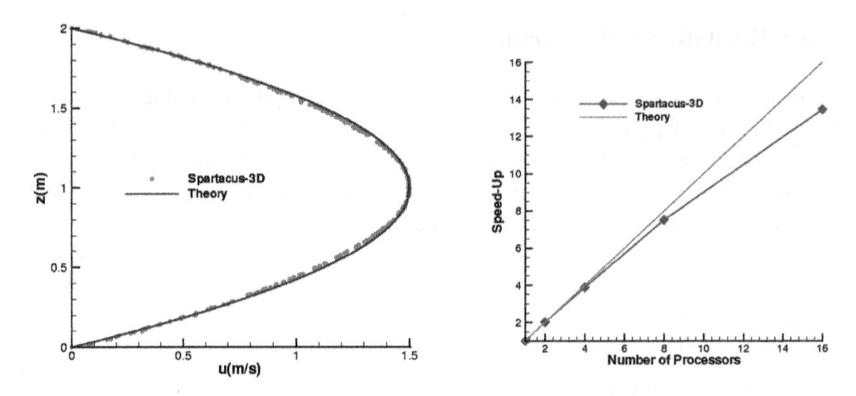

Fig. 1. Left: Poiseuille flow at $Re = 100$. Comparison between SPH and the analytical solution. Right: Speed-up of the previous parallel version of the code.

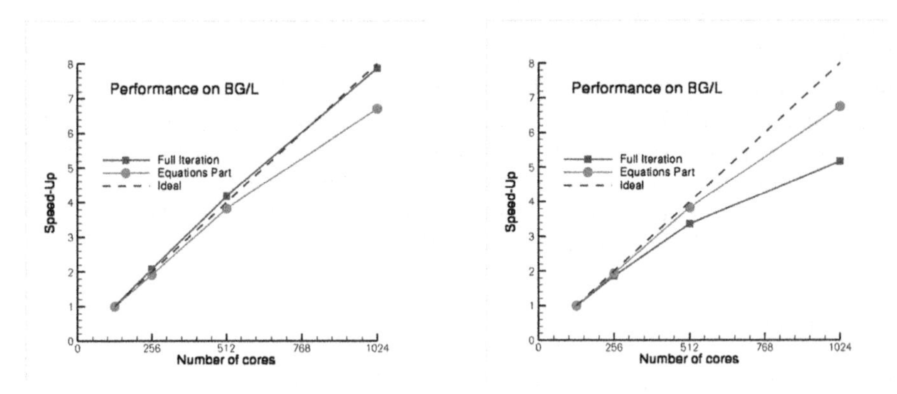

Fig. 2. Speed-up of the code on BGL without optimisation in the sorting of the links (Left) and with optimisation there (Right).

Another important factor of the previous test case of 335,000 particles is that it cannot be run with the former version of the code on the Blue Gene's (BG's) because of insufficient memory per processor. However, the new version, based on using local communications and a smaller amount of memory per processor, can be ran on the BG's and even more particles can be considered. In this work, the case of a 2M particle channel is simulated.

Table 1 shows the CPU time per iteration in 4 cases.
Two versions of the codes are considered, the former without optimisation of the link re-indexing (denoted as P for portable) and the latter with this optimisation (denoted as NPO for non-portable, as the IBM ESSL library is used for that purpose). Figures 3 and 4 show the speed-up of the code from 64 to 1,024 processors (CO mode for

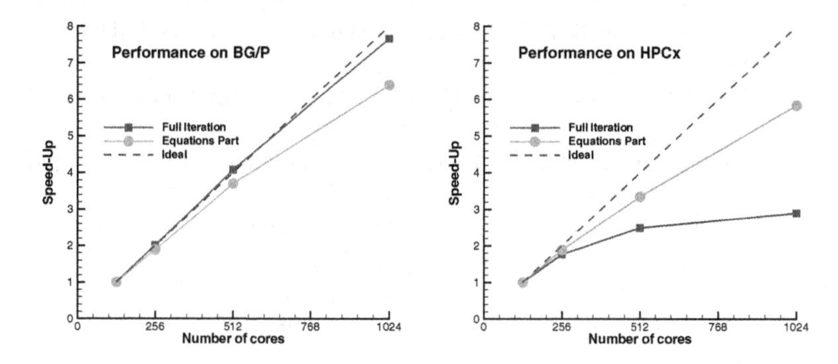

Fig. 3. Speed-up of the code on BGP without optimisation in the sorting of the links (Left) and on HPCx with optimisation there (Right).

	BGP (P)	BGL (P)	BGL (NPO)	HPCx (NPO)
CPU Time/Iteration	1.2894s	1.6919s	0.9689s	0.5732s

Table 1. CPU Time/Iteration on $1,024$ processors on various machine, with a portable (P) version and a non-portable (NPO) one.

BG/L and SMP mode for BG/P). The scaling is generally good on both BG's, with or without optimisation of the link sorting. On the other hand, if the calculation part scales well on HPCx, almost no gain is observed from 512 to 1,204 processors. It might be necessary to increase the number of particles, and also to localise the last global array to increase performance.

Note that the code reaches 6.3% of peak of HPCx system (IBM AIX), which is in the range for CFD codes (the peak computational speed of the HPCx system is 15.3 Tflops and the test has been carried out on 96 processors).

6 Concluding remarks

This paper showed that the new version of Spartacus-3D for HPC can efficiently handle very large simulations. This is due to the localisation of most of the arrays, and the use of local communications instead of global ones, wherever possible. The performance for such an amount of particles is good in terms of calculation time per iteration, but relatively poor for the total calculation time per iteration. The reason is that the particle search is managed through the construction of a global array for the particle list. As a result, some global communications are still required, which impacts on the performance. This is an area we aim to look at in the future.

[1] R. Issa, Numerical assessment of the Smoothed Particle Hydrodynamics gridless method for incompressible flows and its extension to turbulent flows, PhD thesis, UMIST, (2004).
[2] J. J. Monaghan, Particle Methods for Hydrodynamics. Comput. Phys. Rep., 3 (1985) 71.
[3] J.J. Monaghan, Smoothed Particle Hydrodynamics, Ann. Rev. of Astronomy and Astrophysics. 30 (1992) 543.
[4] J. P. Morris and P. J. Fox and Y. Zhu, Modelling low Reynolds Number Incompressible Flows Using SPH, J. of Comp. Phys., 136 (1997) 214.
[5] C. Moulinec, R. Issa, J.C. Marongiu and D. Violeau, Parallel 3-D SPH Simulations. Comp. Mod. in Engineering and Science 25 (2008), 133.
[6] http://www.hpcx.ac.uk.

Part VIII

software Framework and Component Architecture

An integrated object-oriented approach for parallel CFD

Dominique Eyheramendy[1],David Loureiro[2], Fabienne Oudin-Dardun[3]

[1] Ecole Centrale Marseille LMA-CNRS, Technopôle de Château-Gombert 38 rue Frédéric Joliot Curie, 13451 Marseille, France
dominique.eyheramendy@ec-marseille.fr
[2] LIP/ENS-LYON/INRIA/CNRS/UCBL, 46 Allée d'Italie, 69364 Lyon, France
david.loureiro@ens-lyon.fr
[3] Université de Lyon, Université Lyon 1, CNRS UMR5208, Institut Camille Jordan, CDCSP-ISTIL, 15 Blvd Latarget, 69622 Villeurbanne, France
fabienne.oudin@univ-lyon1.fr

Abstract. In this paper, we present a global computer science approach to deal with parallel computations. The proposed approach consists in managing at the same level either multithreading or distributed strategies, whatever the computation may be. The integration of the concept is held in a Java framework which proposes both, a pure object-oriented paradigm and, convenient libraries to deal with threads management and communications schemes. The approach is illustrated on a domain decomposition method for a Navier-Stokes flow.
Keywords: Finite elements, object-oriented programming, domain decomposition, multi-threaded computing, distributed computing.

1 Introduction

In this paper, we present the key ideas of a Java application to a finite element code. This approach is based on the reusability of code and the portability of application in the case of parallel application. The key idea of the developments is that the parallel algorithms are programmed within the application using a single programming concept and a single language. In classical approaches, the programming language (C, C++, Fortran) is associated to additional libraries such as PVM or MPI. Existing libraries in the Java environment are used here. Thus, the maintenance of the application is made easier, which is important from an industrial point of view. The proposed approach offers the advantage of a high level of reusability of the different parts of the code whatever the computational strategy is. The consequence of it is a high reliability of the software for domain decomposition methods. This reliability is obtained through a classical object-oriented paradigm which allows the programmer

D. Tromeur-Dervout (eds.), *Parallel Computational Fluid Dynamics 2008*,
Lecture Notes in Computational Science and Engineering 74,
DOI: 10.1007/978-3-642-14438-7_29,

to separate the management of finite elements data, the solution algorithms, the management of different processes and the communication schemes.

In section 2, we give a state of the art for Java applications in Computational Mechanics that most often includes parallel computations. In section 3, we described two different implementations of an overlapping Schwarz domain decomposition method within the same finite element code. In section 4, we give a numerical application that has been run using the same code on different systems.

2 Java in Computational Mechanics

Until now, most of the developments in Java computational mechanics have been considered by the computational science community, in general concerned by parallel computations. In the domain of numerical computations, Java retained some attention for its networking capabilities and its Internet easy portability. E.g. in [1], a trivial application based on a boundary element method has been developed. Similar developments may be found directly on the Internet, including basic finite element applications, such as in [2] for an application of fracture mechanics. In the computational mechanics community, Java is often considered as a simple tool to produce applications on the Internet and/or to effectuate computation on the network. For example, Java kind technologies are often used to couple and manage traditional codes written in C/C++/Fortran. This permits the developers to use ancient codes or part of code in coupled applications. A consequence of it is to keep a real computational efficiency. E.g. in [3], an interactive finite element application based on a coupled C++/Java is described. Comparative tests with FORTRAN and C are conducted on small problems using direct solvers based on tensor computations; this aims at illustrating the high efficiency computational potential of Java. In [4], the development of GUIs is put in prominent position on an unstructured mesh generator. In [5], the Java code CartaBlanca is presented. It is an environment for distributed computations of complex multiphase flows. Based on a finite volume approach, a solution scheme based on a Newton-Krylov algorithm is described. The code exhibits good performances. A similar environment has been developed to simulate electromagnetisms problems in [6]. Both applications show the high potential of the approach to design more complex and general computational tools in mechanics for example. These developments exhibit the networking facilities provided in Java. A large number of publications shows the interest of Java and its efficiency: direct solution of linear systems [7], FFT and iterative and direct linear systems solvers on Euler type flows [8], solution of Navier-Stokes flows [9] and [10]. Again, in all these papers, multiprocesses' management and networking capabilities of Java are put forward. In [8], [9] and [11], performances of Java are tested on simple matrix/vector products. Compared to C/C++ code, only 20 to 30% of efficiency is lost. More recently, the description of a finite element code in Java was proposed in [12]. The proposed design remains rather similar to the existing ones based on classical object-oriented approaches. Nevertheless, this paper shows that it is possible to develop a global code based on a pure Java approach. The latter can be applied to design large complex applications in computational mechanics taking into account complexity in modern computational mechanics: multiscale, multiphysics and multiprocesses applications. Firstly, enhanced and homogeneous data organization schemes to deal with complexity are proposed in Java. Secondly, developing a global application in a uniform environment including all the libraries needed is somehow an attractive idea for scientists and industrials. It is worth noting that similar approaches exist in C# (e.g. see [15], [16]). This brief analysis shows that even if performances do not achieve the one of C/C++/Fortran but get close to a certain extent, Java or similar approaches can be used in computational mechanics including CFD. These strategies offer the main advantage to design the simulation tools such a way that the technical strategies and the numerical algorithms are integrated in a seamless way.

3 Schwarz multiplicative multithreaded and distributed applications

The Schwarz multiplicative algorithm is described in [11]. In Fig. 1, the same decomposed domain is shown to be solved in a multi-threaded application or in a distributed application. Solving both problems is made through the same basic code for the physics and numerical algorithm. A description of the code goes beyond the scope of this paper and can be found in [13], [17] and [18]. The description of the multithreaded application is given in [17]. As shown in Fig. 1, the general structure of the code organization is the same. In the distributed version, the global Schwarz algorithm is managed through a thread located on the computer 1. The algorithm lies in two points: the management of the domains located on alternative computers and the communication schemes between these domains located on various computers. The global management is held by the way of distributed objects. The Java RMI package is used for that purpose. It permits the programmer to keep a natural and homogeneous organization of classes to deal with the domain decomposition algorithm. The consequence of it is to keep exactly the same structure as in both algorithms, managing real objects in the one case, and distributed objects in the other case. The later remains seamless for the programmer. The Schwarz algorithm is given in both cases as follows (programming details are omitted in the following code):

```
public void solve( )
{
    // ... INITIALIZING OF THE SOLVER
    for(int it = 0; (it < maxIteration) && (! iteration.converged() ); it++)
    {
        for( int color = 0 ; color < 2 ; color++ )
        {
            int anteColor = ( color + 1 ) % 2 ;
            //**************** solution *****************
            // SOLUTION BLOCK OMMITED
            //*************** exchanges ****************
            // EXCHANGES BLOCK OMMITED
```

```
            //*****************************************
        }
        // FINALIZING THE ITERATIONS
        //...
    }
}
```

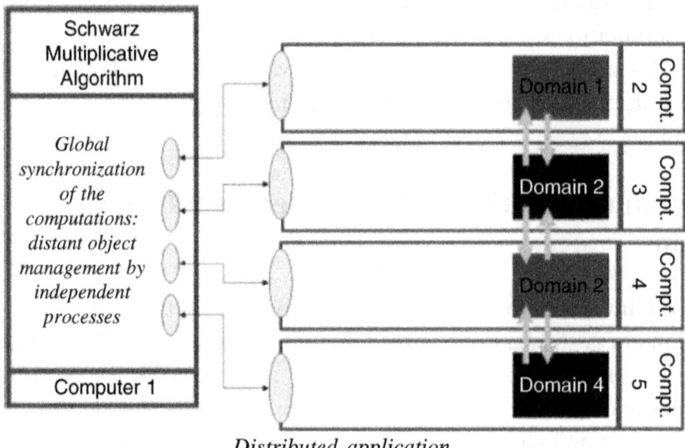

Distributed application

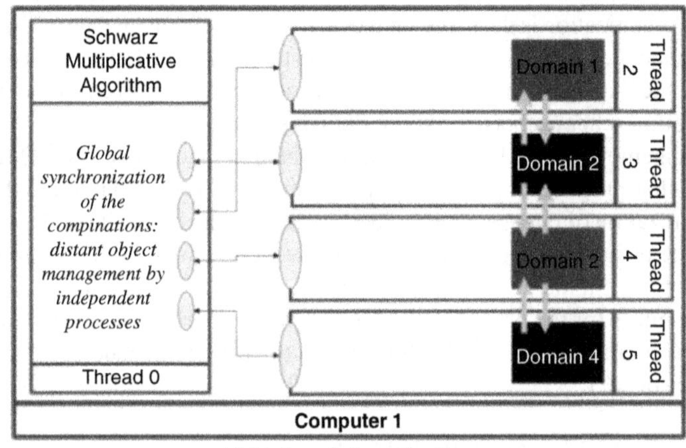

Multi-threaded application

Fig. 1. Multi-threaded application versus distributed application.

The major differences between both approaches remain the initialization phases in the different threads (solution, exchanges,...). We give here the example of the thread creation to solve the physical problem for a given domain:

```
Thread threadii = new Thread ( new Runnable ()
    {
```

```
        int number = Ni ;
        public void run()
        {
            domains[number].solveSchwarz () ;
        }
    });
```

The variable called domains[number] represents the domain (numbered number) to be solved at a given time. In the distributed version, the same piece of code becomes:

```
Thread threadii = new Thread( new Runnable()
    {
        public void run() {
            try {
                ((ServerDomain) listOfServerDomain.get(indice0a)).solveSchwarz();
            } catch (RemoteException e) { }
        }
    });
```

In this piece of code, the equivalent domain is called using the code: listOfServer-Domain.get(indice0a) but the global scheme remains the same. In the first approach, the domain is located in the same memory space, as in the second approach, it is located on a separate computer. The lack of efficiency in communication schemes is the main drawback of the use of RMI distributed objects. The latter is not a problem for the management of the global Schwarz algorithm but produces a bottleneck when exchanging data. This is the reason why communications are programmed using the Socket Java class which a point to point communication scheme. This scheme is implemented at the level of boundary conditions object where the information provided by the neighbor domain id needed. Once again, only a local modification is needed within the global framework. For illustration purposes, both schemes are applied to a Navier-Stokes flow. For programming details in Java, the reader may refer to [19].

4 Numerical application

To illustrate both approaches, we study a Galerkin Navier-Stokes formulation stabilized by adding least-squares type terms. The equations of the problem and the formulation are presented in Fig. 2. Linearization of the problem is introduced in a Newton like scheme in the Schwarz multiplicative scheme. A direct linear system solver based on a Crout decomposition is used to solve the linear system at each iteration. Note, the same code is compiled once and run on all systems. The multi-threaded version of code is run on a SGI Altix 16-processors Itanium2 1.3Ghz, 32 Go of RAM. The distributed version is run on a set of simple cluster of PC linked by a classical network. The code used for both applications is exactly the same. The global solution algorithm and the boundary conditions in charge of the communication scheme are of course not the same. The formulation is applied to the computation of a flow through a set of cylinders shown in Fig. 3. The domain is a periodic layer of cylinders. Numerical results are given in Fig. 3. First, the pressure contour on a typical cell is plotted. Secondly, the mean velocity computed overall the domain is given with respect to the gradient of pressure over the cell in the direction of the main flow. The latter represent the homogenized flow through the cylinders. For low velocities, the relation between the mean velocity and the gradient of pressure is linear. From a global point of view this can be assimilated to a Darcy's flow. But as far as the mean velocity is increasing the linearity disappears. The advection term is no more negligible and the global flow cannot be assimilated to Darcy's flow. It shows the influence of the advection term on the homogenized flow, that cannot any more considered as a linear Darcy's flow. Filtration laws could be established such a way. The same computation has been held using both approaches. From a practical point of view, we show that the same code can be run on heterogeneous systems. As both systems are different, we cannot compare the efficiency of both algorithms. This goes beyond the scope of our qualitative test. We show here that we can easily switch from a computer to another one without any problem using exactly the same code and without compiling it. This feature can be very interesting from an industrial point of view.

$$
\begin{array}{ll}
\sigma_{ij,j} + f_i = \rho(u_i u_{i,j} + u_{i,t}) & \text{on } \Omega \times T \\
u_{i,j} = 0 & \text{on } \Omega \times T \\
\sigma_{ij} n_j = F_i & \text{on } \partial_2\Omega \times T \\
u_i = \overline{u_i} & \text{on } \partial_1\Omega \times T \\
\sigma_{ij} = -p\delta_{ij} + 2\mu\varepsilon_{ij}(u) & \text{on } \Omega \times T \\
\varepsilon_{ij}(u) = \frac{1}{2}(u_{i,j} + u_{j,i}) & \text{on } \Omega \times T \\
u_i(0) = u_{i0} & \text{on } \Omega \text{ at } t = 0
\end{array}
$$

$u = \overline{u}$ $\partial_1\Omega$ Ω $\partial_2\Omega$ f F

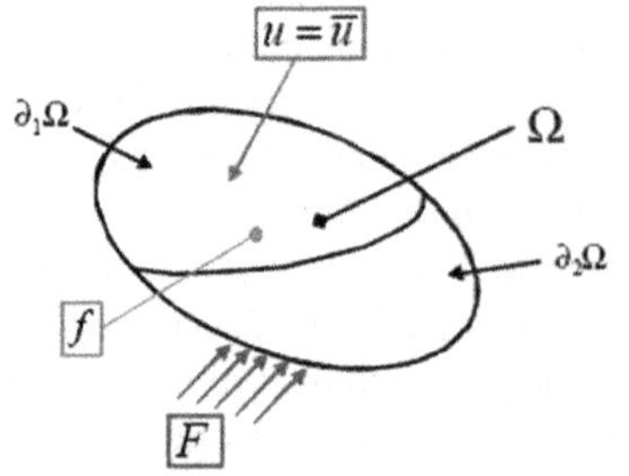

• Given f, find $(u^h, p^h) \subset ((\mathcal{S}^h)_n \times (\mathcal{P}^h)_n)$ such that for each $(w^h, q^h) \subset ((\mathcal{V}^h)_n \times (\mathcal{P}^h)_n)$, one has :

$$
\int_\Omega \rho u_j^h u_{i,j}^h + \rho u_{i,t}^h dv - \int_\Omega 2\mu\varepsilon_{ij}(u^h)\varepsilon_{ij}(w^h) dv + \int_\Omega p_h w_{i,i}^h dv + \int_\Omega u_{i,i}^h q^h dv - \int_\Omega f_i w_i^h dv
$$

$$
+ \sum_{\Omega^e \in \Omega^h} \left[\int_{\Omega^e} (\rho u_j^h u_{i,j}^h + \rho u_{i,t}^h - 2\mu\varepsilon_{ij,j}(u^h) + p_{,i}^h - f_i)\tau_{mom}(\rho u_j^h w_{i,j}^h + q_{,i}^h) dv \right] = 0
$$

Fig. 2. Navier-Stokes model. Initial-boundary value problem and stabilized finite elements formulation.

5 Conclusion

In this paper, we have presented two computational approaches based on the same numerical algorithm, a Schwarz overlapping domain decomposition method. The first approach is a multithreaded approach of the algorithm; the second one is a distributed version of it. In the first one, which is to be run on a share memory system, no communication between the domains is needed. The second one is based on two different communication schemes. The first one permits a master to manage the global Schwarz algorithm based on a classical object-oriented approach for distributed objects. Communications for data exchanges are implemented using classical sockets. This point to point communication scheme allows us to achieve efficiency. The implementation is held in Java which ensures the code to be run directly on all the systems. The same code is run for both applications. The advantages of such an approach are:

1. from a programmer point of view: a single language and a single approach for all the algorithms which means simplicity and reliability due to the fact that the finite element core remains identical.
2. From a user point of view: the same code can be used on heterogeneous systems depending on the availability of different systems.

We advocate that such an approach may simplify a lot the use of complex systems for single applications. This opens new tracks in the design of code that can be used in the context of either a shared memory system or a distributed memory system. In this context, mixing both approaches within a single application, taking advantage of the heterogeneous computers systems available at a given time is made possible in a simple way.

[1] M. Nuggehally, Y.J. Lui, S.B. Chaudhari and P. Thampi, An internet-based computing plateform for the boundary element method, Adv. In Engrg. Software, 34 (2003) 261-269.

[2] G.P. Nikishkov and H. Kanda, The development of a Java engineering application for higher-order asymptotic analysis of crack-tip fields, Advances in Engineering Software 30 (1999) 469-477.

[3] G.R. Miller, P. Arduino, J. Jang and C. Choi, Localized tensor-based solvers for interactive finite element applications using C++ and Java, Comp. & Struct. 81 (2003) 423-437.

[4] R. Marchand, M. Charbonneau-Lefort and M. Dumberry, ARANEA, A program for generating unstructured triangular meshes with a Java Graphics User interface, Comp. Phys. Communications, 139 (2001) 172-185.

[5] N.T. Padial-Collins, W.B. VanderHeyden, D.Z. Zhang, E.D. Dendy and D. Livescu, Parallel operation of CartaBlanca on shared and distributed memory computers, Concurrency and Computation: Practice and Experience 16 (2004) 61-77.

[6] L. Baduel, F. Baude, D. Caromel, C. Delb, N. Gama, S. El Kasmi and S. Lanteri, A parallel object-oriented application for 3-D electromagnetism, ECCOMAS 2004, Jyvskyl, Finland (2004).

[7] G.P. Nikishkov, Y.G Nikishkov and V.V Savchenko, Comparison of C and Java performance in finite element computations, Computer & Structures, 81 (2003) 2401-2408.

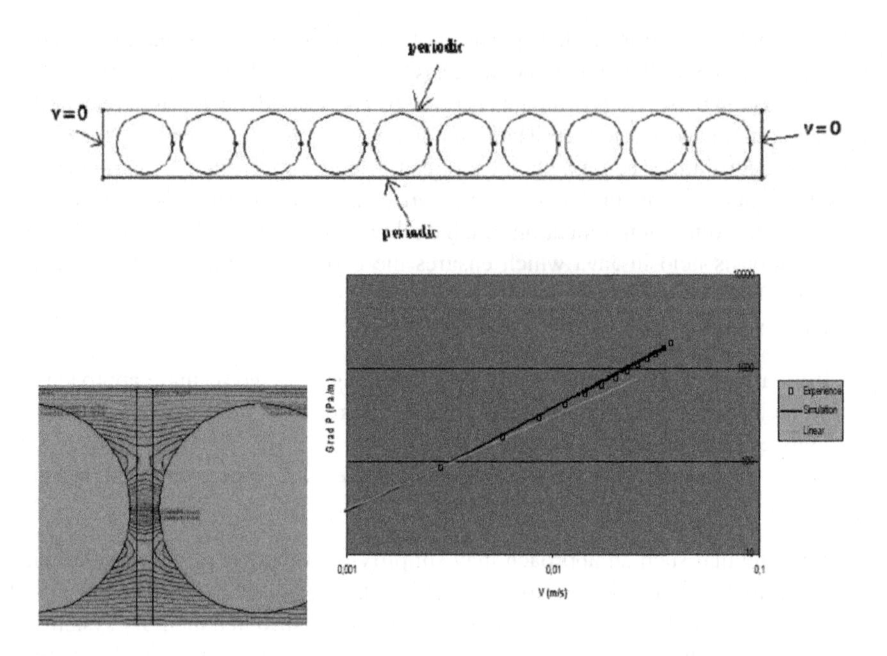

Fig. 3. Flow through cylinders. Computational domain, decomposed domain, numerical results: detail of iso-pressures, curve velocity/gradient of pressure (Experiment and simulation).

[8] J.M. Bull, L. A. Schmith, L. Pottage and R. Freeman, Benchmarking Java against C and Fortran for Scientific Applications, Joint ACM JavaGrande - ISCOPE 2001 Conference, Stanford Universtity, June 2-4, 2001.

[9] J. Huser, T. Ludewig, R.D. Williams, R. Winkelmann, T. Gollnick, S. Brunett and J. Muylaert, A test suite for high-performance parallel Java, Advances in Engineering Software, 31 (2000) 687-696.

[10] C.J. Riley, S. Chatterjee and R. Biswas, High-performance Java codes for computational fluid dynamics, Concurrency and Computation: Practice and Experience 15 (2003) 395-415.

[11] D. Eyheramendy, Object-oriented parallel CFD with JAVA, 15th International Conference on Parallel Computational Fluid Dynamics, Eds. Chetverushkin, Ecer, Satofuka, Priaux, Fox, Ed. Elsevier, (2003) pp. 409-416.

[12] D. Eyheramendy, Advanced object models for mathematical consistency enforcement in scientific computing, WSEAS Transactions on Mathematics, vol. 4, N° 4, (2005), pp. 457-463.

[13] D. Eyheramendy, High abstraction level frameworks for the next decade in computational mechanics, Innovation in Engineering Computational Technology, Eds. B.H.V. Topping, G. Montero and R. Montenegro, ©Saxe-Cobourg Publications, Chap. 3, (2006) pp. 41-61.

[14] G.P.Nikishkov, Object oriented design of a finite element code in Java. Computer Modeling in Engineering and Sciences 11 (2006) pp. 81-90.
[15] R.I. Mackie, Object-oriented design of pre-conditionned iterative equation solvers using .NET, Proceedings of 5th Int. Conf. on Engineering Computational Technology, Las Palmas de Gran Canaria, Spain, 12-15 Sept. 2007.
[16] R.I. Mackie, Lessons learnt from using .NET for distributed finite element analysis, Proceedings of 11th Int. Conf. on Civil, Structural and Environmental Engineering Computing, St. Julians, Malta, 18-21 Sept. 2007.
[17] D. Eyheramendy and F. Oudin, Advanced object-oriented techniques for coupled multiphysics, In Civil Engineering Computation: Tools and Techniques, Ed. B.H.V. Topping, ©Saxe-Cobourg Publications, ISBN 978-1-874672-32-6 Chap. 3, (2007) pp. 37-60.
[18] D. Eyheramendy, Advanced object models for mathematical consistency enforcement in scientific computing, WSEAS Transactions on Mathematics, vol. 4, N° 4, (2005), pp. 457-463.
[19] D. Flanagan, Java in a Nutshell, Fourth edition, Ed. O'reilly (2002).

Fast Multipole Method for particle interactions: an open source parallel library component

F. A. Cruz[1], M. G. Knepley[2], and L. A. Barba[1]

[1] Department of Mathematics, University of Bristol,
University Walk, Bristol BS8 1TW, United Kingdom
[2] Mathematics and Computer Science Division, Argonne National Laboratory,
9700 S. Cass Avenue, Argonne, IL 60439 U.S.A.

Abstract. The fast multipole method is used in many scientific computing applications such as astrophysics, fluid dynamics, electrostatics and others. It is capable of greatly accelerating calculations involving pair-wise interactions, but one impediment to a more widespread use is the algorithmic complexity and programming effort required to implement this method. We are developing an open source, parallel implementation of the fast multipole method, to be made available as a component of the PETSc library. In this process, we also contribute to the understanding of how the accuracy of the multipole approximation depends on the parameter choices available to the user. Moreover, the proposed parallelization strategy provides optimizations for automatic data decomposition and load balancing.

1 INTRODUCTION

The advantages of the fast multipole method (FMM) for accelerating pair-wise interactions or N-body problems are well-known. In theory, one can reduce an $\mathcal{O}(N^2)$ calculation to $\mathcal{O}(N)$, which has a huge impact in simulations using particle methods. Considering this impact, it is perhaps surprising that the adoption of the FMM algorithm has not been more widespread. There are two main reasons for its seemingly slow adoption; first, the scaling of the FMM can really only be achieved for simulations involving very large numbers of particles, say, larger than $10^3 - -10^4$. So only those researchers interested in solving large problems will see an advantage with the method. More importantly, perhaps, is the fact that the FMM requires considerable extra programming effort, when compared with other algorithms like particle-mesh methods, or treecodes providing $\mathcal{O}(N \log N)$ complexity.

One could argue that a similar concern has been experienced in relation to most advanced algorithms. For example, when faced with a problem resulting in a large system of algebraic equations to be solved, most scientists would be hard pressed to have to program a modern iterative solution method, such as a generalized minimum residual (GMRES) method. Their choice, in the face of programming from scratch, will most likely be direct Gaussian elimination, or if attempting an iterative method,

D. Tromeur-Dervout (eds.), *Parallel Computational Fluid Dynamics 2008*,
Lecture Notes in Computational Science and Engineering 74,
DOI: 10.1007/978-3-642-14438-7_30,

the simplest to implement but slow to converge Jacobi method. Fortunately, there is no need to make this choice, as we nowadays have available a wealth of libraries for solving linear systems with a variety of advanced methods. What's more, there are available parallel implementations of many librarires, for solving large problems in a distributed computational resource. One of these tools is the PETSc library for large-scale scientific computing [1]. This library has been under development for more than 15 years and offers distributed arrays, and parallel vector and matrix operations, as well as a complete suite of solvers, and much more. We propose that a parallel implementation of the FMM, provided as a library component in PETSc, is a welcome contribution to the diverse scientific applications that will benefit from the acceleration of this algorithm.

In this paper, we present an ongoing project which is developing such a library component, offering an open source, parallel FMM implementation, which furthermore will be supported and maintained via the PETSc project. In the development of this software component, we have also investigated the features of the FMM approximation, to offer a deeper understanding of how the accuracy depends on the parameters. Moreover, the parallelization strategy involves an optimization approach for data decomposition among processors and load balancing, that should make this a very useful library for computational scientists.

This paper is presented as follows, in section §2 we present an overview of the original FMM [2], in section §3 we present our proposed parallelization strategy, and in §4 scaling results of the parallel implementation are presented.

2 CHARACTERIZATION OF THE MULTIPOLE APPROXIMATION

2.1 Overview of the algorithm

The FMM is a fast summation method that accelerates the multiple evaluation of functions of the form:

$$f(y_j) = \sum_{i=1}^{N} c_i \mathbb{K}(y_j, x_i) \tag{1}$$

where the function $f(\cdot)$ is evaluated at a set of $\{y_j\}$ locations. In a single evaluation of (1) the sum over a set of sources $\{(c_i, x_i)\}$ is carried out, where each source is characterized by its weight c_i and position x_i. The relation between the evaluation and source points is given by the the kernel $\mathbb{K}$, in general, the kernel function is required to decay monotonically.

In order to accelerate the computations, the FMM uses the idea that the influence of a cluster of sources can be approximated by an agglomerated quantity, when such influence is evaluated far enough away from the cluster itself. The method works by dividing the computational domain into a *near-domain* and a *far-domain*:

Near domain: contains all the particles that are near the evaluation point, and is usually a minor fraction of all the N particles. The influence of the near-domain is computed by directly evaluating the pair-wise particle interactions with (1). The computational cost of directly evaluating the near domain is not dominant as the near-domain remains small.

Far domain: contains all the particles that are far away from the evaluation point, and ideally contains most of the N particles of the domain. The evaluation of the far domain will be sped-up by evaluating the approximated influence of clusters of particles rather than computing the interaction with every particle of the system.

The approximation of the influence of a cluster is represented as Multipole Expansions (MEs) and as Local Expansions (LEs); these two different representations of the cluster are the key ideas behind the FMM. The MEs and LEs are Taylor series (or other) that converge in different subdomains of space. The center of the series for an ME is the center of the cluster of source particles, and it only converges outside the cluster of particles. In the case of an LE, the series is centered near an evaluation point and converges locally.

The first step of the FMM is to hierarchically subdivide space in order to form the clusters of particles; this is accomplished by using a tree structure, illustrated in Figure 1, to represent each subdivision. In a one-dimensional example: level 0 is the whole domain, which is split in two halves at level 1, and so on up to level l. The spatial decomposition for higher dimensions follows the same idea but changing the number of subdivisions. In two dimensions, each domain is divided in four, to obtain a quadtree, while in three dimensions, domains are split in 8 to obtain an oct-tree. Independently of the dimension of the problem, we can always make a flat drawing of the tree as in Fig. 1, with the only difference between dimensions being the number of branches coming out of each node of the flat tree.

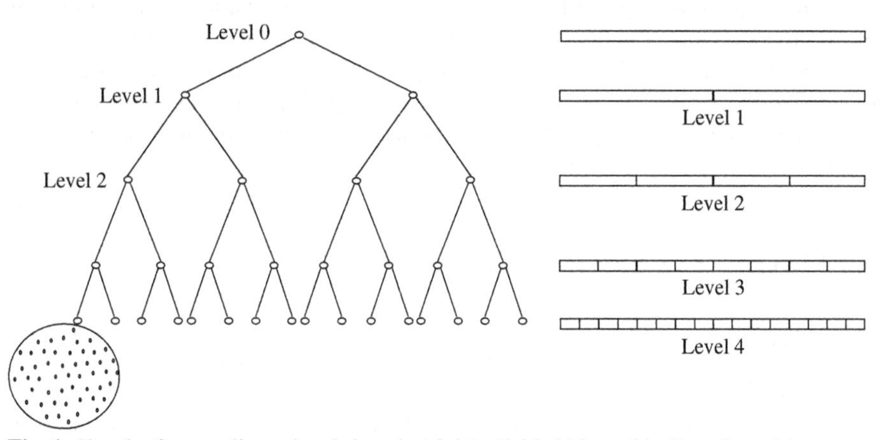

Fig. 1. Sketch of a one-dimensional domain (right), divided hierarchically using a binary tree (left), to illustrate the meaning of levels in a tree and the idea of a final leaf holding a set of particles at the deepest level.

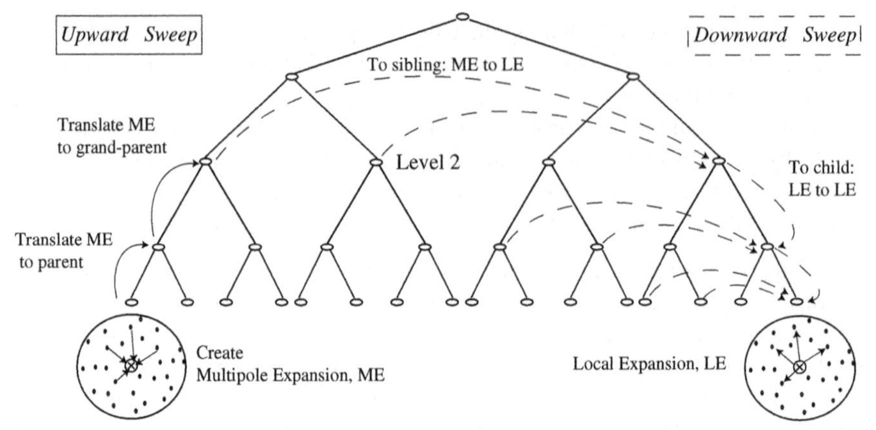

Fig. 2. Illustration of the *upward sweep* and the *downward sweep* of the tree. The multipole expansions (ME) are created at the deepest level, then translated upwards to the center of the parent cells. The MEs are then translated to a local expansion (LE) for the siblings at all levels deeper than level 2, and then translated downward to children cells. Finally, the LEs are created at the deepest levels.

After the space decomposition, the FMM builds the MEs for each node of the tree in a recursive manner. The MEs are built first at the deepest level, level l, and then translated to the center of the parent cell, recursively creating the ME of the whole tree. This is referred to as the *upward sweep* of the tree. Then, in the *downward sweep* the MEs are first translated into LEs for all the boxes in the *interaction list.* At each level, the interaction list corresponds to the cells of the same level that are in the far field for a given cell, this is defined by simple relations between nodes of the tree structure. Finally, the LEs of upper levels are added up to obtain the complete far domain influence for each box at the leaf level of the tree. The result is added to the local influence, calculated directly with Equation (1).These ideas are better visualized with an illustration, as provided in Figure 2. For more details of the algorithm, we cite the original reference [2].

3 PARALLELIZATION STRATEGY

In the implementation of the FMM for parallel systems, the contribution of this work lies on automatically performing load balance, while minimizing the communications between processes. To automatically perform this task, first the FMM is decomposed into more basic algorithmic elements, and then we automatically optimize the distribution of these elements across the available processes, using as criteria the load balance and minimizing communications.

In our parallel strategy, we use sub-trees of the FMM as the basic algorithmic element to distribute across processes. In order to partition work among processes, we cut the tree at a certain level k, dividing it into a *root* tree and 2^{dk} *local* trees,

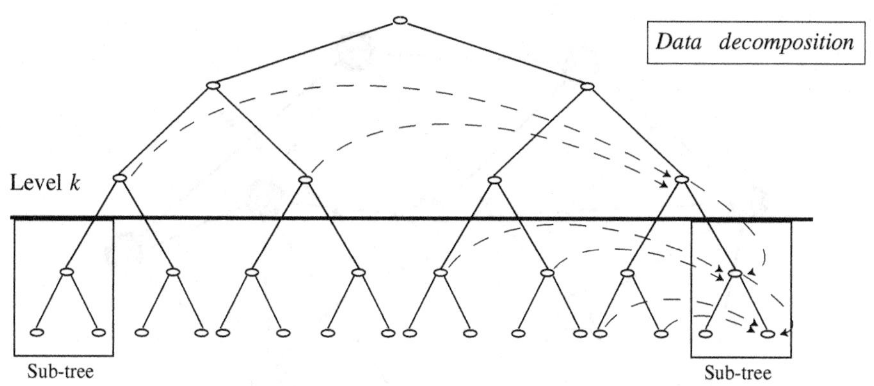

Fig. 3. Sketch to illustrate the parallelization strategy. The image depicts the FMM algorithm as a tree, this tree is then "cut" at a chosen level k, and all the resulting sub-trees are distributed among the available processes.

as seen in Figure 3. By assigning multiple sub-trees to any given process, we can achieve both load balance and minimimal communication. A basic requirement of our approach is that when decomposing the FMM, we produce more sub-trees than available processes.

To optimally distribute the sub-trees among the processes, we change the problem of distributing the sub-trees into a graph partitioning problem, that we later partition in as many parts as processes available. First, we assemble a graph whose vertices are the sub-trees, with edges (i, j) indicating that a cell c in sub-tree j is in the interaction list of cell c' in sub-tree i, as seen in Figure 4. Then weights are assigned to each vertex i, indicating the amount of computational work performed by the sub-tree i, and to each edge (i, j) indicating the communication cost between sub-trees i and j.

One of the advantages of using the sub-trees as a basic algorithmic elements, is that due to its well defined structure, we can rapidly estimate the amount of work and communication performed by the sub-tree. In order to produce these estimates, the only information that we require is the depth of the sub-trees and neighbor sub-tree information.

By using the work and communication estimates, we have a weighted graph representation of the FMM. The graph is now partitioned, for instance using ParMetis [3], and the tree information can be distributed to the relevant processes.

Advantages of this approach, over a space-filling curve partitioning for example, include its simplicity, the reuse of the serial tree data structure, and reuse of existing partitioning tools. Moreover, purely local data and data communicated from other processes are handled using the same parallel structure, known as Sieve [4]. The Sieve framework provides support for parallel, scalable, unstructures data structures, and the operations defined over them. In our parallel implementation, the use of

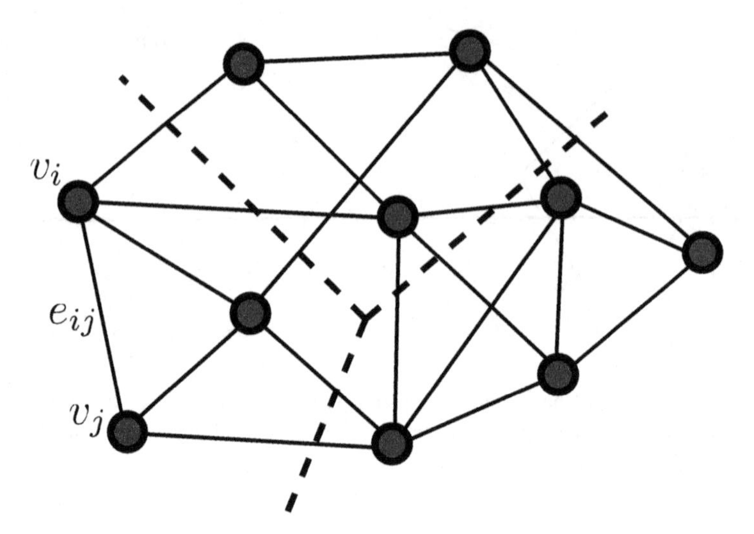

Fig. 4. Sketch that illustrates the graph representation of the sub-trees. The FMM sub-trees are represented as vertex in the graph, and the communication between sub-trees is represented by the edges between corresponding vertices. Each vertex has an associated weight v_i, that represents an estimate of the total work performed by the sub-tree. Each edge has an associated weight e_{ij}, that represents an estimate of the work between sub-trees v_i and v_j. By finding an optimal partitioning of the graph into as many section as processes available, we achieve load balance while minimizing communications. In the figure the graph has been partitioned in three sections, the boundaries between sections are represented by the dashed lines, and all the sub-trees that belongs to the same section are assigned to the same process.

Sieve reduces the parallelism to a single operation for neighbor and interaction list exchange.

4 RESULTS

In this section we present results of the parallel implementation on a multicore architecture, a Mac Pro machine, equipped with two quad-core processors and 8GB in RAM. The implementation is being currently integrated into the open source PETSc library. In Figure 5, we present experimental results of the timings obtained from an optimized version of PETSc library using its built-in profiling tools. The experiments were run on a multicore setup with up to four cores working in parallel, while varying the number of particles in the system. We performed experiments from three thousand particles up to system with more than two million particles. All the experiments were run for a fixed set of parameters of the FMM algorithm, in all cases we used parameters that ensure us to achieve high accuracy an performance (FMM expansion terms $p = 17$, and FMM tree level $l = 8$).

In the experimental results, we were able to evaluate a problem with up to two million particles in under 20 seconds. To put this in context, if we consider a problem

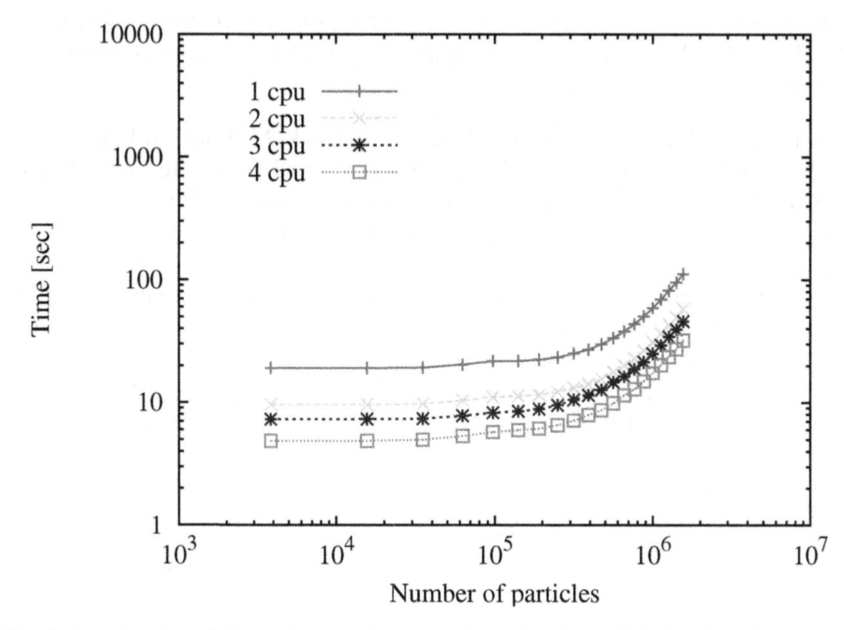

Fig. 5. Log-log plot of the total execution time of the FMM parallel algorithm for fixed parameters (expansion terms $p = 17$, and FMM tree level $l = 8$) vs. problem size (number of particles in the system). In the figure, each line represents the total amount of time used by the parallel FMM to compute the interaction in the system when using 1, 2, 3, and 4 processors.

with one million particles ($N = 10^6$), the direct computation of the N^2 operations would be proportional to 1 Tera floating point operations (10^{12} operations), in order to compute all the particles interactions of the system. Instead, the FMM algorithm accelerates the computations and achieve an approximate result with high accuracy in only 10 Giga floating point operations, meaning 100 times less floating point operations.

5 CONCLUSIONS AND FUTURE WORK

In this paper we have presented the first open source project of a parallel Fast Multipole Method library. As second contribution in this work, we also introduced a new parallelization strategy for the Fast Multipole Method algorithm that automatically achieves load balance and minimization of the communication between processes. The current implementation is capable of solving N-body problems for millions of unknowns in a multicore machine with good results. We are currently tuning the library and running experiments on a cluster setup.

With this paper we expect to contribute to widespread the use of fast algorithms by the scientific community, and the open source characteristic of the FMM library is the first step towards this goal. The open source project is currently being incorporated into the PETSc library [1] and a first version of the software is going to be released by October 2008.

[1] S. Balay, K. Buschelman, W. D. Gropp, D. Kaushik, M. Knepley, L. Curfman-McInnes, B. F. Smith, and H. Zhang. PETSc User's Manual. Technical Report ANL-95/11 - Revision 2.1.5, Argonne National Laboratory, 2002.

[2] L. Greengard and V. Rokhlin. A fast algorithm for particle simulations. *J. Comput. Phys.*, 73(2):325–348, 1987.

[3] George Karypis and Vipin Kumar. A parallel algorithm for multilevel graph partitioning and sparse matrix ordering. *Journal of Parallel and Distributed Computing*, 48:71–85, 1998.

[4] Matthew G. Knepley and Dmitry A. Karpeev. Mesh algorithms for pde with sieve i: Mesh distribution. *Scientific Programming*, 2008. to appear.

Hybrid MPI-OpenMP performance in massively parallel computational fluid dynamics

G. Houzeaux[1] *, M. Vázquez[1], X. Sáez[1] and J.M. Cela[1]

Computer Applications in Science and Egineering Department
Barcelona Supercomputing Center Edificio C6-E201, Campus Nord UPC, Jordi Girona 1-3, Spain

Abstract. We analyze in this paper the parallel performance of a computational mechanics code, based on a hybrid MPI-OpenMP parallelization. The specific problems considered are an explicit compressible flow solver and an implicit incompressible flow solver. Their performance using a fully MPI approach and a Hybrid approach are compared. The performance obtained on different architextures are also compared.

1 Introduction

In a previous paper [4],the authors presented the parallel performance of a computational mechanics code, which parallelization was based on a mesh partitioning technique using MPI as a communication library. The physical problems treated were the following: the incompressible Navier-Stokes(NS) equations; the compressible NS equations; the wave equation. The latter was solved using an implicit monolithic approach where as the last two problems were solved using an explicit time integration scheme. This paper presents the developments carried out on the incompressible and compressible NS equations. In the first set of equations, a projection method was developed and parallelized. In the second set of equations, OpenMP was implemented in the code. The objective of this work consists in assessing the parallel performance obtained on these sets of equations, on different architectures.

2 Incompressible flow equations

Given ρ the density of the fluid, μ its viscosity, u and p its velocity and pressure, the incompressible flow equations are:

$$\begin{aligned} \rho \tfrac{\partial u}{\partial t} + \rho(u \cdot \nabla)u - \nabla \cdot [2\mu\varepsilon(u)] + \nabla p = \rho f \text{ in}\,\Omega, \\ \nabla \cdot u = 0 \text{ in}\Omega. \end{aligned} \quad (1)$$

* This work has been carried out in the framework of the Spanish Project OPTIDIS (ENE2005-05274). The research of Dr. Houzeaux has been partly done under a Ramon y Cajal contract with the Spanish Ministerio de Educación y Ciencia.

D. Tromeur-Dervout (eds.), *Parallel Computational Fluid Dynamics 2008*,
Lecture Notes in Computational Science and Engineering 74,
DOI: 10.1007/978-3-642-14438-7_31,

where f is a force term. This system must be provided with appropriate initial and boundary conditions. Time discretization is achieved with the trapezoidal rule with constant θ and a time step δt. Space discretization is carried out in the finite element context together with a variational multiscale approach [2]. Let superindex i denote the linearization loop counter and n the time loop counter. When unknowns are considered at current linearization step $i+1$ or time step $n+1$, superscript are omitted. We end up solving the following algebraic system:

$$\begin{bmatrix} A_{uu} & A_{up} \\ A_{pu} & A_{pp} \end{bmatrix} \begin{bmatrix} u \\ p \end{bmatrix} = \begin{bmatrix} b_u \\ b_p \end{bmatrix} \tag{2}$$

Matrix A_{pp} comes from the stabilization of the pressure term (GLS like term), enabling the use of equal order interpolation. In a variational multiscale context, this term comes from the algebraic approximation of the subgrid scale equation [2]. Picards method is used to linearize the convection term. The direct solution of (2) is referred to as the monolithic solution of the equations. It is well known that the convergence properties of classical iterative solvers like GMRES is very poor, due to the coupling of the velocity and pressure. Therefore, another approach is envisaged, similar to projection methods [5] [3] and extensively described in [1]. It consists in solving a preconditioned Richardson iteration for the Schur complement of the pressure. In addition, this iteration is coupled to the linearization loop. The preconditionner is a slightly modified Laplacian matrix, $\delta t\tilde{L}$. The modification is necessary to take into account the Neumann part of the boundaries. In algebraic words, at each time step $n+1$, given an initial u^0 and p^0, solve for $i=0,1,2...$ until convergence

$$A_{uu}u^{i+1} = b_u - A_{up}p^i, \tag{3}$$

$$(A_{pp}+\delta t\tilde{L})p^{i+1} = b_p - A_{pu}u^{i+1} + \delta t\tilde{L}p^i. \tag{4}$$

This predictor-corrector scheme is similar to classical incremental projection methods. The differenceis thatweiterate within each time step to obtain the solution ofthe monolithic scheme. In fact, note that when $p^{i+1}=p^i$, we recover the monolithic solution. A *not strict* continuous representation of this scheme is the following system:

$$\rho\frac{u^{i+1,n+1}-u^n}{\theta\delta t} + \rho(u^{i,n+1}\cdot\nabla)u^{i+1,n+1} - \nabla\cdot[2\mu\varepsilon(u^{i+1,n+1})] = \rho f - \nabla p^{i,n+1}, \tag{5}$$

$$\delta t\Delta p^{i+1,n+1} = -\nabla\cdot u^{i+1,n+1} + \delta t\Delta p^{i,n+1}. \tag{6}$$

3 Compressible flow equations

Let U be the linear momentum, T the temperature, E the total energy and k the thermal conductivity. The compressible flow equations in its conservative form are [7]:

$$\begin{aligned} &\partial_t U + \nabla\cdot(U\otimes u)_\nabla\cdot[2\mu\varepsilon(u)] + \nabla p = \rho g, \\ &\tfrac{\partial\rho}{\partial t} + \nabla\cdot U = 0, \\ &\tfrac{\partial E}{\partial t} + \nabla\cdot(uE - k\nabla T - u\cdot\sigma) = \rho u\cdot g. \end{aligned} \tag{7}$$

This system is closed with a state equation, typically the ideal gas'one. Additionally, laws for the viscosity and the thermal conductivity can be used for some particular regimes or materials. Boundary and initial conditions must be provided. Across the literature, different sets of unknowns have been proposed to properly solve the compressible flow equations. The conservative set $\Phi = (U, \rho, E)$ is one of the most favored due to its properties face to the discretization process. This is the unknowns' set chosen to run the compressible flow benchmarks below.

After space and time discretization and linearization, we obtain the following system

$$\begin{bmatrix} A_{UU} & A_{U\rho} & A_{UE} \\ A_{\rho U} & A_{\rho\rho} & A_{\rho E} \\ A_{EU} & A_{E\rho} & A_{EE} \end{bmatrix} \begin{bmatrix} \\ \Phi \\ \\ \end{bmatrix} = \begin{bmatrix} b_U \\ b_\rho \\ b_E \end{bmatrix} \tag{8}$$

that can be compactly written as:

$$A\Phi = b \tag{9}$$

where A is the system matrix that can be decomposed in blocks according to the coupling among the different unknowns. Matrix A structure is heavily dependent on the unknowns set, the linearization, the stabilization procedure, the shock capturing technique or the time and space discretizations. In this paper, when dealing with compressible flow we follow an explicit formulation, meaning tha the system is solved using a preconditioned Richardson equation at each time step as follows:

$$\begin{aligned} U^{n+1} &= u^n + P^{-1}[b_U - A_{U\rho}\rho^n - A_{UE}E^n) - A_{UU}U^n], \\ \rho^{n+1} &= \rho^n + P^{-1}[b_\rho - A_{\rho U}U^n - A_{\rho E}E^n) - A_{\rho\rho}\rho^n], \\ E^{n+1} &= E^n + P^{-1}[b_E - A_{EU}U^n - A_{E\rho}\rho^n) - A_{EE}E^n], \end{aligned} \tag{10}$$

and in compact form,

$$\Phi^{n+1} = \Phi^n + P_c^{-1}R^n \tag{11}$$

The preconditionner P is taken as

$$P = \frac{1}{\delta t}M, \tag{12}$$

where M is the diagonal mass matrix, computed using a closed integration rule for which the integration points are located on the nodes. P_c is its extension in compact notation, considering that each equation can have its own time step.

Additionally, in order to deal with low-Mach regions in an accurate and stable way, fractional step techniques are used. They are so-called fractional because the Navier-Stokes equations are solved by seggregating the operators, leading to a solution scheme of succesive stages. In this paper,we follow the strategy proposed by the CBS algorithm [6], where the fractional step is taken on the pressure gradient term in the momentum equation. In compact form, the fractional explicit algorithm here used is: at each time step solve

$$\begin{aligned} \tilde{U} &= U^n + P^{-1}R_U^n \\ \Phi^{n+1} &= \tilde{\Phi}^n + P_c^{-1}\tilde{R}^n \end{aligned} \tag{13}$$

where $\tilde{U}$ is the fractional momentum, R_U^n is the space residual of the linear momentum equation with the pressure gradient term seggregated[2] and $\tilde{\Phi}^n = (\tilde{U}, \rho, E)$.

4 Parallelization strategy

Concerning the MPI communication, the parallelization strategy was already presented in [4]. As long as the explicit compressible flow solver is concerned, OpenMP directives were implemented to the loop elements. Criticial sections are only located at the scatter of the local residual to the global residual.

5 Performance

The speedup for the incompressible flow solver is shownin Figure 1(Left). The data were obtained on a BlueGene/L and aBlueGene/P.The number of elements is 5M hexahedra, which is relatively low with respect to the number of CPUs used. On 4000 CPUs the average number of elements per CPU is a bit higher that 1000. However, the speedup is around 85%.

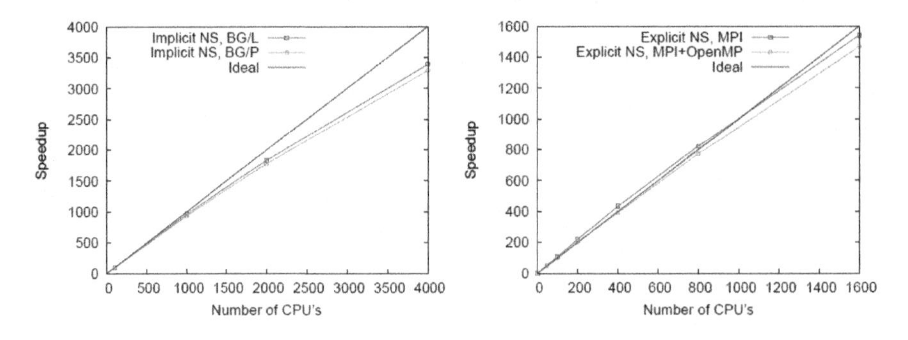

Fig. 1. Speedup, incompressible flow solver. Comparison BlueGene/L and/P.

Figure 1(Right) shows the speedup obtained with a full MPI approach and an hybrid MPI-OpenMP approach. The mesh is composed of 2.6M hexahedra elements. Both give similar speedup results, with a slightly lower speedup for the hybrid approach. Note that not all the loops were parallelized in this case.

[1] G. Houzeaux and M. Vazquez, Parallel implementation of an incompressible Navier-Stokes solver, in preparation.

[2] G. Houzeaux and J. Principe, A variational subgrid scale model for transient incompressible flows, Int. J. CFD, (2008), In press.

[2] Seggregated means that the pressure term is partially considered, multiplied by a incremental factor ranging from 0 to 1.

[3] S. Badia and R. Codina, Algebraic pressure segregation methods for the incompressible Navier-Stokes equations, Arch. Comp. Meth. Eng., (2008), Inpress.
[4] G. Houzeaux, M. Vázquez, R. Grima, H. Calmet and J.M. Cela, Experience in Parallel Computational Mechanics on MareNostrum, PARCFD2007, (2007), Antalya(Turkey), May 21-24.
[5] A. Quarteroni, F. Saleri and A. Veneziani, Factorization methods for the numerical approximation of Navier-Stokes equations, Comp. Meth. Appl. Mech. Eng.,(2000), 188, 505-526.
[6] O.C. Zienkiewicz, P. Nithiarasu, R. Codina, M. Vazquez and P. Ortiz. The characteristic based- split procedure: an efficient and accurate algorithm for fluid problems. Int. J. Numer. Meth. Fluids, (1999), 31, 359-392.
[7] L.D. Landau and E.M. Lifshitz.Fluid Mechanics. Butterworth - Heinemann(1987).

[8] [illegible] and R. Codina. Algebraic [illegible] projection methods for the incompressible Navier-Stokes equations. Arch. Comp. Meth. Eng. (2008) [illegible]
[illegible] M. Vázquez, R. [illegible] and J.M. Cela. Exp[illegible] Computational Mechanics on MareNostrum. PRACE-ECCOMAS 2007, [illegible] May 2007.
[illegible] [illegible] E. Salvador and A. [illegible] Navier-Stokes equations. [illegible] Comp. Meth. Appl. [illegible] [illegible]-526.
[illegible] [illegible] P. [illegible] [illegible]
[illegible] [illegible]

Hierarchical adaptive multi-mesh partitioning algorithm on heterogeneous systems

Y. Mesri[1], H. Digonnet[1], T. Coupez[1]

CIM group, Centre For Material Forming, ENSMP,
Rue Claude Daunesse BP 207 06904 Sophia Antipolis cedex, France

Abstract. Heterogeneous systems are increasingly being used as platforms for resource-intensive distributed parallel mesh applications. A critical contributors to the performance of such applications is i) the scheduling of constituent application tasks ont he system and ii) the mesh partitioning. Since often application submissions from users cannot improve the total turnaround time, the application must be scheduled automatically by the scheduler using information about its scalability characteristics. To obtain the best performance, it is imperative to take into account both application-specific and dynamic system information in developing i) a schedule which meets his performance criteria and ii) a mesh partitioning scheme that takes into account this system heterogeneity. In [1], we have presented a mesh partitioning scheme, that takes into account the heterogeneity of CPU and networks. Load balancing mesh partition strategy improves the performance of parallel applications running in a heterogeneous environment. In this work, we present a new hierarchical adaptive scheme that allows to optimize the scheduling of N parallel mesh applications simultaneously on a heterogeneous system. The new scheme consists in selecting the partition size of each application and then partitioning each application on the allocated heterogeneous ressources. We illustrate our scheduling approach with a detailed description and results for a distributed 3D CFD application on a heterogeneous platform.

1 Introduction

Computational fluid dynamics(CFD) applications usually operate on a huge set of application data associated to unstructured meshes. For this reason CFD problems represent a significant part of high performance supercomputing applications. Finite element and finite volume methods use unstructured meshes. We represent the application as a weighted undirected graph $W = (V(W), E(W))$, which we will call the workload graph . Each vertex v has a computational weight $\omega(v)$, which reflects the amount of the computation to be done at$v$. An edge between vertices u and v, denoted

D. Tromeur-Dervout (eds.), *Parallel Computational Fluid Dynamics 2008*,
Lecture Notes in Computational Science and Engineering 74,
DOI: 10.1007/978-3-642-14438-7_32,

$\{u,v\}$, has a computational weight $\omega(\{u,v\})$, which reflects the data dependency between them. Partitioning applications onto heterogeneous architecture such as a Grid environment requires a special model architecture that reflects both heterogeneous resource characteristics and also non homogeneous communication network between these different resources. The machine architecture can be represented as a weighted undirected graph $A = (V(A), E(A))$, which we will call the architecture graph. It consists in a set of vertices $V(A) = \{p_1, p_2, ..., p_n\}$, denoting processors, and a set of edges $E(A) = \{\{pi,\} \mid pi, pj \in P\}$, representing communication links between processors. Each processor p has a processing weight s_p, modeling its processing power per unit of computation. Each link has a link weight v_{pq}, that denotes the communication bandwidth per unit of communication between processors p and q. In [1] we have proposed a heterogeneous re-partitioning algorithm, called MeshMigration that takes into account the architecture characteristics. MeshMigration generates a high quality partition and provides a load balance on each processor of the heterogeneous architecture. In this paper, we develop an iterative scheme that performs automatically a heuristic based search for a schedule that minimizes average turnaround time coupled with the mesh partitioning scheme described in [1]. Figure 1 represents a diagram of the coupled multi-mesh partitioning algorithm. This paper is organized as follows: In section 2, we define formally an iterative scheduling algorithm, followed by a description of the partitioning problem in section 3. Section 5 discusses the hierarchical mesh partitioner MeshMigration. Finally in section 6, we will describe preliminary experimental results for a finite element code executed on the Grid.

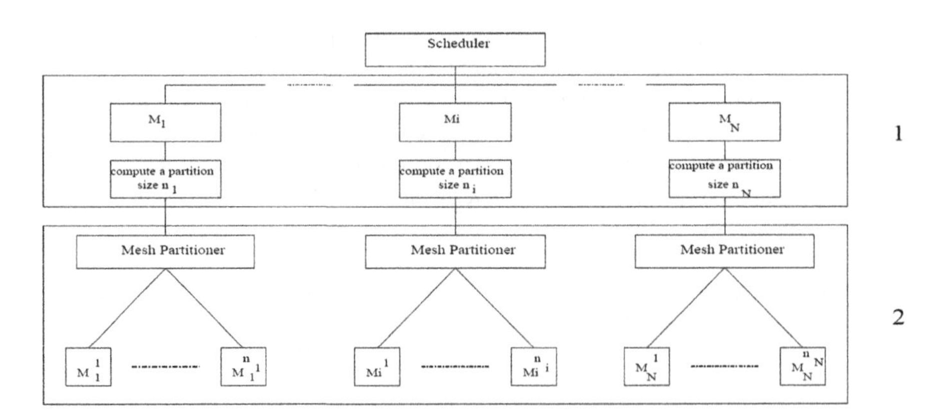

Fig. 1. A diagram of the global multi-mesh partitioning approach

2 The iterative scheduling algorithm

An optimal moldable scheduling strategy would inherently take into account the efficiency, job size and system load without the need to tune parameters. In this section, we describe an iterative scheme that does so. Our iterative algorithm begins by giving each job an initial minimal partition of one processor. A conservative schedule is generated; this schedule is iteratively modified by giving a processor to the most worthy job, the job that,if given an additional processor,has the greatest decrease in runtime. If the addition of a processor to the most worthy job decreased the average response time of the schedule, the addition is accepted, otherwise not.

1. Input List of reserved jobs
2. unmark all jobs and set partition sizes to one
3. While unmarked jobs exist
   a) find unmarked candidate job j(see Algorithm 2)
   b) add one to partition size of job j
   c) create a conservative schedule for all jobs
   d) If average turnaround time did not improve
      i. mark job j
      ii. decrement partition size of candidate job j
      iii. create a conservative schedule for all jobs
   e) end if
4. end while

Algorithm 1: The iterative scheduling algorithm

Algorithm 1 shows the algorithm for the iterative algorithm. Initially each job is assigned one node. This allocation results in optimal per job efficiency, but may result in poor average turn around and system utilization. The next step(lines3to4) searches for a schedule with an improved average turn around time. Step3-(a) chooses the job which will benefit the most of receiving an extra processor. This job is a good candidate to try to increasing its processor allocation. Step 3-(b) to 3-(e) determine if the increased allocation results in a better schedule. If the increase produces a worse schedule,the job is marked as a bad choice and the remaining jobs are considered.

1. Input List of reserved jobs
2. set *bestImprovement* to zero
3. for each unmarked job j in the reserved job list
   a) let n be the current node assignment if job j
   b) let i be the expected runtime on n processors
   c) i be the expected runtime on n+1 processors
   d) if $(i - i' > bestImprovement)$
      i. set best Improvement to $i - i'$
      ii. set best Job to j
   e) end if
4. end for

5. return best Job

Algorithm 2: The unmarked candidate search algorithm

This approach takes all the aspects discussed previously into account: load, scalability, job size, and utilization. If a job is small,the improvement from adding a processor will be minimal, and thus it will be less likely to receive an increased allocation. Likewise, if a job scales poorly,it will benefit less from receiving more processors, and will be less likely to be chosen as the candidate. If the load is low, wider jobs will result in a better average turnaround time, and wider allocations will be given. If the load is high, increasing the allocation of poorly scalable jobs will increase average turnaround time. Finally, the system achieves a good utilization, as processors will not be wasted unless there is no work to be done or using the processor reduces the average turnaround time.

We evaluate our iterative algorithm via simulation. The simulation is based on the workload logs archive of the machine SP2 of th eCornell Theory Center(CTC).The CTC workloads are available on the web at `http://www.cs.huji.ac.il/labs/parallel/workload/`. We use Downeys model [4] of the speedup ($S(n) = t_1/t_n$) of parallel jobs to derive execution time for a partition size n. Downeys speedup model uses two parameters: A (the average parallelism) and σ (an approximation of the coefficient of variance in parallelism). The speedup of a job j is then given by:

$$S(n) = \begin{cases} \frac{An}{A+\sigma(n-1)/2} & \sigma \leq 1, 1 \leq n \leq A \\ \frac{An}{\sigma(A-1/2)+n(1-\sigma/2)} & \sigma \leq 1, A \leq n \leq 2A-1 \\ A & \sigma \leq 1, n \geq 2A-1 \\ \frac{nA(\sigma+1)}{\sigma(n+A-1)+A} & \sigma \geq 1, 1 \leq n \leq A+A\sigma-\sigma \\ A & \sigma \geq 1, n \geq A+A\sigma-\sigma \end{cases} \quad (1)$$

Figure 2 shows the results comparison between the rigid workload extracted from the log files of the CTC machine and our moldable scheduling algorithm. In these experiments, we supposethatjobsscaletothe size of the system(A = system size). Results show that our algorithm is able to reduce the turn-around time to about half of that obtained by the user request.

3 Partitioning Problem

We consider a workload graph $W(V(W),E(W))$which represents the application, and a architecture graph $A(V(A),E(A))$ which represents the Grid. On a computational grid, the machine architecture is heterogeneous both for network and processors. So we consider the characteristics of architecture graph to define the partitions. A mapping of a workload graph onto a architecture graph can be formally described by:

$$m : V(W) \rightarrow V(A) \quad (2)$$

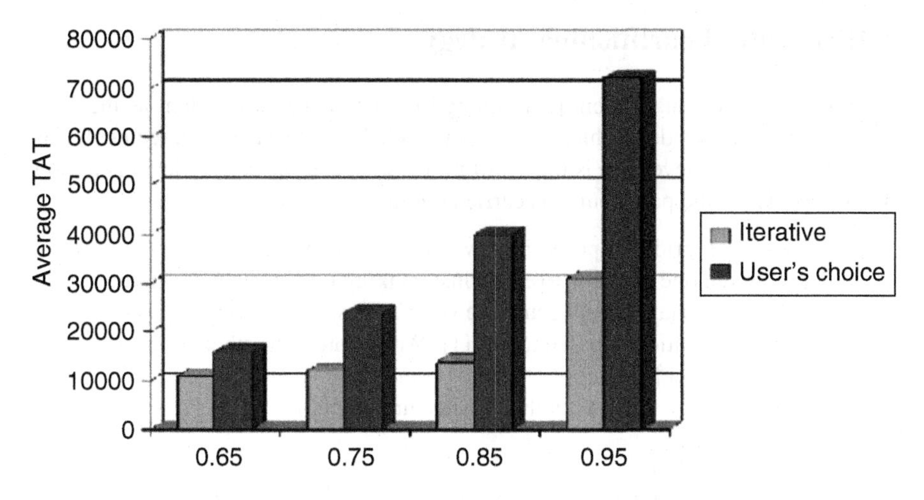

Fig. 2. Effective load (EL= (Number of jobs)/Number of processors)

where $m(v) = p$, if the vertex v of W is assigned to a processor p of A. In order to evaluate the quality of a mapping,we define two cost models: one for estimating the computational cost and the other one for the communication cost evaluation. For each mapping of the workload graph onto the architecture graph we can estimate the computational cost as follows: If a vertex v is assigned to a processor p, the computational cost is given by $t_p^v = \omega(v)/s_p$, that is the ratio of the computational weight of v per the processing weight of p. Computational cost estimates the time units required by p to process v. The communication cost is introduced when we have a data communication transfer between two different nodes in the target graph. Suppose $\{u,v\} \in E(W)$ and $u \in V(W)$ is assigned to processor p and $v \in V(W)$ is assigned to processor q. The data is transferred from the local memory of p to the local memory of q via message passing. In this case, the communication cost is given by $c_{pq}^{u,v} = \omega(\{u,v\})/v_{pq}$, that is the ratio of the communication weight of edge $\{u,v\}$ per the link weight between p and q. The communication cost represents the time units required for data transfer between the vertices u and v. As in [1], we define a cost function as follows: $\Phi(W,A,m) := T + C$ where

- $T = (1,...,t_{card(V(A))})^t$, $t_p = \frac{C(p,m)}{s_p}$ the computational time of p, and $C(p,m)$ is the weight of subgraph assigned to p.
- $C = (C(1,m),...,C(card(V(A)),m)^t$ and $C(p,m) = \sum_{q \in V(A) \neq p} C(\{p,q\},m)$ the communication time associated to a processor p with his neighbors.

The definition of the graph partitioning problem is to find a partition (mapping m) which minimizes the cost function $\Phi(W,A,m)$. Clearly, the problem is extensible to the classical graph partitioning and task assignment problem, and it is well known that this problem is NP-complete. In the next section, we describe the iterative algorithm chosen to minimize this cost function and find the efficient partitioning.

4 Hierarchical partitioning strategy

In order to provide an efficient partitioning in a heterogeneous system as the Grid environment, we introduce a hierarchical approach. We define two architecture levels in the target graph: level one is the set of processors, and the second level is the set of clusters. Then, the partitioning is carried out in two stages:

1. We consider a grid composed by several clusters. We decompose the mesh in order to assign one partition to each cluster taking into account the bandwidth of communications layers between these various clusters. The decomposition takes place via the relationship defined in [1]. We denote N numbers of clusters and $G = \{C0,...,C_{N-1}\}$ the set of clusters. Formally:
Let p and q two processors in the architecture graph, $p \in C_i$ and $q \in C_j$:

$$\begin{cases} \text{if} C_i \neq C_j \\ Friendship(p,q) = \max(F_{0|_{pq}} - F^p_{min}, F_{0|_{pq}} - F^q_{min}) \\ \text{else} \, Friendship(p,q) = 0. \end{cases} \quad (3)$$

2. We denote $\Pi = \{\pi_0,...,\pi_{N-1}\}$, the set of the sub-domains mapped on G. For every partitionπ_i assigned to the cluster $C_i = \{p_0,...,p_{I-1}\}$, where I is the number of processors of C_i, we re-partitioned π_i on the set of processors of C_i: For every p and q two processors on the architecture graph, $p \in C_i$ and $q \in C_j$:

$$\begin{cases} \text{if} C_i = C_j \\ Friendship(p,q) = \max(F_{0|_{pq}} - F^p_{min}, F_{0|_{pq}} - F^q_{min}) \\ \text{else} \, Friendship(p,q) = 0. \end{cases} \quad (4)$$

5 Experimentation of a finite element code on the grid

5.1 Pour of a jerrican

This application depicts the flow of a fluid by the neck of a jerrican (see [2] for more details). At the beginning of the simulation, the fluid is completely in the jerrican which is reversed; its open neck is directed downwards. The fluid then starts to flow by the opening neck, while the air is engulfed in the jerrican in the form of bubbles. Thus, the cavity empties little by little its initial contents and filled with air, until the fluid is completely poured. The flow isgoverned by Navier-Stokes equations and the evolution of the two involved phases (liquid and air) is ensured by the transport of a Level Set function which enables the capture of theinterface [2]. The mesh discretizing the domain contains 500,000 nodes and 2,750,000 elements. The parameters used are as follows: the fluid has a viscosity of 10 Pa.s and a density of $1000 kg/m^3$, air has a viscosity of 1 Pa.s and a density of $1 kg/m^3$. The boundary conditions are of sticking type, i.e. a null speed is imposed on all surfaces of the geometry. Figure 3 represents the interface between the fluid and the air at times $t = 0s$ and $t = 18s$. Whereas air bubbles go up in the jerrican (the top cavity), the fluid flows in the bottom cavity through the neck of the jerrican. A total of 15 processors, and 3 clusters (NINA,PF and IUSTI) of MecaGrid[1] are used to carry out this application. It chosen to allocate 4 NINA and 5 PF and 6 IUSTI. This choice is justified by the calculated power of the processors: for both NINA and IUSTI clusters,the power is approximately $125Mflops/s$, and for the PF, it is $44Mflops/s$, that is to say 2.5 times lower. On the INRIA site, 2 NINA are replaced by 5 PF. Thus, with 4 NINA, 5 PF and 6 IUSTI, the computing power distributed on the two sites is almost the same ones. Measurements of following times are raised after 100 time increments, with homogeneous partitions, than optimized partitions. In Table 1, we use a ILU(1) preconditioning and a MinRes solver to solve the linearized flow system. The first observation is that the gain of timeprovided by the optimized partitioning is very important; the execution time passed from 2 days and 19 hours to one day and 8 hours, which makes a gain of 53%. Furthermore distributing the workload equitably, the hierarchical partitioning algorithm allows to decrease considerably the very penalizing inter sites communication while cutting the mesh in a place where there is few nodes. Finally, the hierarchical partitioning approach saved almost 53% of the computing time of MecaGrid.

4 PF-5 NINA-6 IUSTI	Homogeneous partition	optimized partition
Nb procs	15	15
NSAssembling(s)	3680	1904
NSResolution(s)	235504	107899
Alphaassembling(s)	2357	2896
Alpha resolution(s)	258	113
Total(s)	241799	113364

Table 1. run time on Grid with ILU(1) preconditionner

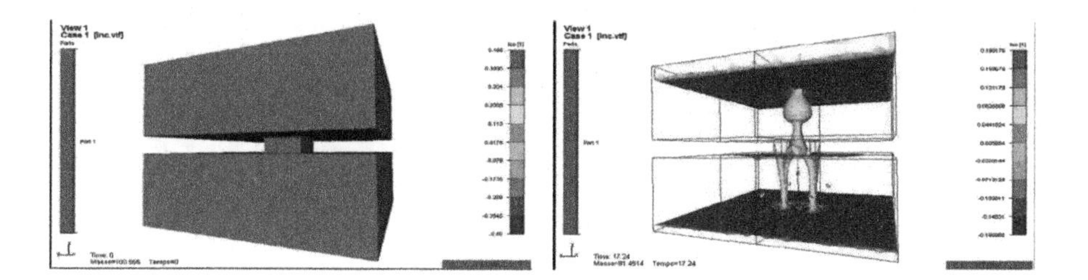

Fig. 3. simulation of jerrican poor: sequences at t=0s (left) and t=18s (right)

[1] www-sop.inria.fr/smash/mecagrid

6 Conclusion

In this paper, we have presented a new adaptive multi-mesh partitioning approach, for partitioning workloads graph onto heterogeneous architecture graph. This algorithm runs in parallel and have shown that optimized load balancing strategies improve the performance of the applications executed on a heterogeneous environment. One note that the results presented here are associated to just a single mesh application, a work-in progress consists in applied this approach to multi-mesh applications.

[1] Y. Mesri, H. Digonnet and H. Guillard, Mesh partitioning for parallel computational fluid dynamics applications on a grid, in Finite volumes for complex applications IV, Hermes Science Publisher, 2005, p. 631-642

[2] O. Basset, H. Digonnet, T. Coupez and H. Guillard Multi-phase flow calculations with interface tracking coupled solution, in Coupled Problems 2005, barcelona, Spain, CIMNE, 2005

[3] A. Basermann and al., (2000), Dynamic load balancing of finite element applications with the DRAMA library Appl. Math. Modelling, 25(2).83-98

[4] A. Downey. A model for speedup of parallel programs. U.C.Berkely Technical report CSD-97-933, January 1997.

Part IX

Parallel Performance

Towards Petascale Computing with Parallel CFD codes

A.G. Sunderland[1], M. Ashworth[1], C. Moulinec[1], N. Li[2], J. Uribe[3], Y. Fournier[4]

[1] STFC Daresbury Laboratory, Warrington, UK
[2] NAG Ltd, Wilkinson House, Jordan Hill Road, Oxford, UK
[3] University of Manchester, UK
[4] EDF R&D, Chatou, France

Keywords: Petascale, High-End Computing, Parallel Performance, Direct Numerical Simulations, Large Eddy Simulations

1 Introduction

Many world leading high-end computing (HEC) facilities are now offering over 100 Teraflops/s of performance and several initiatives have begun to look forward to Petascale computing[5] (10^15 flop/s). Los Alamos National Laboratory and Oak Ridge National Laboratory (ORNL) already have Petascale systems, which are leading the current (Nov 2008) TOP500 list [1]. Computing at the Petascale raises a number of significant challenges for parallel computational fluid dynamics codes. Most significantly, further improvements to the performance of individual processors will be limited and therefore Petascale systems are likely to contain 100,000+ processors. Thus a critical aspect for utilising high Terascale and Petascale resources is the scalability of the underlying numerical methods, both with execution time with the number of processors and scaling of time with problem size. In this paper we analyse the performance of several CFD codes for a range of datasets on some of the latest high performance computing architectures. This includes Direct Numerical Simulations (DNS) via the SBLI [2] and SENGA2 [3] codes, and Large Eddy Simulations (LES) using both STREAMS_LES [4] and the general purpose open source CFD code Code_Saturne [5].

2 Parallel CFD codes

We analyse the parallel performance of several parallel CFD codes on the target high-end computing systems. The codes have been chosen to reflect a range of appli-

[5] Petascale assumes 10s of Petaflop/s Peak Performance and 1 Petaflop/s Sustained Performance on HEC applications

D. Tromeur-Dervout (eds.), *Parallel Computational Fluid Dynamics 2008*,
Lecture Notes in Computational Science and Engineering 74,
DOI: 10.1007/978-3-642-14438-7_33,

cations (e.g. turbulence at the shock/boundary layer interaction, combustion) using both DNS-based and LES-based computational methods. All codes are written in Fortran with MPI [6] for data transfer between processors. The Code_Saturne package also has modules written in the C programming language and the Python scripting language.

2.1 SBLI

Fluid flows encountered in real applications are invariably turbulent. There is, therefore, an ever-increasing need to understand turbulence and, more importantly, to be able to model turbulent flows with improved predictive capabilities. As computing technology continues to improve, it is becoming more feasible to solve the governing equations of motion, the Navier-Stokes equations, from first principles. The direct solution of the equations of motion for a fluid, however, remains a formidable task and simulations are only possible for flows with small to modest Reynolds numbers. Within the UK, the Turbulence Consortium (UKTC) has been at the forefront of simulating turbulent flows by direct numerical simulation (DNS). UKTC has developed the parallel code SBLI to solve problems associated with shock/boundary-layer interaction. SBLI [2] is a compressible DNS code based on finite difference method using high-order central differencing in space and explicit Runge-Kutta for time marching. A grid transformation routine enables this code to simulate relatively complex-geometry flows. The parallel version is under active development and its parallel performance has been fine-tuned. A set of test cases, some with complex geometry involving multiple Cartesian-topology blocks, have been specified for its testing and benchmarking on a range of HPC platforms.

2.2 SENGA2

The SENGA2 [3] code has been developed at The University of Cambridge and has been designed to facilitate combustion DNS with any desired level of chemistry, from single-step Arrhenius mechanisms through all classes of reduced reaction mechanisms up to fully detailed reaction mechanisms. The Navier-Stokes momentum equations are solved in fully compressible form together with the continuity equation and a conservation equation for the stagnation internal energy, as well as any required number of balance equations for species mass fraction. Each component of the reacting mixture is assumed to obey the equations of state for a semi-perfect gas. Boundary conditions are specified using an extended form of the Navier-Stokes Characteristic Boundary Condition formulation, and available boundary conditions include periodic as well as several types of walls, inflows and outflows. The numerical framework is based on a finite-difference approach for spatial discretisation together with a Runge-Kutta algorithm for time-stepping. High-order explicit schemes are preferred due to their speed of execution and ease of parallel implementation, and a 10th order explicit scheme is standard for interior points. The code is fully parallel using domain decomposition over a cubic topology. Current HEC architectures permit 3D DNS of the turbulent flow fields but with only limited representation of the combustion chemistry and a highly simplified representation of the geometry. At the Petascale it will be possible to move towards more complex configurations that are much closer to industrial requirements.

2.3 STREAMS-LES

STREAM-LES [4] is a CFD package developed at Imperial College, London, for Large Eddy Simulations (LES) of incompressible flow. Its numerical framework rests on a general structured, multi-block, collocated-storage finite volume method with non-orthogonal mesh capability. The spatial scheme is second-order central and the time-matching is based on a fractional-step method in which a provisional velocity field is made divergence-free through the solution of the pressure-Poisson equation. The code is fully parallelized using MPI through standard domain decomposition and runs on several high-end computing platforms.

2.4 CODE_SATURNE

Code_Saturne [5] is an open source general purpose computational fluid dynamics software package developed by EDF [7]. It is based on a co-located Finite Volume approach that accepts meshes with any type of cell, including tetrahedral, hexahedral, prismatic, pyramidal, polyhedral and any type of grid structure, including unstructured, block structured, hybrid, conforming or with hanging nodes. Its basic capabilities enable the handling of either incompressible or expandable flows with or without heat transfer and turbulence (mixing length, 2-equation models, v2f, Reynolds stress models, Large Eddy Simulations etc.). Dedicated modules are available for specific physics such as radiative heat transfer, combustion (e.g. gas, coal), magneto-hydro dynamics, compressible flows, two-phase flows (Euler-Lagrange approach with two-way coupling) with extensions to specific applications (e.g. for atmospheric environment: code Mercure_Saturne).

3 High-End Computing Platforms

3.1 HPCx

HPCx [8] is the UK's National Capability Computing service, located at the Computational Science and Engineering Department at STFC Daresbury Laboratory [9] and comprising of 160 IBM eServer 575 nodes. Each eServer node contains 16 1.5 GHz POWER5 processors, giving a total of 2560 processors for the system. The total main memory of 32 GBytes per node is shared between the 16 processors of the node. The nodes in the HPCx system are connected via IBM's High Performance Switch. The current configuration has a sustained Linpack performance of 12.9 Tflop/s.

3.2 HECToR

HECToR [10] is the UKs latest high-end computing resource, located at the University of Edinburgh and run by the HPCx consortium. It is a Cray XT4 system comprising 1416 compute blades, each of which has 4 dual-core processor sockets. This amounts to a total of 11,328 cores, each of which acts as a single CPU. The processor is an AMD 2.8 GHz Opteron. Each dual-core socket shares 6 GB of memory, giving a total of 33.2 TB in all. The Linpack performance of the system is 54.6 Tflops/s, positioning the system at number 46 in the current (Nov 2008) TOP500 list. We have extended the available range of processor counts with runs on the Jaguar[6] Cray XT4 at ORNL [11]. Jaguar has quad-core nodes running at 2.1 GHz but is otherwise similar to the HECToR system.

3.3 BlueGene

STFC Daresbury Laboratory operates single rack Blue Gene/L and Blue Gene/P systems. Both systems contain 1024 chips, with 2 processor cores per chip in the L system and 4 processor cores per chip in the P system, giving a total of 2048 cores and 4096 cores respectively. Memory is provided at 512 Mbytes per core in both systems. The basic processor in the L system is the Power440 running at 700 MHz, whilst the P system uses a processor from the Power450 family running at 850 MHz. Inter-processor communications take place via two different networks: a 3-D torus for general communications and a tree network for collective communication operations. The philosophy behind the Blue Gene design is that the speed of the processor is traded in favour of very dense packaging and low power consumption. As a consequence of these features, Blue Gene systems have featured prominently in the Green500 supercomputer list [12].

4 Performance Results

Performance results are presented in this section for the codes and platforms under investigation.

4.1 SBLI

The benchmark case for the SBLI code is a simple turbulent channel flow benchmark run for 100 timesteps using grid sizes of 360^3 ($360 \times 360 \times 360$), 480^3, 600^3 and 1024^3. The most important communications structure is a halo-exchange between adjacent computational sub-domains. Providing the problem size is large enough to give a small surface area to volume ratio for each sub-domain, the communications costs are small relative to computation and do not constitute a bottleneck and we see

[6] This Jaguar system was the older system available to general users in late 2008, not the Petascale system of the same name from the Nov 2008 TOP500 list.

almost linear scaling from all systems out to 1024 processors. Performance for the three smaller grid sizes is shown in Figure 1. In order to be able to compare different problem sizes the performance is defined as the number of grid points times the number of timesteps divided by the execution time. Some cache effects are seen up to around 1000 processors where the smaller grid sizes perform better as the sub-domains on each processor fit into cache at lower processor counts. However, above 1000 processors we see that, as expected, the scalability is better for larger problem sizes due to the better compute/communication ratio. This is confirmed by profiling measurements which show that the time spent in MPI routines reaches 42% for the smallest grid size on 6144 processors.

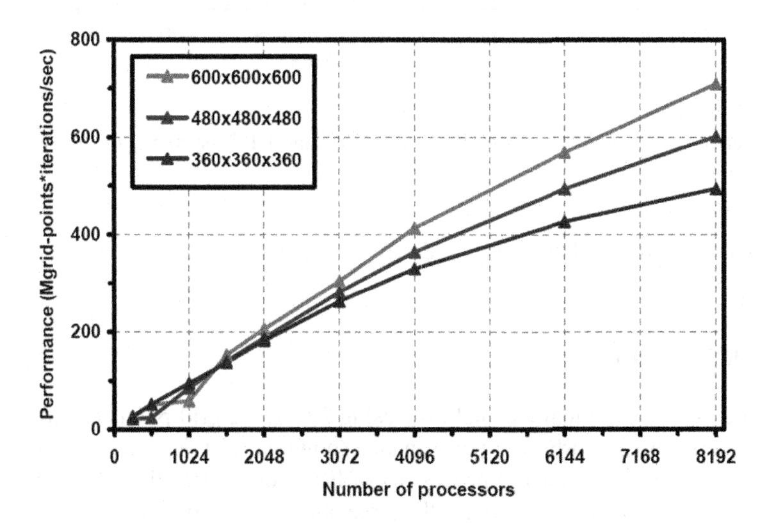

Fig. 1. Performance of SBLI on up to 8192 processors of the HECToR Cray XT4 for three grid sizes.

Parallel performance on jaguar up to 24576 processors, on HECToR up to 8192 and on HPCx on up to 1536 is shown in Figure 2 for the largest problem size of 1024^3 grid points. Scalability is good right out to 24576 processors with parallel efficiency of 61% at this point relative to 1024 processors, the smallest practical configuration for this data size due to memory constraints. HPCx has very similar per-processor performance to jaguar. HECToR is faster per processor than jaguar by about 16%, rather less than is expected from the 33% faster clock speed (2.8 GHz vs. 2.1 GHz). There are many factors e.g. dual-core vs. quad-core, memory architecture and OS, compiler and library version differences which could be responsible for this. Hardware profiling studies of this code have shown that its performance is highly dependent on the cache utilization and bandwidth to main memory [13]. This is confirmed by under-populating the nodes (using only one core per node on HECToR and only one or two cores per node on jaguar), which shows a speed-up of 33%

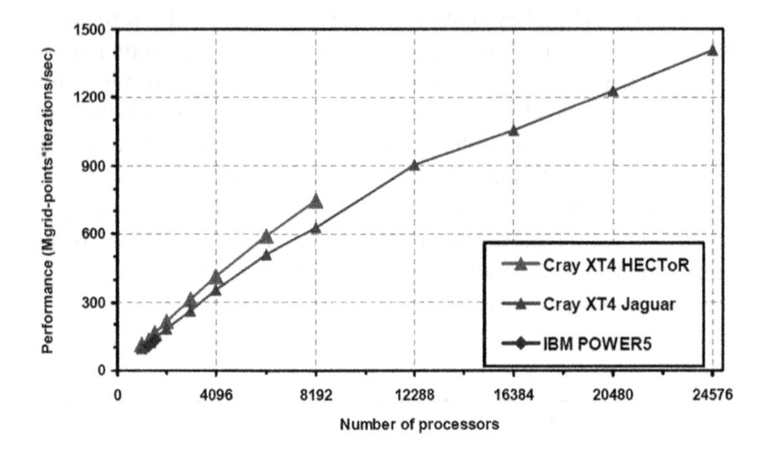

Fig. 2. Performance of SBLI on the HECToR, jaguar and HPCx systems for the 1024^3 grid size.

using only half the available cores and 50% when using only one core of the quad-core nodes.

4.2 STREAMS-LES

The test case is a turbulent channel flow using 2,097,152 grid points calculated for 75000 time steps. The arbitrary performance shown is relative to the time taken to complete the simulation on 128 processors of BlueGene/P. Parallel efficiency is between around 76% to around 96% for the three platforms up to 256 processors. However from 256 to 512 processors the parallel efficiency falls to approximately 45% (BG/P), 53% (Cray XT4) and 64% (IBM Power5). Currently, further investigation of the parallel performance of STREAMS-LES is being undertaken, including the analysis of larger datasets where possible.

4.3 Code_Saturne

The parallel performance of Code_Saturne on the IBM PWR5, BG/P and BG/L systems is shown in Figure 4. This test case consists of a LES at Reynolds number based of the friction velocity of 395 in a channel. The dimension of the box is $6.4 \times 2 \times 3.2$ and a $256 \times 256 \times 256$ (16M cells) mesh is considered. Periodic boundary conditions apply in stream-wise and span-wise directions and no-slip conditions in the wall normal ones. Van Driest damping is used at walls and the flow is driven by a constant pressure gradient. The parallel scaling properties up to 1024 processors is very good for all the IBM platforms considered here, with the PWR5-based architecture performing around twice as fast as the BG/P, which in turn is around 40% faster than the BG/L. Figure 5 shows the parallel performance of Code_Saturne on the Hector

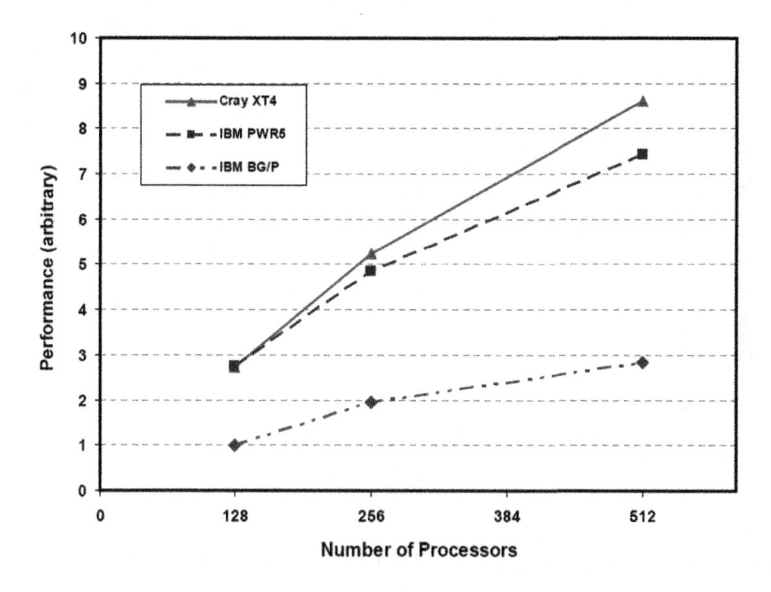

Fig. 3. Performance of STREAMS-LES on high-end systems.

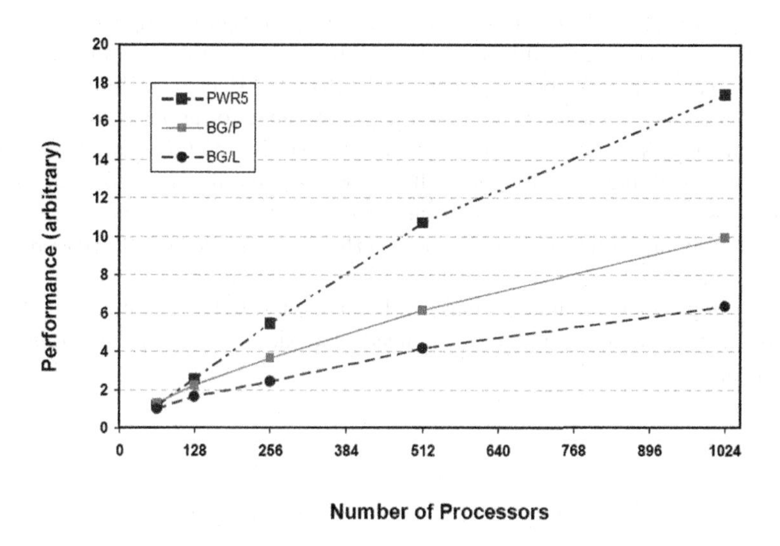

Fig. 4. Performance of Code_Saturne for a 16M cell open channel dataset on IBM platforms.

Cray XT4 platform. Here a larger problem involving an unstructured grid and 100 million cells is chosen as the dataset for the parallel performance appraisal. The simulation represents the flow of around a bundle of tubes in a nuclear reactor, where the geometry is too complex to be represented by structured gridding. The problem

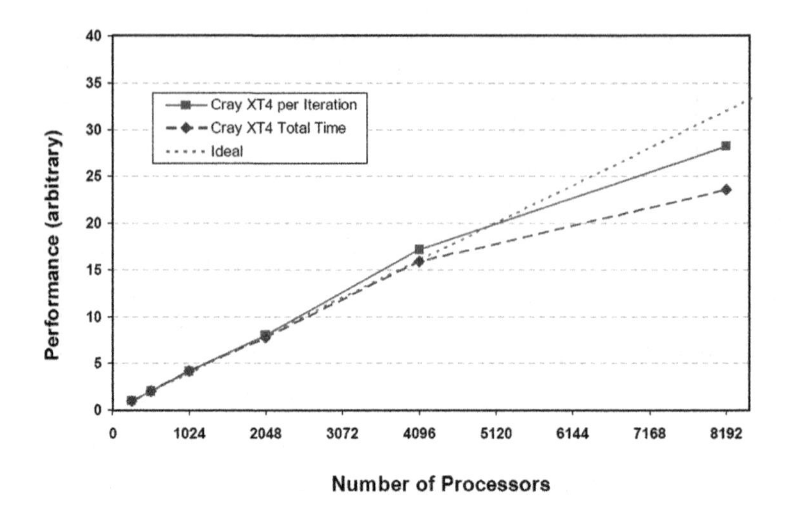

Fig. 5. Performance of Code_Saturne for a 100M cell "Mixing Grid" dataset on Hector (Cray XT4).

represents a very large-scale computational challenge (requiring high-end systems) as the flow is strongly three dimensional, with secondary vortices existing between pipes. Two sets of parallel performance results from the Hector machine are shown in Figure 5: one data series shows how the performance for an individual iteration scales with processor count, the other data series shows the total time based on a calculations involving 1000 iterations. It can be seen that the time taken per iteration in the solver stage scales near-linearly up to the maximum processor count of 8192. However the total time is scaling less perfectly as overheads in the partitioning stage and the I/O stages increase relative to solver time on larger processor counts. Evidently large-scale simulations that involve many thousands of time-steps and/or iterations will suffer less from these overheads than simulations involving relatively few steps in the solver.

4.4 SENGA2

The two datasets, whose performance is examined here, are outlined below.

1. A 4-step calculation, which is an early example of a reduced methane-air mechanism due to Peters and Williams. It uses 4 forward steps (two of them reversible) and 7 species, and is notoriously stiff. It is there to test the ability of the code to cope with stiff chemistry. The benchmark calculation is undertaken with each processor holding 32^3 grid points. Therefore the global grid size is expanded linearly with the number of processors used and is used to assess the weak scaling properties of the code.

2. A 1-step calculation is a simple global 1-step generic non-reversible Arrhenius-type mechanism. It is representative of the previous generation of combustion DNS, and is inexpensive to run. A volume of 1cm3 of air with periodic boundary condition is simulated over 10 time steps of 1 ps. Initial conditions are 300K and atmospheric pressure with an initial turbulent filed. Snapshot data is dumped every 5 steps (twice over the simulation length).

In common with other Direct Numerical Simulation codes, memory bandwidth is expected to be the dominant constraint on performance. Communications within Senga2 are dominated by halo-exchange between adjacent computational sub-domains. Providing the problem size is large enough to give a small surface area to volume ratio for each sub-domain, the communications costs are small relative to computation and do not constitute a bottleneck. This is exemplified in Figures 6, 7, and 8, below where both weak scaling and strong scaling properties are very good on the target platforms. Scaling on the Blue Gene /P is generally good, but overall speed is around 2.5 times slower than the Cray XT4. However it should be noted that this performance ratio betters the relative clock speeds (2800/800 MHz) of the machines' underlying processors.

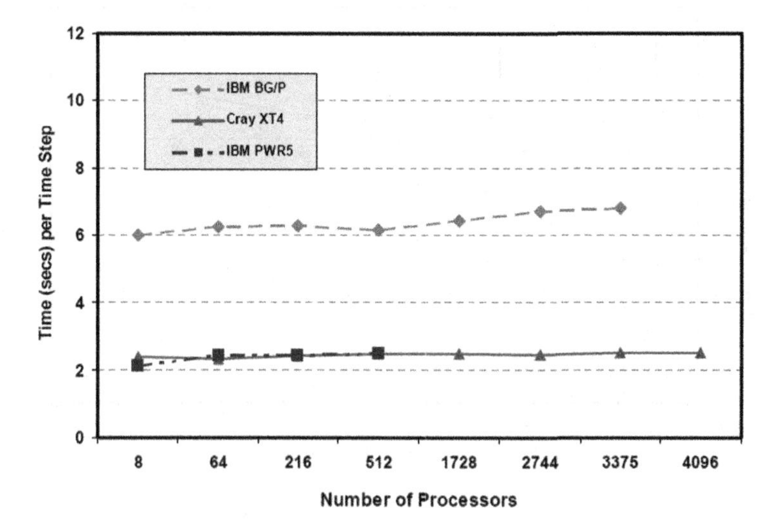

Fig. 6. Weak scaling performance of Senga2-4 Step reduced-air methane mechanism 32^3 grid points per processor.

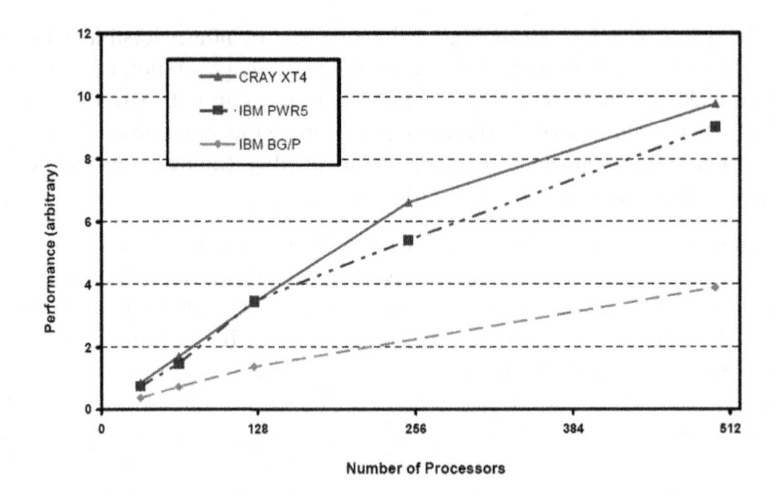

Fig. 7. Strong scaling performance of Senga2-1 Step Arrhenenius-type Mechanism with 500^3 global grid points.

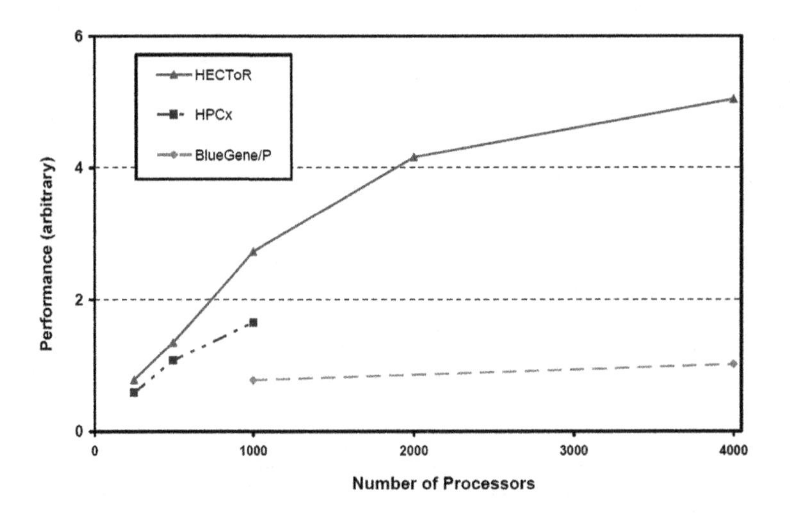

Fig. 8. Strong scaling performance of Senga2-1 Step Arrhenius-type Mechanism with 1000^3 global grid points.

5 Summary

The parallel performance of all four CFD codes analyzed here is encouraging as we head towards future Petascale computing environments. It is demonstrated that three of the codes - SBLI, Code_Saturne and Senga2 - perform well up to many thousands of processing cores, with the performance of SBLI improving right out to 12288 cores on the Jaguar Cray XT4 at Oakridge National Laboratory. All but one of the tests measured the generally more challenging strong scaling properties of the code where global problem size is constant for all runs, rather than weak scaling properties where problem size is scaled up as processor count increases. In practice it is likely that CFD researchers will be interested in both properties when running leading-edge calculations. It is becoming clear that new challenges will arise when computing at the Petascale. In the past much attention has been paid to applying state-of-the-art parallel algorithms to the solving of the mathematical equations at the heart of the simulation. For effective use of available computing power when utilizing many hundreds of thousands of cores code optimizers may now have to attention to new areas, such as the application of highly effective, load-balanced algorithms for mesh-partitioning and efficient parallel I/O methods.

Acknowledgements

This research used resources of the National Center for Computational Sciences at Oak Ridge National Laboratory, which is supported by the Office of Science of the Department of Energy under Contract DE-ASC05-00OR22725. The authors would also like to express their gratitude to Dr Stewart Cant at the University of Cambridge for his input.

[1] Top 500 Supercomputer sites, http://www.top500.org/
[2] N.D. Sandham, M. Ashworth and D.R. Emerson, Direct Numerical Simulation of Shock/Boundary Layer Interaction, , http://www.cse.clrc.ac.uk/ceg/sbli.shtml
[3] http://www.escience.cam.ac.uk/projects/cfd/senga.xsd
[4] L. Temmerman, M.A. Leschziner, C.P. Mellen, and J. Frohlich, Investigation of wall-function approximations and subgrid-scale models in large eddy simulation of separated flow in a channel with streamwise periodic constrictions, International Journal of Heat and Fluid Flow, 24(2): 157–180, 2003.
[5] F. Archambeau, N. Méchitoua, and M. Sakiz, Code_Saturne: a finite volume code for the computation of turbulent incompressible flows industrial applications, Int. J. Finite Volumes, February 2004.
[6] MPI: A Message Passing Interface Standard, Message Passing Interface Forum 1995, http://www.netlib.org/mpi/index.html
[7] EDF Research and Development, http://rd.edf.com/107008i/EDF.fr/Research-and-Development/softwares/Code-Saturne.html
[8] HPCx -The UK's World-Class Service for World-Class Research, www.hpcx.ac.uk
[9] STFC's Computational Science and Engineering Department, http://www.cse.scitech.ac.uk/
[10] HECToR UK National Supercomputing Service, http://www.hector.ac.uk
[11] Supercomputing at Oak Ridge National Laboratory, http://computing.ornl.gov/supercomputing.shtml.
[12] The Green Top 500 List, http://www.green500.org/lists/2007/11/green500.php.

[13] Single Node Performance Analysis of Applications on HPCx, M. Bull, HPCx Technical Report HPCxTR0703 2007, http://www.hpcx.ac.uk/ research/ hpc/technical_reports/HPCxTR0703.pdf.
[14] J. Bonelle, Y. Fournier, F. Jusserand, S.Ploix, L. Maas, B. Quach, Numerical methodology for the study of a fluid flow through a mixing grid, Presentation to Club Utilisateurs Code_Saturne, 2007, http://research.edf.com/fichiers/fckeditor/ File/EDFRD/Code_Saturne/ClubU/ 2007/07-mixing_grid_HPC.pdf.

Scalability Considerations of a Parallel Flow Solver on Large Computing Systems

Erdal Yilmaz, Resat U. Payli, Hassan U. Akay, Akin Ecer, and Jingxin Liu[1]

Computational Fluid Dynamics Laboratory, Dept. of Mechanical Engineering, Indiana University-Purdue University Indianapolis, Indianapolis, Indiana, 46202 USA
http://www.engr.iupui.edu/cfdlab

Abstract. In this paper, we present scalability characteristics of a parallel flow solver on two large computing systems. The flow solver is based cell-centered finite volume discretizations along with explicit and implicit time integration methodologies. It has capability to solve moving body problems using Overset grid approach. Overset option is yet in sequential form. This solver is compared with another in-house flow solver for the parallel performance on two large-scale parallel computing platforms up to 2048 number of processors. Parallel timing performance of the solver was analyzed using the Vampir timing tool for DLR-F6 wing body configuration with 18 million elements. Timing of the Overset component was tested for a butterfly valve flow problem in a channel.

Key words: Parallel CFD, Overset Grid, unstructured grid, parallel performance.

1 Introduction

In today's code development environment with several software and hardware resources at the disposal of the researchers in academic and research institutions, developing a new Computational Fluid Dynamics (CFD) code may not seem to be as challenging as in the past couple of decades. However, making it well-performing on scattered and constantly upgraded computing platforms for variety of problems would be an art form researchers sometimes feel challenged. It could be an elaborate process beyond dealing with the nuts and bolts CFD codes and parallel computers. Ultimate goal would be making the code as scalable as possible for changing sizes of the computing resources as well as the problem itself.

If not designed at the developing phase, adjustments would be needed to the parallel codes for better run-time performance. The reason for poor parallel performance can be very complex and developers need to understand correct performance problems. Performance tools can help by monitoring a program's execution and producing data that can be analyzed to locate and understand areas of poor performance. Several commercial and research tools are available to conduct performance analysis such as Vampir [1, 2], Pablo[3], Tau[4], and Paradyn [5]. These performance tools

D. Tromeur-Dervout (eds.), *Parallel Computational Fluid Dynamics 2008*,
Lecture Notes in Computational Science and Engineering 74,
DOI: 10.1007/978-3-642-14438-7_34,

include libraries to time stamp and record the solver sequential and parallel events. Most of these tools do not require any changes in sources codes but links with the solvers at the building phase along with Message Passing Interface libraries. Outcome can be visualized for CPU time, communication time, message sizes, relative comparisons, etc. In this paper, we used Vampir tool to extract timing information.

Performance of flow solvers can also vary with the algorithms and mesh data structures used. For unstructured mesh based finite volume flow solvers, usually two types of mesh data structure are most popular: cell-center based and cell-vertex (or node) based. While cell-center based solvers integrate fluxes on the faces of individual cells, node-based solvers integrates the fluxes over the cells connected to individual nodes. Though numbers of faces are the same in both approaches, the actual finite volume used in the cell-vertex approach is smaller than the node-based one, hence better actual spatial resolution. On the other hand, the node-based approach works with less memory, as it does not need values at cell centers. Usually number of cells for tetrahedral meshes are five-six times higher than nodes. Besides, face values calculated from nodes will be more accurate than from cell values. We compared timing of two different solvers, one with cell-vertex based one with cell-centered, for this purpose.

Besides the algorithms and physics, mesh capabilities to handle complex problems such as overset mesh method for moving components makes performance managements even difficult. As the number of moving blocks increases, complexity of the problem makes parallel computing a challenge. Different than a few solver-embedded overset capabilities for unstructured grids [6, 7], Dirtlib/Suggar [8, 9] overset grid package can be considered as a stand alone pluggable package for unstructured grid solver. However, its parallel performance yet to be evaluated. In this paper, we will give flavor of our overset capability implemented in our cell-center based flow solver, however full parallelization using MPI was not demonstrated yet. In the following two sections, we will present algorithmic details of our parallel solvers, called PACER3D and SunFlo3D, overset grid methodology employed in SunFlo3D, parallel performance and timing studies of both solvers, and performance with the overset version. Finally we will draw our conclusions.

2 Flow Solver

2.1 Features of the Flow Solver

Our flow solver is based on earlier versions of an existing code, USM3D, developed at NASA [10], an Euler version of which was parallelized and extended to dynamically deforming meshes at IUPUI [11]. The fluid motion is governed by the time-dependent Navier-Stokes equations for an ideal gas, which express the conservation of mass, momentum, and energy for a compressible Newtonian fluid in the absence of external forces. The equations are non-dimensionalized with the freestream reference values for density and speed of sound. Three-dimensional turbulent flow equations are solved on unstructured grids. Spatial discretization is accomplished by a cell-centered finite-volume formulation using an accurate linear reconstruction scheme and upwind flux differencing. Inviscid flux quantities are computed across each cell face using the Roe flux-difference splitting approach, [12] or the Van Leer flux-vector splitting technique [13]. Spatial discretization is accomplished by a novel cell reconstruction process, which is based on an analytical formulation for computing solution gradients within tetrahedral cells. The viscous fluxes are approximated at centroids of cell faces by linear reconstruction, which provides a continuous representation of the solution variables across the cell faces. The viscous computations are advanced to steady state by the implicit time integration algorithm of reference [14]. The scheme uses the linearized, backward Euler time differencing approach to update the solution at each time step for the set of equations. Convergence to the steady state solution is accelerated by sacrificing the time accuracy of the scheme, and advancing the equations at each mesh point in time by the maximum permissible time step in that region. Closure of the Reynolds stress is provided by the one-equation Spalart-Allmaras (S-A) turbulence model [15]. The S-A model requires that the distance of each cell to the nearest wall be provided for the near-wall damping terms for cells, which are in proximity to ‘viscous’ surfaces. These distances are determined prior to the code execution for cells in the “viscous” layers and contribute to only a small portion of the overall overhead.

At IUPUI, we have rewritten the original solver code in Fortran 90 language to take advantage of dynamic allocations feature for vectors and arrays and convenience of the efficient structure and syntax of Fortran 90. The flow solver is parallelized using MPI. Single program multiple data parallelization procedure is employed for distributed memory architectures. Solution domain is partition into blocks with one-element overlap interfaces via in-house General Divider program [16]. General Divider program prepares the sending and receiving cells and nodes at the interface which requires exchanging the flow variables between blocks. At each iteration, cell flow values at the block interfaces exchange between neighbor blocks before inviscid and viscous fluxes are calculated. L2 norm of the residuals are calculated for each block and using the reduction function of the MPI maximum value of the calculated norm values distributed to all blocks for both implicit and explicit solvers. For the implicit solver, Gauss Seidel iterative procedure is used to solve a sparse set of equations. 20 sub-iterations are carried out to complete for one iteration. At the end of each sub-iteration flow values of the interfaces are exchanged between neighbor blocks. Also, calculated nodal values at the interfaces are exchanges between neighbor blocks both for implicit and explicit solvers.

2.2 Overset Grid Method

The overset or Chimera grid methodology has been in use as a technique in CFD to simplify grid generation of complex geometries or to enable relative motion between geometry components. An overset grid system utilizes a set of grids to discretize the domain with each component grid generated locally around a portion of the geometry. The grids are allowed to overlap without needing to match point-to-point with other grids like in a traditional multi-block structured grids system. This ability to grid locally greatly simplifies the grid generation process for complex geometries. In addition, the flexibility of overlapping grids allows the grids to move relative to each other to accommodate relative motion. The flow solution uses interpolation at appropriate points to couple the solutions on the different grids. The result is a flexible gridding strategy that allows components of the geometry to be easily added or moved without global re-gridding. There are two major steps to establish communications in this overset method: 1) automatic hole cutting, which essentially involves blanking cells of a grid in regions that overlap with non-flow-domain in the other grids of an overset grid system and identifying the Chimera boundary cells that lie along the hole or fringe boundary surfaces as well as interpolation boundary surfaces and 2) identification of interpolation stencils, which involves searching donor cells and getting interpolation coefficients for all intergrid boundaries cells.

2.3 Comparison of Two Flow Solvers

The SunFlo3D flow solver is compared against another in-house CFD code, Pacer3D [17, 18]. Pacer3D is an explicit cell vertex based flow solver based on Euler flow equations. Therefore, we compared explicit version of the SunFlo3D flow solver with the Pacer3D flow solver. Table 1 summarizes features of both codes for comparison.

PACER3D	SUNFLO3D (explicit time stepping version)
Finite volume spatial discretization for compressible Euler flow equations using unstructured tetrahedral mesh	
Vertex based	Cell-center based *(better spatial resolution than vertex-based*
Central differencing with artificial dissipation of Jameson	Roe's flux difference splitting
Implicit residual smoothing	
Local Time stepping for steady flows	
Explicit 3-stage Runge-Kutta time stepping	
Enthalpy damping	N/A

Table 1. Scheme details of the two flow solvers

3 Numerical Simulations

3.1 Parallel CFD Application

For the parallel performance evaluations we used an external transonic flow over DLR F6 airplane geometry, which is composed of only wing and body. This test case has 18 millions of tetrahedral cell elements. Grid is divided into 256, 512, 1024, and 2048 numbers of partitions using Metis [19].

Parallel computations have been performed on parallel clusters in Indiana University (BigRed) and San Diego Supercomputing Center (SDSC) for the cases. Those two platforms have different architectures for internal communications. Clock speed of BigRed is three times faster than that of SDSC. More details of the parallel clusters are given in Table 2.

Parallel performances studies on BigRed and BlueGene are summarized in Figures 1 and 2, for speedups and elapsed times, respectively. Both codes were run at the same flow conditions as inviscid for fair comparison. SunFlo3D shows linear scalability on both platforms while Pacer3D tips off after 1024 processors in BigRed. This could be associated with the higher CPU speed of BigRed as it would result in lower computation/communication ratio as we increase number of processors. On BlueGene. Pacer3D shows super linear speedup as CPU clock time on this machine is slower compared to BigRed. In addition to CPU speed, architectures and communications patters are much different for these two platforms.

BigRed (IBM e1350) **Indiana University**	**IBM BlueGene/L** **San Diego Supercomputing Center**
768 JS21 Bladeserver nodes	It has 3072 compute an 384 I/O nodes
PowerPC 970MP processors,	PowerPC processors
2 x 2.5GHz dual-core	2 x 700MHz
8GB 533MHz DDR2 SDRAM	32KB, 32-byte line, 64 way L1 cache; 16 128-byte lines L2 cache act as prefetch buffer 4MB 35 cycles shared L3 cache
1 x Myricom M3S-PCIXD-2-I (Lanai XP)	3D torus for point-to point message passing; Global tree for collective message passing

Table 2. Features of the computing platforms

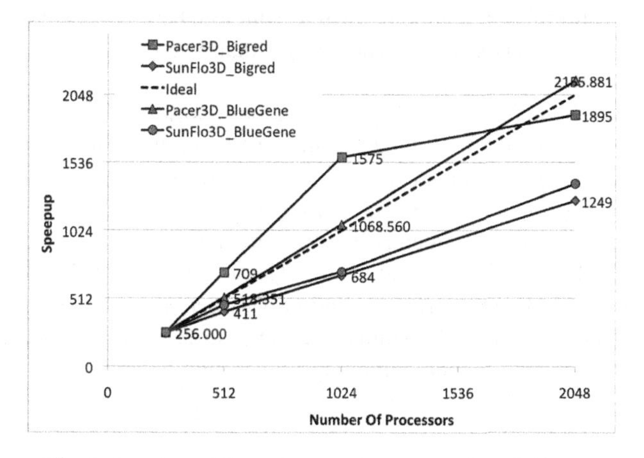

Fig. 1. Speedup of the solvers on two computing platforms

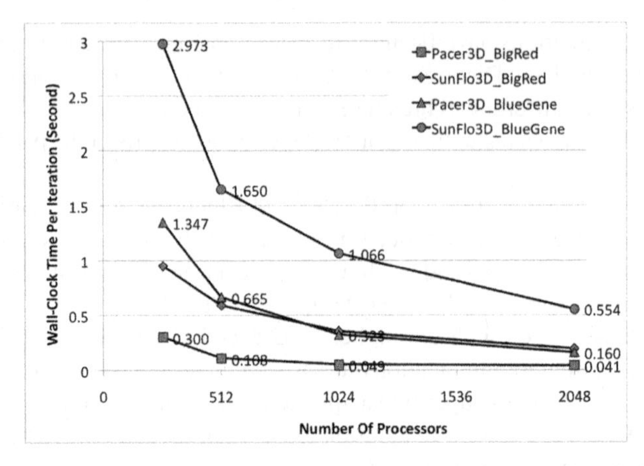

Fig. 2. Timing comparison of the solvers on two parallel computing platforms

3.2 Timing Analysis by Vampir Tool

We used Vampir tool to analyze timing of different functions in both solvers. Vampir provides detailed execution times breakdown of individual routines and functions of the applications. It also provides timing information regarding MPI related activities in the solvers. Figure 3 shows timing of Pacer3D and SunFlo3D obtained by the Vampir tool. Pacer3D takes 49 seconds and SunFlo3D takes 113 seconds to complete 100 explicit time steps. MPI related time includes initialization of the parallel environment, message send and receive between the processors, reduction and gathering of the shared parameters, and finalization of the parallel environment. Pacer3D uses 29 seconds for the parallelization related activities while SunFlo3D uses 67 seconds. Application time includes all the computations including I/O processes. Pacer3D uses 19 seconds for the application part while SunFlo3D uses 44 seconds. Therefore, the SunFlo3D solver performs 2.3 times slower than Pacer3D for both computation and communication.

Table 3 lists breakdown of the timing for both solvers. The SunFlo3D solver needs to update residuals for all blocks. This puts additional burden in the communication of the blocks. The update of the residuals takes significant amount of time – almost 60 percent of total communication time. However, SunFlo3D performs better for exchange of the flow parameters between the neighbors since it uses cell center values while Pacer3D uses nodal values. This is due to the fact that the number of cells at the overlapping interface would be less than the number of nodes. Flux calculations for both codes take almost the same time, though each uses different methodology. However, residual smoothing takes significantly more time in SunFlo3D.

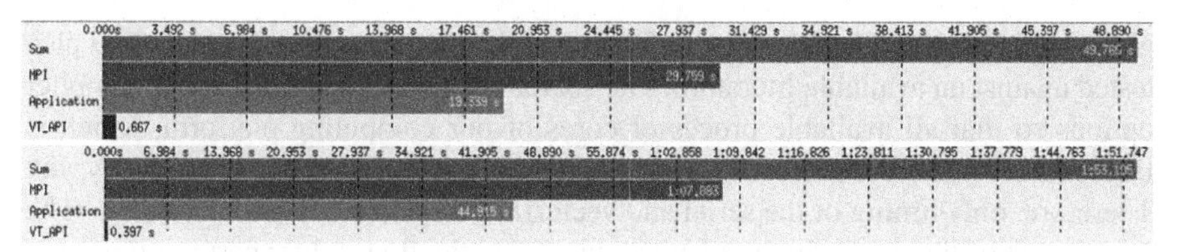

Fig. 3. Timing for Pacer3D and SunFlo3D flow solvers for 100 time steps obtained by Vampir tool, respectively

Components	Pacer3D	SunFlo3D
Communication-1: send and receive at the interfaces	14 seconds	8 seconds
Communication-2: share residuals and other common parameters	N/A	44 seconds
Parallel initialize/finalize	Same (14.8 seconds)	Same (14.6 seconds)
Flux calculation	Central differencing and artificial dissipation of Jameson (11 seconds)	Roe's Flux difference splitting(11 seconds)
Residual smoothing	Operates on edge of the cell and nodes (3.4 seconds)	Operates over cells and faces of the cell. Usually # of cells is 5-6 folds of # of nodes (19 seconds)
Local time stepping	1. 7 seconds	2.3 seconds
I/O	1.0 seconds	2.2 seconds
Others (main, Runge-Kutta, pre-proc, etc)	3 seconds	20 seconds

Table 3. Breakdown of timing for Pacer3D and SunFlo3D by Vampir tool

3.3 Performance of the Overset Version

Overset component of our flow solver have been tested for various flow problems including internal and external flows. Details of our overset research will be given in a separate study. However, here only a flavor of it will be presented. Test case we present here is a valve problem based on the geometry and the flow boundary conditions from experiments of Morrison and Dutton [20]. Steady-state simulations at various angles of the attack of the valve relative to the flow channel are considered. Numerical simulations are made for cases including fully opened status (0 degree valve-position angle), 5-degree position angle, 10-degree valve-position angle and even near-closed sate of 60-degree position. The valve disk angle of 30-degree is approximately a mid-way between the wide open and closed valve conditions. When compared with available experimental data, pressure distributions of the numerical results on the valve disk surfaces are close to that of the experiment. Valve grid positions and static pressure comparison for 30 degrees of the valve position are given in Figure 4 and 5, respectively. Timing of the developed overset method is also tested against an available literature. The solver code was vectorized using compiler options so that all available processor cores of our computing platform, which is BigRed cluster of IU, are used. At this time, MPI parallelization is not ready yet. Therefore, only timing of the serial and vectorized code is compared in Table 4. The timing cost of the overset component compared to the flow solver part per time step is found to be 2.7 times, which much better than reported in reference [7], which is 23.2.

	Overset grid only	Flow Solver	Total Time	Overset/Solver
Sequential	165.41	19.03	184.44	8.7
Vectorized	41.53	15.30	56.83	2.71
Speedup	3.98	1.24	3.25	

Table 4. Timing (in seconds) of the overset and flow solver (one node with 4 cores)

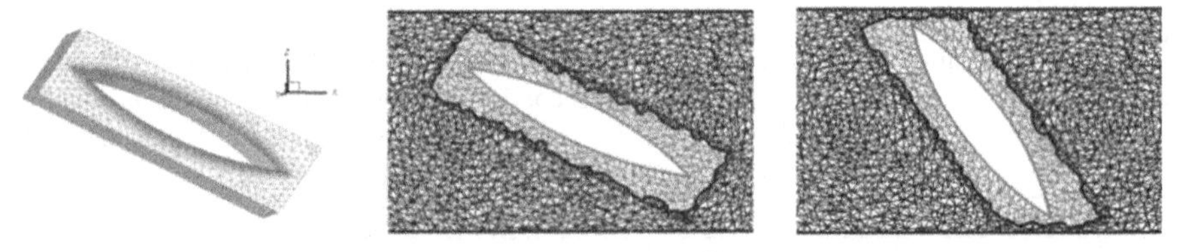

Fig. 4. Valve viscous mesh at 30 degrees and overset mesh at 30 and 60 degrees

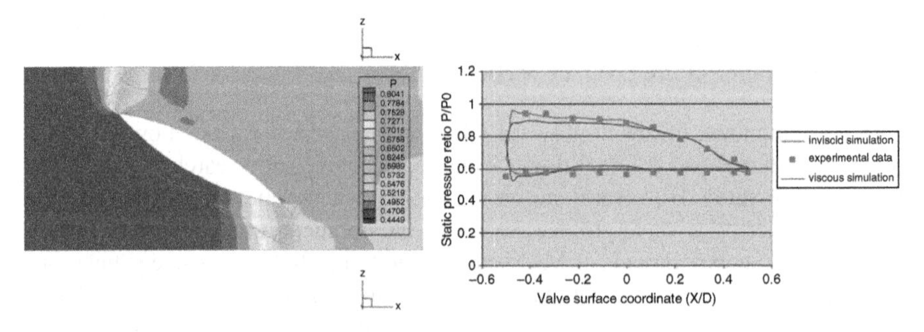

Fig. 5. Pressure contours and comparison with the experiment at 30 degrees of valve position

4 Conclusions

Detailed parallel performance of two finite volume codes are evaluated using a commercially available parallel performance tool. Both codes were found scalable on two parallel systems. While the node-based finite volume code has better computation timing and memory usage and super linear speedup in one of the parallel clusters, cell-vertex based code demonstrated linear speedup up to 2048 processors. Vampir performance tool was found very effective to get detailed timing of the events and components of the code for parallel runs. Overset component of the solver performs very well on multi cores of a compute node. Parallel MPI implementation of the overset component is yet to be done.

Acknowledgments

This research was supported in part by the National Science Foundation under Grants No. ACI-03386181, OCI-0451237, OCI-0535258, and OCI-0504075 and was supported in part by the Indiana METACyt Initiative. The Indiana METACyt Initiative of Indiana University is supported in part by Lilly Endowment, Inc. and by the Shared University Research grants from IBM, Inc. to Indiana University.

[1] Vampire performance tool: http://www.vampir.eu.
[2] Brunst, H. Winkler, M., Nagel, W.E., and Hoppe, H. C.: Performance Optimization for Large Scale Computing: The Scalable VAMPIR Approach. International Conference on Computational Science (ICCS2001) Workshop on Tools and Environments for Parallel and Distributed Programming, San Francisco, CA (2001).
[3] Reed, D.A. Roth, P.C. Aydt, R.A. Shields, K.A. Tavera, L.F. Noe, R.J. Schwartz, B.W.: Scalable Performance Analysis: the Pablo Performance Analysis Environment. Proceedings of Scalable Parallel Libraries Conference (1993).
[4] Shende, S., and Malony, A.D.: TAU: The TAU Parallel Performance System. International Journal of High Performance Computing Applications, Volume 20, Number 2, Pages 287-331 (2006).
[5] Miller, B. P., Callaghan, M.D., Cargille, J. M., Hollingsworth, J. K., Irvin, R. B., Karavanic, K. L., Kunchithapadam, K., and Newhall, T.: The Paradyn Parallel Performance Measurement Tool. IEEE Computer 28, 11: 37-46. Special issue on performance evaluation tools for parallel and distributed computer systems (1995).
[6] Meakin, R.L. and Suhs, N.E.: Unsteady Aerodynamic Simulation of Multiple Bodies in Relative Motion. AIAA-89-1996-CP (1989).
[7] Togashi, F., Ito, Y., and Nakahashi, K.: Extension of Overset Unstructured Grid to Multiple Bodies in Contact. AIAA 2002-2809 (2002).
[8] Noack, R.W.: DiRTLiB: A Library to Add an Overset Capability to Your Flow Solver. AIAA Paper 2005-5116, 17th AIAA Computational Fluid Dynamics Conference, Toronto, Ontario, Canada (2005).

[9] Noack, R.W.: SUGGAR: A General Capability for Moving Body Overset Grid Assembly. AIAA Paper 2005-5117, 17th AIAA Computational Fluid Dynamics Conference, Toronto, Ontario, Canada (2005).
[10] Frink, N.T.: Assessment of an Unstructured-Grid Method for Predicting 3-D Turbulent Viscous Flows. AIAA Paper 96-0292 (1996).
[11] Uzun, A., Akay, H.U., and Bronnenberg, C.: Parallel Computations of Unsteady Euler Equations on Dynamically Deforming Unstructured Grids. Parallel CFD '99 Proceedings, Edited by D. Keyes, et al., Elsevier Science, pp. 415-422 (2000).
[12] Roe, P.L.: Characteristic Based Schemes for the Euler Equations. Annual Review of Fluid Mechanics, Vol. 18, pp. 337-365 (1986).
[13] Van Leer, B.: Flux-Vector Splitting for the Euler Equations. Eighth International Conference on Numerical Methods in Fluid Dynamics, E. Krause, ed., Volume 170 of Lecture Notes in Physics, Springer-Verlag, pp. 507-512 (1982).
[14] Anderson, W.K.: Grid Generation and Flow Solution Method for Euler Equations on Unstructured Grids. NASA TM- 4295 (1992)
[15] Spalart, P.R. and Allmaras, S.R.: A One-Equation Turbulence Model for Aerodynamic Flows. AIAA Paper 92-0439 (1992).
[16] Bronnenberg, C.E.: GD: A General Divider User's Manual - An Unstructured Grid Partitioning Program. CFD Laboratory, IUPUI (1999).
[17] Yilmaz, E., Kavsaoglu, M.S., Akay, H.U., and Akmandor, I.S.: Cell-vertex Based Parallel and Adaptive Explicit 3D Flow Solution on Unstructured Grids. International Journal of Computational Fluid Dynamics, Vol. 14, pp. 271-286 (2001).
[18] Payli, R.U., Yilmaz, E., Akay, H.U., and Ecer, A.: Impact of the TeraGrid on Large-Scale Simulations and Visualizations. Parallel CFD 2007, Antalya, Turkey, May 21-24 (2007).
[19] Karypis, G. and Kumar, V.: Multilevel k-way Partitioning Scheme for Irregular Graphs. J. Parallel Distrib. Comput. 48: 96–129 (1998).
[20] Morrris, M., Dutton, J., and Addy A.L.: Peak Torque Characteristics of Butterfly Valves. Proceedings of the Forum on Industrial Applications of Fluid Mechanics, ASME FED-Vol. 54, pp. 63–66 (1987).

Large Scaled Computation of Incompressible Flows on Cartesian Mesh Using a Vector-Parallel Supercomputer

Shun Takahashi[1], Takashi Ishida[1], Kazuhiro Nakahashi[2], Hiroaki Kobayashi[3], Koki Okabe[4], Youichi Shimomura[5], Takashi Soga[6], and Akihiko Musa[7]

[1] *Ph.D. candidate, Dept. of Aerospace Engineering, Tohoku University, Japan*
[2] *Professor, Dept. of Aerospace Engineering, Tohoku University, Japan*
[3] *Professor, Cyberscience center, Tohoku University, Japan*
[4] *Researcher, Cyberscience center, Tohoku University, Japan*
[5] *Engineer, NEC Software Tohoku, Japan*
[6] *Engineer, NEC, Japan*
[7] *Engineer, NEC Corporation, Japan*

Abstract. Present incompressible Navier-Stokes flow solver is developed in the framework of Building-Cube Method (BCM) which is based on a block-structured, high-density Cartesian mesh method. In this study, flow simulation around a formula-1 car which consists of 200 million cells was conducted by vector-parallel supercomputer NEC SX-9. For exploiting the performance of SX-9, the present flow solver was highly optimized for vector and parallel computation. In this paper, the computational result from the large scale simulation and the parallel efficiency in using flat-MPI or hybrid-MPI are discussed.

1 Introduction

Today's supercomputers have a large number of processors in the system. In order to utilize the large scale parallel computers, one of the authors proposed a block-structured Cartesian mesh method named Building-Cube Method (BCM). [2] It basically employs an equally-spaced Cartesian mesh bacause of the simplicities in the mesh generation, in introducing a spatially higher-order solution algorithm, and in the post processing. These simplicities of Cartesian mesh for all stages of a flow computation as well as the less memory requirement per cell will become more important for large scale computations.

One of the issues of the Cartesian mesh, however, is how to fit the mesh spacing to the local flow scale without introducing algorithm complexities. In the present method, a flow field is described as an assemblage of building blocks of cuboids, named 'cube'. Each cube is a sub-domain which has the same number of cells in it. It is useful for local refinement strategy which is difficult to perform in uniform Cartesian mesh system.

D. Tromeur-Dervout (eds.), *Parallel Computational Fluid Dynamics 2008*,
Lecture Notes in Computational Science and Engineering 74,
DOI: 10.1007/978-3-642-14438-7_35, © Springer-Verlag Berlin Heidelberg 2010

In this study, large scale simulations on BCM are demonstrated by using the latest NEC vector-parallel supercomputer, SX-9, which was installed at the end of March 2008 at Cyberscience center of Tohoku University. This system consists of 16 nodes of SX-9 and each node includes 16 vector processors. In this paper, the parallel efficiency of the present BCM solver was investigated from large scale simulation by using the multi-node system of SX-9.

2 Computational methods

2.1 Building-Cube Method

The BCM is aimed for high-resolution flow computations around real geometries using high-density mesh. In this method, entire flow field is described as an assemblage of cuboids as shown in Fig. 1. Each cube is a sub-domain which includes the same number of equally-spaced Cartesian cells, by which an object is represented with staircase pattern. In BCM, local flow characteristic can be captured easily by simple refinement of cubes with keeping practical computational resource. Moreover, it has advantages about fast and robust mesh generation [6] around complicated geometries and easy introducing spatially higher-order scheme from the character of Cartesian mesh method.

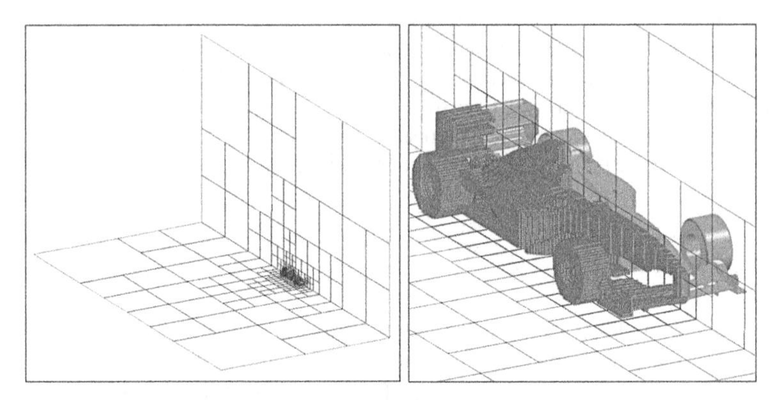

Fig. 1. Example of computational mesh around formula-1 car

In the present flow solver, 3D incompressible Navier-Stokes equations (Eq.(1)) are solved by fractional-step method with staggered arrangement. [7, 8, 5]

$$\begin{cases} \frac{\partial \mathbf{u}}{\partial t} + (\mathbf{u} \cdot \nabla)\mathbf{u} = -\nabla p + \frac{1}{Re}\nabla^2 \mathbf{u} \\ \nabla \cdot \mathbf{u} = 0 \end{cases} \tag{1}$$

In the fractional-step method, three processes are implemented sequentially at each time step. In this paper, second order accurate Adams-Bashforth explicit time integration is implemented in solving the temporal velocity field $\mathbf{u}^*$ in Eq. (2).

$$\frac{\mathbf{u}^* - \mathbf{u}^n}{\Delta t} = -\left(\frac{3}{2}\mathbf{A}^n - \frac{1}{2}\mathbf{A}^{n-1}\right) + \left(\frac{3}{2}\mathbf{B}^n - \frac{1}{2}\mathbf{B}^{n-1}\right) \tag{2}$$

Here the convection term $\mathbf{A}$ and diffusion term $\mathbf{B}$ are discretized by third order upwind finite difference scheme [1] and second order central finite difference scheme respectively. When larger CFL number is needed like a moving boundary problem, second order accurate Crank-Nicolson implicit time integration with dual time stepping method [4] in Eq. (3) is available. In this process, the inner iteration about fictitious time τ is implemented to keep the second order time accuracy. When the Eq. (3) was converged by $m \rightarrow \infty$ ideally, the first term of Eq. (3) is vanished with $\Delta \mathbf{u}^m \rightarrow 0$. Finally, the original equation is satisfied.

$$\frac{\Delta \mathbf{u}^m}{\Delta \tau} + \frac{\mathbf{u}^{n,m} - \mathbf{u}^n}{\Delta t} = -\left(\frac{1}{2}\mathbf{A}^{n,m} + \frac{1}{2}\mathbf{A}^n\right) + \left(\frac{1}{2}\mathbf{B}^{n,m} + \frac{1}{2}\mathbf{B}^n\right) \tag{3}$$

The pressure filed p^{n+1} is solved by Poisson equation of Eq. (4). In solving incompressible Navier-Stokes equations, most of computational cost is paid for the procedure. Therefore hyper-plane method is used with SOR method to exploit vector processing of SX-9 by eliminating data dependency. Detail of the technique is described below.

Moreover pressure perturbation in incompressible flow field should be propagated to far field at once ideally. Then flow information of each cube is exchanged between adjacent cubes because of the multi-block structure of the BCM computational mesh. This process is also highly optimized for vector processing by MPI PUT/GET function of MPI-2.

$$\nabla^2 p^{n+1} = \frac{1}{\Delta t} \nabla \cdot \mathbf{u}^* \tag{4}$$

Finally the real velocity field $\mathbf{u}^{n+1}$ is solved by the temporal velocity field and the pressure as shown in Eq. (5).

$$\frac{\mathbf{u}^{n+1} - \mathbf{u}^*}{\Delta t} = \nabla p^{n+1} \tag{5}$$

In the present computations, a wall object is expressed by simple staircase pattern. Hence nonslip velocity boundary condition and non pressure gradient boundary condition along with normal direction are applied to staircase cells directly.

2.2 Vectorization and parallelization

The present code is optimized to exploit the vector-parallel supercomputer NEC SX-9 shown in Fig. 2.

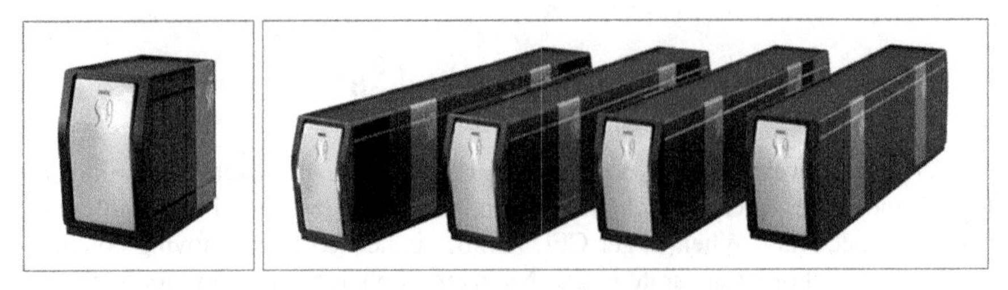

Fig. 2. Single node and 16 multi-node system of NEC SX-9

One of techniques for the vectorization is loop fusion. It is used to extend the length of the vector processing by fusing multiple loops into single loop. In BCM, it is easy to implement because most of loops of the equally-spaced Cartesian cells in each cube consist of simple IJK indices.

As mentioned above, the hyper-plane method as shown in Fig. 3 is applied to SOR method in solving the Poisson equation. Though vector processing is not be able to be utilized in usual SOR method because of the data dependency, it is eliminated by using the hyper-plane method and elements on the same hyper-plane are calculated at once by the vector processing. As a result, about 99.6 percent vector ratio is successfully achieved.

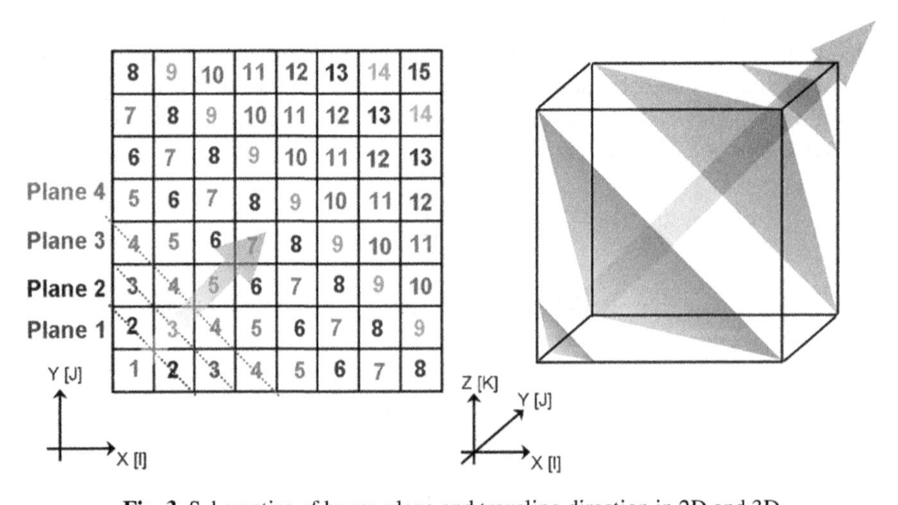

Fig. 3. Schematics of hyper-plane and traveling direction in 2D and 3D

The parallelization is implemented by MPI and OpenMP. In this paper, flat-MPI which consists of only MPI parallelization and hybrid-MPI which consists of MPI and OpenMP parallelization are investigated. For efficient and fast communication, MPI PUT/GET functions of MPI-2 are implemented in this study.

In BCM, the load balance can be kept easily if only dividing whole computational domain into several sub-domains of the same number of cubes. Figure 4 shows an example of the decomposition into 4 domains. The load balance is kept almost same even though the geometrical size of each domain is different and some islands are created as Fig. 4.

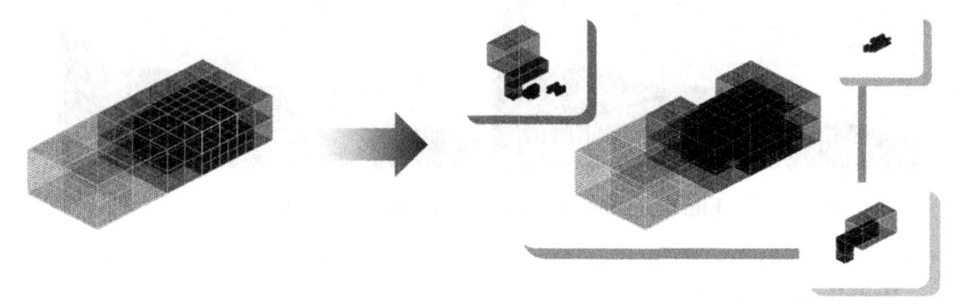

Fig. 4. Example of 4 decomposed domains

3 Results and discussion

3.1 Computational mesh and result

In this chapter, the parallel efficiency of the present flow solver is investigated from large scale incompressible flow simulation around formula-1 car model [3] by using SX-9. The computational mesh around the model consists of 5,930 cubes, in which 32^3 cells are included. The total number of cells is 194,314,240. Minimum mesh spacing is 7.3×10^{-4} based on the overall length of the model which corresponds to 3.5×10^{-3} meter in the real scale. Reynolds number is 2.6 million based on the overall length.

The computational mesh and results are visualized in Figs. 5 and 6 that are coarsened eighth part of the original data to compress the data size. The robustness of the present flow solver could be confirmed from the stable computation around the complicated geometry.

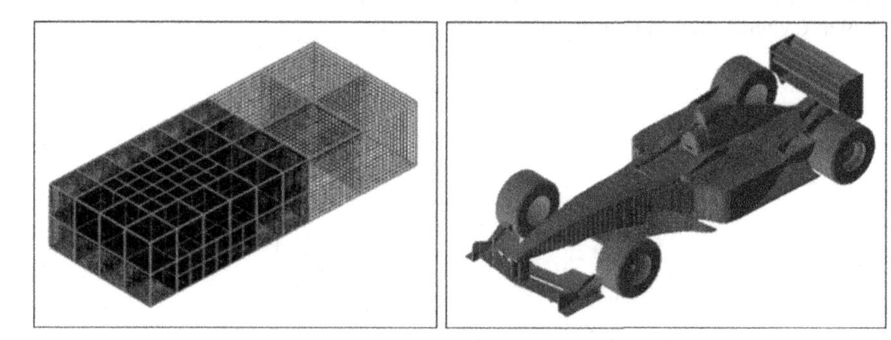

Fig. 5. Computational domain and surface representation

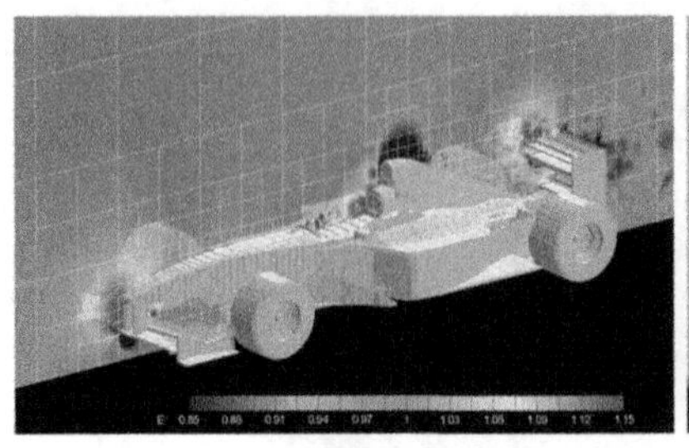

Fig. 6. Instantaneous pressure and velocity field

3.2 Parallel efficiency

The speed-up ratio in using 1, 2 and 4 nodes of SX-9 is discussed here. As mentioned above, 16 CPU are included in 1 node of SX-9. In this study, flat-MPI and hybrid-MPI computations were conducted. In the case of flat-MPI, whole computational domain is decomposed to the same number of domains with the number of PE (CPU). In the case of hybrid-MPI, on the other hand, whole computational domain is decomposed to the same number of domains with the number of nodes, and OpenMP parallel computation is implemented in each node at the same time.

Figure 7 and 8 show the speed-up ratio and communication time in each case of 1, 2 and 4 nodes, respectively. In using single node, hybrid-MPI shows better performance than flat-MPI because the hybrid-MPI computation in single node is just OpenMP parallel computation without any data exchange. In using several nodes, on the other hand, flat-MPI shows better performance than hybrid-MPI with the reduction of communication time as shown in Fig. 8. The flat-MPI parallel computation is conducted with more sub-domains than hybrid-MPI one. For example, in the case of 32 parallel computation by 2 nodes, the numbers of sub-domains are 32 and 2 in flat-MPI and hybrid-MPI respectively. Therefore, the amount of data which is exchanged with adjacent cubes in flat-MPI becomes smaller than one in hybrid-MPI. Now, 130 times speed-up ratio and 99.6 percent parallel ratio are accomplished as shown in Fig. 9 by using flat-MPI on 16 multi-node system (including 256 CPU) of SX-9 at Cyberscience center of Tohoku University.

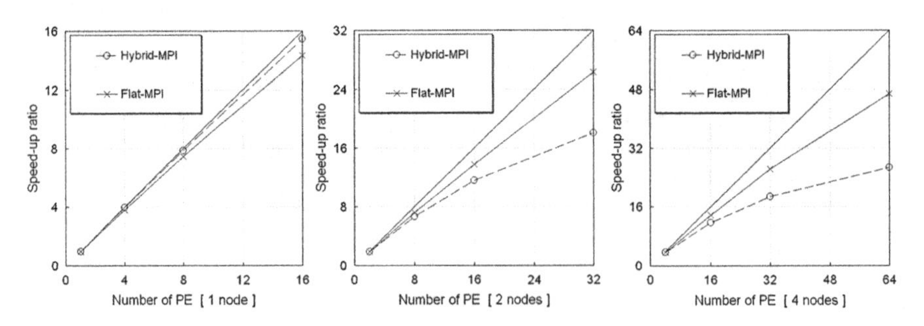

Fig. 7. Speed-up ratio

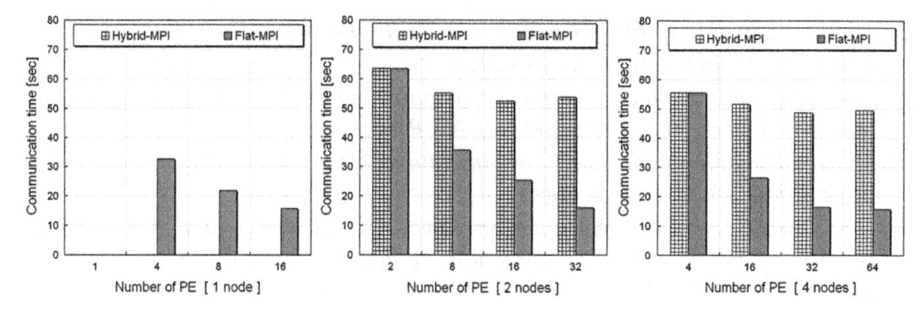

Fig. 8. Communication time

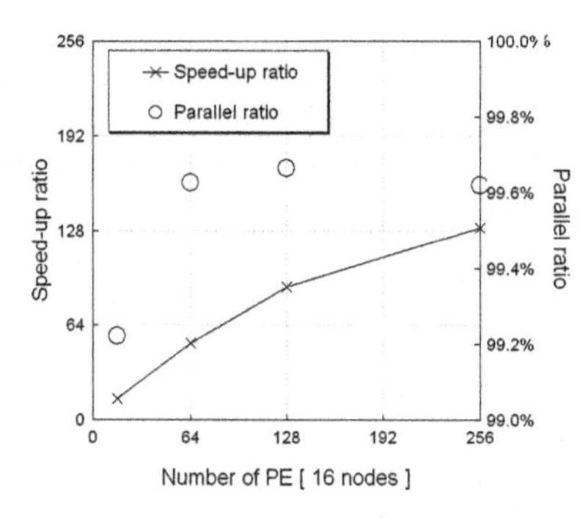

Fig. 9. Parallel perfornamce in 16 multi-node system of SX-9

Parallel overhead commonly disturbs the speedup in the highly vectorized computation because the vector processing is implemented so rapidly. However the present flow solver is achieved good parallel performance with highly vectorized computation. For more efficient parallel computation, appropriate MPI function and parallel algorithm that are suitable for MPP should be considered.

4 Conclusions

The parallel performance of the present flow solver was investigated from large scale incompressible flow simulation on the latest vector-parallel supercomputer NEC SX-9. The robustness of the present solver was confirmed by the simulation around a formula-1 car of 200 million cells. By the comparison between flat-MPI and hybrid-MPI parallel computation, it was confirmed that the speed-up ratio of hybrid-MPI was greater than one of flat-MPI in the single node computation. On the other hand, the speed-up ratio of flat-MPI was greater than one of hybrid-MPI in the several nodes computation. Moreover, the maximum parallel performance was 130 times speed-up in using 16 multi-node system including 256 CPU with flat-MPI parallel computation. From now, more appropriate MPI function and parallel algorithms should be considered to handle MPP in near future.

[1] *Computation of high Reynolds number flow around circular cylinder with surface roughness*, number 84-0340, 1984. AIAA Paper.

[2] *High-Density Mesh Flow Computations with Pre-/Post-Data Compressions*, number 2005-4876, 2005. AIAA Paper.
[3] *Turbulent Flow Simulation around Complex Geometries with Cartesian Grid Method*, number 2007-1459, 2007. AIAA Paper.
[4] *Dynamic Load Balancing for Flow Simulation Using Adaptive Refinement*, number 2008-920, 2008. AIAA Paper.
[5] J. K. Dukowicz and A. Dvinsky. Approximate factorization as a high order splitting for the implicit incompressible flow equations. *Journal of Computational Physics*, 102:336–347, 1992.
[6] T. Ishida, S. Takahashi, and K. Nakahashi. Efficient and robust cartesian mesh generation for building-cube method. *Journal of Computational Science and Technology*, 2, 2008.
[7] J. Kim and P Moin. Application of a fractional-step method to incompressible navier-stokes equations. *Journal of Computational Physics*, 59:308–323, 1985.
[8] J. B. Perot. An analysis of the fractional step method. *Journal of Computational Physics*, 108:51–58, 1993.

Dynamic Load Balancing on Networked Multi-core Computers

Stanley Chien[1], Gun Makinabakan[1], Akin Ecer[1], and Hasan Akay[1]

Purdue School of Engineering and Technology Indiana University-Purdue University Indianapolis 723 W. Michigan Street, Indianapolis, IN 46202, USA
Mechanical, Aerospace and Structural Engineering Department, Washington University in St. Louis, MO 63130 schien@iupui.edu

Abstract. As multi-core computers becoming the main computation force of the computer network, dynamic load balancing for parallel applications on multi-core computers needs to be studied. The paper describes a method for extending the dynamic load balancing tool to support multi-core computers. The method scales all cost function information into per CPU core unit and performs dynamic load balancing in terms of CPU cores. The parallel processes are assigned to each computer (not each CPU core) assuming that the operating system assigns the parallel processes equally among all CPU cores. The proposed tool has been successfully tested for supporting parallel CFD applications.

Keywords: parallel computing, multi-core computer, dynamic load balancing.

1 Introduction

The fast improvement of the computer speed has been partially achieved by increasing CPU clock speed. However, as the computer clock speed reaches Giga Hertz range, the CPU clock speed is approaching its physical limitation under the current fabrication technology [1]. The high CPU clock frequency creates a lot of heat during operation that results low energy efficiency and causes CPU package and system design problems. Therefore, the new trend in CPU designs have been emphasized on the multi-core CPUs each of which is a single integrated circuit that consists several CPU cores. Multi-core CPUs enable increased productivity, computation power and powerful energy efficient performance. There are two different types of multi-core CPUs. The first type has homogeneous CPU cores. Most new PCs with Intel processors use two or more CPU cores which are identical and share the same memory. The second type has heterogeneous CPU cores. An example is IBM cell processor design which has a master processor core and a set of slave processor cores [2]. The slave processor cores are different from the master processor core and each slave processor core has its own local memory space. The computer architecture experts

D. Tromeur-Dervout (eds.), *Parallel Computational Fluid Dynamics 2008*,
Lecture Notes in Computational Science and Engineering 74,
DOI: 10.1007/978-3-642-14438-7_36,

envision that computers with multi-core CPU will be the main computation force in the near future.

How to take the advantage of the multi-core CPUs is an interesting subject. Several directions are currently being explored. One direction is to write parallel code explicitly to assign the process to multi-cores. This approach is required for using IBM Cell multi-core processors. The second direction is to rely on operating systems to assign multiple tasks to different CPU cores. The advantage of the second direction is that the existing software can be easily running on the multi CPU core computers.

Most existing parallel CFD applications have been developed based on domain decomposition approach in which the data are segmented into many blocks and the application code for processing all blocks are the same. These type applications can be naturally adapted to multi-core computers based on the second direction. Although the operating system can assign parallel processes to multiple CPU cores of a computer, a load-balancing tool is still needed for assigning parallel processes to each multi-core computer.

In this paper, we will discuss the expansion of our dynamic load-balancing tool, DLB [3], to automatically assign parallel CFD processes to networked heterogeneous computers with multi homogeneous CPU cores. We will start with the information gathering needed for load balancing, and then describing the algorithm used for the dynamic load balancing. An experimental example will demonstrate the effectiveness of the method.

2 Dynamic Load Balancing

The goal of load balancing of a parallel application on networked computers is to make all parallel processes finish at the shortest time. In most cases (when communication time between parallel processes is low), this goal can be translated as making all computers busy all the time before the parallel application finishes. In order to make all computers busy, the speed of the computer, the workload on each computer, and the amount of new work to be assigned onto each computer need to be known qualitatively so that the execution time (or cost) of each parallel process on each computer can be calculated. Then an optimization method can be adopted to minimize the cost and hence generate the optimal load balancing. This approach has been successfully used in for load balancing on networked computers with single CPU core [4]. Here we extend the same idea for dynamic load balancing of networked multi-core computers.

The steps for achieving dynamic load balancing include: (1) Find the relative computation speed of each multi-core computer. (2) Find the extraneous workload on each computer. (3) Find the workload of every process of a parallel job. (4) Use a cost function to predict the effective computation time of each parallel process under a given process distribution. (5) Find a load distribution that minimizes the elapsed program execution time.

To find the relative computation speed of each multi-core computer, we run a small single thread benchmark program at the system start up time on each computer and measure the

execution CPU time on each computer. Since the benchmark program can be executed on only one CPU core, the measured time provides the speed information of only one CPU core. The ratios of the measured CPU times of the benchmark program on different computers essentially provide the information of the relative speed of the CPU cores on different computers. This method for obtaining the CPU core computation speed frees the user from guessing the computer speed for parallel process distribution.

The computation speed of a multi-tasking computer to a particular user is not only the CPU speed, but also the number of extraneous processes running on the computer. Therefore, it is essential to know the extraneous load on the computer before performing load balancing. To find the extraneous workload on each computer, operating system tools can be used. For example, *ps* command in UNIX or *PsTools* from Microsoft, can be used to measure the percentage of CPU time used by each process. However, the number of running process measured by these tools is not usable for load balancing. The reason is that the measured load on a computer also includes the processes of the parallel applications to be dynamically balanced, processes of other parallel jobs, and demon processes that do not use much computer power. We solve this problem by letting each computer remember the parallel process running on it. Therefore, by running *ps* command on UNIX or *PsTools* program on Windows, we can find how many single load and how many parallel load are running on the computer. During load counting, only the jobs that run over 5% of one CPU core time are counted as the extraneous load. The processes run under 5% of one CPU core time are usually system demons and other small processes that are negligible.

Finding the computation required for each process of a parallel job is essential for the proper load balancing. To release the burden of the user for providing this information, we developed a profiling library to obtain this information during the execution of the application code [5]. The library supports MPICH2 [6] and needs to be linked with the application program.

The profiling library routines are executed during each MPICH2 library access in each parallel application process and provide the elapsed execution time, the elapsed communication time, the execution CPU time, and communication topology information of all application process. Since all information is measured, it already takes into account of the effect of the compiler efficiency, the memory sizes on each computer, the cache size, and computer configuration differences.

Once the information of the computer speed, the extraneous computation load, the workload of user's parallel processes, and the communication cost between computers are obtained, all these information are scaled to per CPU core basis, such as CPU execution time per CPU core and extraneous load per CPU core. This approach essentially treats a multi-core computer as multiple computers for load balancing, e.g., treat a dual-core computer as two computers. All information are used to construct a cost (time) function for parallel process distribution on networked computers.

Then the greedy algorithm is used to move the parallel processed among the computers in the network for minimizing the cost (time) of parallel job execution [4].

In order to take the advantage of dynamic load balancing, the parallel application must be able to utilize the new load redistribution dynamically suggested by DLB. In order to let the parallel application know that there is a newly suggested load distribution, two simple DLB interfacing library functions *check_balance()* and *application_stopped()* are inserted into the parallel application program (see Figure 1). Before each check pointing (to be done at the end of every N time steps), the library function *check_balance()* is called to find if load balancer suggested a better load distribution. If there is a better load distribution, the parallel program stores the intermediate information, stops the program and calls the library function *application_stopped()*. Once DLB detects that the parallel application program stopped execution, DLB restarts the parallel application program according to the proposed new load distribution. If it is not possible to include the DLB interfacing library to the application program, the application program needs to be stopped and restarted periodically.

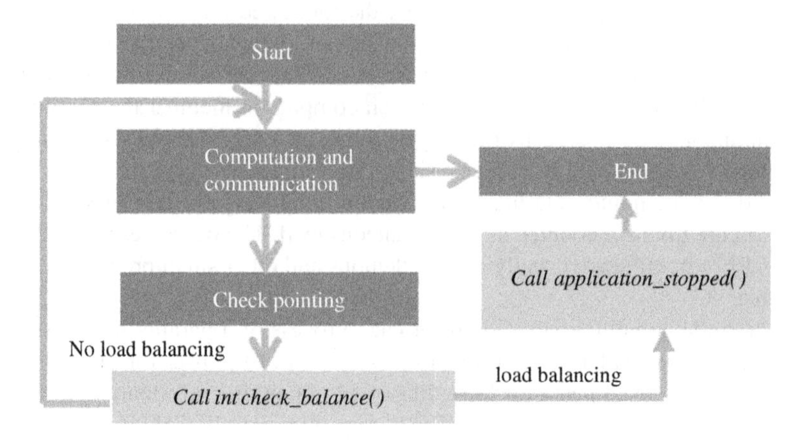

Fig. 1. Inserting profiling library function into application code.

To describe how much computation power is used by the processes on a multi-core computer, we define CPU usage percentages as follows:

$$\text{CPU usage percentage} = \frac{\sum_{i=1}^{\#\text{ of processes}} CPU elapsedtimeofi^{th}\text{bloc}}{(\#\text{ of CPU}) \times \text{Wall clock time}} 100 \quad (1)$$

In Equation (1), it was assumed that all the CPUs installed in the computer are identical, which is the case with the currently developed popular multi-core processors. Therefore, if only one process utilizes a CPU, the actual CPU usage percentage of the process is calculated by dividing the processor usage percentage of the process by the number of total CPU cores on the CPU. For example, if a process uses forty seconds of CPU time in an elapsed one hundred second period in a dual core PC, this process is treated as only uses twenty percent of the total computational power of this PC.

To evaluate the system utilization of all hardware used for a parallel job, *System Efficiency* (*SE*) is defined as follows

$$SE = \frac{\sum_{i=1}^{n} CP^i \times CS^i}{\sum_{i=1}^{n} CS^i} \tag{2}$$

where CP^i is the CPU utilization percentage of the ith computer and CS^i is the relative CPU speed of ith PC. *SE* indicates the percentage of the maximum computational power used by all processes running on the networked multi-core computers. Notes that *SE* calculation also gives higher weight for high speed computer in efficiency calculation since faster computer has high *CS*. *CP* for each computer can be dynamically measured during program execution.

3 DLB Implementation

The proposed load balancing method for multi CPU core computers are incorporated to DLB tool [4]. DLB is a middleware that has three major components, the Dynamic Load Balancing Agent (DLBA) Job Agent, and System Agent (SA). DLBA is a user/system program that supports users to optimally distribute the parallel processes to available computers. Job Agent is responsible to start one parallel job, monitor the execution health of the job and, providing fault tolerance. SA is a program installed on every computer that provides the computer load information and communication speed information. During parallel process dispatching, DLBA requests the computer load and speed information and communication speed information from SA of each PC and then determines an optimal load distribution based on greedy algorithm. To support multi-core computers, SA is also responsible to communicate with the operating systems to get the number of CPU core information. This information is passed to DLBA and be used in the cost function for load distribution optimization. Although DLBA does load balancing based on each CPU core, the operating system does not allow DLBA to submit the job to each CPU core. Therefore, the DLBA submit the load of all CPU core of a computer to the computer and expect the operating systems to distribute the job evenly.

4 Experiments

The following experiment demonstrates the effectiveness of the proposed approach. The application code is a parallel CFD test solver which implements a 3D transient heat equation on structured grids with forward time central space differentiation method. The code can be downloaded from `http://www.engr.iupui.edu/`

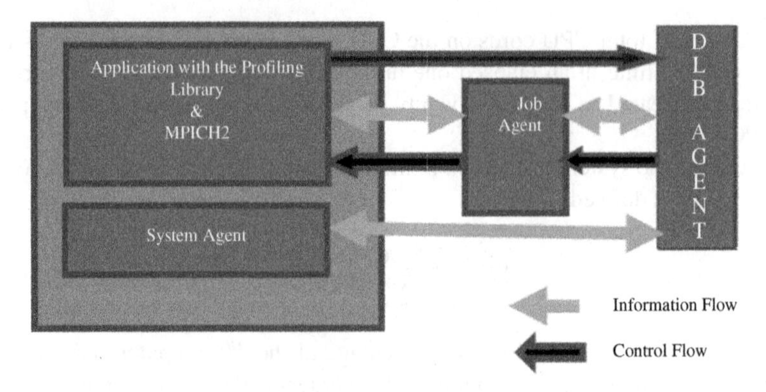

Fig. 2. The components of DLB.

me/newmerl/cfdl_software.htm. The data grid was divided into twelve equal sized blocks with 2 blocks in X direction, 2 blocks in Y direction, and 3 blocks in Z direction. The block size is approximately 8MB (15,625 nodes/block).

The test code was run on four Windows based computers. Two of these computers had two CPU cores, while the other two PCs had only one CPU core. The execution CPU time of a small benchmark program on each computer during system start up is listed in Table 1. Even though the benchmark code execution CPU time for the dual CPU core PC is about the same as the single CPU core PC, only one CPU core of the dual CPU core computer was used to run the benchmark program. Therefore, the dual core PCs were approximately two times faster than the single CPU core PCs.

PC Name	Benchmark Execution CPU Time (sec)	Number of CPUs
in-engr-sl11133	8.234375	1
in-engr-sl11132	7.984375	1
In-engr-sl11134	8.465625	2
In-engr-sl11135	8.578125	2

Table 2. PCs used in the experiment

Twelve data blocks were initially distributed evenly on all PCs (3 on each PC). The execution time for each process is measured by the profiling library. The CPU usage percentages calculated based on the execution time of all processes are shown in Table 2. The system efficiency for the initial load distribution is 68.64%. Since the speeds of the PCs were not identical, this result was expected, which demonstrates that the PCs with two CPU cores were underutilized.

PC Name	TotalLoad	Blocks	Measured CPU Usage Percentage
in-engr-sl11133	3	1 2 3	98.024
in-engr-sl11132	3	4 5 6	98.265
in-engr-sl11134	3	7 8 9	52.127
in-engr-sl11135	3	10 11 12	54.168

Table 4. Experimental results for the initial distribution.

PC Name	Total Load	Blocks	Measured CPU Usage percentage
in-engr-sl11133	2	1 2	87.014
in-engr-sl11132	2	3 4	87.352
in-engr-sl11134	4	5 6 7 8	94.914
in-engr-sl11135	4	9 10 11 12	97.299

Table 6. Experimental results of the final distribution

Based on the measured timing data, the DLB tool suggested to move blocks from the two single CPU core PCs to the dual CPU core PCs in order to balance the load. In the new block distribution, the computers with two CPU cores had twice the load of the computers with one CPU core. After the parallel application restarted with the new process distribution, the experimental results (Table 3) showed that the DLB tool responded properly for the dual CPU core PCs. The CPU usage percentage of the PC's with dual CPU core went up from the 50% range to the middle 90% range and the efficiency percentage increased by twenty five percent to 93.03%. Figure 3 shows the side by side comparison of CPU usage percentage between initial load distribution and DLB suggested balanced load distribution. This result demonstrates that new approach is capable of working in environments which contain multi core PCs and single processor PCs.

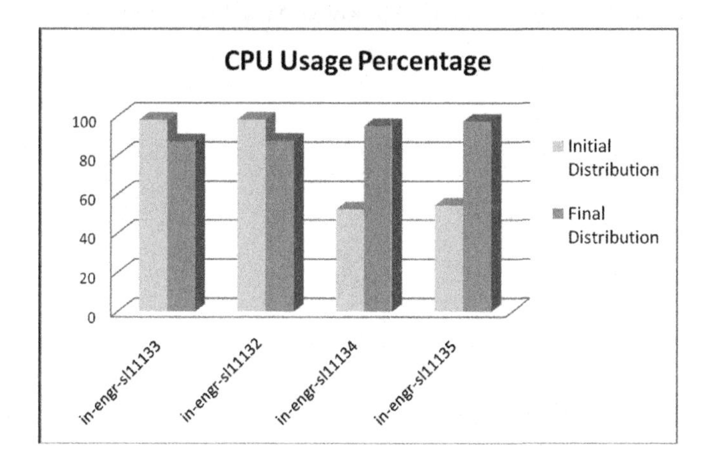

Fig. 3. Comparison of CPU usage percentage between initial load distribution and DLB suggested balanced load distribution.

5 Conclusion

Multi-CPU core computers will be widely used in the heterogeneous computer network for parallel applications. The proposed method shows that the load balancing method can use CPU core as a basic unit for parallel process load balancing if the multi cores are homogeneous on the PC. The CPU usage percentage and system efficiency definitions described in this paper can be used as a measure of the system performance. The dynamic load balancing tool was augmented to support multi-core UNIX based and Microsoft Windows (XP and Vista) based computers. The tool is successfully tested for supporting a parallel CFD code on multi core computers.

[1] L. Chai, Q. Gao, Panda, K. Dhabaleswar , Understanding the Impact of Multi-Core Architecture in Cluster Computing: A Case Study with Intel Dual-Core System, Seventh IEEE International Symposium on Cluster Computing and the Grid, pp. 471-478 Electronic Edition.
[2] The cell project at IBM research, `http://www.research.ibm.com/cell`
[3] Y.P. Chien, J.D. Chen, A. Ecer, H.U. Akay, and J. Zhou, DLB 2.0 – A Distributed Environment Tool for Supporting Balanced Execution of Multiple Parallel Jobs on Networked Computers, *Proceedings of Parallel Computational Fluid Dynamics 2001*, May 2001, Amsterdam, Holland.
[4] S. Secer, Genetic Algorithms and Communication Cost Function for Parallel CFD Problems, M.S.E.E. Thesis, Purdue University, May 1997.
[5] CFD test solver, `http://www.engr.iupui.edu/me/newmerl/cfdl_software.htm`
[6] MPICH 2, `http://www-unix.mcs.anl.gov/mpi/mpich2/`
[7] S. Chien, G. Makinabakan, A. Ecer, and H.U. Akay, Non-Intrusive Data Collection for Load Balancing of Parallel Applications, *Proceedings of Parallel Computational Fluid Dynamics 2006*, Busan, Korea, May 13-15, 2006.

On efficiency of supercomputers in CFD simulations

Andrey V. Gorobets[1], Tatiana K. Kozubskaya[1], Sergey A. Soukov[1]

Institute for Mathematical Modelling of Rus. Ac. Sci.
4A, Miusskaya Sq., Moscow, 125047, Russia
e-mail: cherepock@mail.ru, web page: http://www.imamod.ru

Abstract. The present work is reporting the experience on different aspects of using supercomputers for CFD applications. Several problems that appear with new supercomputers built of multi-core nodes are considered. In particular, it is a problem of increased influence of irregular memory access, problems with efficient use of such a big numbers of processors that are available in present time. A hybrid parallelization using MPI and OpenMP technologies is also considered to improve efficiency of computations. Results of performance tests where the hybrid approach outperformed the MPI-only parallelization are shown. Finally, a performance comparison of Marenostrum and MVS-50000 supercomputers is presented, illustrative numerical results on the DNS of turbulent flows are also included.

Keywords: supercomputers; parallel algorithms; multi-core architecture; hybrid parallelization

1 Introduction

A further progress in the CFD applications is directly connected with a possibility of performing giant-size DNS computations in order to show the advantages of computational experiments together with (and sometimes instead of) physical ones. Another important goal of such DNS computations is a calibration of semi-empirical turbulence models (RANS, LES and the hybrid RANS/LES models). The only way to compute such problems is to use high-performance supercomputers which are being rapidly developed nowadays. Many new machines with Rmax above 100TFlops can be found in the TOP500 list. Several Petaflop computer systems are being built. But those new machines bring new problems. Of course, a primary problem is how to use supercomputers efficiently. An efficient use of even one CPU is not a trivial task with all this memory cache hierarchy, multithreading, vector operations etc. Each node of supercomputer may have several multi-core CPUs and this set of cores shares RAM memory, network channel etc. So an efficient use even of one node which is itself a shared memory parallel system is not straightforward, especially considering that

D. Tromeur-Dervout (eds.), *Parallel Computational Fluid Dynamics 2008*,
Lecture Notes in Computational Science and Engineering 74,
DOI: 10.1007/978-3-642-14438-7_37,

cores can easily be too many for the hardware they installed in (limited RAM memory bandwidth, etc.). And, finally, there are hundreds and thousands of nodes that must work together efficiently within a huge DNS or LES computations. Is sounds like a challenging problem.

The most common CFD approach of parallelization is MPI with geometric parallelism for MMP machines with distributed memory. It also works well on SMP machines with shared memory and can even outperform its inherent OpenMP parallelization. Supercomputer is in fact a hybrid - it is an MMP machine built of SMP nodes. It used to be no problem when single-processor nodes were replaced with twin CPU nodes. Then multi-core CPUs came and now the number of cores is becoming too much for RAM memory bandwidth and size, network channels etc. A general tendency is that a number of cores per node grows much faster than the memory size and bandwidth (Imagine a 16-core node with the memory that can work at full speed only with say 4 cores. If all the 16 processors use the memory intensively the performance may go down 4 times). This brings new problems that should be considered. OpenMP parallelization in addition to MPI should be also taken into account when developing parallel algorithms. And the requirement of scalability to such a big number of CPUs is also not easy to be satisfied. This leads to the situation which can be often seen on supercomputers - mostly they are loaded with plenty of small 10-100 CPU tasks. They are dragged apart into small pieces instead of running full power to solve really big cases they are designed for.

This work is devoted to investigation of ways to use efficiently the huge power of supercomputers for CFD applications. It is based on particular examples of several research codes and DNS for compressible and incompressible flows performed on different supercomputers.

2 Supercomputers and codes under consideration

Following supercomputers were used for the DNS simulations and efficiency tests within this work:

1. **MVS-50000** (Hewlett-Packard) in Joint SuperComputer Center of Russian Academy of Science. It has about 450 nodes interconnected with Infiniband network. Each node has 2 quad-core CPUs Intel Xeon 3.0 GHz, 8 cores in total, that share 4 Gb of RAM memory.
2. **MareNostrum** (IBM) in Barcelona Supercomputing Center. It has about 2500 nodes interconnected with Myrinet network. Each node has 2 double-core CPUs IBM Power PC 970MP 2.3 GHz , 4 cores in total, that share 8 Gb of RAM memory.

The following in-house CFD codes are involved:

1. In-house code **NOISEtte** [2]. It is designed for solving 2D and 3D CFD and aeroacoustics problems on compressible flows using unstructured triangular and tetrahedral meshes and higher-order explicit algorithms [1]. The parallel algorithm is well-scalable due to the explicit time integration.

2. In-house code **WOMBAT** [3]. It is designed for 3D CFD problems, in particular DNS of compressible flows using unstructured meshes. Parallelization is similar in general to the first code.
3. In-house code **KSFD** from CTTC lab of Technical University of Catalonia [5]. It is designed for large DNS of incompressible flows. It uses structured meshes, high-order numerical scheme and is well scalable on parallel systems. It can use efficiently up to at least a thousand of CPU of supercomputer.

The first code is used in efficiency tests and as a playground for hybrid parallelization MPI with OpenMP. The second code also implements both the MPI and OpenMP technologies and is used to evaluate the performance with data reordering for efficient memory access. The third code is used to compare the performance of two supercomputers Marenostrum vs. MVS-50000.

3 Efficiency of multi-core nodes

The first test shows the problem of performance loss due to multi-core configuration of nodes on MVS-50000. It has 8 CPU cores per node and it is not straightforward to use them all efficiently. The DNS test case was performed with the NOISEtte code using unstructured tetrahedral mesh of 10^6 nodes decomposed into 40 subdomains. This 40-cpu case can run on 5 nodes using all 8 cores of each node. It can also run on 10 nodes using only 4 cores, on 20 nodes using 2 cores and on 40 nodes using only 1 core as well.

A wall clock time spent on computation of 100 time steps is measured. The efficiency shown in figure 1 is simply given by

$$E = T_1/T_p \times 100\% \tag{1}$$

where T_p - time for 40 CPU case using P cores per node.

The execution on 40, 20, 10, 5 nodes corresponds respectively to P=1, 2, 4, 8. T_1 - the time for execution on 40 nodes having only one process per node. It is clear that there is a significant decrease of efficiency when all 8 cores are in use. The comparison of 4 and 8 processes per node shows 55% performance loss (which is $(T_8 - T_4)/T_4 \times 100\%$).

The same behavior was observed with In-house code KSFD [5], when the 200CPU case with mesh of 27 millions of nodes was executed first having 4 processes per node, then 8. A performance loss on 8 cores was 60% comparing with 4 cores per node used. A similar test was executed for WOMBAT code and showed the same problem but in less scale. The results for all three codes are represented in the figure 1.

The test with NOISEtte code was repeated with all communications switched off to ensure that it was not a problem of network operations. The results remained the same. This leads to the conclusion that 8 cores are too much for RAM memory bandwidth. When all 8 cores use the memory intensively a significant slowdown happens. This motivates the use of approaches considered further which allow to improve efficiency and performance.

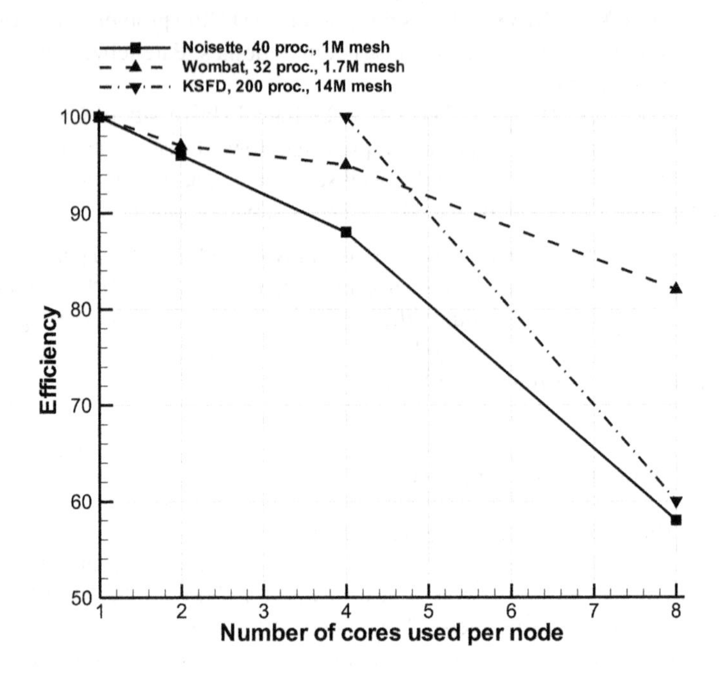

Fig. 1. Efficiency of using cores in 40 CPU case running on 8-core nodes

4 The two-level preprocessing model

Due to computer memory hierarchy an efficiency of parallel implementation of finite-volume and finite-element methods on unstructured meshes strongly depends on a numeration of the mesh nodes. An order of numbering of the mesh elements determines the arrangement of data needed for the calculation of fluxes on the cell faces, variables gradients, etc. in RAM. The calculations per mesh elements are usually not of high computational costs, so the efficiency is mainly dependent on the memory access. A non-optimal data disposition in RAM memory leads to frequent losses in cache. This results in a significant performance decrease, especially for multi-core nodes.

To optimize the memory access, a special algorithm of two level mesh preprocessing is implemented in the in-house code WOMBAT. At the first level the mesh is decomposed into a large number of small micro-subdomains and a coarse graph of mesh is built. At the second level the coarse graph is decomposed into the required number of subdomains. This approach has the several advantages:

- Decomposition of mesh at the second level can be done sequentially. The coarse graph is small enough to be decomposed by single CPU, then each CPU can combine its fragments

- Better load balance. The comparison with ParMetis for the decomposition of $2.1 * 10^8$ nodes mesh into 1280 subdomains shows that ParMetis results in load imbalance 52%, while with 2-level model (1st level decomposition into 20000 fragments) it is only 6%.
- Reordering of the mesh elements by micro-subdomains improves memory access. The size of micro-subdomains is chosen small enough to fit its data in cache minimizing the cache losses and stabilizing the computation time for different variants of original mesh numeration.

5 Hybrid parallelization based on MPI and OpenMP

The hybrid MPI + OpenMP parallelization becomes more popular due to a fast growth of the number of cores per node. OpenMP is used for the parallelization within multi-core nodes. A number of cores grows faster than RAM memory and network performance hence the use of OpenMP gives several advantages reducing a number of MPI processes:

- More RAM memory per MPI process. RAM per core becomes too small when only MPI is used. In case of MVS-50000 it is less than 0.5Gb per each of 8 MPI processes running on a node.
- Reduction of communications. Without OpenMP multiple MPI processes share limited network resources of the node which results in a slowdown.
- Load on file system is reduced since less processes use it simultaneously.

One of the problems with OpenMP parallelization is an intersection of the data the threads work with. Presence of the critical sections and atomic operations lead to a substantial slowdown. In some cases it can be avoided by a straightforward replication of output arrays when the threads write to their own arrays and the results are joined afterwards.

A more universal approach is the further decomposition of subdomain belonging to the MPI process into smaller blocks. Elements of the mesh are reordered in a way that all inner elements of the blocks are grouped in memory and a set of interface elements (belonging to more than one block) is separated and also grouped compactly. OpenMP threads process inner elements of the blocks without intersections and then one thread processes the interface elements.

6 Performance and illustrative results

Speedup results for the code NOISEtte with the hybrid MPI+OpenMP parallelization are represented in figure 2. The test is done on MVS-50000 using a small mesh of only 10^6 nodes and the higher-order numerical scheme based on the extended space stencil which requires large data exchange. Apart from the speedup the test shows the comparison of MPI only (8 MPI processes per node) and MPI+OpenMP (4 MPI processes with 2 OpenMP threads each). MPI+OpenMP goes as fast as MPI until 640

CPU but providing twice more memory and at 1280CPU MPI+OpenMP outperforms MPI on about 20%. Further they both decay.

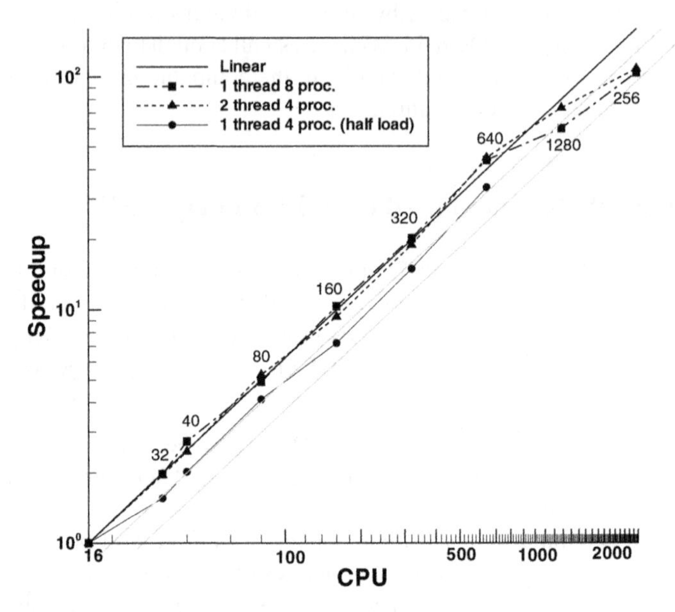

Fig. 2. Comparison of MPI and MPI+OpenMP parallelization in a speedup test using a small mesh

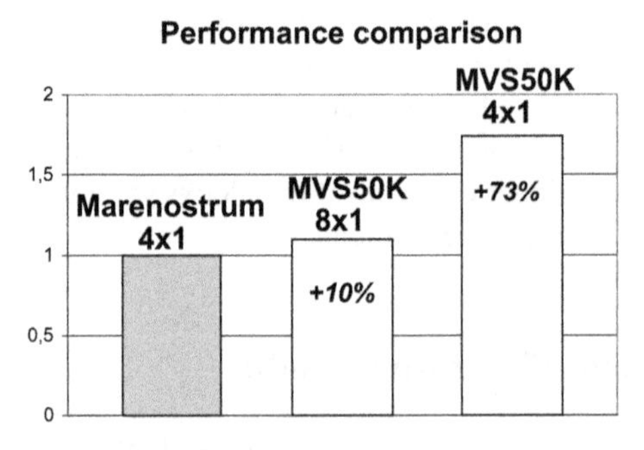

Fig. 3. Performance comparison of MareNostrum and MVS-50000 on a relatively small parallel task

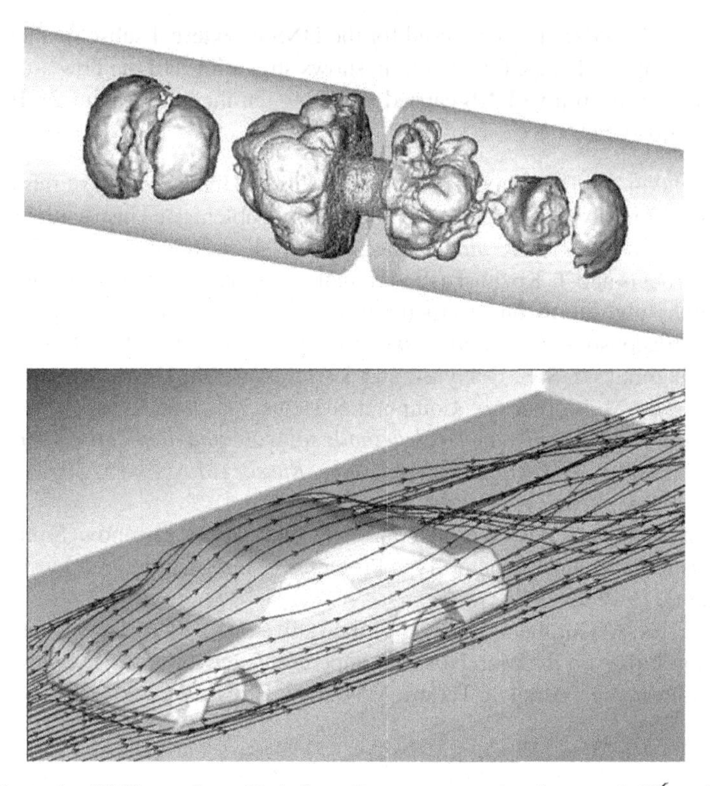

Fig. 4. Illustrative DNS snapshots. Turbulence in a resonator chamber, mesh 10^6 nodes (top); flow around a car body, mesh $2 * 10^8$ (bottom)

Another test was carried out to compare performance of the two supercomputers MVS-50000 and MareNostrum on a relatively small parallel task of 200 MPI processes (overall performance of supercomputers is available in Top500 list).

In the test the KSFD code [5] was used to perform DNS of incompressible natural convection turbulent flow in an open cavity. The mesh size is $14 * 10^6$ nodes, the scheme is of the 4-th order of accuracy. The results are represented in figure 3. On MareNostrum the test was performed with 4 parallel processes per node, on MVS-50000 - with 4 and 8 processes. At the same load with 4 processes per node, MVS-50000 is found substantially faster, but at its full load with 8 processes per node MVS-50000 appears only 10% faster than MareNostrum. It also should be noted that despite MareNostrum is slower, it offers 4 times more memory per core.

Finally, examples of DNS performed with NOISEtte and WOMBAT codes are shown in figure 4.

Fig. 4 top shows a vortex structure in a resonator mouth. It is a snapshot from the DNS of acoustic wave absorption in the resonator chamber installed in the impedance tube. The simulation was performed using the code NOISEtte within a series of numerical experiments with acoustic liners [4].

The code WOMBAT is now used for the DNS of external subsonic flow around bodies of complex shapes. Fig. 4 bottom shows streamlines of the flow around a car body from the illustrative DNS carried out using tetrahedral mesh of $2 * 10^8$ nodes and up to 3000 CPU.

[1] I. Abalakin, A. Dervieux, and T. Kozubskaya. High Accuracy Finite Volume Method for Solving Nonlinear Aeroacoustics Problems on Unstructured Meshes. *Chinese Journal of Aeroanautics*, pages 97–104, 2006.
[2] A. Gorobets and T. Kozubskaya. Technology of parallelization of explicit high-accuracy algorithms on unstructured meshes in computational fluid dynamics and aeroacoustics. *Matematicheskoe modelirovanie*, 19 (2):68–86, 2007.
[3] M.V. Iakobovski, S.N. Boldyrev, and S.A. Sukov. Big Unstructured Mesh Processing on Multiprocessor Computer Systems. *Parallel Computational Fluid Dynamics: Advanced numerical methods software and applications. Proc. of the Parallel CFD 2003 Conference Moscow, Russia (May 13-15, 2003), Elsevier, Amsterdam*, pages 73–79, 2004.
[4] T. Kozubskaya, I. Abalakin, A. Gorobets, and Alexey Mironov. Simulation of Acoustic Fields in Resonator-Type Problems Using Unstructured Meshes. *AIAA Paper 2519-2006*, 2006.
[5] F. X. Trias, A. Gorobets, M. Soria, and A. Oliva. DNS of Turbulent Natural Convection Flows on the MareNostrum Supercomputer. In *Parallel Computational Fluid Dynamics*, Antalya, Turkey, May 2007. Elsevier.

Part X

Environment and biofluids applications

Numerical Study of Pulsatile Flow Through Models of Vascular Stenoses with Physiological Waveform of the Heart

Jason D. Thompson[1], Christian F. Pinzn[1] and Ramesh K. Agarwal[1]

Mechanical, Aerospace and Structural Engineering Department, Washington University in St. Louis, MO 63130 rka@wustl.edu

Abstract. Recently an in-vitro experimental investigation on axisymmetric models of stenotic arteries was conducted by Peterson and Plesniak to determine the influence of three fundamental disturbances on stenotic flows: a geometric perturbation resulting in asymmetry of stenosis; a skewed mean inlet velocity profile; and flow downstream of a bend (skewed mean inlet velocity profile plus secondary flow due to bend). The goal of this paper is to numerically simulate the flow fields in the experiments of Peterson and Plesniak and compare the computed results with the experimental data. A commercially available CFD flow solver FLUENT is employed in the numerical simulations. The stenosis is modeled as an axisymmetric 75% area reduction occlusion. The actual physiological waveform of the heart is employed at the inlet in both the simulations and the experiments. Computations are in good agreement with the experimental data for flow in an axisymmetric stenosis with 75% area reduction occlusion. Computations for flow in an asymmetric stenosis (due to small geometric perturbation to axisymmetric configuration) are also in reasonable agreement with the experimental data.
Keywords: Vascular Stenosis, Pulsatile Blood Flow in Arteries

1 Introduction

Heart and other circulatory system diseases are among the leading causes of death in the adult population worldwide. Vascular and aortic valve stenoses are diseases that occur when there is narrowing of a blood vessel or valve due to formation of plaque. Vascular stenosis is caused by the accumulation of intravascular artherosclerotic plaques that build up along the vessel wall and extend into the vessel interior; also the impingement of extravascular masses sometimes causes stenosis. Since vascular stenosis occurs inside the vessel, it is not easy to diagnose it clinically since it cannot be inspected visually. Using invasive pressure drop measurements from catheters inserted upstream and downstream of the blockage, an estimate of the throat area of the stenosis can be obtained by performing a simple fluid dynamics analysis.

D. Tromeur-Dervout (eds.), *Parallel Computational Fluid Dynamics 2008*,
Lecture Notes in Computational Science and Engineering 74,
DOI: 10.1007/978-3-642-14438-7_38,

Using X-ray contrast angiography, images of severe stenosis can be obtained, but these images provide little or no information about flow properties such as pressure, velocity and wall shear stress. Doppler ultrasound techniques can also be used to measure velocities and waveforms in stenotic vessels and to determine the estimates of the throat area, however this method also has limitations in providing accurate predictions of the valve area because of several simplifying assumption involved. In recent years, the application of Phase Contrast Magnetic Resonance Imaging (PC-MRI) has become popular in the measurement of velocity field in vascular flows, and has shown some promise as a tool for diagnosing the vascular disease [1].

It is now well established that several flow related phenomenon play a critical role in the progression of the vascular disease. These are oscillating and low wall shear stress, high blood pressure, and flow phenomenon such as recirculating flow regions and turbulence that may occur after the onset of stenosis. In the 1997 review paper, Ku [2] has described in detail the fluid dynamics issues related to the flow of blood in healthy and diseased arteries, including some of the analytical models and measurement techniques used to diagnose the vascular disease. Berger and Jou [3] have reviewed the state of the art in analytical and computational techniques in analyzing the blood flow in stenotic vessels, with emphasis on flow through bifurcations and junctions including studies of steady and pulsatile flows in two and three dimensions. They also discuss the fluid dynamics related factors that may be responsible for triggering the buildup of plaque and subsequent formation of stenoses. Young [4] has also provided a very detailed study of the fluid mechanics of flow through vascular stenoses for both steady and pulsatile flow. Kim et al. [5] have performed numerical studies to simulate the local hemodynamics in the human circulatory system. In their CFD simulation, they employed non- Newtonian flow models for blood flow, an analytical model to describe the arterial wall motion due to fluid-wall interactions, a vascular bed model based on lump parameters for outflow boundary conditions, and a model for auto-regulation to account for systemic circulation in the entire cardiovascular system.

In a recent paper, Okpara and Agarwal [6] presented the results of simulations of steady and sinusoidal pulsatile flow in axisymmetric and 3D concentric phantoms and their comparison with experimental data. The goal of this paper is to extend the work reported in [6] to compute the pulsatile flows with actual physiological waveform of the heart in models of vessels with varying degrees of stenoses (mild to severe) for different flow rates to analyze and understand the details of the flow field such as pressure and velocity distributions, wall shear stress etc. The key objective of the study is to validate the experimental data recently obtained by Peterson and Plesniak [7] for actual physiological heart waveform for a 75% area reduction occlusion. Peterson and Plesniak recently conducted an experimental study to determine the influence of three fundamental disturbances on stenotic flows: a geometric perturbation resulting in asymmetry of stenosis; a skewed mean inlet velocity profile; and flow downstream of a bend (skewed mean inlet velocity profile plus secondary flow due to bend). The goal of this paper is to numerically simulate the flow fields in the experiments of Peterson and Plesniak and compare the computed results with the experimental data. Unsteady Reynolds-Averaged Navier-Stokes (URANS) equations in conjunction with several turbulence models are solved using the commercially available CFD flow solver FLUENT. The stenosis is modeled as an axisymmetric 75% area reduction occlusion. The actual physiological waveform of the heart is employed at the inlet in both the simulations and the experiments.

2 CFD Fluent solver

FLUENT is a commercially available numerical flow solver package, which is employed in this paper to compute the flow fields through vascular stenoses. This CFD software package solves the governing equations of incompressible, Newtonian or Non-Newtonian fluid using a finite-volume method. It has several numerical algorithms for both steady and unsteady flow calculations on structured as well as unstructured grids. The software also has several zero-, one-, and two-equation turbulence models. GAMBIT is the pre-processing grid generation software that is provided with the FLUENT package; it is used to create the geometry as well as to generate the appropriate structured or unstructured meshes.

To mimic the blood flow, we use a density of 1030 kg/m3 and viscosity of 0.00255 kg/ ms in the calculations assuming the blood to be a Newtonian fluid. Inlet velocity profiles are specified with appropriate User Defined Functions (UDFs) to mimic the actual physiological waveform of the heart. The flow is assumed to be laminar and fully developed upstream of the stenosis. An extrapolation boundary condition is applied at the outlet, which assumes zero normal gradients for all flow variables except pressure. In our calculations we have employed a second-order upwind solver in FLUENT 6.2.16 for solution of the momentum equations. Pressure-velocity coupling in incompressible flow is solved using the Pressure-Implicit with Splitting of Operators (PISO) scheme, and the pressure is computed using the standard Poisson solver. The reduction in area at the throat of the axisymmetric stenosis results in Reynolds numbers in the transitional/turbulent flow regimes. Therefore a turbulence model was employed; the turbulence model was modified to compute both the

transitional and fully developed turbulent flow. Three different turbulence models: $k-\varepsilon$, $k-\omega$, and full Reynolds Stress model were tested. The computed results using the Reynolds stress model compared best with the experimental data. Computations were performed on a coarse grid and a fine grid to ensure that the solutions were grid independent.

3 Results

3.1 Pulsatile Flow in Axisymmetric Stenosed Phantoms

In Reference [6], we computed the pulsatile flow in axisymmetric stenosed phantoms assuming the inlet flow to be sinusoidal. Here we consider the inlet flow to be the physiological waveform of the heart. The physiologically forcing waveform at the inlet of the stenosis was obtained by digitizing the data from a graph from Sean Petersons thesis [7]. This data was then fit to a curve using a third-order polynomial spline as shown in Figure 1. The spline was used to acquire a constant time interval. From the spline, an 8th order Fourier series was fit to the blood flow data to remove high frequencies. The average radius of an aortic stenosis was assumed to be 0.0064 meters. The density and viscosity of blood were assumed to be 1060 kg/m3 and 0.0035 kg/ms respectively. To match the experimental data (the experiments were performed in water), the blood flow and period were scaled to water. From this, the Reynolds number over the waveform was determined to be in the range 28.7 - 1402.7 and the Womersley number was calculated to be 4.64. The density and viscosity of water were assumed to be 1000 kg/m3 and 0.0001 kg/ms respectively. The period was scaled to 11.5972 seconds.

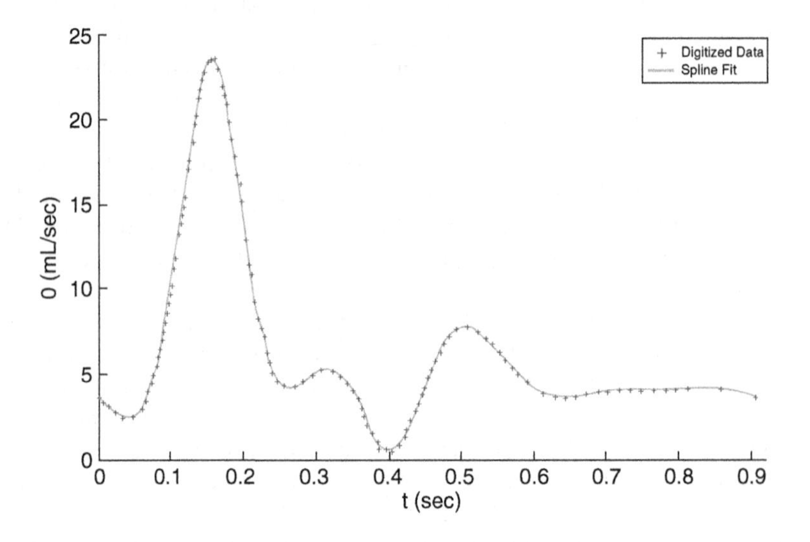

Fig. 1. Spline fit of the digitized data for the physiological waveform of the heart.

A bench mark computational experiment was first performed. A tube of diameter 0.0126 meters and 1.15 meters long was constructed within GAMBIT. The inlet boundary condition was set to be the scaled flow rate by using a user defined function within FLUENT. A laminar flow solver in FLUENT was employed and the velocity profile one meter downstream was recorded. The theoretical solution is well known Womersley solution which is given as:

$$\frac{u}{u_c} = (1-(\frac{r}{R})^2) + REAL\left\{\sum_{n=1}^{12} \frac{1}{2}(\frac{Q_n}{Q} F_{0,n} e^{i\omega_n t})\right\}, \text{where} \tag{1}$$

$$F_{0,n} = \frac{1 - \frac{J_0(i^{3/2}\alpha_n \frac{r}{R})}{J_0(i^{3/2}\alpha_n)}}{1 - \frac{2J_1(i^{3/2}\alpha_n)}{i^{3/2}\alpha_n J_0(i^{3/2}\alpha_n)}}$$

In equation (1), Q_n are the Fourier coefficients determined before. The computed velocity (using FLUENT) and the theoretical velocity (given by equation (1)) at various radial locations z ($z = 0$ being the centerline) of the tube at different time are compared in Figure 2; excellent agreement between the two is obtained. It should be noted that the graphs in Figure 2 also match perfectly with the experimental data obtained by Sean Peterson [7]. The Womersley solution was used to develop an inlet boundary condition. A transformation was needed due to a limit of 16 digits double precision within FLUENT. This Womersley inlet profile was used as a user defined function at the inlet of a tube with a symmetric stenosis. The uniform inlet region was 5D long, where D is the diameter of the tube. The stenosis was 2D long and the outlet tube was of uniform diameter 23D long. The equation for the stenosis geometry in both the experiment and computation is as follows:

$$S(x) = \frac{D}{2}(1 - s_0(1 + \cos(frac2 * \pi(x - x_0)L))) \tag{2}$$

The computations inside the 75% area reduction stenotic occlusion were performed using the FLUENT with inlet profile given in Figure 2. The centerline velocities were computed at various points downstream of the stenosis and are shown below in Figure 3. It should be noted that the computed values in Figure 3 are in excellent agreement with the experimental data. The three-dimensional plots of velocity profiles at select postions downstream of the stenosis specifically at a distance 2D and 4D from the beginning of the stenosis region are shown in Figure 4. Figure 4 shows the corresponding experimental velocity plots at a distance 2D and 4D downstream from the beginning of the stenosis [7]. Figures 4 and 5 are in reasonably good agreement; there are some small discrepancies in the reattachment region of the flow. These calculations were performed using the full Reynolds stress model (RSM) of turbulence.

For this axisymmetric case (Figure 6), pressure plots upstream and downstream of the stenosis are shown in Figures 6 and 7. Figures show the expected pressure drop downstream of the stenoses. The peak value of the pressure drop is an important measure of the area reduction of the occlusion and is used by the physician in assessing the severity of the stenosis.

3.2 Pulsatile Flow in Asymmetric Stenosed Phantoms

As mentioned before, Peterson and Plesniak [7] introduce a small geometric perturbation in the stenosed region of axisymmetric stenoses given by equation (2) and Figure 6. They study the effect of this geometric perturbation on the flow field. In this section, we simulate this effect. Figure 9 shows the geometry of the asymmetric stenosed phantom. It can be seen in from this figure that there is asymmetry in the throat region. It should be noted that the cross-section of the phantom in Figure 9 is symmetric about the z-axis and is asymmetric about the y-axis. 3D grid employed in Figure 9 has 76,572 cells and 236,257 faces. Computations are performed with the second-order accurate URANS solver with RSM in FLUENT. Three-dimensional simulations are compared at various stream-wise stations downstream of the stenosis

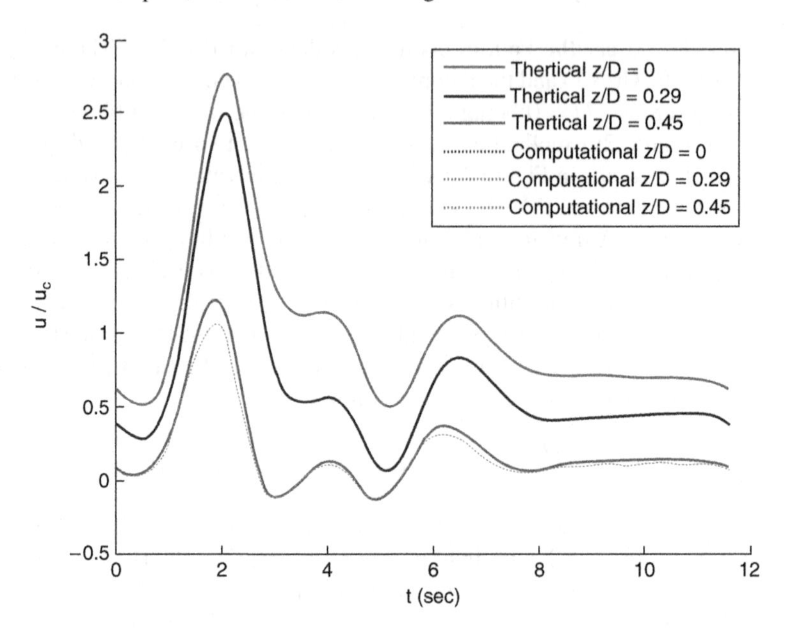

Fig. 2. Comparison of computed and theoretical Womersley solution for velocity at various radial locations.

along the symmetric zaxis and asymmetric y- axis of this eccentric stenosis against the experimental results of Peterson and Plesniak. Figures 10-13 compare the results of computations with experiments at stream-wise locations 2D and 4D downstream of the stenosis along both the symmetric z-axis and the asymmetric y-axis. CFD simulations shown in Figures 10-13 compare reasonably well with the experimental data although some disagreement exists for the results along the symmetric z-axis. The simulations overestimate the magnitude of the non-dimensional velocity u/uc; however, this discrepancy could be attributed to the R.M.S. or variance in the velocity measurements represented by the yellow and red colors in the experimental data. However, the three dominant peaks labeled as primary, secondary and tertiary are clearly present in both the computations and the experiments, and their corresponding locations (at t/T) are in good agreement.

Figures 14 and 15 present a direct comparison between the experimental and numerical jet centerline velocity in the eccentric stenosis at different locations downstream. A reasonably good agreement between the experimental and numerical jet centerline velocities can be observed for most locations downstream, especially the velocities at the primary, secondary and tertiary peaks occurring at $t/T \approx 0.2, 0.35 and 0.6$. Numerical simulations of the jet centerline velocity appear to have the same overall shape at each location downstream just like in the experimental results.

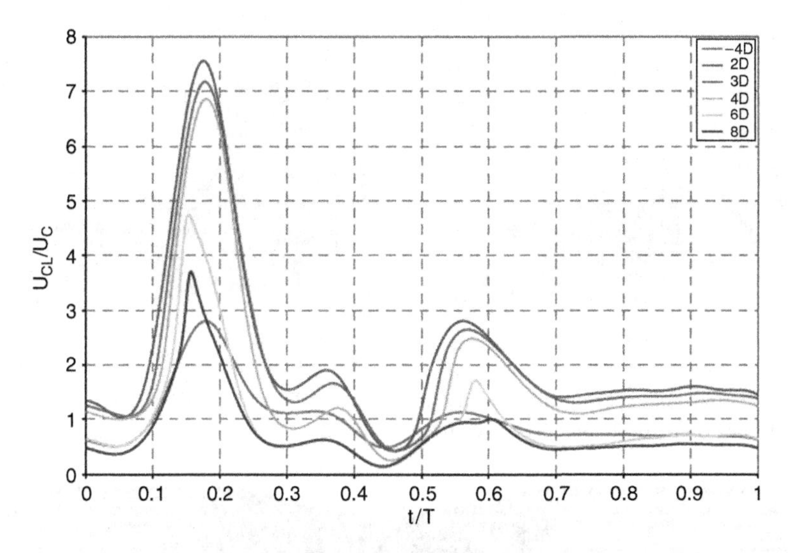

Fig. 3. Computed values of the centerline velocity at various distances downstream of the beginning of the stenosis given by equation (2).

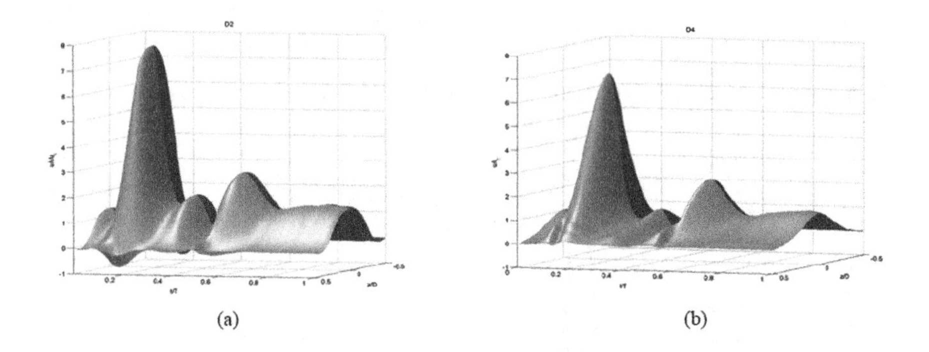

Fig. 4. 3D plots of computed velocity profiles at (a) 2D and (b) 4D distance downstream from the beginning of the stenosis given by equation (2).

Overall the computations and the experiments are in acceptable agreement given the uncertainty in the computations due to turbulence modeling as well as the uncertainty in the measurements.

4 Parallelisation

Computations were performed on a 126-processor SGI Origin 2000 parallel supercomputer. Only 16 processors were used in the calculations because the number of

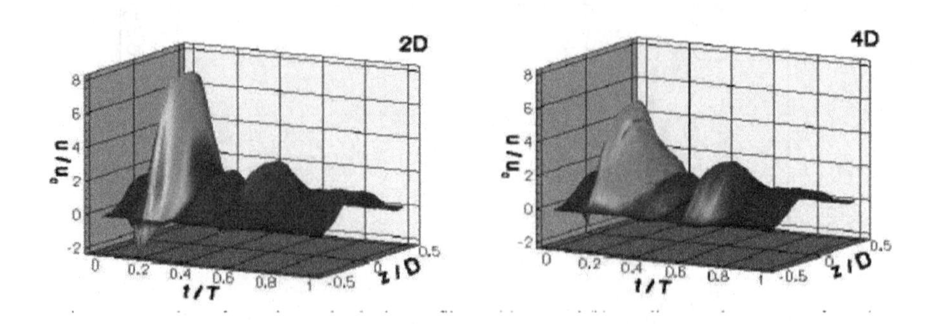

Fig. 5. 3D plots of experimental velocity profiles at (a) 2D and (b) 4D distance downstream from the beginning of the stenosis given by equation (2).

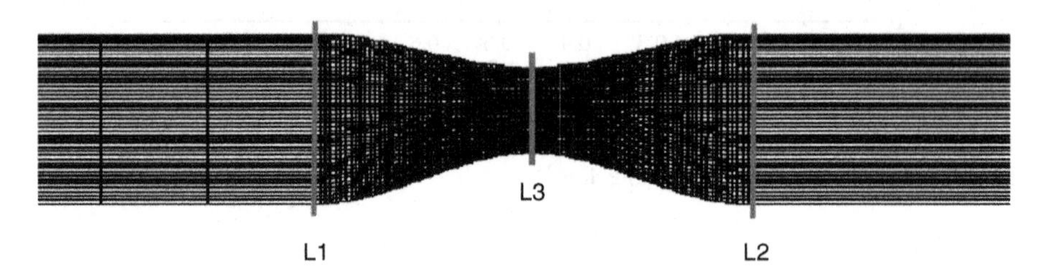

Fig. 6. Axisymmetric stenosis (Equation 2) with specific locations at which pressure was computed.

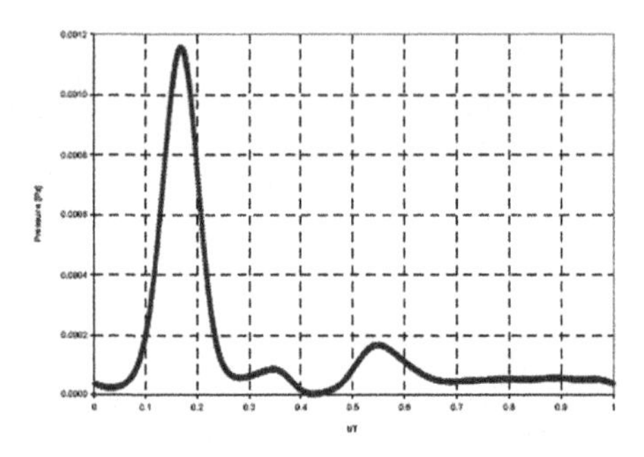

Fig. 7. Average pressure at location L1 upstream of the stenosis.

grid points (76,572) employed were relatively small. SGI Origin 2000 is a cache-coherent non-uniform access multiprocessor architecture. The memory is physically distributed among the nodes but is globally accessible to all processors through interconnection network. The distribution of memory among processors ensures that memory latency is achieved. Parallelization was achieved by MPI. For 3D calculations, parallel speedup efficiency of 94.6% was obtained.

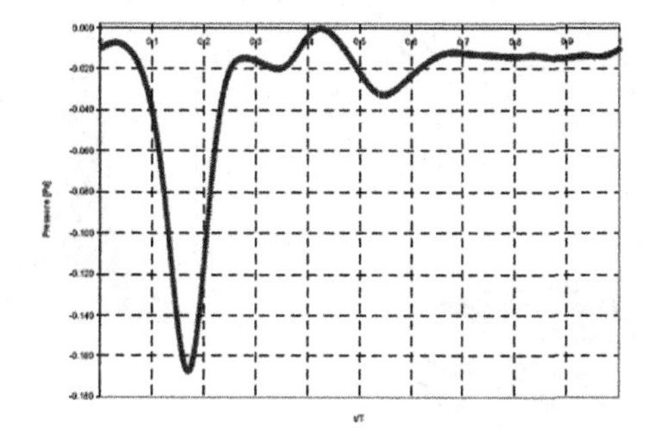

Fig. 8. Average pressure at location L2 downstream of the stenosis.

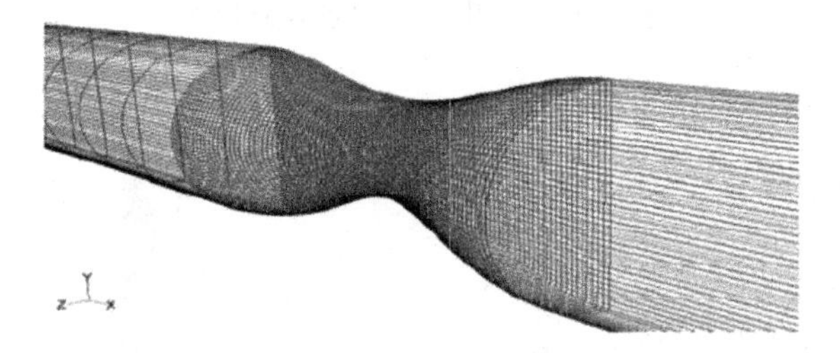

Fig. 9. Geometry of the stenosed asymmetric phantom [7].

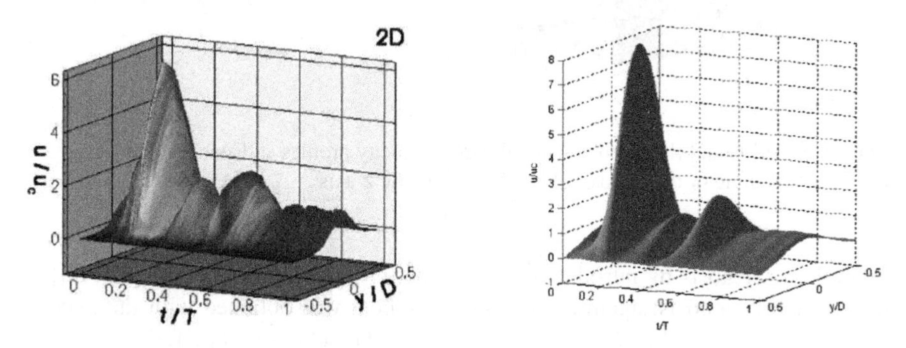

Fig. 10. Experimental (left) and numerical (right) velocity profiles at distance 2D downstream from the beginning of the stenosis along the symmetric z-axis.

5 Conclusions

Computations were performed for pulsatile flow in an axisymmmetric 75% area reduction occlusion with the actual physiological waveform of the heart at the inlet. Good agreement was obtained with the experimental data of Peterson and Plesniak [7] for all the details of the flow field. Computations were also performed for an

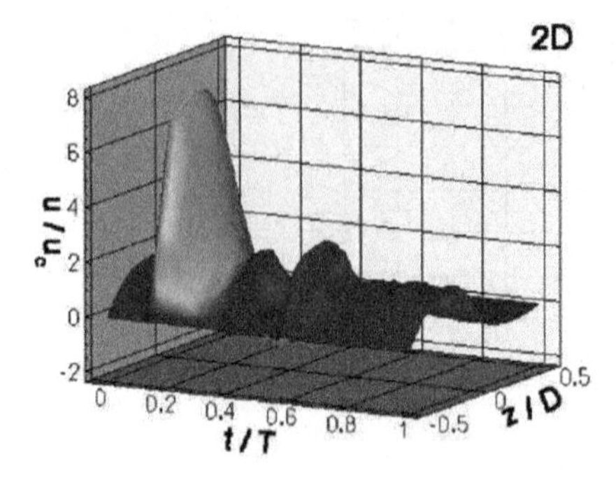

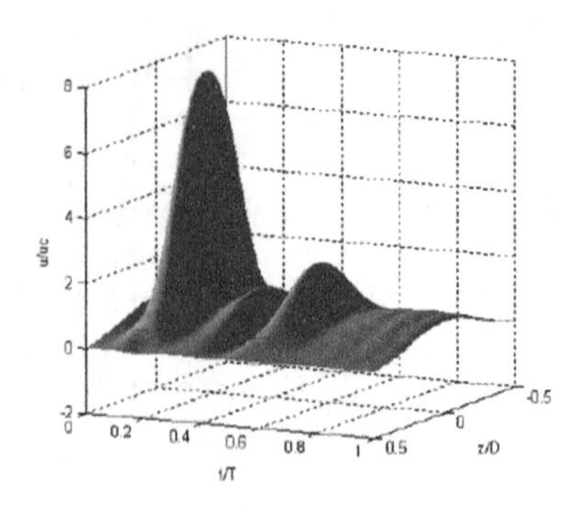

Fig. 11. Experimental (left) and numerical (right) velocity profiles at distance 2D downstream from the beginning of the stenosis along the asymmetric y- axis.

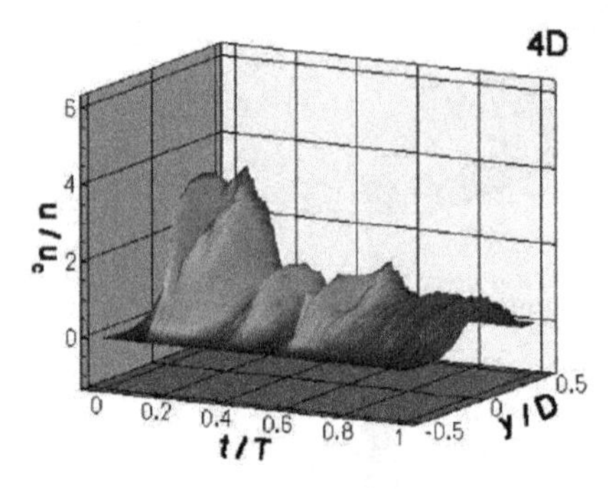

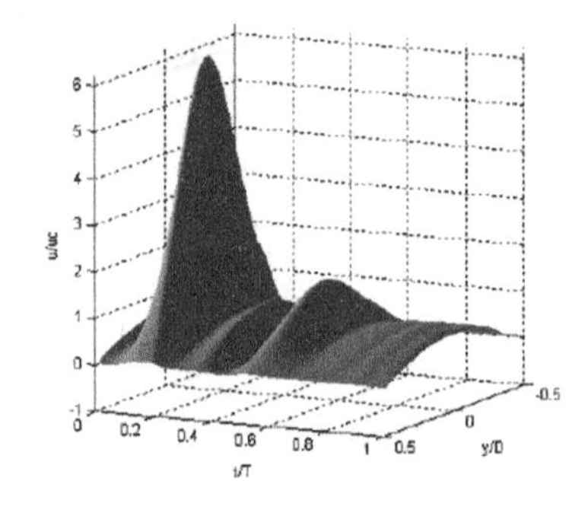

Fig. 12. Experimental (left) and numerical (right) velocity profiles at distance 4D downstream from the beginning of the stenosis along the symmetric z-axis .

asymmetric stenosis by introducing a small perturbation in the throat region of the axisymmetric stenosis; again reasonable agreement was obtained with the experiments of Peterson and Plesniak [7]. It is concluded that the small perturbation in the stenosed geometry does not cause significant change in the velocity downstream and therefore the pressure drop. The small geometric perturbation is likely to have larger effect for severely stenosed vessel (e.g. with 90% area reduction) or the geometric perturbation has to be large enough which will then reduce the area of the throat significantly.

[1] Moghaddam, A.N., Behrens, G., Fatouraee, N., Agarwal, R.K., Choi, E.T., and Amini, A.A., Factors Affecting the Accuracy of Pressure Measurements in Vascular Stenoses from Phase-Contrast MRI, MRM journal, Vol. 52, No. 2, pp. 300-309, 2004.

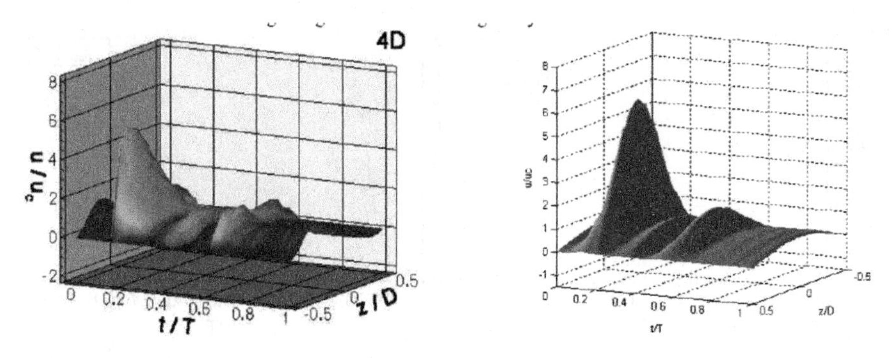

Fig. 13. Experimental (left) and numerical (right) velocity profiles at distance 4D downstream from the beginning of the stenosis along the asymmetric y-axis.

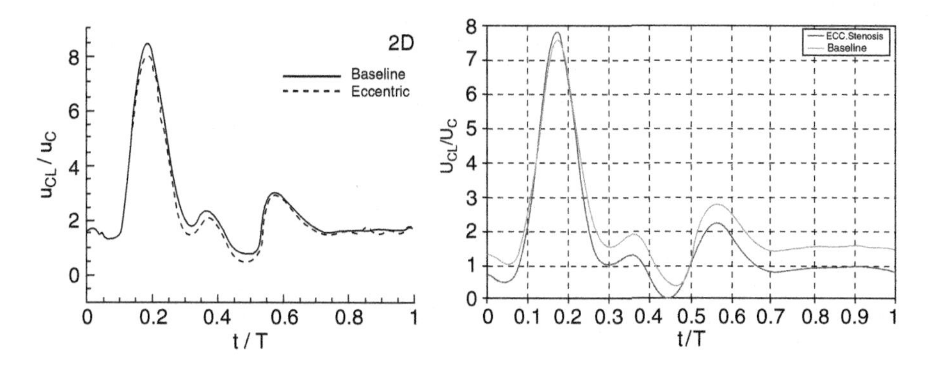

Fig. 14. Experimental (left) and numerical simulation (right) comparison of the jet centerline velocity in the eccentric stenosis at 2D downstream; the baseline case is for the axisymmetric stenosis.

[2] Ku, D.N., Blood Flow in Arteries, Annu. Rev. Fluid Mech., Vol. 29, pp. 399-434, 1997.

[3] Berger, S.A. and Jou, L-D., Flows in Stenotic Vessels, Annu. Rev. Fluid. Mech., Vol. 32, pp.347-382, 2000.

[4] Young, D.F., Fluid Mechanics of Arterial Stenoses, J. of Biomed. Eng., Vol. 101, pp. 157-175, 1979.

[5] Kim, C.S., Kiris, C., Kwak, D., and David, T., Numerical Models of Human Circulatory System under Altered Gravity: Brain Circulation, AIAA Paper 2004-1092, 42nd AIAA Aerospace Science Meeting, Reno, NV, 5-8 January 2004.

[6] Okpara, E. and Agarwal, R. K., Numerical Simulation of Steady and Pulsatile Flow Through Models of Vascular and Aortic Valve Stenoses, AIAA Paper 2007-4342, AIAA Fluid Dynamics Conference, Miami, FL, 25-28 June 2007.

[7] Peterson, S., On the Effect of Perturbations on Idealized Flow in Model Stenotic Arteries, Ph.D. Thesis (supervisor: M. Plesniak), Purdue University, December 2006.

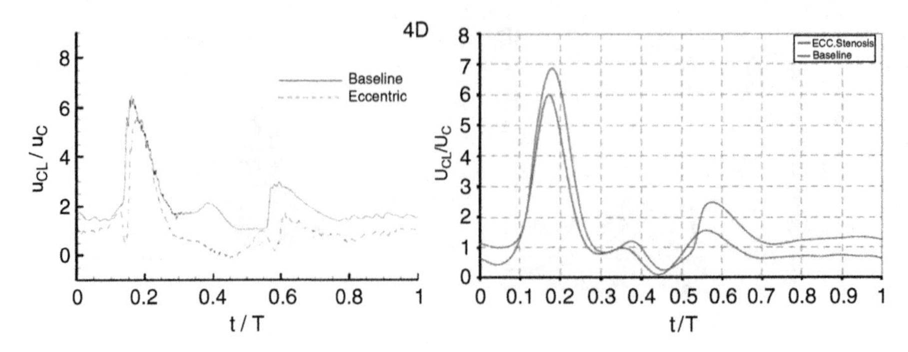

Fig. 15. Experimental (left) and numerical simulation (right) comparison of the jet centerline velocity in the eccentric stenosis at 4D downstream; the baseline case is for the axisymmetric stenosis.

Fluid Flow - Agent Based Hybrid Model for the Simulation of Virtual Prairies

Marc Garbey[1], Cendrine Mony[2], and Malek Smaoui[1]

[1] Department of Computer Science, University of Houston, Houston, Texas
garbey@cs.uh.edu, msmaoui@cs.uh.edu
[2] UMR CNRS 6553 ECOBIO, University of Rennes 1, Rennes, France
cendrine.mony@univ-rennes1.fr

1 Problem Background

For a long time, natural prairial ecosystems have been considered to be supports for primary production of agricultural needs. Therefore, research focused on the evaluation of prairies' productivity and the effect of management on their agronomical values.

Starting from the 50's, other roles of prairies in ecosystems functioning have been detected, while prairies were degraded to be converted into croplands (through erosion, eutrophication, and biodiversity loss) [8]. International policies defined new goals for prairial ecosystems to serve new ecological functions. They can be used to face the two major environmental challenges for the following decades which are (1) the availability of fresh, unpolluted irrigation water, (2) the regulation of carbon emission.
Consequently, we tend now to create new natural systems, as surrogates of the degraded ones, to provide these ecological services. Recent works have demonstrated, for example, the capacity/capability of natural prairies to provide alternative biofuels (carbon negative biofuels) [7], or their role in carbon storage [6].

These new systems are elaborated by sowing mixed-species seeds. Then, questions are raised on the temporal evolution of these plants with different lifestrategies and constantly interacting with one another. Proposing precise design of these systems need to take into account all these complex interactions. The urgent need for short term responses makes it impossible to respond to this sociological demand through the only classical experimental approach which may necessitate long-term surveys.

Consequently, our goal is to use computational modelling to achieve realistic virtual experimentations at a broader scale, a cheapest cost and within a shortest time. Parallel computing is a key tool because these simulation are still compute-intensive. Indeed, such a model usually has a very large parameter space and it is almost impossible to know apriori which properties will emerge from such complex system.

D. Tromeur-Dervout (eds.), *Parallel Computational Fluid Dynamics 2008*,
Lecture Notes in Computational Science and Engineering 74,
DOI: 10.1007/978-3-642-14438-7_39,

We introduce in this paper an hybrid model of a prairial system (modelled through multi-agent IBM), coupled with the environment (modelled through PDEs), taking into account biological complexity of ecological ecosystems. Such model may provide an original and efficient tool for testing ecological hypotheses on realistic complex systems. In the next section we are going to detail our hybrid model.

2 Hybrid Model

Our hybrid model schematic is given in Fig. 1.

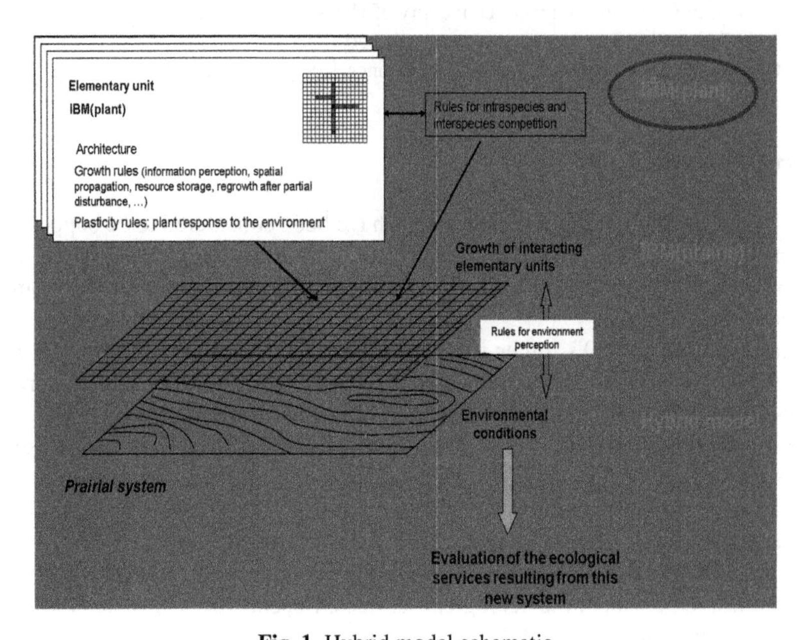

Fig. 1. Hybrid model schematic

First, we need to describe the representation and processing of individual clonal plants. So, we use an individual based model (IBM) [4, 3] which is a set of parameterized rules representing the metabolism process (carbohydrate production, ramet and connection costs of production), the plant architecture (ramification, elongation processes) and its resource strategy (resource sharing and storage).
The plant is represented by a set of cells or modules. Each module is either a ramet module or a connection module (c.f. Fig. 2).

At t_0, the plant has only one initial ramet module (considered also as a ramet module). Plant spatial colonization is directly linked with branching and elongating patterns. We have chosen to work with an hexagonal grid to facilitate the control on the plant growth. There are only six equivalent directions of potential growth on an hexagonal grid. The creation of a new cell is made either by elongating an existing

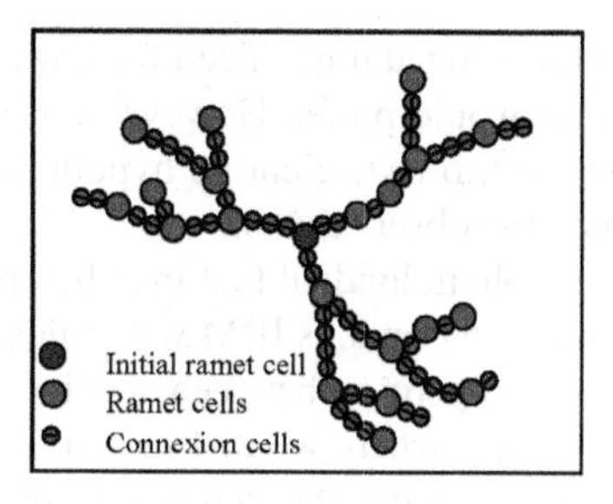

Fig. 2. Clonal plant model

connection or by branching at an existing ramet node in order to initiate another connection. Thus, at each time step, the IBM decides either to add a connection module (elongation), a ramet module (ramification or branching) or no module at all, according to several factors:

- The order of the branch (primary, secondary, tertiary).
- The length of the branch.
- The available resources.

The placement of the new module is chosen according to whether it is an elongation process or a branching process. For elongation process, the location of the new module follows the direction of the branch. For the ramification process, the location is a random choice between the different possible directions. The simulation is done on hundred time steps which corresponds roughly to one season. At a given time t_1 and before adding a new module, all the ramet modules found on the grid accumulate biomass by photosynthesis.
All the rules of the model obey parameterized probability distributions. This set of rules is based on the literature [3, 4], but also on a set of controlled experiments that we have done. Fig. 3 shows the experimental set up.

Fig. 3. Experiment companion by the image segmentation of the clone

Experiments with various species and environment were run and the structure of the clone was extracted assisted by image analysis. This type of experiments is however very tedious and time consuming. Therefore, we cannot afford to have as many individual plants as we would prefer. However, our modeling, guided by this set of experiments, can be used to test scientific hypothesis qualitatively and/or do reverse engineering to match the observations.

Prairies are made of multiple individual that may belong to different species. In a second step, we supplement the previous IBM with rules representing the interaction between plants and the competition between species. In fact, we consider this time a significantly larger surface where we distribute randomly a given number of initial modules (hundreds to thousands). At each prairie-scale time step, we grow the plants one by one using the plant-scale time steps. The order followed for growing the plants is determined randomly at every prairie-scale time step. Additional IBM rules exclude for example the superposition of ramets of different plants, though connexions can overlap. Fig. 4 is an example of the graphic output of the application for a prairie with one hundred individuals.

Additionally, the clonal plants feed on soil resources. Then, our hybrid model describes the coupling between the individual plants growth and ground flow of plant nutrient. We are interested in analyzing the effect of spatio-temporal variations in resource distribution on plant growth and space colonization. The modifications of nitrate concentration in the soil will be taken as an example. Nitrate is a key element determining plant production. Simulating the complex feedbacks between the soil (nitrate concentration) and the prairie may play a key role in the understanding of the impact of a prairie on water purification. The nitrate transport in the soil is simulated by flow models in porous media in the form of PDEs.

We start with a Darcy Law: $u = \eta \nabla p$, which drives the transport of a chemical solute:

$$\frac{\partial S}{\partial t} = \varepsilon \Delta S - u.\nabla S - \sigma \delta_i(C) S, \tag{1}$$

The last term of the rhs is for the consumption of the solute at each ramet location. Inversely, the energy intake per ramet is a function of the local value of the solute concentration. We have therefore a non linear feedback mechanism between the prairie dynamic and the transport of solute. The IBM is stochastic and there is no simple prediction on the outcome of this hybrid model.

Our ongoing research is then to (i) find realistic interactive rules, (ii) analyse the possibility for constructing approximate models in the limit of large population, (iii) test the effect of the initial conditions (density of plants, profiles of individuals, number of different profiles sown) on the nitrate dynamic.

Fig. 4 and 5 show the effect of the non linear coupling between the transport of Solute in the ground and the virtual prairie after one hundred time steps starting at $t = 0$ from a slick of solute. A small fraction of pollutant leaves the field through the upper half of the right boundary and most of it is absorbed by the plants (red: initial concentration, dark blue: no nitrates).

More details on this application will be published in a companion paper [5](in preparation). Now, let's us discuss the implications in term of computation cost of the deployment of our numerical experiments.

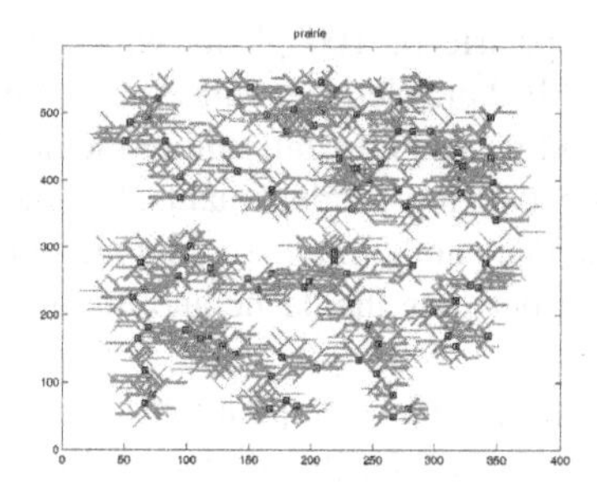

Fig. 4. Darcy Flow

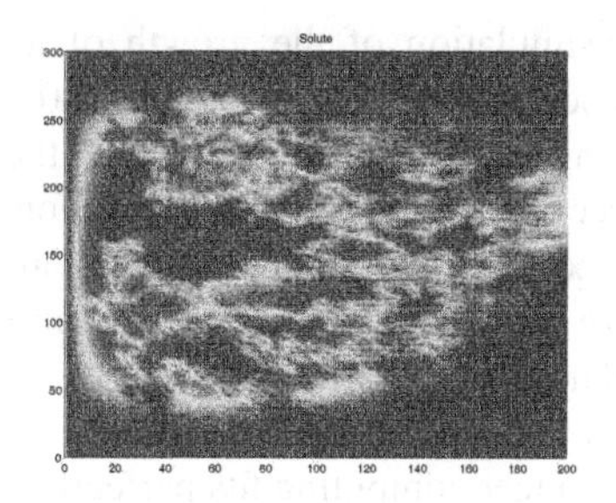

Fig. 5. Concentration of the Solute

3 Computation Needs for Virtual Prairie

As previously mentioned , the IBM obeys a set of parameterized rules. The dimension of the parameter space for one individual plant can reach 16 and even more. This dimension depends on the focus of the ecological study. We need to perform two tasks:

• *Task 1* Get a parameter space response i.e. browse the parameter space with selected values for each parameter and get the simulation results for these parameter combinations.

• *Task 2* Perform an optimization process using for example a genetic algorithm or a particle swarm algorithm.

Both steps are necessary and in that order, since the ecologist (end user) does not know necessarily in advance what emergent properties he is looking for.

Added to that, the IBM is by nature a stochastic algorithm. It relies on random number generation/selection, thus, one simulation of the plant growth is far from being significant. We need to repeat this simulation many times and calculate the mean and the standard deviation of the desired outputs (total biomass accumulated, number of ramet modules, number of connection modules, length of branch ...) obtained at each simulation. Actually, part of testing the application was verifying that the mean and standard deviation on the output values tend to a fixed value as the number of stochastic simulations increases.

The convergence of this Monte Carlo method is very slow. We decided to simulate the growth of a plant with 1000 runs for each parameter setting. Nevertheless, a 100x100 grid simulation with enough time steps to represent a little bit less than a season is still very fast. It takes on an average 45 seconds on an Intel Pentium 1.7GHz with 1GB of RAM and a Windows XP platform . Although this metric depends enormously on the combination of parameter values input of the model, it proved to be a good estimation when we run the computations on the volunteered computers using BOINC [1]. Browsing a parameter space of dimension 16 with only 3 values per parameter takes about 14 million simulations. These simulations will take around 20 years on a single PC. With a 72-nodes dedicated cluster, it may take more than 3 months.

A simulation of the growth of a mono-species prairie of 100 plants with 1000 (stochastic) running in a 600x700 grid and during 100 prairie-scale time steps takes significantly more time. In fact, the application induces significantly more memory accesses to manage the competition for resources between individual plants. Besides, since the size of the problem has also increased, it does not fit anymore in a medium size cache. So, this induces much more swapping between the cache and the main memory.
Overall, we have a huge number of embarrassingly parallel computation jobs. So, volunteer computing fits perfectly this situation and with the computing and storage potential of this resource, we can expect to achieve the computations in a reasonable time. We are going now to describe this technique.

4 Virtual Prairie with BOINC

Volunteer computing is an arrangement in which people (volunteers) provide computing resources to projects, which use the resources to do distributed computing and/or storage.

- Volunteers are typically members of the general public who own Internet-connected PCs. Organizations such as schools and businesses may also volunteer the use of their computers.
- Projects are typically academic (university-based) and do scientific research.

Several aspects of the project-volunteer relationship are worth noting.

- Volunteers are effectively anonymous; although they may be required to register and supply email address or other information, there is no way for a project to link them to a real-world identity.
- Because of their anonymity, volunteers are not accountable to projects.

BOINC [1] is a middleware for volunteer computing developed by the team of Dr. David Anderson at the Berkeley Space Science Lab. Nowadays, there are at least 50 volunteer computing projects in different research fields. All the projects are independently operated, see http://boinc.berkeley.edu/.
BOINC has a server/client architecture. The volunteers install the client on their computers and decide to volunteer their resources to the projects of their choice. The client will then download jobs from the different projects, compute them and report the results back.
The server side hosts a web interface, a data base and a set of functionalities for the distribution of jobs and aggregation of results. The web interface allows the users to get information about the science involved in the project, consult the work status of the project, communicate with each other and with the project administrators via a message board, consult and edit their accounts and profiles, etc.
To handle the work distribution and aggregation, BOINC uses abstractions which are mainly "workunits" and "results". A workunit is a reference to an application and an input file. A result is a reference to a workunit and an output file. Since the resources (volunteered computers) are not necessarily reliable, BOINC uses redundant calculation to ensure the correctness of the results obtained for each workunit. So, for each

workunit, a predefined number of results are sent to different users and then their output files are compared and validated if they agree.
A set of daemons running on the server interacting with a data base work together to ensure all these mechanisms. A feeder is responsible for creating the needed results for each workunit according to its status. A scheduler is responsible for dispatching these jobs (results) to the clients [2]. A "validator" is responsible for comparing the results for every workunit and validating or invalidating them. An "assimilator" is responsible for the post-processing of the validated results on the fly. These demons "communicate" with each other by updating the status of the workunits and the results in the database.

We have used BOINC first to browse the parameter space of the IBM of one individual plant. We ran 22 10^6 parameter combinations in a fairly systematic way to test our model. At this point we have no apriori knowledge on what is the best fitness function, and we are interested in optimum solution as well as failures of our model. This campaign of experiments did growth 22 10^9 individual clonal plants and took 3 weeks with the equivalent of 800 full time individual PCs of our volunteer community. The lessons we learned from data mining this large scale numerical experiment, using clustering technic among other things, are the following. First it seems that there is no trade-offs between plant traits but there can be some positive correlations. Second the properties involved in clonal plant fitness depend on the output parameter considered, such as biomass, number of ramet modules, total length of stolon etc... Third, there are several combinations of traits that promote plant performance. In other words, there is no single strategy but a few different "optimum" strategies. Based on these preliminary results, we are now studying the optimum solutions that emerge from our hybrid complex model of virtual prairies.

5 Conclusion

We have presented here the first green project of BOINC, and shown the potential of this approach to analyze a multi-scale complex adaptive ecosystem. The social aspect of volunteer computing certainly contributes to the awareness of the public on the cyber-world. We have paid particular attention to the design of a web site for this BOINC project - see http://vcsc.cs.uh.edu/virtual-prairie/- to recruit our volunteers. The key challenging problem remains however the careful validation of the model. This is a process that traditionally takes time and manpower indeed, since we depend on seasons. We are looking therefore at new innovative ways of accumulating more experimental data.

Thanks: We would like to thank Dr. David Anderson for his advice and support for the project and Marie Lise Benot for providing data on experimentations describing spacial plant growth. We would like also to acknowledge CNRS & Cemagref for partly financing the project.

[1] David P Anderson. Boinc: A system for public-resource computing and storage. In *Proceedings of the 5th IEEE/ACM International Workshop on Grid Computing*, pages 1–7, Pittsburgh, PA, 2004.
[2] David P Anderson and Eric Korpela. High-performance task distribution for volunteer computing. In *Proceedings of 1st IEEE International Conference on e-Science and Grid Computing*, pages 196–203, Melbourne, Australia, 2005.
[3] Uta Bergera, Cyril Pioua, Katja Schiffersb, and Volker Grimm. Competition among plants: Concepts, individual-based modellingling approaches, and a proposal for a future research strategy. to appear in International Journal on Perspectives in Plant Ecology, Evolution and Systematics.
[4] Volker Grimm and Steven F Railback. *Individual-based Modelling and Ecology*. Princeton University Press, Princeton, NJ, 2005.
[5] Cendrine Mony, Marc Garbey, Malek Smaoui, and Marie Lise Benot. Clonal growth in optimal conditions: an IBM approach. in preparation, companion paper.
[6] T. J. Purakayastha, D. R. Hugginsb, and J. L. Smithb. Carbon sequestration in native prairie, perennial grass, no-till and cultivated palouse silt loam. *Soil Science Society of America Journal*, 72(2):534–540, February 2008.
[7] David Tilman, Jason Hill, and Clarence Lehman. Carbon-negative biofuels from low-input high-diversity grassland biomass. *Science*, 314(5805):1598–1600, December 2006.
[8] Peter M. Vitousek, John Aber, Robert W. Howarth, Gene E. Likens, Pamela A. Matson, David W. Schindler, William H. Schlesinger, and David G. Tilman. Human alteration of the global nitrogen cycle: sources and consequences. *Ecological Applications*, 7(3):737–750, August 1997.

HPC for hydraulics and industrial environmental flow simulations

Réza Issa[1 2], Fabien Decung[1], Emile Razafindrakoto[1], Eun-Sug Lee[2], Charles Moulinec[3], David Latino[4], Damien Violeau[1 2], Olivier Boiteau[5]

[1] EDF R&D, National Hydraulics and Environment Laboratory, 6, quai Watier, 78400, Chatou, France
[2] Saint-Venant Laboratory for Hydraulics (Paris-Est University, Joint research unit EDF R&D CETMEF ENPC), 6, quai Watier, 78400, Chatou, France
[3] Daresbury Laboratory, Computational Engineering Group, Daresbury, Warrington, WA4 4AD, United Kingdom
[4] IBM Systems and Technology Group, IBM Global Engineering Solutions, IBM Middle East, Dubai, P.O. Box 27242, Dubai, United Arab Emirates
[5] EDF R&D, SINETICS, 1, avenue du Gnral de Gaulle,92141, Clamart, France

This paper describes the parallelization needs required by the TELEMAC system, a complete hydroinformatics tool developed by EDF R&D. A focus on three codes part of the TELEMAC system, namely Estel-3D, Telemac-3D and Spartacus-3D, is achieved. For each code, performance in terms of speed up is presented, as well as an industrial application.

Keywords: Groundwater flows Free surface flows Finite elements SPH HPC Spillway

1 HPC through the TELEMAC system

The TELEMAC hydroinformatics system has been developed since 1987 at EDF R&D [1]. It addresses free surface and groundwater flows and is composed of several numerical codes based on the finite element technique except Spartacus-3D code which relies on the SPH formalism. Each code aims at modelling a specific physical phenomenon. Wave effect as well as sediment transport or water quality can be modelled in the context of hydraulics or environmental flows. In this framework, three codes of the Telemac system, namely Estel-3D, Telemac-3D and Spartacus-3D have been selected for HPC due to their respective computational time requirements.

1.1 Historic of HPC in the TELEMAC system

HPC in the TELEMAC system began with the emergence of vector computers, hence inducing many changes in algorithms. Because of possible backward dependence, an

D. Tromeur-Dervout (eds.), *Parallel Computational Fluid Dynamics 2008*,
Lecture Notes in Computational Science and Engineering 74,
DOI: 10.1007/978-3-642-14438-7_40,

elementary vector assembling was not automatically vectorized. Renumbering operation allowed us to force vectorization by a compiler instruction. This first performance development was successfully tested on several vector architectures such as a Cray YMP and Fujitsu computers with respectively 64 and 1024 vector register lengths [1]. However, even with improved algorithms, this kind of parallelization technique suffers from intrinsic limitations. The idea of extracting most parallelism from the codes has been followed in the development by a first implementation of algebraic partitioning based on PVM. This release was tested with success on an ORIGIN 2400 with distributed memory [1]. As MPI became an international standard, PVM was rapidly replaced by a collection of basic MPI routines.

1.2 TELEMAC, a SPMD hydroinformatics system

According to the Flynn classification, the parallelization technique used for the TELEMAC system corresponds to the SPMD (Single Program Multiple Data) technique. For the sake of simplicity and adaptability to the different physics solved in the system, algebraic parallelization has been preferred to the parallelization based on domain decomposition methods. In the case of our implicit algorithms, which led to solve a linear system, specifications for the parallel structures were:

1. Partitioning of the domain and definition of the parallelism structure;
2. Development of a parallel wrapper able to process communications between processors and updating values;
3. Transformation of scalar to scalable algorithms such as vector assembling, dot product

Obviously, the way of partitioning the mesh strongly influences the above mentioned points. According to our finite element method, partitioning is done without overlapping, edge to edge, meaning that elements belong to a unique sub-domain. Then, nodes are shared at the interface between sub-domains. 2D (triangles) and 3D (tetrahedrons) mesh decomposers are both using METIS decomposer [7] for achieving load balancing. Partition of prisms is easily performed by extracting from a 2D triangle partition. As this is done as a - scalar - pre-processing, level of parallelism is very high in the TELEMAC system since MPI runs at the beginning of the simulation.

1.3 Structure of parallelism in the TELEMAC system

A parallel structure is necessary for mapping the message-passing scheme, where most difficulties lie. As dot product is easily computed by a sequence of a global sum reduction and division of the interfacial node value by its multiplicity (meaning the number of sub-domains it belongs to), the number of nodal values has to be exchanged by point-to-point communication in the case of matrix-vector product. Point to point communication needs special mapping of interface nodes provided by a specific parallel mesh structure. So, extra-arrays, storing initial global number of nodes and mapping of nodes belonging to different sub-domains, are provided when partitioning.

1.4 Point to point communication and data exchange

At first, blocking point-to-point communications have been implemented for value updating. This operation is quite complex to achieve but the algorithm, though complex, does not occur any deadlock. The transmission is decomposed into a sequence of 4 tasks, depending on the rank of the processors, as described in figure 1:

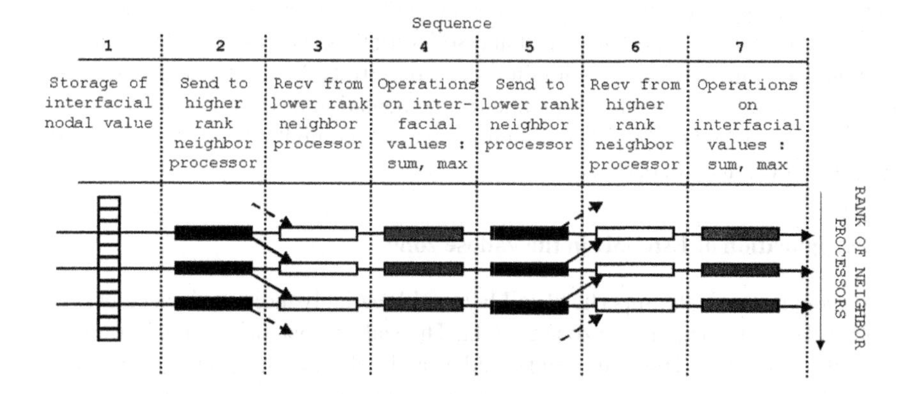

Fig. 1. Point to point communication and data exchange

So, the communication scheme can be summarize as follows:

- Transmission of the data to the higher rank processor;
- Reception by the higher rank processor;
- Transmission of the data to the lower rank processor;
- Reception by the lower rank processor.

Each processor separately performs the operation after its own reception. Since associative addition is not valid in parallel computing, the summing task may compute value with a slight difference on digits on nodes belonging to neighbor processors. This algorithm is being optimized and simplified by using non-blocking communications.

2 HPC through soil pollution with Estel-3D

2.1 Estel-3D code

Estel-3D is developed within the TELEMAC system and is applied to the safety of waste nuclear repository projects in France. It solves the following Richards equation with finite element method:

$$\frac{\partial \theta}{\partial t} = \underline{\nabla} \cdot k_r \underline{\underline{K}}_S \cdot \underline{\nabla}(h+z) + S \tag{1}$$

where h is the pressure head and θ is the saturated moisture content. z denotes the vertical distance from the soil surface downward and $\underline{\underline{K}}_S$ the saturated hydraulic conductivity. k_r is the relative permeability of water to air in the unsaturated regime and S a volumetric source/sink term. Flow velocities are determined by differentiation using the hydraulic head, implying that the spatial accuracy is reduced by one order of magnitude. Obviously, accuracy of velocities is crucial for solving advection-dispersion equation within a safety study. The discretization of the flow domain is obtained by using tetrahedral finite elements and linear and piecewise shape functions, respectively for pressure head and soil variables such as moisture content or hydraulic conductivity. Integration in time is performed by a forward Picard finite-difference scheme. The current parallel implementation based on domain partition has been developed for distributed-memory parallel computer [4]. It is based on the technique described in part 1.

2.2 Application of Estel-3D in the vadose zone

The vadose zone is the superficial soil layer with variably moisture located between land surface and the phreatic zone content. The vadose zone plays an important role in refilling aquifers. However, safety studies and risk assessment are generally performed only with inclusion of the aquifers although contaminant migration goes from the top soil through the vadose zone. The relationships $k_r = k_r(h)$ and $\theta = \theta(h)$ strongly depend on the type of porous medium and some experiments are necessary for calibration. The following test case is taken from [2] and is used to validate the parallel non-linear scheme. A soil column is composed of homogeneous sand and parameterized by the following constitutive relationships, where the parameters are defined in table 1: Although this case is a one-dimensional infiltration, the column

$k_r(h) = \frac{A}{A+\lvert h\rvert^{\lambda}}$			$\theta(h) = \theta_r + \frac{\alpha(\theta_s-\theta_r)}{\alpha+\lvert h\rvert^{\beta}}$				
Soil parameter	A	α	β	y	θ_s	θ_r	$\underline{\underline{K}}_S$
value	1.17510^{-6}	1.61110^{-6}	3.96	4.74	0.287	0.075	9.4410^{-3}cm/s

Table 1. Constitutive relationships and soil parameters.

size is about $10 \times 10 \times 40$ cm. The column infiltration problem is chosen such that pressure at the bottom is -61.5 cm and -20.7 cm at the top. The initial pressure profile is specified as constant and equal to -61.5 cm. The mesh size is about 11149 nodes and 60245 tetrahedrons, meaning that there is less than 200 nodes along the depth. This test case ran on a IBM BlueGene/L with 500 Mb memory per cores. The continuous line of figure 2 (left) represents the solution calculated by a mass-conservative scheme [3] and $\Delta t = 0.5$s while circles represent analytical values taken from [2]. The speed-up performance, represented on figure 2 (right), is satisfactory for this number of nodes.

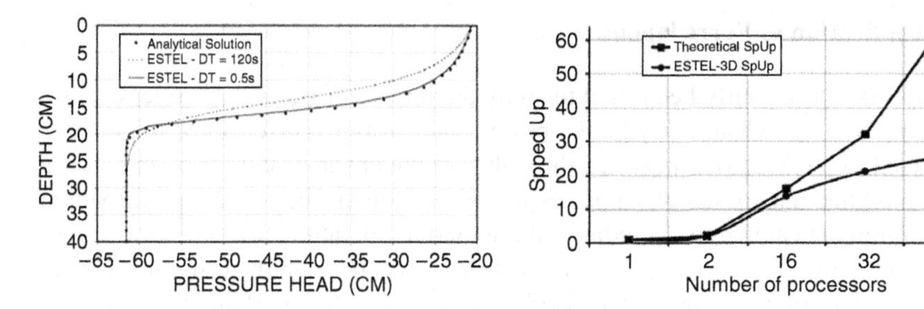

Fig. 2. Comparisons of analytical and numerical profiles at time $T = 360$s for the infiltration column problem: vertical profile of pressure head h in function of depth with different time step (Left). Estel-3D speed-up.

3 HPC for real environmental applications with Telemac-3D

3.1 Telemac-3D code

Telemac-3D is developed for the modeling of complex 3D free surface flows. Many physical phenomena can be taken into account such as turbulence, effect of vertical density resulting from temperature and/or salinity fluctuations, effect of wind surface, Applications involving transport of chemical and radiological species, as well as evolution of thermal plumes and/or salinity stratification are achieved with Telemac-3D. Considering a 3D domain Ω, the following system for a non-hydrostatic free surface model are solved:

$$\frac{d\rho u}{dt} - \underline{\nabla}\cdot(\mu_h \nabla u) - \frac{\partial}{\partial z}(\mu_z \frac{\partial u}{\partial z}) + \nabla(\rho g h) + \nabla \pi = \rho f_u - \nabla(\rho g Z_F) - \nabla p_{atm}, \quad (2)$$

$$\frac{d\rho w}{dt} - \underline{\nabla}\cdot(\mu_h \nabla w) - \frac{\partial}{\partial z}(\mu_z \frac{\partial w}{\partial z}) + \frac{\partial \pi}{\partial z} = \rho f_w, \quad (3)$$

$$\underline{\nabla}\cdot(\rho u) - \frac{\partial \rho w}{\partial z}) = 0, \quad (4)$$

$$\frac{d\rho S_i}{dt} - \underline{\nabla}\cdot(k_h \nabla S_i) - \frac{\partial}{\partial z}(k_z \frac{\partial S_i}{\partial z}) = \rho f_{S_i}, \quad (5)$$

$$\rho = \rho(S_1, \ldots, S_i, \ldots). \quad (6)$$

On 2D horizontal projected domain ω:

$$\frac{\partial \rho h}{\partial t} + \underline{\nabla}\cdot[\int_{Z_F}^{Z_S} \rho u dz] = 0 \quad (7)$$

where u, w(respectively $\mu_h z, \mu_z, k_h, k_z$) are the horizontal and vertical fluid velocity (resp. total dynamic velocity and scalar viscosities, included eddy turbulent viscosities), S_i scalar variables (as salinity or temperature,), f_u, f_w, f_{S_i} source terms, g the gravity, Z_s, Z_f the z levels of free surface and the bottom, $h = Z_f - Z_s$ the water column height, π the dynamic pressure and P_{atm} the atmospheric pressure. The system is completed with appropriate boundaries and initial conditions.

3.2 Application to Berre lagoon

Telemac-3D has recently been used to study the impact of fresh water release coming from the Saint-Chamas hydraulic power plant in Berre lake. Since this lake is connected to the Mediterranean sea through Caronte channel, salinity of the lake is hence modified due to Saint-Chamas exploitation, inducing ecosystem perturbation. A measurement campaign, including salinity and temperature, and a programme of CFD modelling were performed. Velocities and fluxes through the Caronte Channel have been carefully measured at measurement stations SA1, SA2 and SA3 (see figure 3) to assess the salt and water exchanges with the Mediterranean Sea. Wind, air temperature and atmospheric pressure are recorded by a meteorological station. A Telemac-3D numerical model has been constructed: 12 000 000 prisms are used for spatial dicretization of the entire 3D domain. For temporal discretization, θ-semi-implicit scheme is used. Thanks to the large datasets of measurements, the results could be duly validated and show a very complex hydrodynamics, depending on tide, density gradients due to salinity and mistral, the dominant local wind (see figure 4). The speed up performance relative to IBM BlueGene/P when considering 12 million nodes is satisfactory, as displayed on figure 5. 1 hour computing was needed on 8000 processors IBM BlueGene/P in order to simulate 16 hours of physical time.

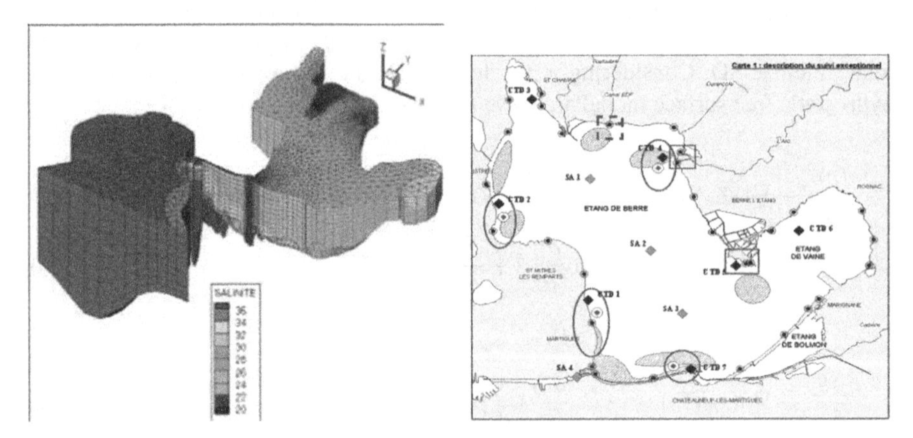

Fig. 3. Berre lake (right) and its Telemac-3D modeling (left).

4 HPC for complex environmental flows with Spartacus-3D

4.1 The SPH method

Smoothed Particle Hydrodynamics (SPH) is a meshfree particle method initially developed for astrophysical applications. Recently adapted to fluid mechanics [5], it

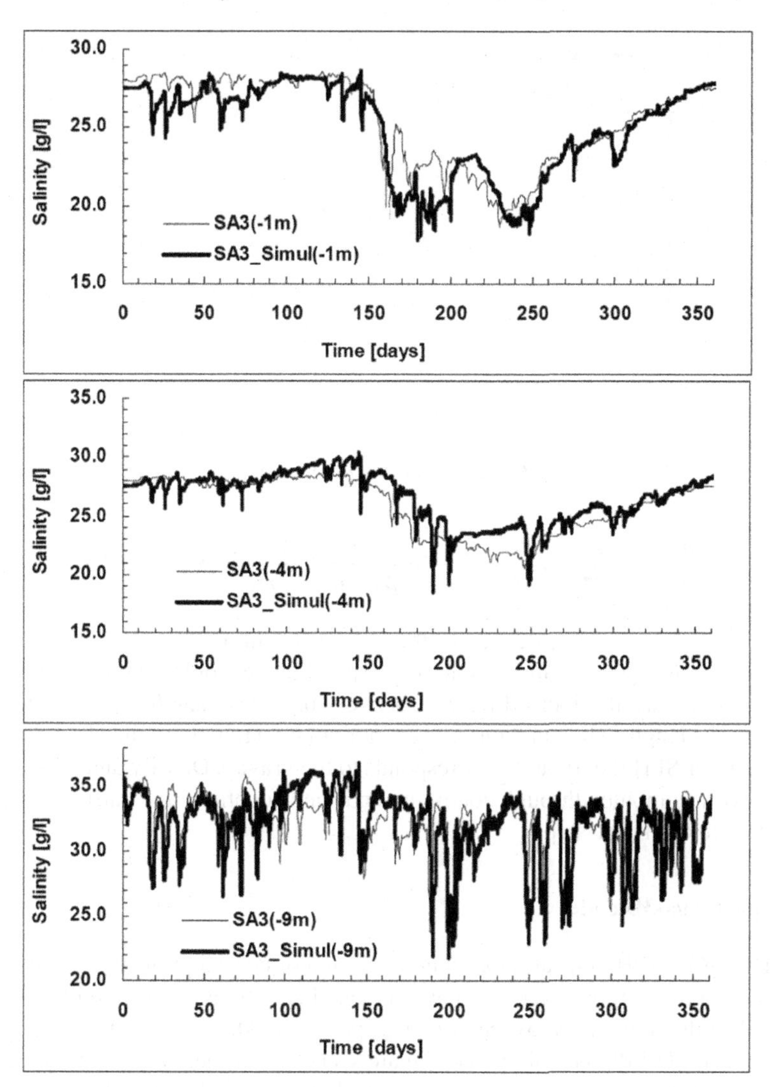

Fig. 4. Salinity comparisons at measurement stations SA1, SA2 and SA3 (bottom) at 3 water column locations. Measurements (respectively simulation) correspond to thick lines (resp. light lines)

has shown its capability to succesfully handle complex flow modeling with free surface, e.g. breaking wave, wave flumes. In the SPH formalism, the fluid is discretized with a finite number of macroscopic particles which are characterized by a constant mass m_a, a velocity vector $\underline{u}_a$, a pressure p_a, a density ρ_a and a position vector $\underline{r}_a$, which are computed at each time step. The momentum equation relative to particle a is usually discretized by:

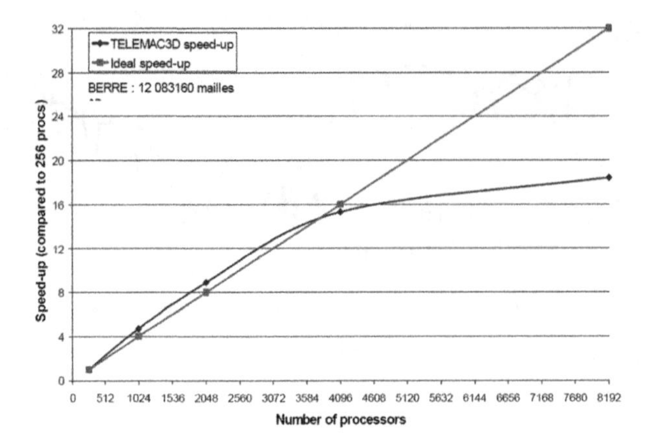

Fig. 5. Telemac-3D speed-up.

$$\frac{D\overline{\underline{u}}_a}{Dt} = -\sum_b m_n\left(\frac{\overline{p}_a+\overline{p}_b}{\rho_a\rho_b} - 8\frac{\nu_{T,a}+\nu_{T,b}}{\rho_a+\rho_b}\frac{\overline{\underline{u}}_{ab}\cdot\underline{r}_{ab}}{r_{ab}^2}\right)\nabla_a w_h(\underline{r}_{ab}) + \underline{g} \tag{8}$$

where the summation is extended to the closest neighbors b relative to particle a. In the previous equation, all variables correspond to Reynolds averaged values. $\overline{\underline{u}}_{ab}$ denotes the averaged velocity difference between particles a and b, $\underline{r}_{ab}$ the vector $\underline{r}_a - \underline{r}_b$ and r_{ab} its length. The function w_h is an interpolating function called kernel, which is the core of SPH theory and $\underline{g}$ corresponds to the gravity. Density and pressure are respectively computed through continuity and state equations and particles are then displaced at each time step.

4.2 Spartacus-3D code

At EDF R&D, SPH aims at improving the design of coastal protections near power plants or optimizing geometry of environmental structures such as spillways, fish passes For this purpose, a 3D SPH code, Spartacus-3D, has been developed since 1999. Since 3D SPH is still an issue, mainly because of heavy CPU time requirements, Spartacus-3D has been massively parallelized, with the collaboration of the French college Ecole Centrale de Lyon [6] and IBM, by using a dynamic particle decomposition technique preserving the load balance. The Spartacus-3D speed up performance when considering approximately 2 million particles is satisfactory, as displayed on figure 6.

4.3 Application to an industrial hydraulic structure

Hydraulic structures are any structures can be used to control the natural flow water. In this section, two industrial structures are considered: the Goulours dam, devoted to hold water in a large reservoir (see figure 7) and the current sky-jump spillway (see figure 7), devoted to evacuate floodwater. A Spartacus-3D model of the entire

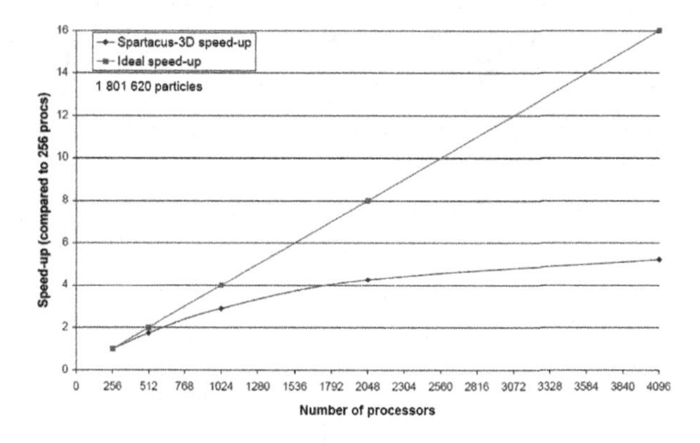

Fig. 6. Spartacus-3D speed-up.

system (involving the real Goulours valley, as well as the dam and the spillway) has been built at EDF R&D. For this purpose, approximately 1 000 000 particles were considered. In order to simulate 16 s of physical time, 5 days of computing were needed by using 1024 IBM BlueGene/L processors. As reveal by figure 8, the simulated flow is realistic and, from a qualitative point of view, match quite well observations achieved on an EDF physical model. More quantitative comparisons will be carried out in 2009.

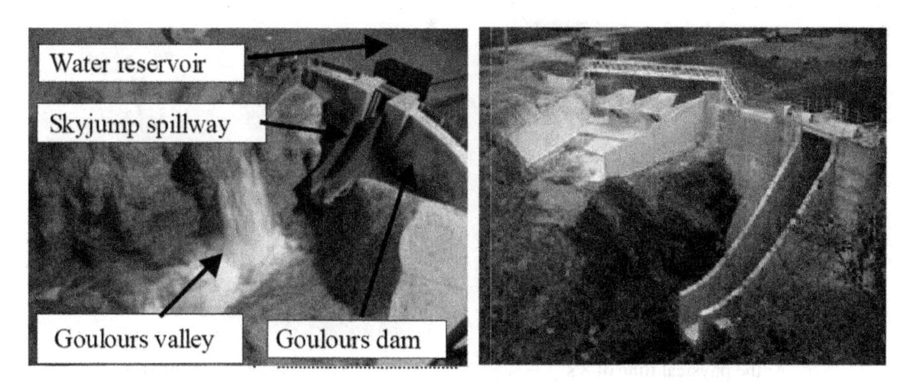

Fig. 7. Left: picture of an EDF physical model of the Goulours structure; right: real Goulours dam and spillway.

5 Conclusion

This paper presented the parallelization needs required by the TELEMAC system, a complete hydroinformatics tool developed by EDF R&D. A focus on three codes part of the TELEMAC system, namely Estel-3D, Telemac-3D and Spartacus-3D, has

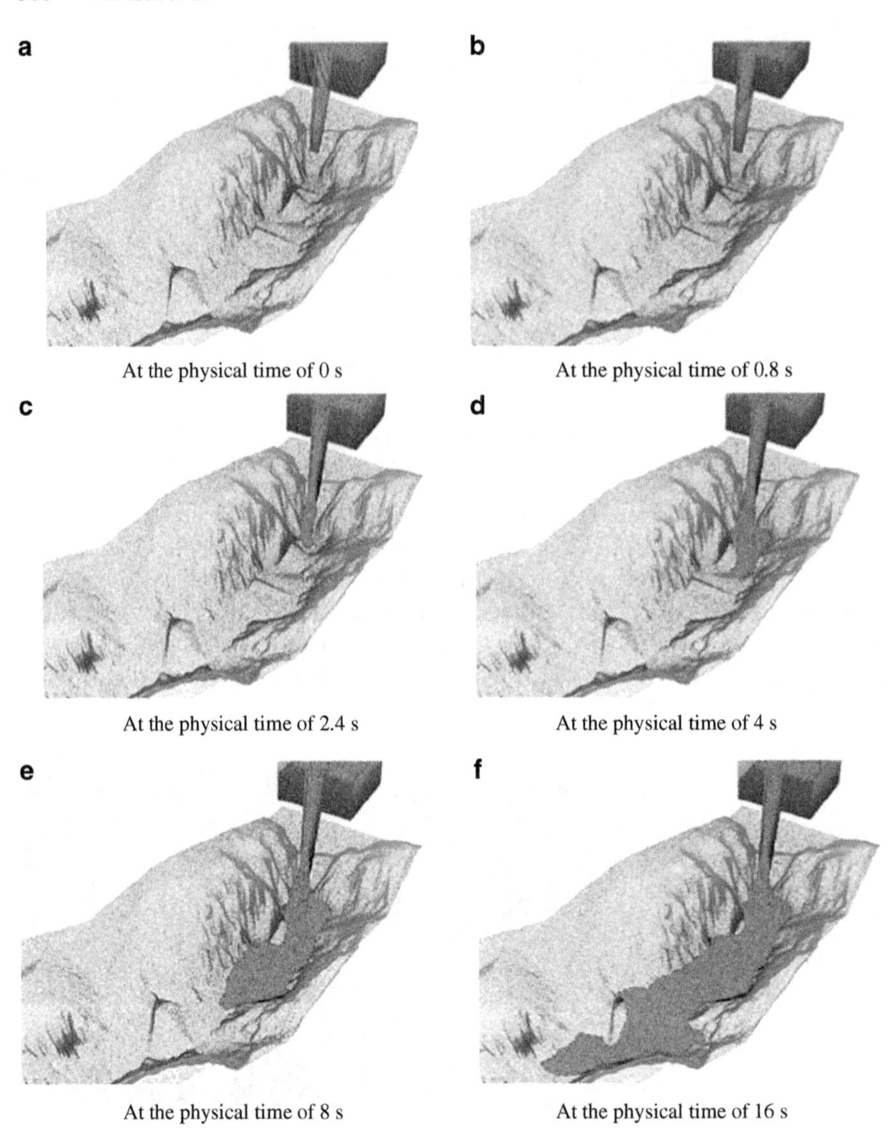

Fig. 8. Snapshots of the simulated flow at different physical time.

been achieved. For each code, performance in terms of speed up is satisfactory and suitable with industrial needs. However, when refining domain discretization, large amount of data has to be handled and a suitable post-processor is required for data analysis.

[1] J.M Hervouet, Hydrodynamics of Free Surface Flows: Modelling with the Finite Element Method Wiley & Sons, 2007.

[2] Haverkamp R., Vauclin M., Tourna J., Wierenga P., Vachaud G: A comparison of numerical solution models for one-dimensional infiltration. Soil Scientific Society America, 41(2), 285-294 (1977)
[3] Celia, M.A., Boutoulas, E.T., Zarba, R.L.: A General Mass-Conservative Numerical Solution for the Unsaturated Flow Equation. Water Resources Research 26(7),1483-1496 (1990)
[4] Hinkelmann, R., Zielke, W.: Parallelization of a Lagrange-Euler model for three- dimensional free surface flow and transport processes. Computers & fluids, Vol. 29, pp. 301-325 (2000)
[5] Monaghan, J. J.: Simulating free surface flows with SPH, J. Comp Phys., Vol 110, pp. 339-406 (1994)
[6] Moulinec, C., Issa,, R., Marongiu, J.-C., Violeau, D.: Parallel 3-D SPH Simulations, CMES, Vol. 25, pp. 133 (2008)
[7] Karypis, G., Kumar, V.: Metis mesh partitioner. Unstructured Graph Partitioning and Sparse Matrix Ordering System, University of Minnesota Technical Report (1995)

[illegible] Haverkamp R., Vauclin M., Touma J., Wierenga P., Vachaud G.: A comparison of numerical simulation models for one-dimensional infiltration. Soil Science Society America J. 41:285–294 (1977)
[illegible] M.A., Bouloutas E.T., Zarba R.L.: A General Mass-Conservative Numerical Solution for the Unsaturated Flow Equation. Water Resources Research [illegible] (1990)
[illegible] R., Zhang W.: Parallel [illegible] of a [illegible] groundwater flow and transport [illegible]
[illegible] Smoothing [illegible] Comp. Phys. [illegible]
[illegible] Vila J.P. [illegible]
[illegible] Methods [illegible]

Multi-parametric intensive stochastic simulations for hydrogeology on a computational grid

J. Erhel[1] *, J.-R. de Dreuzy[2], E. Bresciani[1]

[1] INRIA,35042 Rennes Cedex, France
[2] Geosciences Rennes, UMR CNRS 6118, France

Abstract. Numerical modelling is an important key for the management and remediation of groundwater resources. Numerical simulations must be performed on domains of a large size, at a fine resolution to take into account the scale of geological heterogeneities. Numerical models are based on probabilistic data and rely on Uncertainty Quantification methods (UQ). In this stochastic framework, non intrusive methods require to run multiple simulations. Also, each simulation is governed by multiple parameters and a complete study requires to carry out analysis for more than 50 sets of parameters. We have identified three levels of distributed and parallel computing: subdomain decomposition in one simulation, multiple simulations for UQ methods, multiparametric studies. Our objective is to use the computing and memory resources of computational grids to deploy these multiple large-scale simulations. We discuss our implementation of these three levels, using an object-oriented approach. We present some preliminary results, with a strategy to choose between the first and second level.

1 Introduction

Numerical modelling is an important key for the management and remediation of groundwater resources. Natural geological formations are highly heterogeneous, leading to preferential flow paths and stagnant regions. The contaminant migration is strongly affected by these irregular water velocity distributions. In order to account for the limited knowledge of the geological characteristics and for the natural heterogeneity, numerical models are based on probabilistic data and rely on Uncertainty Quantification methods. In this stochastic framework, non intrusive methods require to run multiple simulations. Also, numerical modelling aims at studying the impact of various physical parameters, such as the Peclet number. Therefore, each simulation is governed by multiple parameters and a complete study requires to carry out analysis for more than 50 sets of parameters. The hydraulic simulations must be performed on domains of a large size, at the scale of management of the groundwater resource or at the scale of the homogeneous medium type in terms of geology. This domain must be discretized at a fine resolution to take into account the scale

* this work was partly funded by the Grid'5000 grant from the French government

D. Tromeur-Dervout (eds.), *Parallel Computational Fluid Dynamics 2008*,
Lecture Notes in Computational Science and Engineering 74,
DOI: 10.1007/978-3-642-14438-7_41,

of geological heterogeneities. Characterization of transport laws requires simulating advection and dispersion on very long times and in turn in very large domains. Our objective is to use the computing and memory resources of computational grids to deploy these multiple simulations.

A first level of parallelism is used in each simulation. Indeed, in order to reach the target of large scale domains, it is necessary to run each simulation on a parallel computer with enough memory and with enough computing power. A second level of parallelism comes from Uncertainty Quantification. A third level of parallelism is the study of different sets of parameters. These multiparametric simulations are clearly independent and are thus very well-suited to techniques inspired from peer-to-peer. However, it should be kept in mind that each study is in itself a heavy computation involving a large number of random simulations, requiring high performance computing for each simulation. Our objective is to use current middleware developed for grid architectures, in order to make the most of the three levels of parallelism. Several difficulties arise, ranging from basic software engineering (compatibility of systems, libraries, compilers) to scheduling issues.

2 Existing work

2.1 Parallel Monte-Carlo

The Monte-Carlo method is heavily used in physical simulations, either to evaluate integrals or in the framework of stochastic models. In general, a run is composed of more or less independent simulations, so that a run is embarassingly parallel. The main difficulty is to generate random numbers correctly. Also, since the flowchart of a Monte-Carlo run is identical for many applications, it is natural to design a generic software. For example, the ALPS project (Algorithms and Libraries for Physics Simulations) is an open source effort aiming at providing high-end simulation codes for strongly correlated quantum mechanical systems as well as C++ libraries for simplifying the development of such code (http://alps.comp-phys.org/). The ALPS software contains a module for parallel Monte-Carlo runs and parallel multiparametric simulations [14]. It is based on a distributed memory paradigm and uses MPI, with clusters as target computers. Currently, each simulation is sequential.

2.2 Distributed multiple simulations and grid applications

Multiparametric experiments arise in many application domains, for example in biology, chemistry and earth science. Computational grids provide interesting power and memory resources. Many initiatives of grids are built around the world. For example, DEISA is a grid gathering several supercomputing centers in Europe (http://www.deisa.org/). Other grids build an infrastructure with several partners to create an integrated, persistent computational resource. Some examples are Teragrid in USA (http://www.teragrid.org/about/), EGEE in Europe (http://www.eu-egee.org/)

and Grid'5000 in France (http://www.grid5000.fr). These infrastructures aim at developing e-science applications using global resources provided by the grid. For example, the GEON web portal (http://www.geongrid.org) provides tools to a network of partners in earth science, like SINSEIS, a synthetic seismogram computation toolkit, built as a grid application. Some applications are multiparametric scientific simulations; for example, in earth science, the footprint project (http://www.eu-footprint.org/) develops tools for pesticide risk assessment and management, based on Monte-Carlo and multiparametric simulations for dealing with uncertainty [10]. Whereas most grid initiatives use computing resources of research laboratories or computing centers, another approach rely on Internet to run scientific software on desktop computers. The platform BOINC [2] is a distributed computing tool which allows to run computationally expensive projects by using the aggregate power of desktop computers. The project climateprediction.net [13] uses BOINC to run millions of simulations of a climate model coupling atmosphere and ocean.

2.3 Middleware on grids

Computational grids are often built as a network of several clusters located in different geographical places. Multiple simulations can in principle run on these clusters by using the grid infrastructure. However, scheduling tools are required to distribute the simulations and to communicate data between the clusters. Some projects enable scientists and engineers to seamlessly run MPI-conforming parallel application on a Computational Grid, such as a cluster of computers connected through high-speed networks or even the Internet. For example, MPICH-Madeleine [6] is a free MPICH-based implementation of the MPI standard, which is a high-level communication interface designed to provide high performance communications on various network architectures including clusters of clusters. Another solution is to provide a tool specifically devoted to distributed computing. Nimrod/G is an example of such software and has been successfully used with different grids [1]. The Grid'5000 project provides the software Oaregrid [7] and Taktuk [12], which can also help to deploy parametric simulations. Other approaches are based on a software component paradigm [8].

3 Our work

We are developing a scientific platform H2OLab for hydrogeology. Our platform is designed in order to ensure integration of new modules and to facilitate coupling of existing modules. We use C++ development environments and software engineering tools. We pay a lot of attention to test generation, non regression issues and numerical validation. Modularity and genericity are essential for a scientific platform of this size. These requirements are satisfied by a rigorous design inspired from UML techniques and by an object-oriented approach. We have identified three levels of distributed and parallel computing. At the simulation level, we choose to define distributed memory algorithms and to rely on the MPI library for communications between processors. These parallel deterministic simulations are operational in the software H2OLab and we are investigating scalability issues [9]. The intermediate level is the Uncertainty Quantification non intrusive method, currently Monte-Carlo. We apply a paragim similar to the software ALPS and have developed a generic Monte-Carlo module. It differs from ALPS in a number of ways including the use of Monte-Carlo, random number generation, the physical application, the memory and CPU intensive simulations and the development tools used. Our objective is to design a facility for running this run of Monte-Carlo by choosing either a parallel approach with MPI or a distributed approach with a grid middleware. We use a specific random number generator in order to guarantee independent simulations. At the multiparametric level, we choose only the distributed approach as is done in most projects on computational grids.

3.1 Generic tools in H2OLab

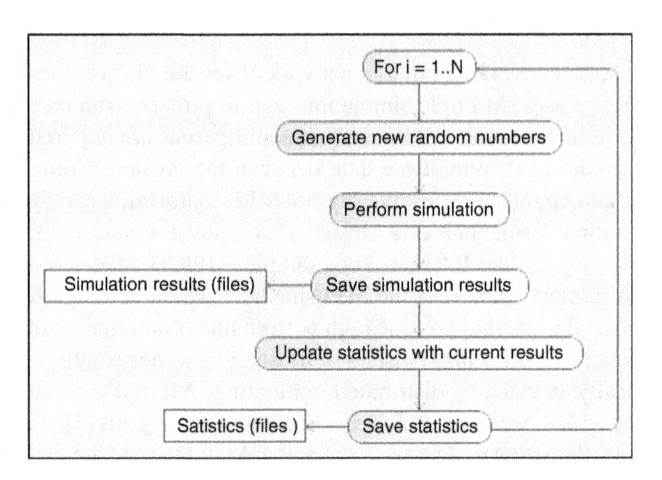

Fig. 1. Monte-Carlo simulations, saving results and collecting statistics.

We have implemented a generic Monte Carlo method where a loop performs N simulations and computes first and second statistical moments of the results. This loop contains a checkpoint at every iteration with a restore point. This recovery facility allows to resume simulations in case of failure or to complete already existing statistics with new results. The generic loop is depicted in Figure 1. Some operations are always done by any executable program : reading inputs, creating results directories, initializing random number generators, launching the simulation, writing the results, initializing visualization tools, etc. All those operations are factorized and performed by a Launcher class. This is lot-of-time saving for the user. In order to be generic to any application, the Launcher and the Monte Carlo modules need a common interface. We thus developed a Simulation virtual class, which owns all necessary objects to perform a simulation: parameters, results, random number streams.

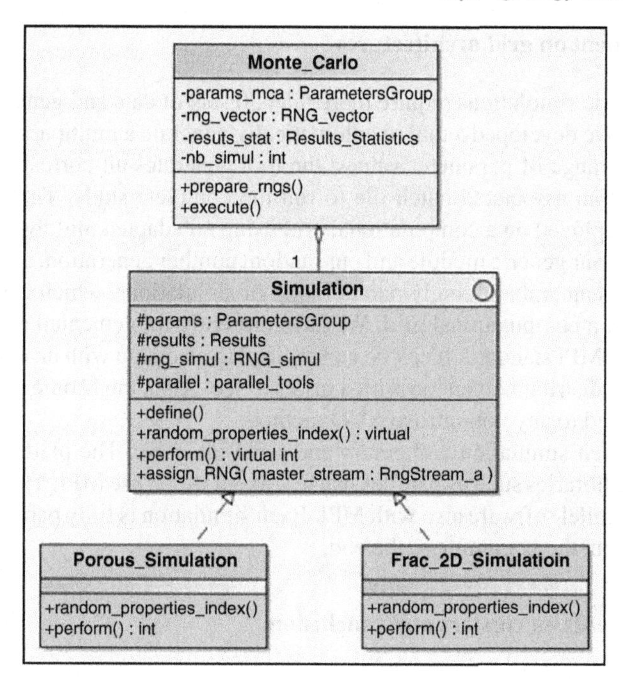

Fig. 2. The virtual class simulation, and its use for Monte-Carlo with various applications.

Practically, the user only has to override two functions to define the specific random properties associated to the simulation and to write the very job of the simulation. The virtual class is depicted in Figure 2.

We have opted for the XML standard language to define a generic structure for input parameters. The use of this standard has allowed us to easily define an associated user interface and develop C++ read/write methods thanks to already existing tools. In our scheme, the parameters are defined by four fields: name, value, default value and description, and can be organized in groups in a recursive manner. This structure facilitates the development of non-conflicting C++ packages.

The simulation results are stored in a generic structure (C++ class) which can contain scalars, vectors and matrices, organized in categories. This class also provides a method to write the results in files in an appropriate format. Application-specific results and categories are defined in XML files.

Regarding random number generation, we have to deal with several difficulties: a simulation has several random properties, a random property can require several random numbers and this quantity is not fixed in advance, depending on the studied medium or phenomenon. Moreover, the run must be reproducible for validation and composed of independent simulations for distributed computing. We have designed a set of classes, based on the RngStream package [11], to generate random numbers streams that solve these difficulties. These classes are generic to any application.

3.2 Deployment on grid architectures

Multiparametric simulations require more than 50 sets of data and generate as many results. We have developed a tool to automatically generate a multiparametric study: from a given range of parameter values, the tool generates all corresponding input data files and an associated batch file to run the complete study. This tool is now ready to be deployed on a computational grid using an adapted middlware.

Thanks to our generic module and our random number generation, a run of Monte Carlo contains an embarassingly parallel loop of simulations, which can be readily distributed on a computational grid. We have currently implemented a parallel version using the MPI standard. It can be generalized to a version with an extended MPI library or to a distributed version with a grid service. Also, the Monte Carlo module can be extended to any non intrusive UQ method.

Finally, each simulation is memory and CPU intensive. The platform relies on free software libraries such as parallel sparse solvers which use MPI. Thus we choose to develop parallel software also with MPI. Each simulation is fully parallel with data distributed from the beginning to the end.

3.3 Experiments on clusters and conclusion

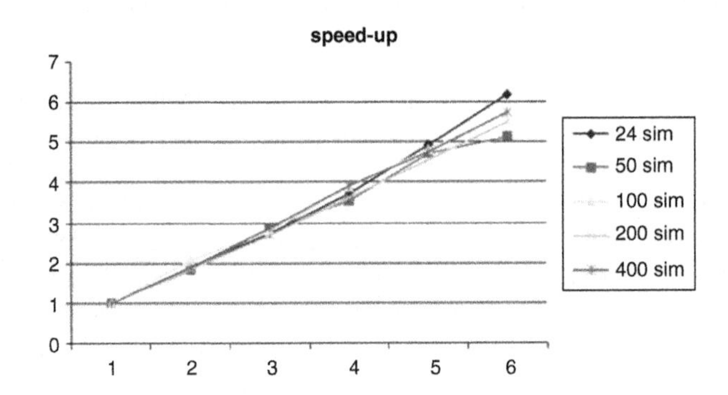

Fig. 3. Speed-up versus the number of nodes (2 subdomains per node).

We use the different clusters available thanks to the french grid project Grid'5000. We have made a thorough performance analysis of our parallel simulations applied to flow and solute transport in heterogeneous porous media [3, 5]. We have also done some experiments with parallel simulations applied to flow in Discrete Fracture of Networks [4]. This analysis allows us to find out in advance the number of processors necessary for a given set of data. Thus we can rely on a static scheduling

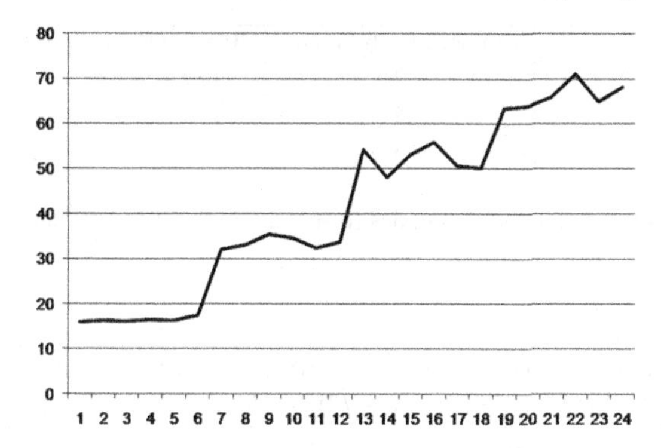

Fig. 4. CPU time versus the number of simulations, with 6 nodes.

of resources for each simulation. For a run of Monte Carlo, we can also define the number of simulations and use a static scheduling.

Here, we give the results for several experiments combining parallel Monte Carlo runs of parallel simulations. In a first step, we run parallel Monte-Carlo simulations of moderate size so that each simulation can run on one node. We use a cluster of nodes with a Myrinet network where each node is one-core bi-processor, with 2GB memory. We apply our method to flow and solute transport in heterogeneous porous media, with a mesh of 1024 times 1024 cells. We have done several measures with varying parameters of the model with similar results, so we plot here the results with one set of parameters. In Figure 3, we plot the speed-up in function of the number of processors. For a small number of nodes, the speed-up is almost linear, as could be expected since parallel Monte-Carlo does not induce communication. In Figure 4, we plot the CPU time in function of the number of simulations, using 6 nodes of the same cluster. As could also be expected, we get a stairwise function, due to the distribution of simulations which is obtained simply by dividing the number of simulations by the number of processors.

In a second experiment, we use a two-level parallelism. We run the same application on a cluster of four nodes and try three configurations, distributing both the subdomains of one simulation and the Monte-Carlo simulations. Results are given in Table 3.3. Clearly, Monte-Carlo parallelism is more efficient since subdomain decomposition involves communications. Therefore, the limiting resource is here the memory available and the best strategy is to choose the smallest number of subdomains so that each subdomain fits in the memory of one core, defining chunks of processors. Then Monte-Carlo simulations are distributed among the different chunks.

These results are preliminary but show clearly that what we get is what we expect. So we can now adopt the same strategy for larger computational domains with a larger number of nodes. Also, in a next future, we plan to use middleware available on grid'5000, in order to run the three levels of distributed computing.

[1] D. Abramson. Applications development for the computational grid. In *Applications Development for the Computational Grid*, volume 3841 / 2006 of *Lecture Notes in Computer Science*, pages 1–12, 2006.

Number of subdomains	CPU time
2	129
4	199
8	299

Table 1. Two-level parallelism with 4 nodes (bi-processors): CPU time according to the number of subdomains.

[2] David P. Anderson. Boinc: A system for public-resource computing and storage. In *5th IEEE/ACM International Workshop on Grid Computing*, Pittsburgh, USA, November 2004.

[3] A. Beaudoin, J-R. de Dreuzy, and J. Erhel. An efficient parallel particle tracker for advection-diffusion simulations in heterogeneous porous media. In A.-M. Kermarrec, L. Boug, and T. Priol, editors, *Euro-Par 2007, LNCS 4641*, pages 717–726. Springer-Verlag, Berlin, Heidelberg, 2007.

[4] A. Beaudoin, J-R. de Dreuzy, J. Erhel, and H. Mustapha. Parallel simulations of underground flow in porous and fractured media. In G.R. Joubert, W.E. Nagel, F.J. Peters, O. Plata, P. Tirado, and E. Zapata, editors, *Parallel Computing: Current and Future Issues of High-End Computing*, volume 33 of *NIC Series*, pages 391–398, Jlich, Germany, 2006. NIC.

[5] A. Beaudoin, J. Erhel, and J.-R. de Dreuzy. A comparison between a direct and a multigrid sparse linear solvers for highly heterogeneous flux computations. In *Eccomas CFD 2006*, volume CD, 2006.

[6] Darius Buntinas, Guillaume Mercier, and William Gropp. Data Transfer in a SMP System: Study and Application to MPI. In *Proc. 34th International Conference on Parallel Processing(ICPP 2006)*, Colombus, Ohio, August 2006.

[7] Nicolas Capit, Georges Da Costa, Yiannis Georgiou, Guillaume Huard, Cyrille Marti n, Grgory Mouni, Pierre Neyron, and Olivier Richard. A batch scheduler with high level components. In *Cluster computing and Grid 2005 (CCGrid05)*, 2005.

[8] A. Denis, C. Prez, and T. Priol. Achieving portable and efficient parallel corba objects. *Concurrency and Computation: Practice and Experience*, 15(10):891–909, August 2003.

[9] Jocelyne Erhel, Jean-Raynald de Dreuzy, Anthony Beaudoin, Etienne Bresciani, and Damien Tromeur-Dervout. A parallel scientific software for heterogeneous hydrogeology. In *Parallel CFD*, Antalya (Turkey), May 2007. Invited plenary talk.

[10] Stenemo F. and Jarvis N.J. Accounting for uncertainty in pedotransfer functions in vulnerability assessments of pesticide leaching to groundwater. *Pest Management Science*, 63(9):867–875, 2007.

[11] Pierre L'ecuyer, Richard Simard, E. Jack Chen, and W. David Kelton. An object-oriented random-number package with many long streams and substreams. *Operations Research*, 50(6):1073 – 1075, 2002.

[12] Vincent Martin, Jrme Jaffr, and Jean E. Roberts. Modeling fractures and barriers as interfaces for flow in porous media. *SIAM Journal on Scientific Computing*, 26:1667–1691, 2005.

[13] N. Massey, T. Aina, M. Allen, C. Christensen, D. Frame, D. Goodman, J. Kettleborough, A. Martin, S. Pascoe, and D. Stainforth. Data access and analysis with distributed federated data servers in climateprediction.net. *Advances in Geosciences*, 8:49–56, 2006.

[14] Matthias Troyer, Beat Ammon, and Elmar Heeb. Parallel object oriented monte carlo simulations. In *ISCOPE '98: Proceedings of the Second International Symposium on Computing in Object-Oriented Parallel Environments*, pages 191–198, London, UK, 1998. Springer-Verlag.

Parallel computation of pollutant dispersion in industrial sites

J. Montagnier[1,2], M. Buffat[2], and D. Guibert[1,3]

[1] Modelys, 66, Boulevard Niels Bohr, BP 2132, 69603 Villeurbanne Cedex France julienmontagnier@modelys.net
[2] Université de Lyon, CNRS, Ecole Centrale de Lyon, Université de Lyon 1, Laboratoire de Mécanique des fluides et d'Acoustique, bat Omega 721, 43 bvd du 11 nov. 1918, 69622 Villeurbanne Cedex France marc.buffat@univ-lyon1.fr
[3] ICJ CNRS UMR 5208, Center for the Developement of Scientific Parallel Computing, University Claude Bernard Lyon 1, ISTIL 15, Boulevard Latarjet, 69622 Villeurbanne Cedex France davidguibert@modelys.net

Understanding the pathway of toxic air pollutants from their source is essential to government agencies that are responsible for the public health. CFD remains an expansive tool to evaluate the flow of toxic air contaminants and requires to deal with complex geometry, high Reynolds numbers and large temperature gradients. To perform such simulations, the compressible Naviers Stokes equations are solved with a collocated finite volume method on unstructured grid and the computation speed is improved as a result of parallelism.

The developed computer code is written in an object oriented language (C++) using a 3D finite element library libMesh ([5]) associated with PETSc ([1]) a general purpose scientific computing library. Parallelization is performed using the domain decomposition and the parallel linear solvers of PETSc.

In this paper, we propose to evaluate Algebraic Multigrid preconditioning methods used in the Navier Stokes solver.

1 Description of the numerical method

1.1 Implicit Navier Stokes Solver

A classical implicit finite volume discretization of the compressible Reynolds averaged Navier-Stokes (RANS) equations may be written as :

$$\rho_i^{n+1} - \rho_i^n + \tau \sum_{j \in V(i)} \rho_i^{n+1} \phi_{ij}^{n+1} = 0 \tag{1}$$

D. Tromeur-Dervout (eds.), *Parallel Computational Fluid Dynamics 2008*,
Lecture Notes in Computational Science and Engineering 74,
DOI: 10.1007/978-3-642-14438-7_42, © Springer-Verlag Berlin Heidelberg 2010

$$(\rho u)_i^{n+1} - (\rho u)_i^n + \tau \sum_{j \in V(i)} \phi_{ij}^{n+1} (\rho u)_i^{n+1}$$
$$= \tau \sum_{j \in V(i)} S_{ij} p_{ij}^{n+1} n_{ij} + \tau \sum_{j \in V(i)} S_{ij} \nabla (\nu_{ij} (\rho u)_{ij}^{n+1}) \quad (2)$$

$$(\rho E)_i^{n+1} - (\rho E)_i^n + \tau \sum_{j \in V(i)} (\rho H)_i^{n+1} \phi_{ij}^{n+1} = 0 \quad (3)$$

with $\tau = \triangle t / V_i$ and $V(i)$ is the set of neighboring cells of the cell i.

The superscripts n and n+1 indicate the old and the new time step, respectively, $\triangle t$ is the time step, V_i the volume of cell i. ρ, P, u, H and E denote respectively the density, pressure, velocity, total enthalpy and total energy. $\partial C_i \cap \partial C_j$ is the interface located between the two cells i and j. S_{ij} and n_{ij} are respectively the surface and normal vector of the face $\partial C_i \cap \partial C_j$.

For the convective fluxes, the cell face velocity $\phi_{ij} = \int_{\partial C_i \cap \partial C_j} u_{ij}.n_{ij} ds$ is classically calculated using a specific interpolation ([10]). On the interface, the transported quantities are interpolated using a first or second order upwind scheme.

Momentum, energy and mass conservation equations are solved implicitly using a projection method ([8]). At each non linear iteration (k), a predictor step (*) for density and momentum is calculated from the density and momentum equations :

$$\rho_i^* - \rho_i^n + \tau \sum_{j \in V(i)} \rho_i^* \phi_{ij}^k = 0 \quad (4)$$

$$(\rho u)_i^* - (\rho u)_i^n + \tau \sum_{j \in V(i)} \phi_{ij}^k (\rho u)_i^* = \tau \sum_{j \in V(i)} S_{ij} p_{ij}^k n_{ij} + \tau \sum_{j \in V(i)} S_{ij} \nabla (\nu_{ij} (\rho u)_{ij}^*) \quad (5)$$

After this prediction step, the pressure correction $p_i^{'}$ (and eventually the temperature correction $T_i^{'}$) is computed from a Laplace equation based on the energy (and eventually the density equation). Correction equation (8) is obtained by replacing in equation (3) the total energy and the total enthalpy by expression (6) and (7).

$$(\rho E)^{n+1} = (\rho E)^* + (\rho e)_p^* p^{'} \quad (6)$$

$$(\rho H)_i^{n+1} \phi_{ij}^{n+1} = (\rho H)_i^* \phi_{ij}^* + h_i^* (\rho \phi)_{ij}^{'} \quad (7)$$

$(\rho \phi)_{ij}^{'}$ is related to the pressure correction $p_i^{'}$ through the momentum equation using a SIMPLE algorithm ([9]).

$$(\rho e)_p^* p_i^{'} + \tau \sum_{j \in V(i)} h_i^* (\rho \phi)_{ij}^{'} = - \left((\rho E)_i^* - (\rho E)_i^n + \tau \sum_{j \in V(i)} (\rho H)_i^* \phi_{ij}^* \right) \quad (8)$$

At each step of this algorithm, we have to solve several large sparse linear systems. The key point of the efficiency of this implicit algorithm is to use an efficient linear solver.

1.2 Iterative methods for sparse linear systems

Krylov subspace methods are widely used to solve for large and sparse linear systems. In the present work, the incomplete LU factorisation algorithm and the Algebraic MultiGrid method (AMG) are used as preconditioners for the GMRES algorithm. The next section brievly describes the multigrid preconditioner used in this paper. The reader is refered to the literature for more information (see [11], [2], [12]).

The central idea of AMG is to remove smooth error not eliminated by relaxation on coarser grids. This is done by solving the residual equation on coarse a grid, then interpolating the error back to the finest grid. Significant and important parts of the AMG algorithm rely on the coarse-grid selection process and the interpolation operation.

Coarsening schemes

Various coarsening schemes exist and most are based on the coarsening heuristics proposed by [11]. However when these traditional algorithms are applied to three-dimensional problems, stencil may grow significantly, leading to computational complexity growth and a loss of scalability. For this reason, the Falgout algorithm ([4]) is the only tested classical method.

Recently, [12] developed new methods to obtain sparser coarser grids (see figure 1) that diminish algorithm complexity : The Parallel Modified Independent Set (PMIS) algorithm and the Hybrid Modified Independant Set (HMIS) algorithm are compared to the more classical Falgout method.

Interpolation

Interpolation is an important operation affecting the convergence factor of the AMG algorithm. The Falgout coarsening scheme is used with a classical interpolation ([11]) : the error of a refine point (F-point) i is done with a subset of the nearest neighbors of grid point i.

Because of the coarser grids, PMIS and HMIS approach require more accurate interpolation : we investigate longer range interpolations F-F and F-F1 (see [2]).

2 Description of the tests

2.1 Laplace equation

To select the best precondioning method for our application, scale-up and speed up tests are performed on a Laplace equation $\Delta\phi = 1$ with a null Dirichlet boundary condition $\phi_\Gamma = 0$. The computational domain is a unit cubic box with tetrahedra cells. Speed up tests used 10^6 cells mesh, scale-up tests investigated different numbers of degree of freedom per processor (dofs) from 12 500 to 400 000 dofs. AMG methods described in 1.2 are compared with ILU preconditioning.

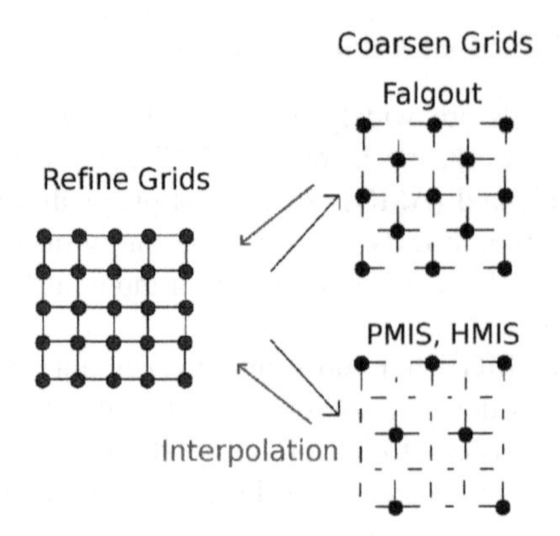

Fig. 1. Coarsening schemes and interpolation

2.2 Flow around a complex industrial site

We are interested in an incompressible isothermal flow around a french factory. The standard k-epsilon model ([6]) with wall function ([7]) was used for assessing turbulent flow. The closure coefficients are obtained from [3] to take into account the atmospheric boundary layer.

The three dimensional unstructured mesh used in the simulations is displayed in figure 2 (b) . It contains approximately 5×10^6 tetrahedra.

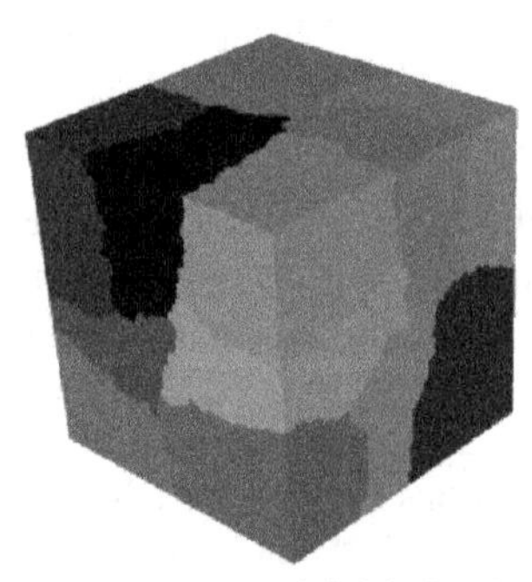

(a) Unstructured Cubic Box

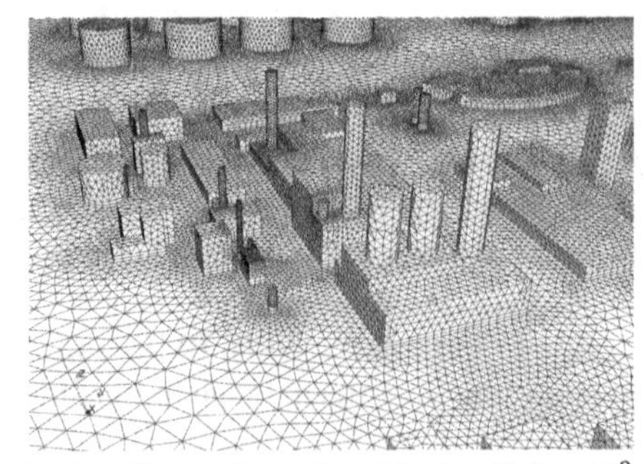

(b) Details of the industrial site (1.5×10^9 cells)

Fig. 2. Meshes

3 Results

The results were obtained at the High Performance Computing Center P2CHPD, Lyon 1 University. The cluster is a mid-range HPC platform. Each node is composed of 2 inter quad-core processor with a CPU speed of 2.9 GHz and are connected with an infiniband network.

3.1 Laplace equation

In figure 3 (a) we show the results of parallel scaling tests. These results immediately highlight the superiority of AMG preconditioning on parallel ILU. Scalability of ILU is poor due to the increase of the number of iterations with the problem size (tab 1).

It is interesting to observe in Tables 2 that low complexity coarsening scheme PMIS with F-F1 interpolation is more efficient than the Falgout coarsening scheme and classical interpolation : complexity is lower and computing time is reduced. HMIS with F-F or F-F1 interpolation show very similar results and are not shown here. AMG with PMIS coarsening schemes and F-F1 interpolation give the best results. Solving time is five hundred times faster than ILU preconditioning on 64 processors. The use of classical interpolation with low complexity scheme leads to an increase of the number of iterations because of sparser grid. Figure 3 (b) displays the influence of the problem size on scalability : Communication times are important and cause a loss in scability.

Table 1. Number of iteration with 200 000 dofs / processor

Method / processor	1	2	4	8	16	32
ILU	226	346	368	500	774	955
PMIS - F-F1	13	14	14	14	16	16
Falgout classical	11	11	11	12	13	13

Table 2. Details of AMG Preconditioning methods with 32 procs, 200 000 dofs

Method	Grid Complexity	Interpolation Complexity	nb it
Falgout classical	2.6	18.5	13
PMIS - F-F1	1.42	3.80	16
PMIS - classical	1.42	2.28	37

3.2 Flow around a complex industrial site

As seen in 3.1, coarsening scheme PMIS with interpolation F-F1 give the best results in term of speed up and computation time for a Laplace equation. This preconditioning method is then evaluated for parallel computation of atmospheric transport

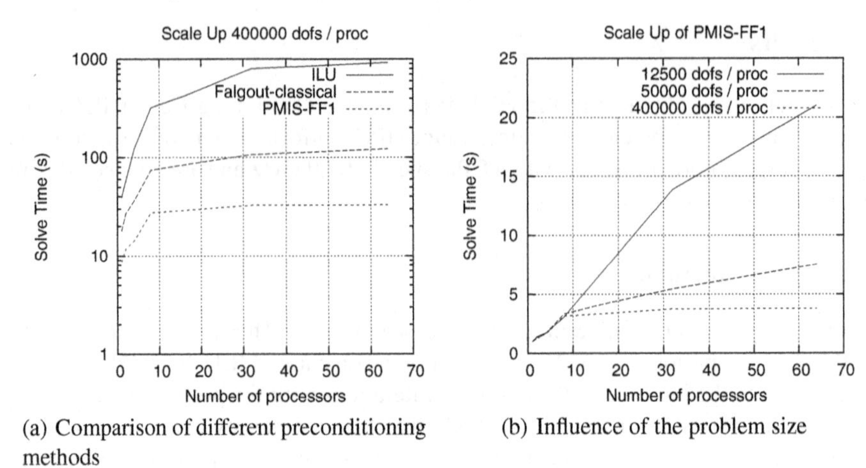

(a) Comparison of different preconditioning methods

(b) Influence of the problem size

Fig. 3. Scale up Results on a Laplace equation

pollutant. Pollutant plume is displayed in figure 5. Figure 4 shows that speed up is excellent with the 5×10^6 cells but is not good with 1×10^6 cells due to communication times.

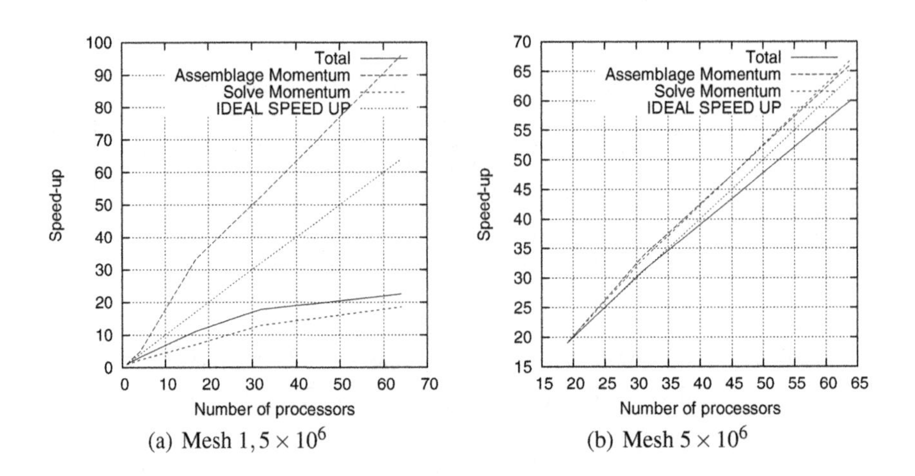

(a) Mesh $1,5 \times 10^6$

(b) Mesh 5×10^6

Fig. 4. Speed up performance of Navier Stokes solver

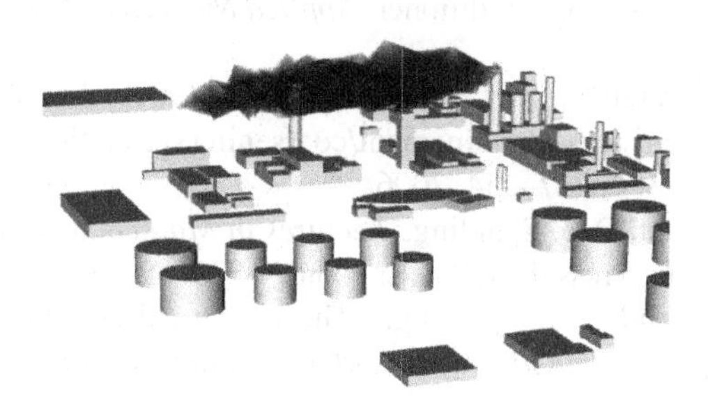

Fig. 5. Pollutant plume around buildings (1.5×10^5 cells)

4 Conclusion

The most challenging aspect of parallelizing a Navier Stokes finite volume solver is the selection of efficient parallel linear solver. We have presented a comparison of different preconditioning methods for Krylov subspace iterative methods.

The preconditioning methods are first tested on a Laplace equation. The tests indicate that AMG preconditioning exhibit better results than ILU preconditioning. The poor scalability of the ILU preconditioning is due to an increase of iteration numbers with the problem size. The Falgout coarsening scheme and classical interpolation have for 3D problems too high complexity algorithm. 3D problems require a low complexity coarsening scheme associated with a long range interpolation. Among the methods tested, PMIS with interpolation F-F1 gives the best results in term of speed up, scale-up and computing time.

Numerical experiments demonstrated good scalability of PMIS coarsening scheme and F-F1 interpolation applied to the Navier Stokes solution of flow around an industrial site. Hence, use of parallel and efficient multigrid preconditioner allows a fast simulation of pollutant dispersion in industrial sites.

Future research will focus on parallelization of high order schemes for Detached and Large Eddy Simulations.

[1] Satish Balay, Kris Buschelman, Victor Eijkhout, William D. Gropp, Dinesh Kaushik, Matthew G. Knepley, Lois Curfman McInnes, Barry F. Smith, and Hong Zhang. Petsc users manual. ANL-95/11, Argonne National Laboratory, 2004.

[2] Butler and Jeffrey. *Improving Coarsening and Interpolation for Algebraic Multigrid.* PhD thesis, Department of Applied Mathematics, University of Waterloo, 2006.

[3] P.G. Dyunkerke. Application of the E-epsilon turbulence closure model to the neutral and stable atmosphere boundary layer. *Journal of the Atmospheric Sciences*, 45:5, 865–880, 1988.

[4] Van Emden Henson and Ulrike Meier Yang. Boomeramg: A parallel algebraic multigrid solver and preconditioner. *Applied Numerical Mathematics*, 41:155–177, 2000.

[5] B. Kirk, J.W. Peterson, R.H. Stogner, and G.F. Carey. libmesh: A c++ library for parallel adaptive mesh refinement/coarsening simulations. *Engineering with Computers*, 22:3–4, 237–254, 2006.

[6] B.E. Launder and D.B. Spalding. *Lectures in Mathematical Models of Turbulence.* Academic Press, London, England,, 1972.

[7] B.E. Launder and D.B. Spalding. The numerical computation of turbulent flows. *Computer Methods in Applied Mechanics and Engineering*, 3:2, 269–289, 1974.

[8] K. Nerinckx, J. Vierendeels, and E. Dick. Mach-uniformity through the coupled pressure and temperature correction algorithm. *Journal of Computational Physics.*, 206:597–623, 2005.

[9] S.V. Patankar and D.B. Spalding. A calculation procedure for heat, mass and momentum transfer in three-dimensional parabolic flows. *International Journal of Heat and Mass Transfer*, 15:1787, 1972.

[10] C.M. Rhie and W.L. Chow. A numerical study of the turbulent flow past an isolated airfoil with trailing edge separation. *AIAA J*, 21:1525–1532, 1983.

[11] J. W. Ruge, K. Stüben, and S. F. McCormick. *Algebraic multigrid (AMG)*, volume 3 of *Frontiers in Applied Mathematics*. SIAM, Philadelphia, 1987.

[12] De Sterck, Hans, Yang, Ulrike M., Heys, and Jeffrey J. Reducing complexity in parallel algebraic multigrid preconditioners. *SIAM J. Matrix Anal. Appl.*, 27(4):1019–1039, 2006.

Part XI

General fluid

3D Numerical Simulation Of Gas Flow Around Reentry Vehicles

S.V. Polyakov[1], T.A. Kudryashova[1], E.M. Kononov[1], and A.A. Sverdlin[1,2]

[1] Institute for Mathematical Modeling, Russian Academy of Sciences, 125047, Miusskaya Sq. 4a, Moscow, Russia
[2] E-mail: alex@gai.ru

Abstract. Mathematical model of 3D gas flow around reentry vehicle is considered. This model is based on QGD (Quasy-GasDynamics) equations. Numerical methods for solving of these equations on tetrahedral meshes are presented. Parallel realization of these methods is based on MPI and OpenMP technologies. Parallelization is performed using domain decomposition principle.

1 Introduction

Modeling of the gas flow is very important for construction of any spacecraft and in particular reentry vehicle. Latest experiments show that protection of space vehicles designed decades ago is superfluous, and the destructive factors such as temperature do not reach such high values the protection designed against.

Computer modelling allows to save a lot of money and predict parameters of gas-dynamic processes which cannot be obtained by ground test facilities. Computational Fluid Dynamics (CFD) is the best approach for calculation of aerodynamic heating. The main factors typical for CFD algorithms are: computational complexity, high amounts of memory required, necessity for carrying out all the calculations in case of alteration in one of the flow parameters (such as Mach number, altitude, angle of attack). Therefore using this technique for calculation of the heat flow throughout space vehicle trajectory becomes difficult and time consuming task. All the mentioned factors force the use of high performace parallel systems for modelling hypersonic flows.

In the branch where one kilogram means tens and hundreds thousands dollars it is very important to calculate all the values precisely. The introduced approach allows to make more accurate calculations in comparison to well known and widely used Navier-Stokes equations system. Also improved stability of numerical methods allows to take into account additional factors such as radiative energy withdrawal that has not been considered before.

This work presents 3D flow calculation approach around reentry vehicle in Earth atmosphere using multiprocessor computational clusters. The mathematical model is

D. Tromeur-Dervout (eds.), *Parallel Computational Fluid Dynamics 2008*,
Lecture Notes in Computational Science and Engineering 74,
DOI: 10.1007/978-3-642-14438-7_43,

introduced and completed with numerical implementation. Parallel algorithm implementing this numerical method is designed.

2 Mathematical Model

QGD (Quasy-GasDynamic) equations ([4, 7, 5]) and a set of boundary conditions are used for description of the gas flow. QGD equations are generalize well known Navier-Stokes system and differ from it in additional dissipative terms with a small parameter.

This system binds together gasdynamic variables density, velocity, energy. Additional variables pressure and temperature are calculated according with corresponding gas state equations. QGD system has a form of conservation laws (of mass, impulse and energy) and in general terms is written as:

$$\frac{\partial \rho}{\partial t} + \mathrm{div}\mathbf{j}_m = 0, \tag{1}$$

$$\frac{\partial (\rho \mathbf{u})}{\partial t} + \mathrm{div}\left(\mathbf{j}_m \otimes \mathbf{u}\right) + \nabla p = \mathrm{div}\Pi, \tag{2}$$

$$\frac{\partial}{\partial t}\left[\rho\left(\frac{\mathbf{u}^2}{2} + \varepsilon\right)\right] + \mathrm{div}\left[\mathbf{j}_m\left(\frac{\mathbf{u}^2}{2} + \varepsilon\right)\right] = \mathrm{div}\mathbf{A} - \mathrm{div}\mathbf{q}. \tag{3}$$

In present work the computation domain is an area inwards closed piecewise linear triangulable boundary. There are no other limitations on the form of computation domain. The geometry of space object is specified as combination of piecewise linear elements and spline surfaces. Example of computation domain is presented on Fig. 1(a). Arrows show the direction of gas flow. X, Y, Z are the axes of cartesian space. Boundary conditions representing oncoming flow are held on one side of computational domain (marked as *ABCD* on Fig. 1(a)). All other bounds allow the gas to run freely out of the computational area. Wall boundary condition specified for the object surface, with no mass flux through the object under consideration. Figure 1(b) shows the example of real computation domain truncated in according to the symmetry of problem with boundary conditions marked on the faces (1 - inlet boundary, 2 - outlet boundary, 3 - planes of symmetry, 4 - surface of the object).

The flow is considered to be quasy-stationary and calculated until establishing. Parameters of the model problem are shown in Table 1.

3 Numerical Implementation

For numerical solution of QGD equations system (1)–(3) we use tetrahedral mesh ([2, 1]). Specially designed mesh generator allows to create locally condensing adaptive meshes in 3D space. This local condensing allows to balance optimal the calculations precision and computational complexity. The number of nodes in the

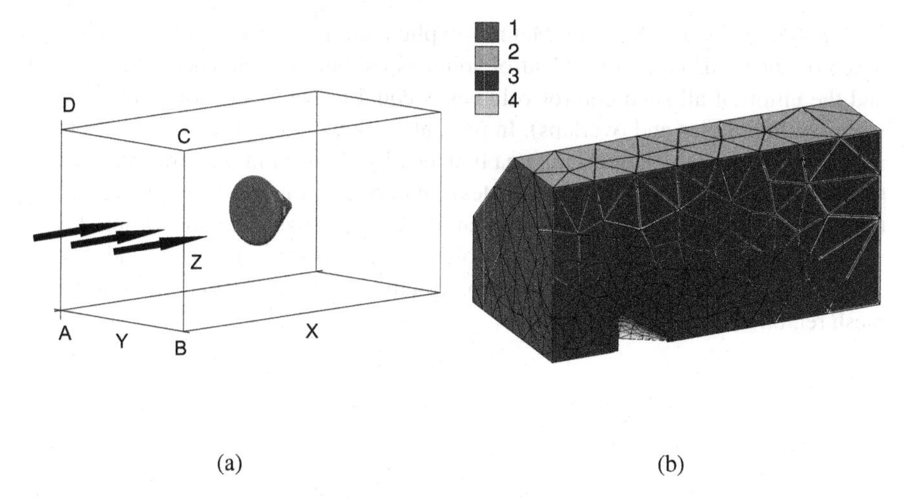

(a) (b)

Fig. 1. Computation domain examples

Table 1. Model problem parameters

Background gas temperature	T_0 (K)	266
Temperature of head shield	T_1 (K)	$2000-3300$
Peak shock wave front temperature	T_2 (K)	12000
Pressure	P_0 (Pa)	$4.3 \cdot 10$
Density	ρ_0 $(\frac{kg}{m^3})$	$5.63 \cdot 10^{-4}$
Characteristic size of problem	L_0 (m)	3.9
Mach number	M_∞	12
Reynolds number	Re	1000

used meshes is vary from 10^5 to $2 \cdot 10^6$. Example of tetrahedral mesh is presented on Fig. 1(b).

Adaptive mesh generator was developed using TetGen ([8]) library that allows to tetrahedralize piecewise linear complexes and a set of vertices in 3D space. In accordance with predefined distribution function additional vertices are added to computation domain for local mesh condesing. This approach allows to generate calculation meshes satisfying the following parameters:

- minimal number of mesh vertices;
- predefined maximal volume of calculation cell in areas of solution peculiarity (for example areas of high density);
- necessary mesh condensing for boundary layer modelling;
- maximal volume of mesh tetrahedra;
- maximal relation of circumsphere radius to minimal cell edge length;
- in addition, the maximal volume of adjacent tetrahedral cells can be limited for each geometry face.

Applying Control Volume Method explicit numerical scheme was designed based on the QGD equations. Control volumes are built around each node of mesh and the union of all such control volumes is equal to the whole computational domain (without holes and overlaps). In present work we use barycentric control volumes. Each control volume is an area bounded by faces build on tetrahedra centers (arithmetic mean of its nodes coordinates), centers of tetrahedra faces and centers of tetrahedra edges in all tetrahedra adjacent to referred node. Example of tetrahedral cell is shown on Fig. 2(a) with control volume elements (faces) built around vertices of this tetrahedra. One complete control volume around a calculation node inside mesh region is shown on Fig. 2(b).

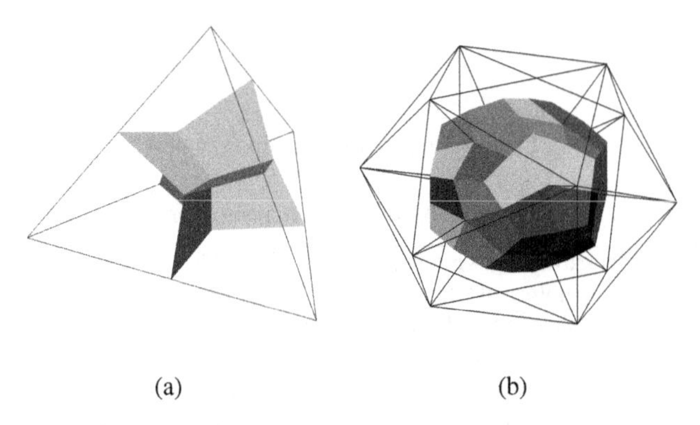

Fig. 2. Tetrahedral cell with control volume elements (a) and tetrahedral mesh region with control volume (b)

The solution of initial boundary problem is calculated by time-explicit finite-difference scheme. Spatial derivatives are approximated with second order of accuracy while time derivatives are approximated with first order.

Stability of numerical method is provided by QGD-additives with a small multiplier as parameter. This parameter has the dimension of time and its order is about free path of the molecule. Thus it can be treated as artificial (or numerical) viscosity.

4 Parallelization

Parallelization of QGD calculations code is done using domain decomposition technique. The computational domain is divided into subdomains with near to equal numbers of nodes. The number of subdomains is equal to number of processors. Number of connections between subdomains (number of shared calculation nodes) is minimized to achieve low communications overhead. The code uses MPI (Message Passing Interface) library and is highly portable being written in C++. Also for hybrid systems (where one node of distributed memory system actually is a shared memory parallel system) the OpenMP technology is applied.

Used parallel systems:

- Smaller cases were run on the cluster in IMAMOD (Institute for Mathematical Modeling). This consists of 48 CPU cores (Intel Xeon 3.06GHz microprocessors). Peak performance is 300 GFlops and peak transfer rate is 1GBit/sec (Ethernet).
- Larger cases were run on the "MVS-100k" cluster in Joint SuperComputer Center (JSCC). This consists of 6144 CPU cores (Intel Xeon 5160 3.0Ghz microprocessors). Peak performance is 45 TFlops and peak transfer rate is 1400 Mbyte/sec (Infiniband DDR).
- The second high performance cluster used for calculations was "SKIF MSU" (Moscow State University - Research Computing Center). This consists of 5000 CPU cores (Intel Xeon E5472 3.0Ghz microprocessors). Peak performance is 60 TFlop/s and peak transfer rate is 1540 Mbyte/sec (Infiniband DDR).

Efficiency graphs for parallel algorithm for shared memory architecture are presented on Fig. 3. This code produces one MPI process per each CPU core. For 1 Million grid nodes the effectiveness is 75% at 128 processors and goes below 40% at 512 processors. For 2 Millions nodes grid configuration the effectiveness is about 76% at 100 processors and falls to 50% at 600 processors.

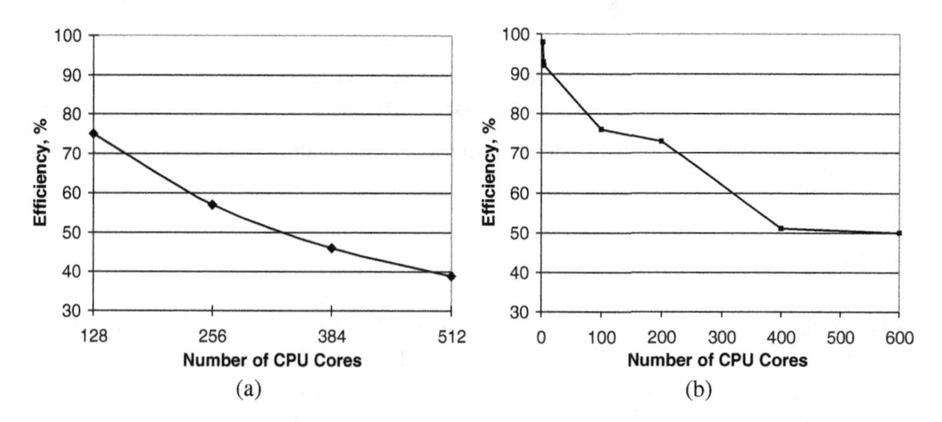

Fig. 3. Parallelization efficiency for 10^6 nodes task, 1 MPI process per core on "SKIF MSU" (a) and efficiency for $2 \cdot 10^6$ nodes, 1 MPI process per core on "MVS-100k" (b)

Systems like "MVS-100k" and "SKIF MSU" that have several CPU cores on one computation node allow to realize hybrid parallelization approach and achieve performance several times higher. In this case we use MPI to distribute tasks among nodes in cluster while applying OpenMP to distribute computational load among cores on one node. Thus we eliminate the communications overhead inside this small group (8 for mentioned systems) of computation cores.

Practically we can use 3-5 times more CPU maintaining the same efficiency and obtain the final result about 3 times faster. This comparison is illustrated on Fig. 4.

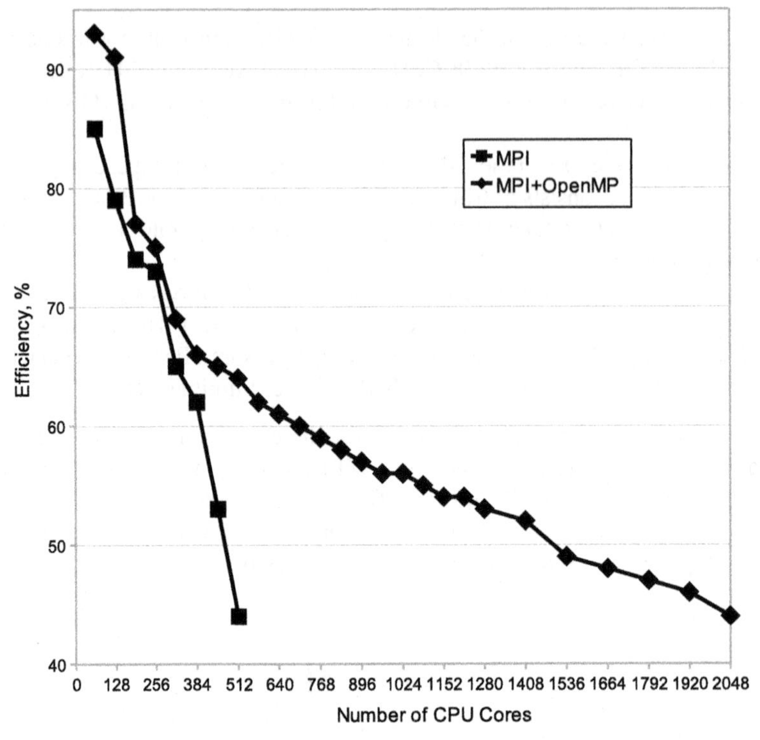

Fig. 4. Comparison of pure MPI parallelization efficiency with hybrid MPI+OpenMP approach ("MVS-100k")

5 Test case

One of the test cases is Apollo project capsule model. The diameter of the capsule is about 4 meters. Note that total mass of the capsule is 5800kg and the thermal protection mass is about 850kg. Considering the model is axisymmetric we can truncate it and calculate only one fourth of the original domain. Another parameters of model problem are shown in Table 1.

The result of simulation is shown on Fig. 5.

6 Conclusions

This paper is intended to show the evolution of previous work ([6, 3]).

Calculated gasdynamic variables can be refined on each time step using, for example, radiative heat transfer model. In this approach QGD calculations module passes energy, pressure and temperature fields to RGD (Radiative GasDynamics) module and receives adjusted energy field back.

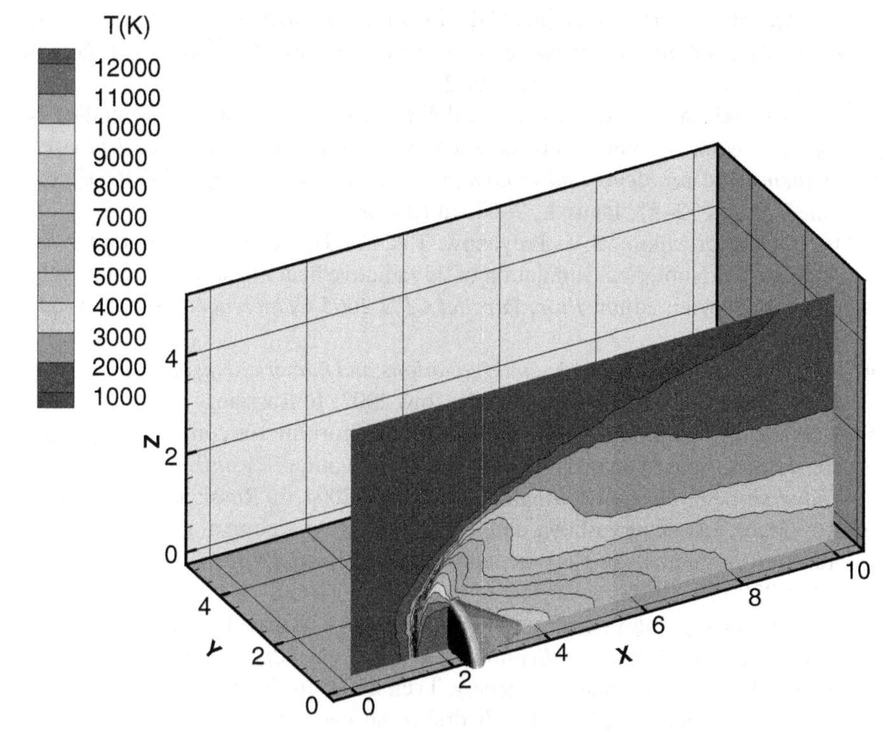

Fig. 5. Temperature distribution (K) around Apollo capsule model at speed 12Ma

Introduced model can be applied to objects with very complicated shape including real space vehicles. Improved stability of numerical method allows to take into account additional factors such as radiation and chemical processes.

Grid generator developed allows to construct adaptive locally condensing tetrahedral grids with high number of nodes (up to 10^9) suitable for parallel computations. The parallel realization of numerical algorithm allows to take advantage of high perfomance computer clusters and obtain results in a reasonable time using very precise grids. The algorithm is designed for distributed and hybrid memory systems and based on geometrical parallelism principle. Parallel algorithm shows good efficiency on different type of parallel systems from low-cost clusters (50-100 CPU cores, with Ethernet transport) to high perfomance parallel systems (2000-6000 CPU cores connected via Myrinet or Infiniband networks).

7 Acknowledgments

The work was performed under research grants: RFBR N06-01-00233, N08-07-00458.

[1] I.V. Abalakin, A. Dervieux, and T.K. Kozubskaya. High accuracy finite volume method for solving nonlinear aeroacoustics problems on unstructured meshes. *Chinese Journal of Aeroanautics*, 19(2), 2006.

[2] I.V. Abalakin and S.A. Sukov. Modeling of external flow around bodies on multiprocessor systems using thetraedral meshes. In *Fundamental physico-mathematical problems and modeling of technico-technological systems*, volume 7, pages 52–57. Janus-K, 2004. In Russian.

[3] B.N. Chetverushkin, S.V. Polyakov, T.A. Kudryashova, E. Kononov, and A. Sverdlin. Numerical simulation of 2d radiative heat transfer for reentry vehicles. In Elsevier, editor, *Proc. Parallel CFD 2005 Conference*, pages 293–299, 2006.

[4] T.G. Elizarova. *Quasi-gasdynamic equations and numerical methods for viscous flow simulation*. Scientific world, Moscow, 2007. In Russian.

[5] T.G. Elizarova and M. Sokolova. Numerical algorithm for computing of supersonic flows is based on quasi-hydrodynamics equations. *Vestnik of Moscow State University, Physics and Astronomy*, (1):10–15, 2004. In Russian.

[6] I.A. Graur, T.A. Kudryashova, and S.V. Polyakov. Modeling of flow for radiative transport problems. In Elsevier, editor, *Proc. Parallel CFD 2004 Conference*, pages 239–245, 2005.

[7] Y.V. Sheretov. Analysis of modified kinetically-consistent scheme stability in acoustic approach. In *Application of functional analisis in approach theory*, pages 147–160. Tver State University, Tver, 2004. In Russian.

[8] Hang Si. Tetgen: A quality tetrahedral mesh generator. `http://tetgen.berlios.de/`.

Effective Parallel Computation of Incompressible Turbulent Flows on Non-uniform Grid

Hidetoshi Nishida[1] and Nobuyuki Ichikawa[2]

[1] Department of Mechanical and System Engineering, Kyoto Institute of Technology, Matsugasaki, Sakyo-ku, Kyoto 606-8585, Japan nishida@kit.ac.jp
[2] Department of Mechanical and System Engineering, Kyoto Institute of Technology, Matsugasaki, Sakyo-ku, Kyoto 606-8585, Japan

In order to improve the parallel efficiency of incompressible turbulent flow solver on non-uniform grid, the multigrid technique with the checkerboard SOR method or the rational Runge-Kutta (RRK) scheme for the elliptic partial differential equation solver is presented. The 3D test problem and the DNS of 3D turbulent channel flow with the Reynolds number $Re_{\tau} = 180$ are considered. The results show that the chackerboard SOR relaxation has the property that the computational time is shorter but parallel efficiency is lower than the RRK relaxation. Then, it is necessary to improve the parallel performance, but the present approach has the possibility of short computational time with high parallel efficiency.

1 Introduction

The incompressible flow simulations are usually based on the incompressible Navier-Stokes equations. In the incompressible Navier-Stokes equations, we have to solve not only the momentum equations but also the elliptic partial differential equation (PDE) for the pressure, stream function and so on. The elliptic PDE solvers consume the large part of total computational time, because we have to obtain the converged solution of this elliptic PDE at every time step. In order to analyze the turbulent flows, the large-scale simulations have to be carried out. In these large-scale incompressible flow simulations, the acceleration of convergence for the elliptic PDE solver is necessary. On the other hand, the wall turbulence is usually simulated on non-uniform grid in the wall direction in order to ensure the resolution. In this case, the convergence of elliptic PDE on non-uniform grid becomes slower than the convergence on uniform grid. Then, for the incompressible flow simulations, especially the large-scale simulations, the efficient elliptic PDE solver on non-uniform grid is very important key technique.

In the parallel computations, the incompressible Navier-Stokes equations, that is, the momentum equations, can be solved with almost theoretical speedup on the parallel platform ([6, 4]). However, the parallel performance of the elliptic PDE

D. Tromeur-Dervout (eds.), *Parallel Computational Fluid Dynamics 2008*,
Lecture Notes in Computational Science and Engineering 74,
DOI: 10.1007/978-3-642-14438-7_44, © Springer-Verlag Berlin Heidelberg 2010

solver becomes lower than the momentum equations solver. Then, for the practical large-scale simulations, the elliptic PDE solver of the multigrid method with high parallel efficiency on the non-uniform grid has high possibility.

In this paper, the multigrid technique with the checkerboard SOR method or the rational Runge-Kutta (RRK) scheme ([8]) on non-uniform grid is presented. For the simple interpolations between grid levels, that is, restriction and prolongation, the elliptic PDE is transformed to the computational plane. The present elliptic PDE solver is applied to the direct numerical simulation (DNS) of 3D turbulent channel flows. The message passing interface (MPI) library is applied to make the computational codes. These MPI codes are implemented on personal cluster system with Pentium D(3.2GHz) processors. The cluster network is constructed by the gigabit ethernet. The computational MPI codes are compiled by Intel fortran 9.1.

2 Computational Technique

The incompressible Navier-Stokes equations in the Cartesian coordinates can be written by

$$\frac{\partial u_i}{\partial x_i} = 0, \tag{1}$$

$$\frac{\partial u_i}{\partial t} + u_j \frac{\partial u_i}{\partial x_j} = -\frac{\partial p}{\partial x_i} + \nu \frac{\partial^2 u_i}{\partial x_i \partial x_i}, \tag{2}$$

where u_i $(i = 1,2,3)$ denotes the velocity, p the pressure and ν the kinematic viscosity.

2.1 Variable Order Method of Lines

The solution procedure of the incompressible Navier-Stokes equations (1) and (2) is based on the fractional step approach on the collocated grid system. In the fractional step approach, first, the fractional step velocity u_i^* is estimated by

$$u_i^* = u_i^n + \alpha \Delta t \left(-u_j \frac{\partial u_i}{\partial x_j} + \nu \frac{\partial^2 u_i}{\partial x_i \partial x_i} \right)^n. \tag{3}$$

Next, the fractional step velocity at the staggered locations is computed by the interpolation. Then, the velocity at next time step u_i^{n+1} can be obtained by

$$u_i^{n+1} = u_i^* - \alpha \Delta t \frac{\partial p^{n+1}}{\partial x_i}. \tag{4}$$

Substituting the velocity at next time step into the discrete continuity equation, the pressure equation

$$\frac{\partial^2 p^{n+1}}{\partial x_i \partial x_i} = \frac{1}{\alpha \Delta t} \frac{\partial \overline{u_i^*}^{x_i}}{\partial x_i}, \tag{5}$$

is obtained. In these relations, α is the parameter determined by the time integration scheme. The overbar denotes the interpolated value from the collocated location to the staggered location.

In the method of lines approach, the spatial derivatives are discretized by the appropriate scheme, so that the partial differential equations (PDEs) in space and time are reduced to the system of ordinary differential equations (ODEs) in time. The resulting ODEs are integrated by the Runge-Kutta type time integration scheme.

In the spatial discretization, the convective terms are approximated by the variable order proper convective scheme ([4]), because of the consistency of the discrete continuity equation, the conservation property, and the variable order of spatial accuracy. This scheme is the extension of the proper convective scheme ([2]) to the variable order. The variable order proper convective scheme can be described by

$$u_j \frac{\partial u_i}{\partial x_j}|_{\mathbf{x}} = \sum_{\ell=1}^{M/2} c_{\ell'} \overline{\overline{u_j}^{x_j} \frac{\delta_{\ell'} u_i}{\delta_{\ell'} x_j}}^{\ell' x_j} |_{\mathbf{x}}, \quad (\ell' = 2\ell - 1), \tag{6}$$

where M denotes the order of spatial accuracy, and the operators in eq.(6) are defined by

$$\overline{f}^{\ell' x_j}|_{\mathbf{x}} = \frac{1}{2}(f_{\mathbf{x}_j+\ell'/2} + f_{\mathbf{x}_j-\ell'/2}), \tag{7}$$

$$\overline{u_j}^{x_j}|_{\mathbf{x}_j \pm \ell'/2} = \sum_{m=1}^{M/2} \frac{c_{m'}}{2} [u_j|_{\mathbf{x}_j \pm \{\ell'+m'\}/2} + u_j|_{\mathbf{x}_j \pm \{\ell'-m'\}/2}], \tag{8}$$

$$\frac{\delta_{\ell'} u_i}{\delta_{\ell'} x_j}|_{\mathbf{x}_j \pm \ell'/2} = \frac{\pm 1}{\ell' \Delta x_j}(u_i|_{\mathbf{x}_j \pm \ell'} - u_i|_{\mathbf{x}_j}), \tag{9}$$

where $m' = 2m - 1$. In this technique, the arbitrary order of spatial accuracy can be obtained automatically by changing only one parameter M. The coefficients $c_{\ell'}$ and $c_{m'}$ are the weighting coefficients and Δx_j denotes the grid spacing in the x_j direction.

On the other hand, the diffusion terms are discretized by the modified differential quadrature (MDQ) method ([7]) as

$$\frac{\partial^2 u_i}{\partial x_i \partial x_i}|_{\mathbf{x}} = \sum_{m=-M/2}^{M/2} \Phi_m''(\mathbf{x}) u_i|_{\mathbf{x}_i+m}, \tag{10}$$

where $\Phi_m''(\mathbf{x})$ is the second derivative of the function $\Phi_m(\mathbf{x})$ defined by

$$\Phi_m(\mathbf{x}) = \frac{\Pi(\mathbf{x})}{(\mathbf{x} - \mathbf{x}_{i+m})\Pi'(\mathbf{x}_{i+m})}, \tag{11}$$

$$\Pi(\mathbf{x}) = (\mathbf{x} - \mathbf{x}_{i-M/2}) \cdots (\mathbf{x} - \mathbf{x}_i) \cdots (\mathbf{x} - \mathbf{x}_{i+M/2}). \tag{12}$$

The coefficients of the variable order proper convective scheme, $c_{\ell'}$, can be computed automatically by using the MDQ coefficients. Then, the incompressible Navier-Stokes equations are reduced to the system of ODEs in time. This system of ODEs is integrated by the Runge-Kutta type scheme.

2.2 Pressure Equation Solver

The pressure equation, eq.(5), is solved by the checkerboard SOR method and the variable order multigrid method ([5]).

In the variable order multigrid method, the unsteady term is added to the pressure equation. Then, the pressure (elliptic) equation is transformed to the parabolic equation in space and pseudo-time, τ.

$$\frac{\partial p^{n+1}}{\partial \tau} = \frac{\partial^2 p^{n+1}}{\partial x_i \partial x_i} - \frac{1}{\alpha \Delta t} \frac{\partial \overline{u_i^*}^{x_i}}{\partial x_i}. \tag{13}$$

Equation (13) can be solved by the variable order method of lines. The spatial derivatives are discretized by the aforementioned MDQ method, so that eq.(13) is reduced to the system of ODEs in pseudo-time,

$$\frac{d\overrightarrow{p^{n+1}}}{d\tau} = \overrightarrow{L}(\overrightarrow{p^{n+1}}). \tag{14}$$

This system of ODEs in pseudo-time is integrated by the rational Runge-Kutta (RRK) scheme, because of its wider stability region. Then, the same order of spatial accuracy as the momentum equations can be specified.

In addition, the multigrid technique, i.e., correction storage algorithm ([1]), is incorporated into the method in order to accelerate the convergence. In the multigrid method, the checkerboard SOR method and RRK scheme are interpreted as the relaxation method.

In this paper, the multigrid method on non-uniform grid is considered. We prepare two approaches. One is that the values on coarse and fine grids are interpolated on the corresponding non-uniform grid. Another is that the pressure equation is transformed to the computational plane, so that the values are interpolated on the computational plane with uniform grid.

3 Numerical Results

3.1 3D Test Problem

We consider the following 3D test problem.

$$\frac{\partial^2 p}{\partial x_i \partial x_i} = -3sin(x_1)cos(x_2)sin(x_3). \tag{15}$$

The periodic boundary condition is imposed in $x_1(=x)$ and $x_3(=z)$ directions. In $x_2(=y)$ direction, the Neumann boundary condition, $\frac{\partial p}{\partial y} = 0$, is imposed. The computational domain is set $(0,0,0) < (x,y,z) \leq (2\pi,\pi,2\pi)$. The convergence criteron is $L_2 < 1.0 \times 10^{-6}$, where L_2 denotes the L2-residual of eq.(15). The non-uniform grid defined by

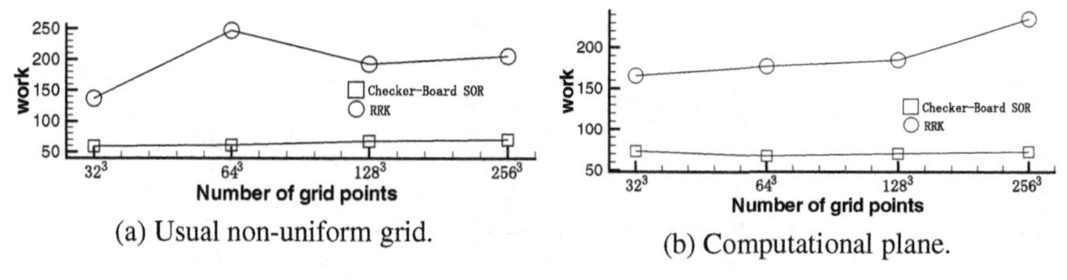

(a) Usual non-uniform grid.

(b) Computational plane.

Figure 1. Multigrid property on non-uniform grid.

$$y_j = \frac{tanh[\gamma(2j/N-1)]}{tanh(\gamma)}, \tag{16}$$

is used in y direction only. The stretching parameter γ is 1.80.

Figure 1 shows the comparison of work unit until convergence. Figure 1(a) is the result on usual non-uniform grid and (b) denotes the result on computational plane. The square and circular symbols are the work unit obtained by the checkerboard SOR and RRK relaxation, respectively. In the checkerboard SOR relaxation, the multigrid convergence is confirmed in both cases. On the other hand, the RRK relaxation shows the dependency on number of grid points. Also, the work unit until convergence obtained by the checkerboard SOR relaxation is less than that obtained by the RRK relaxation in all cases.

Next, in order to investigate the parallel efficiency, Fig.2 shows the parallel efficiency in this test problem. The parallel efficiency is defined by

$$Efficiency = \frac{T_{single}}{N \cdot T_{parallel}} \times 100\,(\%)\,, \tag{17}$$

where T_{single} and $T_{parallel}$ denote the computational time on single processor element (PE) and on N parallel PEs, respectively. In Fig.2 the present parallel efficiency is compared with the usual parallel efficiency on non-uniform grid. On 2PEs the high parallel efficiency in which is larger than 80% can be obtained in the large-scale computations, that is, 128^3 and 256^3 grid points. However, the parallel efficiency becomes lower on 4PEs. In this case, the performance of the RRK relaxation on computational plane presents the highest efficiency.

3.2 DNS of 3D Turbulent Channel Flow

In order to simulate the 3D turbulent channel flow, the incompressible Navier-Stokes equations, eq.(1) and eq.(2), are nondimensionalized by the friction velocity $u_\tau(=\sqrt{\tau_{wall}/\rho})$ and half length of channel h. Table 1 shows the computational conditions. In the table, the conditions of reference simulation ([3]) are shown, too. The periodic boundary conditions in x and z directions and the Neumann boundary condition in y direction are imposed. In y direction, the non-uniform grid with $\gamma = 1.80$ in eq.(16) is adopted. The DNS of 3D turbulent channel flow with $Re_\tau = 180$ is computed on 32^3, 64^3, and 128^3 grid points. The present results are compared with reference solution obtained by Moser et al.. Figure 3 shows the comparison of the mean streamwise

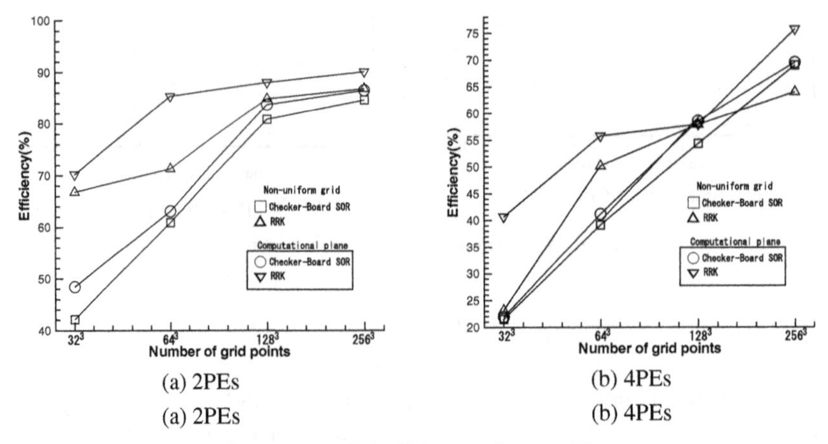

(a) 2PEs (b) 4PEs

(a) 2PEs (b) 4PEs

Figure 2. Parallel efficiency of test problem.

Table 1. Compuational conditions for 3D turbulent channel flow.

	Present			Moser et al.
Grid points	32^3	64^3	128^3	$128 \times 129 \times 128$
Region	$2\pi h \times 2h \times \pi h$	$\rightarrow$	$4\pi h \times 2h \times \pi h$	$4\pi h \times 2h \times 1.3\pi h$
Δx^+	35.3	17.7	17.7	17.7
Δy^+	2.47∼21.3	1.17∼10.7	0.57∼5.4	0.16∼4.4
Δz^+	17.7	8.8	4.4	5.9
Scheme		2nd order FDM		spectral method

velocity, turbulent intensity, and iso-surface of high and low speed regions of u' and second invariant of velocity gradient tensor. The present DNS results, especially high resolution results, are in very good agreement with the reference solution. It is clear that the present instantaneous flow field, Fig.3(c), shows the characteristic features of turbulent channel flow.

On the parallel efficiency, Table 2 shows the present parallel efficiency with checkerboard SOR and RRK relaxations on 64^3 and 128^3 grid points. The chackerboard SOR relaxation has the property that the computational time is shorter but parallel efficiency is lower than the RRK relaxation. On the contrary, the RRK relaxation gives the higher performance. The RRK relaxation, however, consumes large computational time.

4 Concluding Remarks

In this work, the multigrid technique with the checkerboard SOR method or the RRK scheme for solving the elliptic PDE, i.e., the pressure equation, is presented, in order to improve the parallel efficiency of incompressible turbulent flow solver on non-uniform grid. The 3D test problem with the periodic and the Neumann boundary conditions and the DNS of 3D turbulent channel flow with the Reynolds number

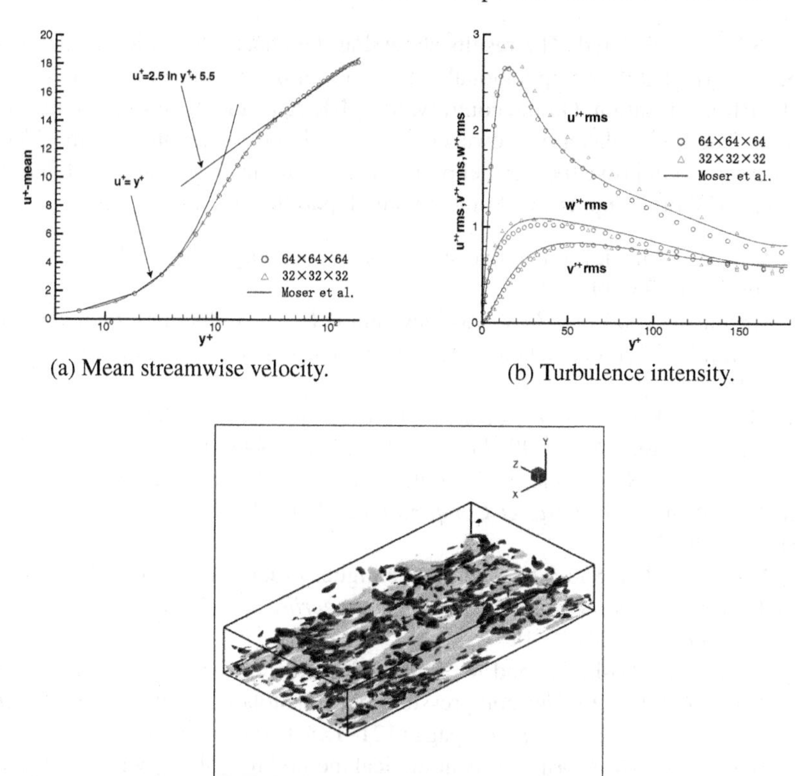

(a) Mean streamwise velocity.

(b) Turbulence intensity.

(c) Iso-surface of high and low speed regions of u' and second invariant of velocity gradient tensor (64^3 grid).

Figure 3. Statistical quantities and instantaneous flow field.

Table 2. Parallel efficiency for 3D turbulent channel flow.

	Checkerboard SOR		RRK	
$64 \times 64 \times 64$	time (sec/step)	efficiency (%)	time (sec/step)	efficiency (%)
single	0.870	-	6.573	-
parallel(2PEs)	0.708	61.47	4.937	66.58
parallel(4PEs)	0.727	29.93	4.482	36.67
$128 \times 128 \times 128$	time (sec/step)	efficiency (%)	time (sec/step)	efficiency (%)
single	12.678	-	133.594	-
parallel(2PEs)	8.136	77.91	76.205	87.65
parallel(4PEs)	6.794	46.65	59.644	56.00

$Re_\tau = 180$ are considered. The results show that the chackerboard SOR relaxation has the property that the computational time is shorter but parallel efficiency is lower than the RRK relaxation. On the contrary, the RRK relaxation gives the higher performance. The RRK relaxation, however, consumes large computational time. Then, it is necessary to improve the parallel performance, but the present approach has the possibility of short computational time with high parallel efficiency.

[1] A. Brandt. Multi-level adaptive solutions to boundary-value problems. *Math. Comput.*, 31:333–390, 1977.
[2] Y. Morinishi, T.S. Lund, O.V. Vasilyev, and P. Moin. Fully conservative higher order finite difference schemes for incompressible flow. *J. Comput. Phys.*, 143:90–124, 1998.
[3] R. Moser, J. Kim, and N. Mansour. Direct numerical simulation of turbulent channel flow up to $Re = 590$. *Physics of Fluids*, 11:943–945, 1999.
[4] H. Nishida, S. Nakai, and N. Satofuka. Parallel efficiency of a variable order method of lines. In *Parallel Computational Fluid Dynamics*, pages 321–328. Elsevier, 2003.
[5] H. Nishida and N. Satofuka. Automatic higher-order multi grid method and its application. *Memoirs of the Faculty of Engineering and Design, Kyoto Institute of Technology*, 36:24–35, 1987.
[6] H. Nishida, H. Yoshioka, and M. Hatta. Higher order parallel DNS of incompressible turbulence using compressible Navier-Stokes equations. In *Parallel Computational Fluid Dynamics*, pages 121–128. Elsevier, 2005.
[7] N. Satofuka and K. Morinishi. A numerical method for solving vlasov equation. *NASA TM*, 81339, 1982.
[8] A. Wambecq. Rarional Runge-Kutta method for solving system of ordinary differential equations. *Computing*, 20:333–342, 1978.

Secondary flow structure of turbulent Couette-Poiseuille and Couette flows inside a square duct

Hsin-Wei Hsu[1], Jian-Bin Hsu[1], Wei Lo[1] and Chao-An Lin[1,2]

[1] Department of Power Mechanical Engineering, National Tsing Hua University, Hsinchu 30013, Taiwan
[2] calin@pme.nthu.edu.tw, corresponding author

Turbulent Couette-Poiseuille and Couette flows inside a square duct at bulk Reynolds number 10,000 are investigated using the Large Eddy Simulations. The mean secondary flow is observed to be modified by the presence of moving wall where the symmetric vortex pattern vanishes. Secondary flow near the top corner shows a gradual change of vortex size and position as the moving wall velocity increased. It is interesting to note that a linear relation exits between the angle and the parameter $r = \frac{W_w}{W_{Bulk}}$, and a change in slope occurs at $r \sim 1.5$. Near the moving wall due to the reduction of the streamwise velocity fluctuation at the moving wall, turbulence structure gradually moves towards a rod-like axi-symmetric turbulence as r increases. As the wall velocity increases further for $r > 1.5$, the rod like structure disappears, and turbulence reverts to a disk like structure.

1 Introduction

Turbulent Poiseuille, Couette-Poiseuille or Couette flows inside a square or rectangular cross-sectional duct are of considerable engineering interest because their relevance to compact heat exchangers and gas turbine cooling systems. The most studied problem is the turbulent Poiseuille type flow inside a square duct and is characterized by the existence of secondary flow of Prandtl's second kind ([16]) which is not observed in circular ducts nor in laminar rectangular ducts. The secondary flow is a mean circulatory motion perpendicular to the streamwise direction driven by the turbulence. Although weak in magnitude (only a few percent of the streamwise bulk velocity), secondary flow is very significant with respect to momentum and heat transfer.

There are investigations directed to explore the influences of the bounding wall geometry, non-isothermal effect, free surface and system rotation on the secondary flow pattern within turbulent Poiseuille duct flows ([20], [15], [2]). The above investigations have implied that with careful manipulation, the secondary flow is very

D. Tromeur-Dervout (eds.), *Parallel Computational Fluid Dynamics 2008*, Lecture Notes in Computational Science and Engineering 74,
DOI: 10.1007/978-3-642-14438-7_45,

much promising on enhancement of particle transport or heat transfer in different industrial devices. Also, the turbulence anisotropy in non-circular ducts could be modified by bounding wall geometry, heating, free surface and system rotation. Previous studies on turbulent Couette-Poiseuille flows have been conducted on simple plane channels. [19] found negative production of streamwise turbulence near the forward moving wall. [8, 4, 5] identified different turbulence statistics and structures between the stationary and moving wall.

However, little is known about the effect of the moving wall on the secondary flow structure. [11, 12] found that the secondary flow structure correlates with the ratio of the speed of the moving wall and duct bulk flow, albeit the ratio was less than 1.17. In the present study, focus is also directed to the influences of the moving wall on the secondary flow pattern and hence turbulence structure, but at elevated ratio of moving wall velocity and duct bulk velocity.

2 Governing Equations and Modeling

The governing equations are grid-filtered, incompressible continuity and Navier-Stokes equations. In the present study, the Smagorinsky model ([17]) has been used for the sub-grid stress(SGS).

$$\tau_{ij}^{s} = -(C_s\Delta)^2 \frac{1}{\sqrt{2}} \sqrt{(S_{kl}S_{kl})} S_{ij} + \frac{2}{3} \rho k_{sgs} \delta_{ij} \quad (1)$$

where $C_s = 0.1, S_{ij} = \frac{\partial \bar{u}_i}{\partial x_j} + \frac{\partial \bar{u}_j}{\partial x_i}$, and $\Delta = (\Delta x \Delta y \Delta z)^{1/3}$ is the length scale. It can be seen that in the present study the mesh size is used as the filtering operator. A Van Driest damping function accounts for the effect of the wall on sub-grid scales is adopted here and takes the form as, $l_m = \kappa y[1 - exp(-\frac{y^+}{25})]$, where y is the distance to the wall and the length scale is redefined as, $\Delta = Min[l_m, (\Delta x \Delta y \Delta z)^{1/3}]$. Although other models which employed dynamic procedures on determining the Smagorinsky constant (C_s) might be more general and rigorous, the Smagorinsky model is computationally cheaper among other eddy viscosity type LES models. Investigations carried out by [1] on the turbulent Poiseuille flow through a straight and bent square duct have indicated that, the difference between the predicted turbulence statistics using dynamic models and Smagorinsky model is negligible.

3 Numerical and parallel Algorithms

A semi-implicit, fractional step method proposed by [7] and the finite volume method are employed to solve the filtered incompressible Navier-Stokes equations. Spatial derivatives are approximated using second-order central difference schemes. The non-linear terms are advanced with the Adams-Bashfoth scheme in time, whereas the Crank-Nicholson scheme is adopted for the diffusion terms. The discretized algebraic equations from momentum equations are solved by the preconditioned Conjugate Gradient solver. In each time step a Poisson equation is solved to obtain a divergence free velocity field. Because the grid spacing is uniform in the streamwise direction, together with the adoption of the periodic boundary conditions, fast Fourier transform (FFT) can be used to reduce the 3-D Poisson equation to uncoupled 2-D algebraic equations. Here, the fast Fourier transform is performed using FFTPACK ([18]). The algebraic equations are solved by the direct solver using LU decomposition.

As indicated earlier, periodic boundary condition is employed in the streamwise direction, thus the flow field is driven by the combined effects of the moving wall (Couette type) and the prescribed pressure gradient (Poiseuille type) in this direction. It is noted that a constant force is adopted in the momentum equation to represent the driving pressure gradient. In all the cases considered here the grid size employed is (128x96x96) in the spanwise, normal, and streamwise direction, respectively.

In the present parallel implementation, the single program multiple data (SPMD) environment on a distributed memory system using MPI is adopted. The domain decomposition is done on the last dimension of the three dimensional computation domain due to the explicit numerical treatment on that direction. The simulation is conducted on the HP Integrity rx2600 server (192 Nodes) with about 80 percent efficiency when 48 CPUs are employed. Linear speed-up is not reached in present parallel implementation mainly due to the global data movement required by the Fast Fourier Transform in the homogenous direction.

A schematic picture of the flows simulated is shown in Figure 1, where D is the duct width. We consider fully developed, incompressible turbulent Couette-Poiseuille flows inside a square duct where the basic flow parameters are summarized in Table 1. Reynolds number based on the bulk velocity (Re_{bulk}) is kept around 10,000 for all cases simulated and the importance of Couette effect in this combined flow field can be indicated by the ratio $r = (W_w/W_{Bulk})$ and $-\frac{D}{\rho W_w^2}\frac{\partial P}{\partial z}$. Due to the lack of benchmark data of the flow filed calculated here, the simulation procedures were first validated by simulating a turbulent Poiseuille flow at a comparable bulk Reynolds number (case P). The obtained results (see [10]) exhibit reasonable agreement with DNS data from [3]).

4 Results

4.1 Mean secondary flow structure

Mean streamwise velocity distributions from the top wall along the wall bisector, i.e. x/D=0.5, at different mean Couette strain rates are shown in Figure 1. For cases P and CP1, the velocity distributions follow closely the 2D channel flow DNS data of [14] (Re_τ : 395) and [6](Re_τ : 300). However, at higher Couette velocity due to the reduction of shear rate, departures from the logarithmic distributions are observed for cases CP2 - C, which are consistent with the findings of plane Couette-Poiseuille flow of [8]. It should be noted that for all cases considered logarithmic distributions prevail at the bottom wall, except in the vicinity of the side wall.

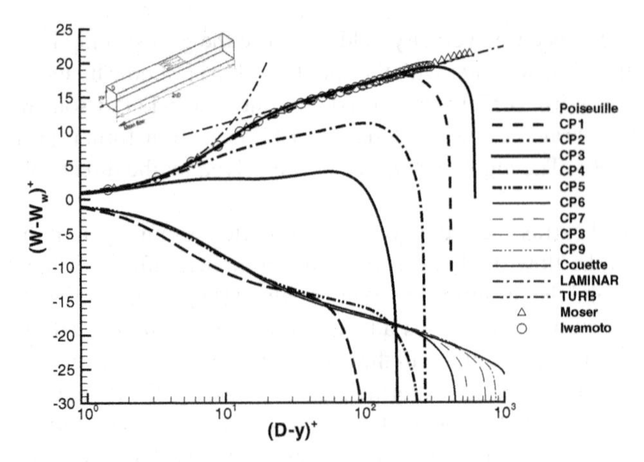

Fig. 1. Geometry and mean streamwise velocity along the wall-bisector.

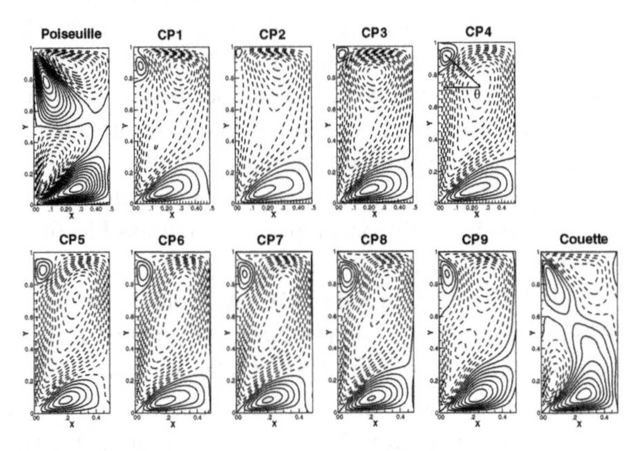

Fig. 2. Streamlines of mean secondary flow; solid lines for counter-clockwise rotation, dashed lines for clockwise rotation.

Streamlines of mean secondary flow for cases CP1 to C are shown in Figure 2. The vortex structure is clearly visible, where solid and dashed lines represent counter-clockwise and clockwise rotation, respectively. The angle formed by the horizonal x axis and the line joining the two vortex cores might become a good representation of the relative vortex positions. This angle is calculated and plotted against the parameter r defined by W_w/W_{bulk} which can be interpreted as the non-dimensional moving wall velocity. It is interesting to note that a linear relation exits between the angle and the parameter r, as shown in Figure 3 and a change in slope occurs at $r \sim 1.5$.

4.2 Anisotropy invariant map

The anisotropy invariant map (AIM) is introduced here in order to provide more specific description of the turbulence structures. The invariants of the Reynolds stress

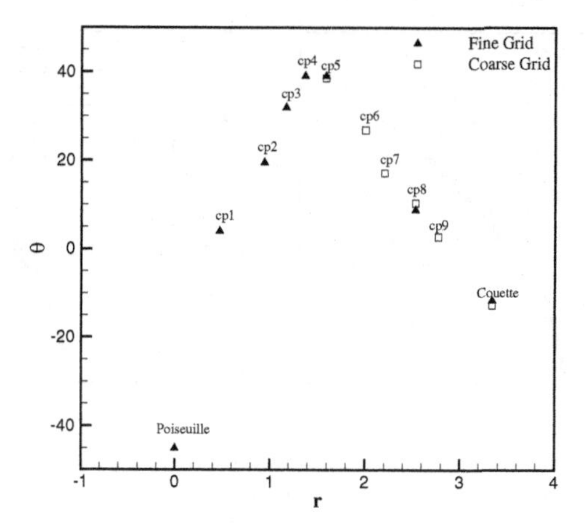

Fig. 3. Angles between vortex cores near the top corners

tensors are defined as $II = -(1/2)b_{ij}b_{ji}$, $III = (1/3)b_{ij}b_{jk}b_{ki}$ and $b_{ij} = < u'_i u'_j > / < u'_k u'_k > -1/3\delta_{ij}$. A cross-plot of $-II$ versus III forms the anisotropy invariant map. All realizable Reynolds stress invariants must lie within the Lumley triangle ([13]). This region is bounded by three lines, namely two component state, $-II = 3(III + 1/27)$, and two axi-symmetric states, $III = \pm(-II/3)^{3/2}$. For the axi-symmetric states, [9] described the positive and negative III as disk-like and rod-like turbulence, respectively. The intersections of the bounding lines represent the isotropic, one-component and two-component axi-symmetric states of turbulence.

The AIM along the horizontal wall bisector for cases CP1 to CP4 is presented in Figure 4. Near the stationary wall ($y/D \leq 0.5$), turbulence behaviors of different Couette-Poiseuille flows resemble those of the Poiseuille flow. In particular, the turbulence structure is similar to the plane channel flow, where turbulence approaches two-component state near the stationary wall due to the highly suppressed wall-normal velocity fluctuation. It moves toward the one-component state till $y^+ \sim 8$ ([20]) and then follows the positive III axi-symmetric branch (disk-like turbulence, [9]) to the isotropic state at the duct center. However, near the moving wall due to the reduction of the streamwise velocity fluctuation at the moving wall, turbulence structure gradually moves towards a rod-like axi-symmetric turbulence (negative III) as r increases. As the wall velocity increases further for $r > 1.5$, the rod like structure disappears, and turbulence reverts to the disk like structure, as is shown in Fig. 5.

5 Conclusion

The turbulent Couette-Poiseuille and Couette flows inside a square duct are investigated by present simulation procedures. Mean secondary flow is observed to be modified by the presence of moving wall where the symmetric vortex pattern vanishes. Secondary flow near the top corner shows a gradual change of vortex size and position as the moving wall velocity increased. The vortex pair consists of a

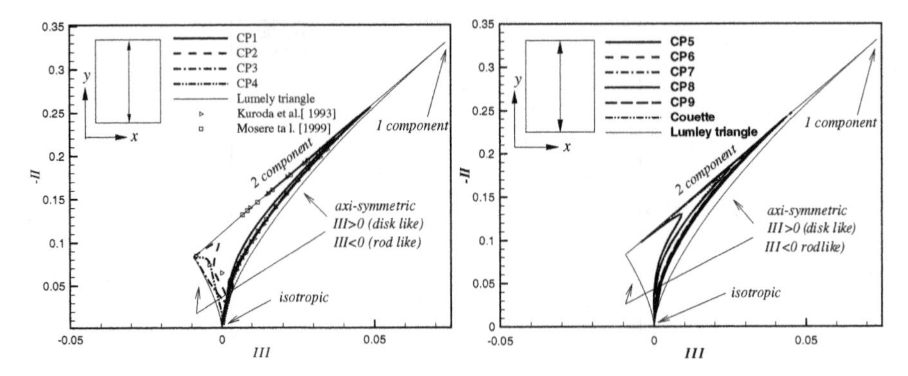

Fig. 4. Anisotropy invariant map - cases CP1-CP4

Fig. 5. Anisotropy invariant map - cases CP5-C

dominate (clock-wise) and relatively smaller (counter-clockwise) vortex. It is interesting to note that a linear relation exits between the angle and the parameter r, and a change in slope occurs at $r \sim 1.5$. Near the moving wall due to the reduction of the streamwise velocity fluctuation at the moving wall, turbulence structure gradually moves towards a rod-like axi-symmetric turbulence (negative III) as r increases. As the wall velocity increases further for $r > 1.5$, the rod like structure disappears, and turbulence reverts to the disk like structure.

6 Acknowledgments

This research work is supported by the National Science Council of Taiwan under grant 95-2221-E-007 -227 and the computational facilities are provided by the National Center for High-Performance Computing of Taiwan which the authors gratefully acknowledge.

[1] M. Breuer and W. Rodi. Large eddy simulation of turbulent flow through a straight square duct and a 180 degree bend. In P. R. Voke et al., editor, *Direct and Large-Eddy Simulation I*, pages 273–285. Kluwer Academic Publishers, 1994.

[2] R. Brogolia, A. Pascarelli, and U. Piomelli. Large eddy simulations of ducts with a free surface. *J. Fluid Mech.*, 484:223–253, 2003.

[3] A. Huser and S. Biringen. Direct numerical simulation of turbulent flow in a square duct. *J. Fluid Mech.*, 257:65–95, 1993.

[4] C. B. Hwang and C. A. Lin. Improved low-reynolds-number k-e model based on direct numerical simulation data. *AIAA J.*, 36:38–43, 1998.

[5] C. B. Hwang and C. A. Lin. Low-reynolds-number k-e modeling of nonstationary solid boundary flows. *AIAA J.*, 41:168–175, 2003.

[6] K. Iwamoto, Y. Suzuki, and N. Kasagi. Reynolds number effect on wall turbulence: toward effective feedback control. *International J. of Heat and Fluid Flow*, 23:678–689, 2002.

Table 1. The flow conditions for simulated cases; Re_τ is defined by mean friction velocity averaged over four solid walls (t=top,b=bottom wall); W_w denotes the velocity of the moving wall and W_{Bulk} is the bulk velocity; $Re_c = \frac{W_w D}{\nu}$; $r = \frac{W_w}{W_{Bulk}}$.

	$Re_{\tau t}$	$Re_{\tau b}$	Re_{Bulk}	Re_c	r	$-\frac{D}{\rho W_w^2}\frac{\partial P}{\partial z}$
Case P	600	600	9708	0	0	∞
Case CP1	441	605	9716	4568	0.47	0.0621
Case CP2	305	591	9760	9136	0.94	0.0138
Case CP3	284	587	9770	11420	1.17	0.0083
Case CP4	363	581	10012	13704	1.37	0.0054
Case CP5	452	576	10034	15998	1.59	0.0037
Case CP6	627	569	10266	20556	2.00	0.0019
Case CP7	712	570	10400	22840	2.20	0.0014
Case CP8	875	574	10812	27408	2.53	0.00077
Case CP9	998	580	11158	30834	2.77	0.00049
Case C	1167	512	10247	34260	3.34	0
Kuroda et al. (1993)	35	308	5178	6000	1.16	0.0026

[7] J. Kim and P. Moin. Application of a fractional-step method to incompressible naviervstokes equations. *J. Comput. Phys.*, 177:133–166, 1987.

[8] A. Kuroda, N. Kasagi, and M. Hirata. Direct numerical simulation of turbulent plane couettevpoiseuille flows: Effect of mean shear rate on the near wall turbulence structures. In *Proc. Turbulent Shear Flows*, page 241. Springer-Verlag, Berlin, 1993.

[9] M. J. Lee and W. C. Reynolds. Numerical experiments on the structure of homogeneous turbulence. Technical Report Report TF-24, Stanford University, Thermoscience Division, 1985.

[10] W. Lo. *Large eddy simulations of Couette-Poiseuille flows within square*. PhD thesis, National Tsing Hua University, Hsinchu, Taiwan, 2006.

[11] W. Lo and C. A. Lin. Mean and turbulence structures of couette-poiseuille flows at different mean shear rates in a square duct. *Phys. Fluids*, 18:068103, 2006.

[12] W. Lo and C. A. Lin. Prediction of secondary flow structure in turbulent couette-poiseuille flows inside a square duct. In *Proc. 2006 Parallel Computational Fluid Dynamics*, pages 173–180. Elsevier S.V., 2007.

[13] J. L. Lumley. Computational modeling of turbulent flows. *Adv. Appl. Mech.*, 18:123–176, 1978.

[14] R. Moser, J. Kim, and N. Mansour. Direct numerical simulation of turbulent channel flow up to retau=590. *Phys. Fluids*, 11:943–945, 1999.

[15] J. Pallares and L. Davidson. Large eddy simulations of turbulent heat transfer in stationary and rotating square ducts. *Phys. Fluids*, 14:2804–2816, 2002.

[16] L. Prandtl. Uber die ausgebildete turbulenz. In *Fur Tech. Mech., Zurich,[English translation NACA Tech. Memo. 435, 62]*, 1926.

[17] J. Smagorinsky. General circulation experiments with the primitive equations. I. The basic experiment. In *Mon. Weather Rev.*, volume 91 of *3*, pages 99–164. 1963.
[18] P. N. Swarztrauber. Fast fourier transforms algorithms for vector computers. *Parallel Computing*, pages 45–63, 1984.
[19] E. M. Thurlow and J. C. Klewicki. Experimental study of turbulent poiseuille-couette flow. *Phys. Fluids*, 4:865–875, 2000.
[20] M. Salinas Vazquez and O. Metais. Large eddy simulation of the turbulent flow through a heated square duct. *J. Fluid Mech.*, 453:201–238, 2002.

Editorial Policy

1. Volumes in the following three categories will be published in LNCSE:

i) Research monographs
ii) Tutorials
iii) Conference proceedings

Those considering a book which might be suitable for the series are strongly advised to contact the publisher or the series editors at an early stage.

2. Categories i) and ii). Tutorials are lecture notes typically arising via summer schools or similar events, which are used to teach graduate students. These categories will be emphasized by Lecture Notes in Computational Science and Engineering. **Submissions by interdisciplinary teams of authors are encouraged.** The goal is to report new developments – quickly, informally, and in a way that will make them accessible to non-specialists. In the evaluation of submissions timeliness of the work is an important criterion. Texts should be well-rounded, well-written and reasonably self-contained. In most cases the work will contain results of others as well as those of the author(s). In each case the author(s) should provide sufficient motivation, examples, and applications. In this respect, Ph.D. theses will usually be deemed unsuitable for the Lecture Notes series. Proposals for volumes in these categories should be submitted either to one of the series editors or to Springer-Verlag, Heidelberg, and will be refereed. A provisional judgement on the acceptability of a project can be based on partial information about the work: a detailed outline describing the contents of each chapter, the estimated length, a bibliography, and one or two sample chapters – or a first draft. A final decision whether to accept will rest on an evaluation of the completed work which should include

- at least 100 pages of text;
- a table of contents;
- an informative introduction perhaps with some historical remarks which should be accessible to readers unfamiliar with the topic treated;
- a subject index.

3. Category iii). Conference proceedings will be considered for publication provided that they are both of exceptional interest and devoted to a single topic. One (or more) expert participants will act as the scientific editor(s) of the volume. They select the papers which are suitable for inclusion and have them individually refereed as for a journal. Papers not closely related to the central topic are to be excluded. Organizers should contact the Editor for CSE at Springer at the planning stage, see *Addresses* below.

In exceptional cases some other multi-author-volumes may be considered in this category.

4. Only works in English will be considered. For evaluation purposes, manuscripts may be submitted in print or electronic form, in the latter case, preferably as pdf- or zipped ps-files. Authors are requested to use the LaTeX style files available from Springer at http://www. springer.com/authors/book+authors?SGWID=0-154102-12-417900-0.

For categories ii) and iii) we strongly recommend that all contributions in a volume be written in the same LaTeX version, preferably LaTeX2e. Electronic material can be included if appropriate. Please contact the publisher.

Careful preparation of the manuscripts will help keep production time short besides ensuring satisfactory appearance of the finished book in print and online.

5. The following terms and conditions hold. Categories i), ii) and iii):

Authors receive 50 free copies of their book. No royalty is paid.
Volume editors receive a total of 50 free copies of their volume to be shared with authors, but no royalties.

Authors and volume editors are entitled to a discount of 33.3 % on the price of Springer books purchased for their personal use, if ordering directly from Springer.

6. Commitment to publish is made by letter of intent rather than by signing a formal contract. Springer-Verlag secures the copyright for each volume.

Addresses:

Timothy J. Barth
NASA Ames Research Center
NAS Division
Moffett Field, CA 94035, USA
barth@nas.nasa.gov

Michael Griebel
Institut für Numerische Simulation
der Universität Bonn
Wegelerstr. 6
53115 Bonn, Germany
griebel@ins.uni-bonn.de

David E. Keyes
Mathematical and Computer Sciences
and Engineering
King Abdullah University of Science
and Technology
P.O. Box 55455
Jeddah 21534, Saudi Arabia
david.keyes@kaust.edu.sa

and

Department of Applied Physics
and Applied Mathematics
Columbia University
500 W. 120 th Street
New York, NY 10027, USA
kd2112@columbia.edu

Risto M. Nieminen
Department of Applied Physics
Aalto University School of Science
and Technology
00076 Aalto, Finland
risto.nieminen@tkk.fi

Dirk Roose
Department of Computer Science
Katholieke Universiteit Leuven
Celestijnenlaan 200A
3001 Leuven-Heverlee, Belgium
dirk.roose@cs.kuleuven.be

Tamar Schlick
Department of Chemistry
and Courant Institute
of Mathematical Sciences
New York University
251 Mercer Street
New York, NY 10012, USA
schlick@nyu.edu

Editor for Computational Science
and Engineering at Springer:
Martin Peters
Springer-Verlag
Mathematics Editorial IV
Tiergartenstrasse 17
69121 Heidelberg, Germany
martin.peters@springer.com

Lecture Notes in Computational Science and Engineering

1. D. Funaro, *Spectral Elements for Transport-Dominated Equations.*
2. H.P. Langtangen, *Computational Partial Differential Equations.* Numerical Methods and Diffpack Programming.
3. W. Hackbusch, G. Wittum (eds.), *Multigrid Methods V.*
4. P. Deuflhard, J. Hermans, B. Leimkuhler, A.E. Mark, S. Reich, R.D. Skeel (eds.), *Computational Molecular Dynamics: Challenges, Methods, Ideas.*
5. D. Kröner, M. Ohlberger, C. Rohde (eds.), *An Introduction to Recent Developments in Theory and Numerics for Conservation Laws.*
6. S. Turek, *Efficient Solvers for Incompressible Flow Problems.* An Algorithmic and Computational Approach.
7. R. von Schwerin, ***M**ulti **B**ody **S**ystem **SIM**ulation.* Numerical Methods, Algorithms, and Software.
8. H.-J. Bungartz, F. Durst, C. Zenger (eds.), *High Performance Scientific and Engineering Computing.*
9. T.J. Barth, H. Deconinck (eds.), *High-Order Methods for Computational Physics.*
10. H.P. Langtangen, A.M. Bruaset, E. Quak (eds.), *Advances in Software Tools for Scientific Computing.*
11. B. Cockburn, G.E. Karniadakis, C.-W. Shu (eds.), *Discontinuous Galerkin Methods.* Theory, Computation and Applications.
12. U. van Rienen, *Numerical Methods in Computational Electrodynamics.* Linear Systems in Practical Applications.
13. B. Engquist, L. Johnsson, M. Hammill, F. Short (eds.), *Simulation and Visualization on the Grid.*
14. E. Dick, K. Riemslagh, J. Vierendeels (eds.), *Multigrid Methods VI.*
15. A. Frommer, T. Lippert, B. Medeke, K. Schilling (eds.), *Numerical Challenges in Lattice Quantum Chromodynamics.*
16. J. Lang, *Adaptive Multilevel Solution of Nonlinear Parabolic PDE Systems.* Theory, Algorithm, and Applications.
17. B.I. Wohlmuth, *Discretization Methods and Iterative Solvers Based on Domain Decomposition.*
18. U. van Rienen, M. Günther, D. Hecht (eds.), *Scientific Computing in Electrical Engineering.*
19. I. Babuška, P.G. Ciarlet, T. Miyoshi (eds.), *Mathematical Modeling and Numerical Simulation in Continuum Mechanics.*
20. T.J. Barth, T. Chan, R. Haimes (eds.), *Multiscale and Multiresolution Methods.* Theory and Applications.
21. M. Breuer, F. Durst, C. Zenger (eds.), *High Performance Scientific and Engineering Computing.*
22. K. Urban, *Wavelets in Numerical Simulation.* Problem Adapted Construction and Applications.

23. L.F. Pavarino, A. Toselli (eds.), *Recent Developments in Domain Decomposition Methods.*

24. T. Schlick, H.H. Gan (eds.), *Computational Methods for Macromolecules: Challenges and Applications.*

25. T.J. Barth, H. Deconinck (eds.), *Error Estimation and Adaptive Discretization Methods in Computational Fluid Dynamics.*

26. M. Griebel, M.A. Schweitzer (eds.), *Meshfree Methods for Partial Differential Equations.*

27. S. Müller, *Adaptive Multiscale Schemes for Conservation Laws.*

28. C. Carstensen, S. Funken, W. Hackbusch, R.H.W. Hoppe, P. Monk (eds.), *Computational Electromagnetics.*

29. M.A. Schweitzer, *A Parallel Multilevel Partition of Unity Method for Elliptic Partial Differential Equations.*

30. T. Biegler, O. Ghattas, M. Heinkenschloss, B. van Bloemen Waanders (eds.), *Large-Scale PDE-Constrained Optimization.*

31. M. Ainsworth, P. Davies, D. Duncan, P. Martin, B. Rynne (eds.), *Topics in Computational Wave Propagation.* Direct and Inverse Problems.

32. H. Emmerich, B. Nestler, M. Schreckenberg (eds.), *Interface and Transport Dynamics.* Computational Modelling.

33. H.P. Langtangen, A. Tveito (eds.), *Advanced Topics in Computational Partial Differential Equations.* Numerical Methods and Diffpack Programming.

34. V. John, *Large Eddy Simulation of Turbulent Incompressible Flows.* Analytical and Numerical Results for a Class of LES Models.

35. E. Bänsch (ed.), *Challenges in Scientific Computing - CISC 2002.*

36. B.N. Khoromskij, G. Wittum, *Numerical Solution of Elliptic Differential Equations by Reduction to the Interface.*

37. A. Iske, *Multiresolution Methods in Scattered Data Modelling.*

38. S.-I. Niculescu, K. Gu (eds.), *Advances in Time-Delay Systems.*

39. S. Attinger, P. Koumoutsakos (eds.), *Multiscale Modelling and Simulation.*

40. R. Kornhuber, R. Hoppe, J. Périaux, O. Pironneau, O. Wildlund, J. Xu (eds.), *Domain Decomposition Methods in Science and Engineering.*

41. T. Plewa, T. Linde, V.G. Weirs (eds.), *Adaptive Mesh Refinement – Theory and Applications.*

42. A. Schmidt, K.G. Siebert, *Design of Adaptive Finite Element Software.* The Finite Element Toolbox ALBERTA.

43. M. Griebel, M.A. Schweitzer (eds.), *Meshfree Methods for Partial Differential Equations II.*

44. B. Engquist, P. Lötstedt, O. Runborg (eds.), *Multiscale Methods in Science and Engineering.*

45. P. Benner, V. Mehrmann, D.C. Sorensen (eds.), *Dimension Reduction of Large-Scale Systems.*

46. D. Kressner, *Numerical Methods for General and Structured Eigenvalue Problems.*

47. A. Boriçi, A. Frommer, B. Joó, A. Kennedy, B. Pendleton (eds.), *QCD and Numerical Analysis III.*

48. F. Graziani (ed.), *Computational Methods in Transport.*

49. B. Leimkuhler, C. Chipot, R. Elber, A. Laaksonen, A. Mark, T. Schlick, C. Schütte, R. Skeel (eds.), *New Algorithms for Macromolecular Simulation.*

50. M. Bücker, G. Corliss, P. Hovland, U. Naumann, B. Norris (eds.), *Automatic Differentiation: Applications, Theory, and Implementations.*

51. A.M. Bruaset, A. Tveito (eds.), *Numerical Solution of Partial Differential Equations on Parallel Computers.*

52. K.H. Hoffmann, A. Meyer (eds.), *Parallel Algorithms and Cluster Computing.*

53. H.-J. Bungartz, M. Schäfer (eds.), *Fluid-Structure Interaction.*

54. J. Behrens, *Adaptive Atmospheric Modeling.*

55. O. Widlund, D. Keyes (eds.), *Domain Decomposition Methods in Science and Engineering XVI.*

56. S. Kassinos, C. Langer, G. Iaccarino, P. Moin (eds.), *Complex Effects in Large Eddy Simulations.*

57. M. Griebel, M.A Schweitzer (eds.), *Meshfree Methods for Partial Differential Equations III.*

58. A.N. Gorban, B. Kégl, D.C. Wunsch, A. Zinovyev (eds.), *Principal Manifolds for Data Visualization and Dimension Reduction.*

59. H. Ammari (ed.), *Modeling and Computations in Electromagnetics: A Volume Dedicated to Jean-Claude Nédélec.*

60. U. Langer, M. Discacciati, D. Keyes, O. Widlund, W. Zulehner (eds.), *Domain Decomposition Methods in Science and Engineering XVII.*

61. T. Mathew, *Domain Decomposition Methods for the Numerical Solution of Partial Differential Equations.*

62. F. Graziani (ed.), *Computational Methods in Transport: Verification and Validation.*

63. M. Bebendorf, *Hierarchical Matrices.* A Means to Efficiently Solve Elliptic Boundary Value Problems.

64. C.H. Bischof, H.M. Bücker, P. Hovland, U. Naumann, J. Utke (eds.), *Advances in Automatic Differentiation.*

65. M. Griebel, M.A. Schweitzer (eds.), *Meshfree Methods for Partial Differential Equations IV.*

66. B. Engquist, P. Lötstedt, O. Runborg (eds.), *Multiscale Modeling and Simulation in Science.*

67. I.H. Tuncer, Ü. Gülcat, D.R. Emerson, K. Matsuno (eds.), *Parallel Computational Fluid Dynamics 2007.*

68. S. Yip, T. Diaz de la Rubia (eds.), *Scientific Modeling and Simulations.*

69. A. Hegarty, N. Kopteva, E. O'Riordan, M. Stynes (eds.), *BAIL* 2008 – *Boundary and Interior Layers.*

70. M. Bercovier, M.J. Gander, R. Kornhuber, O. Widlund (eds.), *Domain Decomposition Methods in Science and Engineering XVIII.*

71. B. Koren, C. Vuik (eds.), *Advanced Computational Methods in Science and Engineering.*

72. M. Peters (ed.), *Computational Fluid Dynamics for Sport Simulation.*

73. H.-J. Bungartz, M. Mehl, M. Schäfer (eds.), *Fluid Structure Interaction II - Modelling, Simulation, Optimization.*

74. D. Tromeur-Dervout, G. Brenner, D.R. Emerson, J. Erhel (eds.), *Parallel Computational Fluid Dynamics 2008.*

For further information on these books please have a look at our mathematics catalogue at the following URL: www.springer.com/series/3527

Monographs in Computational Science and Engineering

1. J. Sundnes, G.T. Lines, X. Cai, B.F. Nielsen, K.-A. Mardal, A. Tveito, *Computing the Electrical Activity in the Heart.*

For further information on this book, please have a look at our mathematics catalogue at the following URL: www.springer.com/series/7417

Texts in Computational Science and Engineering

1. H. P. Langtangen, *Computational Partial Differential Equations.* Numerical Methods and Diffpack Programming. 2nd Edition
2. A. Quarteroni, F. Saleri, P. Gervasio, *Scientific Computing with MATLAB and Octave.* 3rd Edition
3. H. P. Langtangen, *Python Scripting for Computational Science.* 3rd Edition
4. H. Gardner, G. Manduchi, *Design Patterns for e-Science.*
5. M. Griebel, S. Knapek, G. Zumbusch, *Numerical Simulation in Molecular Dynamics.*
6. H. P. Langtangen, *A Primer on Scientific Programming with Python.*
7. A. Tveito, H. P. Langtangen, B. F. Nielsen, X. Cai, *Elements of Scientific Computing.*

For further information on these books please have a look at our mathematics catalogue at the following URL: www.springer.com/series/5151